THE COAST GUARDSMAN'S MANUAL

12TH EDITION

THE COAST GUARDSMAN'S MANUAL

12TH EDITION

PHILLIP N. NULL

Naval Institute Press
Annapolis, Maryland

Naval Institute Press
291 Wood Road
Annapolis, MD 21402

First edition published 1952. Eleventh edition published 2017.

ISBN: 978-1-68247-611-6 (paperback)
ISSN: 0530-045 (print)

♾ Print editions meet the requirements of ANSI/NISO z39.48–1992 (Permanence of Paper).
Printed in the United States of America.

34 33 32 31 30 29 28 27 26 9 8 7 6 5 4 3 2 1
First printing

Unless otherwise noted, all photographs are official U.S. Coast Guard.

SEMPER PARATUS

From Aztec Shore to Arctic Zone, to Europe and Far East,
The Flag is carried by our ships
In times of war and peace;
And never have we struck it yet,
In spite of foemen's might,
Who cheered our crews and cheered again
For showing how to fight.

CHORUS
So here's the Coast Guard marching song,
We sing on land or sea.
Through surf and storm and howling gale,
High shall our purpose be,
"Semper Paratus" is our guide,
Our fame, our glory, too.
To fight to save or fight and die!
Aye! Coast Guard, we are for you.
Surveyor and Narcissus,
The Eagle and Dispatch,
The Hudson and the Tampa,
These names are hard to match;
From Barrow's shores to Paraguay,
Great Lakes' or Ocean's wave,
The Coast Guard fights through storms and winds
To punish or to save.

Aye! We've been "Always Ready"
To do, to fight, or die!
Write glory to the shield we wear
In letters to the sky.
To sink the foe or save the maimed
Our mission and our pride.
We'll carry on 'til Kingdom Come
Ideals for which we've died.

—CAPT F. S. Van Boskerck, USCG

Contents

Creed of the United States Coast Guardsman

VADM Harry G. Hamlet, 1938

I am proud to be a United States Coast Guardsman.

I revere that long line of expert seamen who by their devotion to duty and sacrifice of self have made it possible for me to be a member of a service honored and respected, in peace and in war, throughout the world.

I never, by word or deed, will bring reproach upon the fair name of my service, nor permit others to do so unchallenged.

I will cheerfully and willingly obey all lawful orders.

I will always be on time to relieve, and shall endeavor to do more, rather than less, than my share.

I will always be at my station, alert and attending to my duties.

I shall, so far as I am able, bring to my seniors solutions, not problems.

I shall live joyously, but always with due regard for the rights and privileges of others.

I shall endeavor to be a model citizen in the community in which I live.

I shall sell life dearly to an enemy of my country, but give it freely to rescue those in peril.

With God's help, I shall endeavor to be one of His noblest Works . . .

A UNITED STATES COAST GUARDSMAN

United States Coast Guard Ethos

I am a Coast Guardsman.

I serve the people of the United States.

I will protect them.

I will defend them.

I will save them.

I am their shield.

For them I am Semper Paratus.

I live the Coast Guard Core Values.

I am proud to be a Coast Guardsman.

We are the United States Coast Guard.

1

INTRODUCTION

Since 1790 the U.S. Coast Guard and its predecessor agencies have saved those in peril on the sea, defended the nation from threats delivered by the sea, and protected the sea itself. Today a military, multimission service, the U.S. Coast Guard is one of the six branches of the armed forces of the United States and exists to defend and preserve the U.S. Constitution with civil and military responsibilities that touch every facet of the maritime environment. It is the largest and most powerful coast guard in the world, rivaling the capabilities and size of most navies, and is America's first responder to any maritime incident that demands federal action. We protect and advance important national interests in America's waterways and along its coasts, on international waters and the high seas, and in any other waters where they may be at risk from threats, whether foreign or domestic, natural or man-made. To accomplish these missions 44,000 personnel maintain, operate, and support a multimission, interoperable fleet of over 250 cutters, 200 aircraft, and 1,600 boats. The Coast Guard's distinct blend of authorities, capabilities, competencies, and partnerships provide the president, secretary of homeland security, secretary of defense, and other national leaders with the ability to lead or support a range of operations as national priorities demand. The true value of the Coast Guard is not in its ability to perform any single mission but in its versatile, highly adaptive, multimission character.

All Coast Guardsmen should read and embrace the heritage, history, and enduring hard-won lessons presented in this manual. Anyone interested in the service or the sea will also benefit from its contents and find it a valuable reference on a wide range of topics. Mariners in particular would do well to sail with a copy on board and refer to it often, whether for inspiration or insight, especially in advance of those trying times when a safe return to port is uncertain.

On an average day, the U.S. Coast Guard:
Conducts 42 search and rescue cases;
Saves 12 lives;
Saves over $114,000 in property;
Seizes 1,253 pounds of cocaine and 172 pounds of marijuana;
Conducts 133 waterborne patrols of critical maritime infrastructure;
Interdicts 18 illegal migrants;
Escorts 9 high-capacity passenger vessels;
Conducts 24 security boardings in and around U.S. ports;
Screens 360 merchant vessels for potential security threats prior to arrival in U.S. ports;
Conducts 19 fisheries conservation boardings;
Services 82 buoys and fixed aids to navigation;
Investigates 35 pollution incidents;
Completes 24 safety examinations on foreign vessels;
Conducts 162 marine inspections of U.S. vessels and facilities;
Investigates 17 marine casualties involving commercial vessels; and
Facilitates movement of $15.6 billion worth of goods and commodities through the nation's Maritime Transportation System.

2

VALUES, HISTORY, AND TRADITIONS

As Coast Guardsmen, we share a bond of pride in our rich heritage and a common purpose to uphold our honorable traditions and core values. We are defined by unsung heroism and selfless service. We protect our nation. We risk our lives to save others. We give our utmost when cold, wet, and tired. Countless times we have extended our arm down into the water to rescue those in peril from the sea. We will always extend that same arm to help a shipmate in need. Coast Guardsmen are always ready. We perform our missions with an unrelenting sense of pride. When disaster strikes, whether natural or man-made, we are first on scene. We are often the last to depart.

Coast Guardsmen are agile, adaptable, ready to perform every mission and to answer every call. Born as revenue cuttermen, lighthouse keepers, steamboat inspectors, and surfmen, we have expanded to meet the needs of our maritime nation. We are still the keepers of the lights, but we also now patrol far more distant waters. We readily go wherever there are important, difficult, and dangerous maritime duties to be performed. As we look to the future, we must never forget our past, our history, or our heritage. We revere that long blue line of Coast Guardsmen who by their devotion to duty and sacrifice of self have made it possible for us to be a member of a service honored, respected, and imitated around the globe. We are privileged to be members of a unique service that, due to our collection of missions and legacy agencies, sometimes defies logic when someone attempts to classify us or place a label on us. For this reason, whenever asked to describe what you are, always reply with pride: I am a Coast Guardsman.

Adapted from ALCOAST 234/14—SHIPMATES 29 AND FINAL: WE ARE COAST GUARDSMEN.

CORE VALUES

Distilled from the legacy of service that distinguishes the Coast Guard, our core values provide a common ground, binding us together and guiding our conduct, performance, and decisions both on and off duty, and challenging us to live up to the high standards of excellence exhibited by those who came before. Our core values are honor, respect, and devotion to duty.

Honor

Integrity is our standard. We demonstrate uncompromising ethical conduct and moral behavior in all of our personal and organizational actions. We are loyal and accountable to the public trust.

Respect

We value our diverse workforce. We treat each other and those we serve with fairness, dignity, respect, and compassion. We encourage individual opportunity and growth. We encourage creativity through empowerment. We work as a team.

Devotion to Duty

We are professionals, military and civilian, who seek responsibility, accept accountability, and are committed to the successful achievement of our organizational goals. We exist to serve. We serve with pride.

HISTORY

The history of the Coast Guard includes every remembered story, celebrated date, and recognized name associated with the service. In this history, countless examples of honor, respect, and devotion to duty can be found to provide ready examples to follow and model our own conduct after. By learning the Coast Guard's proud history and heritage, you will be able to understand the Coast Guard and how you can best fit in. Remember, the operations and missions of today will be the history of tomorrow.

Alexander Hamilton

> A few armed vessels, judiciously stationed at the entrances of our ports, might at a small expense be made useful sentinels of the laws.
>
> —Alexander Hamilton, Federalist Paper #12, 27 November 1787

Considered the father of the U.S. Coast Guard, Alexander Hamilton played an integral role in forming the United States and its democratic government. Born on the West Indian island of Nevis on 11 January 1757, Hamilton immigrated to New York in 1772. Although not yet twenty years old, by 1774 he had authored many widely read political publications. Not long after the start of the American Revolution, Hamilton received the captaincy of an artillery unit and fought in the principal campaigns of 1776. In 1777 he advanced to the rank of lieutenant colonel, joined the staff of GEN George Washington as secretary and aide-de-camp, and became Washington's close confidant. Hamilton ended the war as a lieutenant colonel commanding an infantry regiment, which he led with great success during the siege at Yorktown that forced the surrender of British forces in North America.

When the United States government was established in 1789, Hamilton was appointed its first secretary of the treasury. In this role he immediately set about to stabilize national finances, proposing the establishment of a national bank and a fleet of cutters to combat smugglers. Hamilton advised Congress that the cutters could direct ships to specific ports along the East Coast of the United States to ensure import tariffs on the goods they carried were paid to the treasury to fund the federal government. Congress adopted Hamilton's plan on 4 August 1790, which the U.S. Coast Guard now celebrates as its birthday, and the cutters proved the basis for establishment of a Revenue Marine, later known as the Revenue Cutter Service. Since the Continental navy and marines disbanded following the conclusion of the American Revolution, this Revenue Marine was the nation's only sea service in the early years of the new nation.

Today Hamilton's name graces the hull of a national security cutter, CGC *Alexander Hamilton* (WMSL 753), the sixth Coast Guard cutter to bear his name.

Predecessor Services

To refer to the Coast Guard as a long blue line is fitting, but the service could even be described as the longest blue line. The Coast Guard received its name in 1915, but it has existed in some form since the founding of the nation and traces its origins to five distinct federal agencies. These agencies were originally independent but had overlapping authorities and were shuffled around the government. They sometimes received new names and were all finally united under the umbrella of the Coast Guard. The multiple missions and responsibilities of the modern service are directly tied to this diverse heritage and the achievements of all who served in these agencies.

U.S. Revenue Cutter Service (USRCS)

A system of cutters to serve as an armed customs enforcement service was established by an act of Congress on Guard. 4 August 1790, upon the recommendation of Treasury Secretary Alexander Hamilton. As time passed, these cutters gradually gained missions, including those of a military nature. The cutters were generally referred to as the Revenue Marine until 1894, when they were officially organized as the Revenue Cutter Service and operated under the authority of the U.S. Department of the Treasury. On 28 January 1915, the service was merged by an act of Congress with the U.S. Life-Saving Service to form the U.S. Coast Guard.

U.S. Life-Saving Service (USLSS)

The U.S. Life-Saving Service was a government agency formed in 1848 that grew out of private and local humanitarian efforts to save the lives of shipwrecked mariners. The stations of the service fell into three categories: lifesaving, lifeboat, and houses of refuge. Most stations were in isolated areas where crewmen had to launch their boats from the beach into the surf. The service merged with the Revenue Cutter Service to form the U.S. Coast Guard in 1915.

U.S. Lighthouse Service (USLHS)

Boston Light was the first lighthouse built in North America, constructed on Little Brewster Island in Boston Harbor by the Province of Massachusetts Bay in 1716. On 7 August 1789, the ninth act of our first Congress, and the first Public Works Act, provided for the transfer of twelve existing lighthouses, including Boston Light, from the states to the federal government. Over the years lighthouses were placed under the direction of various federal departments. The U.S. Lighthouse Service was a government agency formed in 1910 under the Department of the Treasury and made responsible for the construction, maintenance, and operations of all lighthouses, lightships, and navigation aids in the United States. The service was merged into the U.S. Coast Guard in 1939.

Bureau of Navigation and Steamboat Inspection Service

Following a series of disasters caused by boiler explosions, the government created the Steamboat Inspection Service in 1871 in order to safeguard the lives of passengers on board vessels propelled in whole or in part by steam. The Bureau of Navigation was a separate government agency created in 1884 to enforce laws related to the construction, equipment, operation, inspection, safety, and documentation of merchant vessels and to investigate marine accidents and casualties; collect tonnage taxes and other navigation fees; and examine, certify, and license merchant mariners. These agencies, initially operating under the Department of Treasury and later the Department of Commerce, merged in 1932 to form the Bureau of Navigation and Steamboat Inspection, which, in 1936, was reorganized into the Bureau of Marine Inspection and Navigation. Functions of the bureau were transferred to the U.S. Coast Guard in 1946.

At War

The Coast Guard, beginning as a system of cutters, has fought in every war since the ratification of the U.S. Constitution. The Coast Guard Standard proudly exhibits forty-three battle streamers, tracing the service's wartime service from the Quasi-War with France to the Global War on Terror.

Quasi-War

In the 1790s and early 1800s, Great Britain and France fought a succession of wars that threatened American neutrality on the high seas. Treaty disputes between the United States and the French during this period resulted in the French authorizing privateers to attack American shipping in the Caribbean and off the East Coast in 1796. The naval response of the United States was to employ its revenue cutters in the defense of our merchant fleet, with eight cutters fighting against France alongside the newly formed U.S. Navy. During the war American forces captured twenty French vessels, ten by the cutter *Pickering* and five by the cutter *Eagle*.

War of 1812

After the Quasi-War concluded in 1800, the United States tried to remain neutral in the ongoing conflict between Great Britain and France, but the Royal Navy's practice of boarding American ships and impressing their crews threatened U.S. sovereignty. A series of embargoes and trade restrictions to punish the British only increased tensions, and war was declared on 18 June 1812. Throughout the war the cutters ran blockades, escorted convoys, captured smugglers and privateers, and engaged the Royal Navy, with notable actions including the capture of the British privateer *Dart* by the cutter *Vigilant* near Block Island and the defense of the cutter *Eagle* at Long Island.

Seminole Wars

In 1836 eight revenue cutters were ordered to the coasts of Florida to support Army and Navy forces engaged in the Seminole Wars to displace the Seminole tribe and remove its people to reservations. Actions involving the cutters include the defense of settlements, the blockade of rivers, sea and land expeditions against the Seminoles, and the transport of troops and supplies.

Mexican-American War

Following the annexation of Texas in 1845, the boundary between the United States and Mexico became disputed along the Rio Grande and an attack on U.S. forces camped in the area by the Mexican Army prompted a declaration of war. The U.S. Army and Navy were critically short of the shallow-draft vessels needed for operating in the Gulf of Mexico, and the treasury secretary assigned five cutters to support the war effort. The cutters engaged in convoy escorts; carried troops, dispatches, and supplies; engaged in river and coastal blockades; and landed troops for attacks on Mexican forces at Alvarado and Tabasco.

Civil War

The sympathies of the cutter force were divided between the North and the South during the American Civil War. In a famous dispatch, Treasury Secretary John A. Dix ordered, "If any one attempts to haul down the American flag, shoot him on the spot." Transmitted on the evening of 15 January 1861, the order was to ensure federal control of the cutter *Robert McLelland* at the onset of the war.

The principal duties of cutters during the Civil War were to engage Confederate forces, blockade ports, patrol for smugglers and raiders, and provide transport for troops and government leaders. The cutter *Harriet Lane* fired the first naval shot of the war when it challenged the steamer *Nashville* with a shot across its bow as the ship attempted to enter Charleston harbor without displaying the U.S. flag.

The cutter *E. A. Stevens*, later renamed the *Naugatuck*, was an early ironclad that sailed with the USS *Monitor* in the James River Flotilla, trading shots with the CSS *Virginia* and with a Confederate fort at Drewry's Bluff near Richmond. When President Lincoln was shot in April 1865, cutters were ordered to search for the assassin on all ships on the Potomac.

Spanish-American War

The United States declared war against Spain in April 1898 in support of ongoing struggles by Cubans and Filipinos against Spanish rule and in response to the mysterious explosion of the battleship USS *Maine* in Havana Harbor. Thirteen revenue cutters would serve in the conflict in cooperation with the U.S. Navy, four protecting gold-laden vessels along the West Coast, the cutter *Hugh McCullough* participating in the Battle of Manila Bay, and eight cutters involved in the blockade of Cuba.

In an action at Cardenas, Cuba, the cutter *Hudson* rescued the torpedo boat *Winslow* from imminent destruction, leading Congress to award medals for bravery to all of the involved cuttermen.

During the war the Navy assigned the task of coast watching to the U.S. Life-Saving Service. As a result, approximately two-thirds of the Navy's coastal observation stations were U.S. Life-Saving stations. Although the Spanish fleet never approached the U.S. mainland, this Coast Guard predecessor service dutifully maintained its vigilance throughout the war.

World War I

The outbreak of World War I in Europe in the summer of 1914 caused the establishment of the U.S. Coast Guard, which was immediately transferred to the Navy Department for wartime service. The Coast Guard augmented the Navy with 47 cutters and 279 coastal stations and was made responsible for enforcing U.S. neutrality laws as well as the rules and regulations that governed the anchorage and movements of vessels in American harbors.

The Espionage Act, passed in June 1917, gave the Coast Guard further power to protect merchant shipping from sabotage. This act included provisions for the safeguarding of waterfront property, supervision of vessel movements, establishment of anchorages and restricted areas, and the right to control and remove people from ships.

It was during this same period that Coast Guard aviation was born. LT E. F. Stone, USCG, one of several early Coast Guard fliers, was copilot on the Navy NC-4 flying boat, the first aircraft to cross the Atlantic.

World War II

Immediately after the Japanese attack on Pearl Harbor on 7 December 1941, the Coast Guard was placed under the control of the U.S. Navy and its cutters were ordered to escort duty and into action against Adolf Hitler's U-boat fleet. Coast Guard forces were credited with sinking thirteen U-boats and a Japanese sub. Additionally, two U-boats surrendered to Coast Guard–manned warships. Coast Guardsmen also captured two Nazi surface vessels, the only United States service to do so during the war.

Amphibious operations employed thousands of Coast Guardsmen in both the Atlantic and Pacific theaters, their small-boat experience proving invaluable to landing troops on contested shores. SM1 Douglas Munro, in command of a Higgins boat at the second battle of the Matanikau in September 1942, was posthumously awarded the only Congressional Medal of Honor earned by a member of the Coast Guard for his selfless actions to extract of a force of Marines whose position had been overrun by Japanese forces.

During World War II, 241,093 men and women served in the Coast Guard. Of these, 1,917 died, with a third losing their life in action. Four were held as prisoners of war.

Korean War

When North Korean forces crossed the thirty-eighth parallel to invade the Republic of South Korea on 25 June 1950, the Coast Guard was called upon to support the war effort and establish a presence throughout the Pacific. Coast Guard personnel manned a number of Navy destroyer escorts, and the service formed air detachments in the Philippines and on the islands of Guam, Wake, Midway, Adak, and Oahu to safeguard the tens of thousands of troops that were being airlifted across the Pacific. It also deployed a chain of mobile LORAN (long-range navigation) stations and provided meteorological and communication services to support air and sea navigation by friendly forces.

More than 8,500 Coast Guardsmen were eligible for the Korean Service Medal.

Vietnam War

The United States deployed conventional military forces to Vietnam in 1964 following the Gulf of Tonkin incident in order to openly counter the communist aggression of North Vietnam. At the start of the war the U.S. Navy had an unmet need for shallow-water craft to conduct inshore operations, causing the Coast Guard to deploy twenty-six 82-foot cutters to provide such a capability. Thirty high-endurance cutters were also deployed to blockade the nation's coastline. The cutters were tasked with engaging North Vietnamese and Viet Cong craft, searching junks for contraband, and providing fire support for friendly forces ashore.

Coast Guard forces established and operated multiple LORAN-C stations throughout Southeast Asia, providing for naval and air navigation of friendly forces involved in the war.

The rapid development of deepwater ports in Vietnam to accommodate more than three hundred merchant ships engaged in the sealift of warfighting materials brought an expanded need for navigational aids. South Vietnam's single buoy tender could not meet the demand, and a Coast Guard aids-to-navigation detail was set up in Saigon to maintain the aids and to coordinate work for the service's visiting buoy tenders.

Coast Guard pilots flew combat search-and-rescue missions throughout the war, most assigned to the U.S. Air Force's 37th Aerospace Rescue and Recovery Squadron and flying Sikorsky HH-3F Jolly Green Giants. LT Jack Rittichier (Coast Guard aviator #997) became the first Coast Guardsman to die from enemy action in Vietnam when his helicopter was shot down on 9 June 1968 during an attempt to rescue a downed Marine pilot from enemy-held territory.

Some eight thousand Coast Guardsmen served in Vietnam. Seven lost their lives, and sixty were wounded.

Gulf War

On 2 August 1990 Iraqi forces invaded Kuwait, and a coalition led by the United States responded to defend neighboring Saudi Arabia. Within days Coast Guard law enforcement detachments (LEDET) deployed onto coalition ships as part of Operation Desert Shield where they led multinational boarding teams in intercepting ships going to and from Iraq and Kuwait in violation of a United Nations embargo.

Three port security units (PSU) would also be ordered to the Persian Gulf to conduct waterborne security operations in Saudi Arabia, Kuwait, and Bahrain. Offensive operations to liberate Kuwait commenced in January 1991 under Operation Desert Storm and lasted until April. Coast Guard LEDETs supported these offensive operations by clearing some of Iraq's offshore oil platforms. Coast Guard LEDET personnel on board the USS *Nicholas* (FFG 11) assisted when the frigate cleared eleven Iraqi oil platforms and took twenty-three prisoners. When Iraqi forces set wells ablaze and created concerns of an environmental disaster, an aviation detachment consisting of two HU-25A Falcon and two HC-130 Hercules aircraft was further deployed as part of an interagency oil spill assessment team (USIAT) to map the spills over 40,000 square miles of water.

Operation Uphold Democracy

In July 1994 the United Nations authorized the deployment of a multinational force under American leadership to restore democratic institutions in Haiti following a 1991 military coup d'état that installed a military junta as the island nation's governing body and created a long period of violence in which many thousands crowded into unseaworthy boats to seek asylum in the United States. A last-minute diplomatic approach backed with the threat of military invasion was successful in avoiding offensive operations and the mission changed to one of peacekeeping and nation-building. Coast Guard forces played a significant role in the operation, which lasted from September 1994 to March 1995, deploying two PSUs, a harbor defense unit, five LEDETs, and thirteen cutters.

Operation Allied Force

Operation Allied Force was a seventy-eight-day NATO air campaign from 24 March 1999 to 10 June 1999 against Serbian military targets in the former Federal Republic of Yugoslavia. USCGC *Bear* (WMEC 901) deployed to the Adriatic Sea in support of the operation where it conducted escorts for U.S. Army vessels transporting military cargo between Italy and Albania.

Global War on Terror

On 11 September 2001, terrorists from the Al Qaeda network hijacked four commercial aircraft, crashing two into the World Trade Center in New York and one into the Pentagon in Washington, D.C. The fourth aircraft crashed near Shanksville, Pennsylvania, when passengers on board attempted to regain control from the terrorists. Coast Guard units from Activities New York were among the first military units to respond; with bridges and tunnels closed throughout the city, the Coast Guard coordinated the evacuation of over 500,000 people trapped in Lower Manhattan using a fleet of military, commercial, and recreational vessels.

The Global War on Terror was launched by the United States following the 11 September, or 9/11, attacks. It targets networks of terrorists and the governments that support them throughout the world, with the most prominent being Al Qaeda, the Islamic State, and the Taliban.

Coast Guard Patrol Forces Southwest Asia (PATFORSWA) was established in 2002 at the request of the U.S. Central Command to support combat operations in Operation Iraqi Freedom. In 2003 four 110-foot Island-class patrol boats and associated support elements were deployed to the Arabian Gulf and were present for the commencement of combat operations. These cutters primarily assisted in force protection and the search and seizure of suspected smugglers in Iraqi and international waters, often in proximity to Iran. USCGC *Walnut* (WLB 205) was also deployed at the onset of the war to establish critical aids to navigation in Iraq's river and coastal areas. Two additional patrol boats were deployed to the region in 2004.

PATFORSWA has operated in the region since its establishment and today continues to support maritime security and counter piracy operations with its assigned cutters and a variety of operational, training, and advisory teams. In 2021 the Coast Guard transitioned the six Island-class patrol boats to six 154-foot fast-response cutters specifically appropriated for the mission.

U.S. Naval Station Guantanamo Bay, Cuba, has served as a detention camp and site of military tribunals for terrorist detainees since 2002. The Coast Guard has deployed maritime safety and security teams and PSUs to the naval base in support of Joint Task Force Guantanamo Bay (JTF-GTMO) to guard detainees, provide courtroom security, and ensure maritime security for the naval station's defensive sea area that extends three miles from shore.

Operation Pacific Eagle–Philippines commenced in September 2017 to support the Philippine government and its military forces in countering the Islamic State and other violent extremist organizations operating within its borders. The Coast Guard has played a significant role in this operation, transferring several former high-endurance cutters to the nation and deploying national security cutters and deployable specialized forces to aid the Philippine Coast Guard through joint patrols and training exchanges.

In addition to deploying full units, the Coast Guard brings its unique capabilities and authorities to the war on terrorism through deployable advanced interdiction teams (AIT) that augment Department of Defense and partner nation resources. AITs represent the nation's elite visit, board, search, and seizure capability and have deployed on board U.S. Navy warships for operations in the Arabian Gulf, Arabian Sea, Indian Ocean, and Horn of Africa.

Disasters and Domestic Operations

Closer to home, the Coast Guard provides our nation with tremendous value in service to the public during times of natural and man-made disasters and other contingencies. A sampling of key domestic events and missions for the Coast Guard follows.

Overland Relief Expedition

In November 1897 word that several whaling ships holding 275 sailors were trapped in the ice at Point Barrow, Alaska, reached the federal government and the Overland Relief Expedition was planned. USRC *Bear* was ordered into the Bering Sea, but because she was not an icebreaker and could not sail through pack ice to the trapped whalers' location, the executive officer, LT David Jarvis; the ship's surgeon, Dr. Samuel J. Call, and LT Ellsworth Bertholf were landed near Cape Vancouver with instructions to drive a herd of reindeer overland across roughly 1,500 miles to feed the stranded whalers until the spring thaw when the *Bear* could reach their location. The party reached the whalers with the reindeer in early March 1898, walking most of the distance and enduring temperatures as low as -45 degrees Fahrenheit. In 1902 Jarvis, Call, and Bertholf were awarded Congressional Gold Medals in recognition of their heroic relief efforts.

International Ice Patrol

On her maiden voyage from Southampton, England, bound for New York, the RMS *Titanic* collided with an iceberg in the North Atlantic just south of the Grand Banks on 15 April 1912. Although built with the latest safety features including compartmentation and automatic watertight doors, the ship sank within two-and-a-half hours and more than 1,500 of the 2,224 passengers and crew on board perished. Loss of the *Titanic* gripped the world with a chilling awareness of an iceberg's potential for tragedy and the disaster led to the creation of the International Ice Patrol, which actively patrols the waters of the Grand Banks of Newfoundland during the ice season from 1 February through 31 July.

Today fixed-wing Coast Guard aircraft conduct the primary reconnaissance work, with HC-130 long-range surveillance aircraft making the flights. Iceberg data collected by these patrols are fed into a computer model to predict the drift of icebergs and warn mariners of their position. Except for the years of the two world wars, the International Ice Patrol has conducted its work each season since 1913.

Rum Patrol

On 16 January 1919, the 18th Amendment to the Constitution was ratified, and the following year a national prohibition on the manufacture, sale, transportation, importation, or exportation of intoxicating liquors came into effect. Prohibition gave rise to the smuggling of illicit liquor into the United States, and by 1921 a "Rum Row" of cargo ships loaded with liquor and loitering just outside the three-mile limit existed off New York City and the New Jersey shore as well as near Boston, and the Chesapeake and Delaware bays. Liquor from these mother ships would be moved to small fast boats, known as "rum runners," to make the run to shore under cover of darkness. The Florida coast, New Orleans, Galveston, and San Francisco also saw similar smuggling activity. The Rum Patrol was a Coast Guard operation to interdict vessels smuggling liquor. The service was significantly expanded to deal with the problem, gaining a fleet of refurbished Navy destroyers, newly built picket boats, seaplanes, and intelligence units as well as the personnel to crew them.

Over the thirteen years that prohibition remained in effect the Coast Guard put constant pressure on rum-running syndicates and successfully reduced the volume of illegal liquor coming into the country, cementing the service's reputation as an internationally recognized law enforcement organization and intelligence service.

Operation Deep Freeze

In 1955 Task Force 43, consisting of the 270-foot Wind-class cutter *Eastwind* (WAGB 279) and eight U.S. Navy ships, was formed to provide logistical support for an expedition to the South Pole, considered the least explored area on Earth at the time. Named Operation Deep Freeze, the expedition intended to establish a permanent research station over the Antarctic summer of November 1955 to April 1956, with the goal of advancing world knowledge of Antarctic hydrography, weather, glacial movements, and marine life. The expedition succeeded and on 31 October 1956 landed a Douglas DC-3 aircraft named *Que Sera, Sera* at the South Pole, marking the first time that Americans had ever set foot there. The landing also began the establishment of the first permanent base at the South Pole, which was commissioned on 1 January 1957. Operation Deep Freeze activities were succeeded by Operation Deep Freeze II, and so on, continuing a constant U.S. presence in Antarctica since that date, with the earliest expeditions supported by the cutters *Eastwind* and *Northwind*. Two mountains in Antarctica, Mount Schmidtman and Mount Naab, along with Venzke Glacier, are named after their captains during these early expeditions. Northwind Glacier in the continent's Victoria Land is named for the cutter.

In October 1960 USCGC *Eastwind*, while assigned to Operation Deep Freeze, became the first Coast Guard cutter to ever circumnavigate the Earth, departing Boston and passing through the Panama Canal to reach the Pacific for the transit to Antarctica's McMurdo Sound before returning through the Indian Ocean and the Suez Canal to reach the Mediterranean Sea and cross the Atlantic Ocean.

Operation Deep Freeze has come to be used as a general term for U.S. operations on the continent, and in particular for the regular missions to resupply U.S. Antarctic bases, which have all been supported by the Coast Guard's icebreakers, most recently USCGC *Polar Star* (WAGB 10).

Mariel Boatlift

On 20 April 1980 the communist regime of Fidel Castro, in response to an ailing economy and internal tensions on the island of Cuba, announced that all Cubans wishing to emigrate to the United States were free to board boats at the port of Mariel west of Havana. The first Cuban refugees from Mariel reached Miami the next day. In all, 125,000 Cubans fled to U.S. shores in boats, many overcrowded and barely seaworthy, marking the largest Cuban migration to the United States. Around 25,000 Haitians also entered the United States during the boatlift.

The Coast Guard responded by blockading the flotilla of boats, seizing over 1,400 and removing their occupants to ensure the safety of the refugees. The exodus was finally ended by mutual agreement between the U.S. and Cuban governments in October 1980.

Exxon Valdez

On 24 March 1989 the oil tanker *Exxon Valdez* struck Bligh Reef near Tatitlek, Alaska, and over the next few days spilled 10.8 million gallons of crude oil into Prince William Sound. Coast Guard cutters were some of the first to respond to the spill, quickly establishing a safety zone around the stricken vessel and acting as on-scene command platforms to oversee booming and skimming operations. The Exxon Mobil company assumed full responsibility for the spill and its cleanup, but because the incident took place in navigational waters, the Coast Guard had authority for all activities related to the cleanup effort as the federal on-scene coordinator under the authority under the Clean Water Act.

Considered the worst oil spill worldwide in terms of damage to the environment, the event led to the passage of the Oil Pollution Act of 1990, which advanced tanker construction by requiring double hulls on new tankers, strengthened drug and alcohol testing requirements for licensed mariners, and expanded emergency response capability by creating Coast Guard strike teams. The disaster further caused the International Maritime Organization to introduce comprehensive marine pollution prevention rules (MARPOL) through various conventions for the purpose of achieving safer ships and cleaner oceans.

Hurricanes Katrina and Rita

On 29 August 2005 Hurricane Katrina made landfall as a Category 3 storm near the Louisiana–Mississippi border, bringing winds of 140 miles per hour and storm waters that flooded the city of New Orleans. Tragically, more than 1,800 people lost their lives, and damages across the Gulf Coast topped $108 billion. A few weeks later, on 24 September, Hurricane Rita again battered the area, extending the damage from eastern Texas to western Florida.

In advance of these powerful storms, the Coast Guard anticipated the potential devastation and executed plans to relocate its personnel, aircraft, and vessels, including evacuating eighteen small-boat stations, to an inland military installation so they would not be trapped in flooded areas and unable to respond. As a result, the Coast Guard was able to launch extensive search-and-rescue operations even while Katrina continued to pound the Gulf Coast. Pollution-response strike teams and teams to restore aids to navigation were readied, and they were deployed just as soon as conditions allowed.

The result of this careful preparation is that, during the chaotic days and weeks immediately following the storm, the Coast Guard rescued or evacuated 33,544 people. In that same period the Coast Guard responded to more than 1,100 pollution incidents with a total discharge of more than 7 million gallons of contaminants, and it repaired, replaced, or repositioned over nine hundred navigational aids to reopen the vital waterways. A Presidential Unit Citation was later presented to all active-duty, reserve, auxiliary, and civilian personnel affiliated with the service for their meritorious achievement and outstanding performance in preparation for and response and recovery to the hurricanes.

Haiti Earthquake

On 12 January 2010 Haiti was struck by a devastating 7.0 earthquake. The Coast Guard cutter *Forward* was diverted from a law enforcement patrol in the Caribbean and arrived in Port-au-Prince, Haiti, within seventeen hours of the disaster. The cutter *Forward* provided the world with the first images and reports of the devastation that had struck the impoverished nation. The first helicopters on scene were from Coast Guard units deployed to the Turks and Caicos Islands for counternarcotics operations. Over eight hundred Coast Guard personnel, cutters, and aircraft were involved in response and relief efforts, including operations to evacuate injured personnel and U.S. citizens, and to deliver of aid. Coast Guard cutters and personnel also completed port assessments and developed plans to reconstitute port operations, allowing for safe transit of humanitarian supplies from relief ships to coastal logistics staging points.

Deepwater Horizon

On 20 April 2010 the Coast Guard responded to the distress call following an explosion on the Deepwater Horizon oil rig in the Gulf of Mexico. During the initial search-and-rescue response more than one hundred people were rescued. As the scope and focus of the incident expanded and shifted to the ensuing oil spill, the Coast Guard led the monumental federal response to the first-ever spill of national significance and one of the worst environmental disasters in U.S. history. The scope of the unprecedented response involved over seven thousand Coast Guard personnel, forty-six cutters and boats, and twenty-two aircraft. Throughout the incident, the Coast Guard led a whole-of-government and industry response to secure the well-head, contain and clean up the spill, and protect the Gulf Coast ecosystem.

Legends

- Hopley Yeaton (1739–14 May 1812) was the first officer commissioned under the Constitution of the United States on 21 March 1791. A veteran of the Continental navy, Yeaton was commissioned by George Washington into the Revenue Marine and took command of cutter *Scammel.* Yeaton's mission was to cruise the coast from Nantucket to Passamaquoddy Bay looking for ships attempting to smuggle goods into the United States. He retired from the service in 1809. His tomb lies on the grounds of the U.S. Coast Guard Academy in New London, Connecticut.
- Joshua James (22 November 1826–19 March 1902) was a sea captain, a lifelong volunteer lifesaver with the Massachusetts Humane Society, and the keeper of U.S. Life-Saving Station Point Allerton in Hull, Massachusetts. James was credited with saving over six hundred lives from the age of fifteen, when he first associated himself with the Massachusetts Humane Society, until his death at the age of seventy-five while on duty with the U.S. Life-Saving Service. During his lifetime he was honored with the highest medals of the Humane Society and the United States, including the Gold Lifesaving Medal for the rescue of twenty-nine persons from five vessels during a storm in November 1888. The fourth Legend-class national security cutter, USCGC *James* (WMSL 754), is named in his honor.
- Sumner Increase Kimball (2 September 1834–20 June 1923) was the chief of the Treasury Department's Revenue Marine Division and the organizer and only general superintendent of the U.S. Life-Saving Service. As the civilian head of the Revenue Marine in the post–Civil War period, he made significant changes to advance the service and ensure its economy and in 1877 established the School of Instruction to train young officers, an institution that developed into the modern Coast Guard Academy. In 1878 he convinced Congress to fund and organize the U.S. Life-Saving Service to provide for full-time paid crews under the direction of a keeper and served as its general superintendent for thirty-seven years. The service existed until it was merged with the Revenue Cutter Service to form the U.S. Coast Guard in 1915. The seventh Legend-class national security cutter, USCGC *Kimball* (WMSL 756) is named in his honor.

- Michael Healy (22 September 1839–30 August 1904) was a career officer with the U.S. Revenue Cutter Service. He has been recognized since the late twentieth century as the first man of African American descent to command a ship of the United States. Following the Alaska Purchase in 1867, Healy patrolled the Alaskan coastline in command of several cutters, notably USRC *Bear*, for more than twenty years and was essentially the federal government's law enforcement presence in the vast territory, acting as judge, doctor, and policeman to Alaskan natives, merchant seamen, and whaling crews. Commissioned in 1999, the icebreaker USCGC *Healy* (WAGB 20) is named in his honor.
- Richard Etheridge (16 January 1842–January 1900) was the first African American to command a U.S. government installation when he was appointed keeper of U.S. Life-Saving Station Pea Island. Born into slavery in the Outer Banks of North Carolina, Etheridge enlisted into the 36th United States Colored Troops at the onset of the Civil War and, later, the Ninth Calvary. When the Civil War was over, he returned to the Outer Banks and made his living fishing and serving in the newly formed U.S. Life-Saving Service, first at Oregon Inlet and then Bodie Island before being appointed a keeper at Pea Island in 1880. While in command of the station, Etheridge and his crew were awarded Gold Lifesaving Medals for the rescue of eleven lives from a schooner wrecked during a hurricane. The second of the Coast Guard's 154-foot Sentinel-class fast-response cutters, USCGC *Richard Etheridge* (WPC 1102), is named in his honor.
- Idawalley Zoradia Lewis (25 February 1842–24 October 1911) was the keeper of the Lime Rock Light at Newport, Rhode Island, and the first woman to receive the Gold Lifesaving Medal. In 1857 Ida's father, the first keeper of Lime Rock Light, suffered a stroke, leaving the responsibility of minding the light to his wife and children, as was the custom of the Lighthouse Service. Ida was appointed the official keeper of the light in 1879 and remained so until her death in 1911. Her first rescue occurred in 1854 when, at only twelve years old, she saved the lives of four men whose boat capsized in Newport Harbor. In 1881 she performed her most famous rescue for which she was awarded the Gold Lifesaving Medal when she saved the lives of two soldiers whose boat overturned in icy waters during a blinding snowstorm. During her lifetime, Lewis was called "the Bravest Woman in America" and, with her exploits detailed in the national press, became the world's most famous

lighthouse keeper. In 1995 the Coast Guard named the first of its Keeper-class coastal buoy tenders, USCGC *Ida Lewis* (WLM 551), in honor of her.

- Ellsworth Price Bertholf (7 April 1866–11 November 1921) was the fourth captain-commandant of the U.S. Revenue Cutter Service and the first commandant of the U.S. Coast Guard. As a second lieutenant in the Revenue Cutter Service, Bertholf was awarded the Congressional Gold Medal for his role in the 1897 Overland Relief Expedition. His leadership as commandant of both services was critical to the merger of the Revenue Cutter Service and the U.S. Life-Saving Service in 1915 and to the U.S. Coast Guard's survival at a time when members of Congress debated its creation and then future at the outbreak of World War I. The first Legend-class national security cutter, USCGC *Bertholf* (WMSL 750) is named in his honor.
- Elmer "Archie" Stone (22 January 1887–20 May 1936) was a Coast Guard commander and its first aviator. In early 1915 Stone and another officer, Norman B. Hall, were the first to suggest that the Coast Guard develop an aviation capability for at-sea rescue operations and to participate in experimental flights in a Curtiss Model F flying boat. On 10 April 1917, Stone became the Coast Guard's first aviator upon graduating from flight training at Pensacola and was appointed as Coast Guard Aviator No. 1 and Naval Aviator No. 38. In May 1919 Stone, along with Walter Hinton, made the first successful transatlantic flight in a Curtiss seaplane NC-4, a journey that took twenty-three days, made six stops, and required the support of fifty-three ships broadcasting radio navigation signals. Both pilots were awarded the Navy Cross and the Congressional Medal of Achievement for the fight. On 20 December 1934 Stone made history again when he broke the world seaplane speed record, 191 miles per hour over a 3-kilometer test course. The ninth Legend-class national security cutter, USCGC *Stone* (WMSL 758), is named in his honor.
- Elizebeth Smith Friedman (26 August 1892–31 October 1980) was a Coast Guard cryptanalyst during Prohibition and World War II. In 1927 the Coast Guard Intelligence Division began to monitor the radio communications of rum runners after it was discovered that suspects used encrypted messages to support their operations, presuming they would be able to communicate securely. Friedman's Unit 387, the Coast Guard Cryptanalytic Unit, solved the bulk of intercepts collected by Coast Guard stations and helped to indict high-level criminals such as Al Capone. During World War II, Friedman's unit was transferred to the Navy, where they served as the primary U.S.

code breakers assigned to South America, solving numerous cipher systems used by the Germans and their local sympathizers, including three separate Enigma machines and four thousand messages sent on forty-eight different radio circuits. The eleventh Legend-class national security cutter, USCC *Friedman* (WMSL 760), is named in her honor.

- Frank A. Erickson (6 November 1907–17 December 1978) was a U.S. Coast Guard captain and aviator (#32) who became the service's first helicopter pilot in 1943 when he performed a solo flight on board the Sikorsky XR-4. He commanded the New York Coast Guard Air Station when it was designed as the first U.S. Naval Helicopter Training School in 1943; while there, he developed the first technologies for helicopter flotation, hoists, and baskets that would enable search and rescue. On 3 January 1944, then-commander Erickson performed the first recorded helicopter rescue when he flew an HNS-1 through a winter blizzard to deliver a cargo of blood plasma from Manhattan to Sandy Hook that was used to treat over one hundred sailors injured in the sinking of the USS *Turner*, reversing any perception of helicopters as impractical machines.
- Douglas Albert Munro (11 October 1919–27 September 1942) was a U.S. Coast Guard signalman 1st class who was posthumously decorated with the Medal of Honor for an act of "extraordinary heroism" after leading the withdrawal of a force of Marines whose position had been overrun by Japanese forces at the second battle of the Matanikau in September 1942 during World War II. He died at age 22 of a gunshot wound while using the Higgins boat that he was piloting to shield a landing craft, filled with Marines, from Japanese fire. As of 2021, Munro is the only member of the U.S. Coast Guard to have received the Medal of Honor. The sixth Legend-class national security cutter, USCGC *Munro* (WMSL 755), is named in his honor.
- Alexander Murray Palmer Haley (11 August 1921–10 February 1992) was a U.S. Coast Guard chief journalist. While a mess attendant on board the cargo vessel USS *Murzim* (AK 95) during World War II, Haley began writing what combat was like and was first published in *Coast Guard Magazine* in 1944. Toward the end of the war, Haley was assigned to edit the *Outpost*, an official Coast Guard periodical published by the Coast Guard's Personnel Separation Center. Later he became a reporter and then editor for the Coast Guard publication *Helmsman*. In 1949 the Coast Guard officially recognized his talents, and he was changed ratings and promoted to journalist first class

before being advanced to chief journalist later the same year. Haley went on to a distinguished career as a writer, gaining international fame with his book *Roots: The Saga of an American Family*, first published in 1976 and later made into a television miniseries.

- Bernard Challen Webber (9 May 1928–24 January 2009) was a U.S. Coast Guard boatswain's mate and coxswain. While assigned to Station Chatham, Massachusetts, on 18 February 1952, Webber and a crew of three onboard Motor Lifeboat 36500 performed one of the most notable rescues in the service's history when they saved thirty-two of thirty-three crew members from the stricken T2 tanker SS *Pendleton*, which had broken in half during a storm off Cape Cod. Webber maneuvered the 36-foot lifeboat under Pendleton's stern as the tanker's crew, trapped in the stern section, abandoned the wreck of their ship into the motor lifeboat. For their heroic actions, Webber and his crew were awarded Gold Lifesaving Medals. Webber remained in the Coast Guard for over twenty years, rising to the rank of chief warrant officer. The first Sentinel-class fast-response cutter, USCGC *Bernard C. Webber* (WPC 1101), is named in his honor.
- Colleen A. Cain (18 November 1952–7 January 1982) was a U.S. Coast Guard lieutenant who became the service's third female aviator (#1988) and the first female helicopter pilot in June 1979. In her brief career, Lieutenant Cain flew many rescue missions and completed her qualifications as copilot, first pilot, and aircraft commander on board the Sikorsky HH-52A Seaguard helicopter. In the early morning hours of 7 January 1982, while assigned to Air Station Barbers Point, Hawaii, Lieutenant Cain took flight in severe weather, heavy winds, and limited visibility in response to a distress call from the *Pan Am*, a 74-foot fishing vessel taking on water off the coast of Maui with seven persons on board. While en route to the sinking vessel, the HH-52A she was co-piloting crashed into the side of a mountain in the Wailua Valley of Molokai, Hawaii, killing Lieutenant Cain and her two crew members, LT CDR Buzz Johnson and ASM David Thompson. Lieutenant Cain became the first female Coast Guard member killed in the line of duty. On 25 October 1985, the Coast Guard dedicated Cain Hall, a one-hundred-room residence hall at Training Center Yorktown, Virginia, to her memory.
- Nathan B. "Nate" Bruckenthal (17 July 1979–24 April 2004) was a U.S. Coast Guard damage controlman killed in action during the war on terrorism. On 24 April 2004 Bruckenthal was assigned to Coast Guard Tactical Law

Enforcement Team South and deployed on board the USS *Firebolt* when he and six others on a rigid-hull inflatable boat attempted to stop and board a suspicious vessel within the security perimeter of the Khawr Al Amaya Oil Terminal in the Persian Gulf. As the crew was poised to board the boat, an explosion occurred, killing Bruckenthal and two U.S Navy petty officers. Bruckenthal was posthumously awarded the Bronze Star with Combat Distinguishing Device and the Purple Heart for his actions. The twenty-eighth Sentinel-class fast-response cutter, USCGC *Nathan Bruckenthal* (WPC 1118), is named in his honor.

- Thad William Allen (16 January 1949–) is a former U.S. Coast Guard admiral who served as the 23rd commandant of the Coast Guard. While Commander, Atlantic Area and U.S. Maritime Defense Zone Atlantic, he oversaw all Coast Guard operations on the U.S. East Coast, Gulf Coast, and Great Lakes in the aftermath of the 11 September attacks. He later directed the federal response to hurricanes Katrina and Rita in the Gulf Coast region from September 2005 to January 2006 and served as national incident commander of the Unified Command for the Deepwater Horizon oil spill in the Gulf of Mexico in 2010.

Chiefs

The title "Chief" is given to the top three enlisted ranks: chief petty officer, senior chief petty officer, and master chief petty officer. These are a distinct group of naval elders accorded a high level of respect and admiration due in large part to seniority in their respective ratings and within the organization. First and foremost deckplate leaders of the enlisted workforce, chiefs are unique in that they are empowered to "speak truth to power" and "break up the frozen middle" to influence individuals, commands, and the service as a whole. Chiefs have broad responsibilities that range from command to the development and mentoring of commissioned officers.

The grade of chief petty officer was officially established in the U.S. Navy on 1 April 1893. This change would not be adopted by the Revenue Marine or Revenue Cutter Service and would not be adopted by the Coast Guard until its ratings and pay grades were ordered into alignment with the Navy in 1920. On 18 May of the same year, the secretary of the treasury signed General Order No. 43, which authorized chief petty officers in eleven enlisted ratings. Senior chief petty officer and master chief petty officer (MCPO) rates were created on 1 June 1958. Charles Luther Calhoun became the first master chief petty officer of the Coast Guard (MCPOCG) on 27 August 1969.

SPARS

The Coast Guard Women's Reserve Act was signed into law on 23 November 1942 and provided for a branch of the USCG Reserve called SPARS (from the first letters of the Coast Guard's motto: Semper Paratus—Always Ready), which was intended to staff shore stations with women so the men could be freed for duty at sea.

Dorothy C. Stratton was the first SPAR of more than 10,000 women who would volunteer and serve. SPARS trained on several college campuses until the Coast Guard established its own training center at the Palm Beach Biltmore Hotel in Florida and, near the end of the war, in Manhattan Beach, New York. Most who enlisted earned ratings of yeomen or storekeeper and performed clerical duties, but billets did exist for parachute riggers, chaplain assistants, air control tower operators, radio operators, cooks, and pharmacy and medical assistants. Following the surrender of Japan in 1945, the SPARS were gradually discharged as part of the larger military demobilization effort.

TRADITIONS AND SYMBOLS

The traditions observed in the Coast Guard and the symbols associated with the service have developed over the centuries since the first revenue cutters were commissioned. Coast Guard personnel must learn each of these and take pride in carrying them out properly. Like wearing the uniform, they are a mark of the select group of citizens who have taken a special oath to support and defend their country.

National Ensign

The Stars and Stripes is the flag of the United States of America. On 14 June 1777 the Continental Congress passed an act establishing an official flag for the new nation. The resolution stated: "Resolved, that the flag of the United States be thirteen stripes, alternate red and white; that the union be thirteen stars, white in a blue field, representing a new constellation." Today the flag consists of thirteen horizontal stripes, seven red alternating with six white. The stripes represent the original thirteen colonies and the fifty stars represent the states of the union. The colors of the flag are symbolic as well; red symbolizes hardiness and valor, white symbolizes purity and innocence, and blue represents vigilance, perseverance, and justice.

National Anthem

"The Star-Spangled Banner" is the national anthem of the United States. Its lyrics were written by Francis Scott Key on 14 September 1814 after the bombardment of

Fort McHenry by British ships during the Battle of Baltimore in the War of 1812. Key was inspired by the large U.S. flag, with fifteen stars and fifteen stripes, known as the Star-Spangled Banner, flying triumphantly above the fort during the U.S. victory.

"The Star-Spangled Banner" was recognized for official use by the U.S. Navy in 1889 and was made the national anthem by a congressional resolution on 3 March 1931. Although the poem has four stanzas, only the first is commonly sung today.

O say can you see, by the dawn's early light,
What so proudly we hailed at the twilight's last gleaming,
Whose broad stripes and bright stars through the perilous fight,
O'er the ramparts we watched, were so gallantly streaming?
And the rocket's red glare, the bombs bursting in air,
Gave proof through the night that our flag was still there;
O say does that star-spangled banner yet wave
O'er the land of the free and the home of the brave?

Union Jack

The Union Jack is a flag made exactly like the blue-star-studded field in the corner of the national ensign to represent U.S. nationality. As with the national ensign, each star on the Union Jack represents a state. During its early years, the Revenue Cutter Service flew the canton, or the upper corner of the flag nearest the staff, of the Revenue Cutter Service ensign as its jack. Prior to the Civil War, the Revenue Cutter Service adopted as its new jack the canton of the United States flag, or the Union Jack, and this continues to this day. The jack is flown from the jackstaff on the bow of U.S. vessels that are moored or anchored.

Coast Guard Ensign

The Coast Guard Ensign is a mark of authority and must be displayed whenever a Coast Guard cutter or law enforcement detachment takes active measures related to boarding, examining, seizing, or stopping a vessel for the purpose of enforcing the laws of the United States. The use of the ensign can be traced to the first revenue cutters whose task was to guarantee that the maritime public was not evading taxes. Import taxes were the lifeblood of the new nation, but smuggling had become a patriotic duty during the revolution. If the new nation under the Constitution were to survive, this activity needed to be stopped. Working within a limited budget, cutters needed some symbol of authority. Neither officers nor crew members had uniforms. How could a revenue cutter come alongside a merchant ship during an age of pirates and privateers and order it to heave to? The solution was to create an ensign unique to the revenue cutter to fly in place of the national flag while in American waters.

Nine years after the establishment of the Revenue Cutter Service, the Congress passed the Customs Administration Act on 2 March 1799. This act provided that "the cutters and boats employed in the service of the revenue shall be distinguished from other vessels by an ensign and pendant, with such marks thereon as shall be prescribed and directed by the President of the United States." On 1 August 1799, Secretary of the Treasury Oliver Wolcott issued an order announcing that the distinguishing ensign and pennant would consist of "16 perpendicular stripes, alternate red and white, the union of the ensign to be the arms of the United States in a dark blue on a white field." This ensign soon became very familiar in American waters and served as the sign of authority for the Revenue Cutter Service until the early twentieth century.

The ensign was intended to be flown only on revenue cutters and boats connected with the Customs Service, but over the years it was found flying atop customs houses as well. On 7 June 1910 President William Howard Taft issued an executive order adding an emblem to the ensign flown by the revenue cutters, to distinguish it from the ensign flown from the customs houses. The order read: "I hereby prescribe that the distinguishing flag now used by vessels of the Revenue Cutter Service be marked by the distinctive emblem of that service, in blue and white, placed on a line with the lower edge of the union, and over the center of the seventh vertical red stripe from the mast of said flag, the emblem to cover a horizontal space of three stripes. This change to be made as soon as practicable." When the service adopted the name Coast Guard in 1915, the Revenue Cutter Service's ensign became the distinctive flag on all Coast Guard cutters, as it had been for the revenue cutters.

The ensign is poignant with historical detail: in the canton of the flag, there are thirteen stars, thirteen leaves to the olive branch, thirteen arrows, and thirteen bars to the shield. All correspond to the number of states constituting the union at the time the nation was established. The sixteen vertical stripes in the body are symbolic of the number of states in the union when this ensign was officially adopted. The colors used are all symbolic. The color red stands for our youth and sacrifice of blood for liberty's sake. The color blue not only stands for justice but also for our covenant against oppression. The white symbolizes our desire for light and purity. The ensign has undergone several designs and now includes the Coast Guard emblem centered on the seventh red stripe.

The ensign is authorized only for official Coast Guard use and is never flown at half-mast. On ships having two masts, the ensign is flown at the foremast. On a ship having only one mast, it is flown on the same halyard and just below the commission pennant. As it was intended in 1799, the ensign is displayed as a mark of authority for boardings, examinations, and seizures of vessels for the purpose of enforcing the laws of the United States. The ensign is never carried as a parade or ceremony standard.

Coast Guard Standard

The flag of the U.S. Coast Guard is white with a dark-blue Great Seal of the United States. Inscribed in an arc above the seal is "UNITED STATES COAST GUARD," and below it is the Coast Guard motto, "SEMPER PARATUS," and the numerals "1790." The origins of the Coast Guard Standard are obscure, but an 1840 painting of the revenue cutter *Alexander Hamilton* shows it flying a jack very similar to today's standard. The flag in the painting, like the Union Jack, which is the canton or upper corner of the United States flag, appears to be the canton of the Revenue Cutter Service ensign.

An illustration in 1917 shows thirteen stars in a semicircle surrounding the Great Seal. After 1950 the semicircle of stars was changed to the circle containing thirteen stars. The Coast Guard Standard is used during parades and ceremonies and is adorned by our battle streamers. We are the only branch of the U.S. military to have two official flags, the Coast Guard Standard and the Coast Guard Ensign.

A battle-ax tops the flagstaff on which the Coast Guard Standard flies.

Battle Streamers

Embroidering the names of battles on flags may be traced to the early days of the republic. By the end of the nineteenth century, embroidery was discontinued in favor of inscribed silver bands around the color staffs. This, too, was changed in World War I in favor of small ribbons bearing battle names. The Coast Guard adopted battle streamers in 1968 following the practice established by the U.S. Marine Corps. Streamers are attached to the Coast Guard Standard, replacing cords and tassels. They are carried in all ceremonies representing heroic actions in all naval encounters from 1790 to the present. Only major headquarters commands may display a complete set of battle streamers. Individual units may only display those they have earned. The Coast Guard has authorized a total of forty-three battle streamers:

Unit

Coast Guard Presidential Unit Citation
Navy Presidential Unit Commendation
Department of Transportation—Secretary's Outstanding Unit Award
Coast Guard Unit Commendation
Navy Unit Commendation
Coast Guard Meritorious Unit Commendation
Navy Meritorious Unit Commendation
Army Meritorious Unit Commendation

Campaign

Maritime Protection of the New Republic
French Naval War (Quasi-War with France)
War of 1812
African Slave Trade Patrol
Operations against West Indian Pirates
Indian Wars
Mexican War
Civil War
Spanish-American War
World War I Victory
China Service
Yangtze Service
American Defense Service
American Campaign
European-African–Middle Eastern Campaign
Asiatic-Pacific Campaign
World War II Victory
Navy Occupation Service
Korean Service
National Defense Service
Armed Forces Expeditionary
Vietnam Service
Southwest Asia Service
Kosovo Campaign
Global War on Terrorism Expeditionary
Global War on Terrorism Service

Afghanistan Campaign
Iraq Campaign
Inherent Resolve Campaign

Foreign
Croix de Guerre, French World War II
Philippine Defense
Philippine Liberation
Philippine Independence Philippine Presidential Unit Citation
Republic of Vietnam (RVN) Armed Forces Meritorious Unit Citation Gallantry Cross with Palm
RVN Meritorious Unit Citation Civil Actions Medal First-Class Color with Palm

POW/MIA Flag

The POW/MIA flag is the symbol of our nation's concern and commitment to resolving as fully as possible the fates of Americans still imprisoned, missing, and unaccounted for, thus ending the uncertainty for their families and the nation. It is a rectangular flag with a black background and a silhouette of a prisoner of war camp guard tower overlooking a prisoner of war with the words "POW/MIA YOU ARE NOT FORGOTTEN" inscribed thereon. It is flown at all times that the national ensign is flown. Shore installations fly this flag beneath the national ensign. Ships fly the flag from the inboard halyard, port signal yardarm.

Coast Guard Seal and Service Mark

The creation of an official Coast Guard seal confirmed the existence of a symbol that had evolved over the decades. The Revenue Cutter Service, the Life-Saving Service, and the Lighthouse Service all had their own unique distinguishing devices. The Bureau of Marine Inspection and Navigation used drawings of ships and marine equipment on licenses and stationery. But it wasn't until 1927, after Treasury Secretary Andrew W. Mellon approved a design, that the Coast Guard had its first official seal/emblem. Draftsman Oscar H. Keefor designed the official Coast Guard seal and emblem. At that time the seal and emblem were the same, but over the years became two different devices. The seal is used for official documents and records of the Coast Guard and may also be used for jewelry and stationery at the discretion of the commandant. The Coast Guard service mark, a simplified version of the seal, was created as a visual identifier for the Coast Guard. It not only appears on the Coast Guard ensign but also inside the distinctive slash on our cutters, boats, aircraft, and signage. It is also used on medals and plaques where space is limited.

Seal of the Coast Guard Reserve

Designed in 1991, the crossed anchors supporting the shield are adapted from the U.S. Coast Guard's symbol, and the dark-blue, scarlet, and white colors are traditional with the organization. Gold is emblematic of achievement and excellence. The sword symbolizes defense and the organization's military heritage. The motto "Professionalism, Preparedness, and Patriotism" embodies the U.S. Coast Guard Reserve's total mission to protect our national security. The round shape of the motto scroll alludes to a life preserver and represents the U.S. Coast Guard Reserve's peacetime mission to help others.

Coast Guard Distinguishing Mark

Distinguishing marks were first introduced in the 1905 U.S. Navy uniform regulations and were used to distinguish personnel who had met specific qualifications in addition to qualifications required for their rating. Distinguishing marks were also issued to personnel who were members of crews attaining merit in approved competitions such as gunnery exercises and engineering readiness. Originally authorized in 1920 "to be worn by every enlisted man of the Coast Guard," the Coast Guard distinguishing mark is a shield with thirteen stars and stripes. Today it is worn in several locations on the uniforms of both officers and enlisted Coast Guardsmen alike.

Coast Guard Slash / Racing Stripe

Our familiar and distinctive red slash or "racing stripe" did not appear on cutters, boats, and aircraft until the 1960s. In 1964 an industrial design firm recommended that the Coast Guard adopt a symbol or mark that would be easily distinguished from other government agencies and easily applied to ships, boats, aircraft, stations, vehicles, signs, and printed forms. Their design was a wide red bar to the right of a narrow blue bar, both canted at 64 degrees. Centered on the red bar was a new emblem. Studies were done with experimental markings for their impact on the public as well as for their long-run compatibility with the Coast Guard's mission and traditions. The reaction was overwhelmingly favorable. Three years later, on 6 April 1967, the now-famous slash appeared throughout the Coast Guard. In the final design, only the emblem changed when the traditional Coast Guard emblem was selected for centering on the red stripe.

Coast Guard Commissioning Pennant

The commissioning pennant signifies that the vessel is a U.S. Coast Guard cutter under the command of a commanding officer and is flown until the cutter is decommissioned. The ensign, jack, and commissioning pennant are hoisted directly after the commissioning order and struck as the final act before the captain declares the ship decommissioned, the only exceptions being when either a flag officer or a civilian official is embarked; then the flag officer's or civilian official's personal flag is flown in its place. An elongated pennant with a swallow-tailed fly, the portion near the hoist consists of a white field on which is superimposed thirteen blue five-pointed stars in a row, each of the stars having one point upward. The remaining portion of the flag is vertically divided into red and white stripes with the last few inches split to form a swallow-tail.

When the commanding officer is ashore, the ship will also display the international code flag known as the Third Substitute pennant.

Coast Guard Aircraft Markings / National Insignia

The first military aviation insignia of the United States include a red star painted on the vertical tail of U.S. Army Signal Corps aircraft just over a year before American involvement in World War I began. By May 1917 all branches of the military outside of the Western Front of Europe used an identical tricolor roundel as their national insignia. Lessons learned during World War II found that shapes were more important than colors for visual identification, causing a white bar with a red outline to be added. The current version, adopted in January 1947, roughly nine months before the formation of the U.S. Air Force, consists of a circular dark-blue field containing a single, five-pointed white star bisecting bars of red, white, and blue.

On Coast Guard rotary-wing aircraft the national insignia is applied to both sides of the fuselage. After adopting the Coast Guard slash or "racing stripe" insignia on the fuselages of its fixed-wing aircraft, the Coast Guard, unique among U.S. military services, now uses the roundel on the vertical fin of its fixed-wing aircraft as a fin flash. Fixed-wing aircraft also display the national insignia on the topside of the left wing and the underside of the right wing.

Semper Paratus, "Always Ready"

Semper Paratus is a Latin phrase meaning "Always Ready" and is the Coast Guard's motto and the title of our service song. The earliest recorded use of the phrase in connection to the service was in the New Orleans newspaper *Bee* in January 1836, which used the phrase in an article praising the revenue cutter *Ingham*. The motto was adopted when a new seal containing the phrase appeared on a general order of the Revenue Cutter Service on 21 May 1897.

CAPT F. S. Van Boskerck wrote the words to the song in 1922 while he was commanding officer of USCGC *Yamacraw* (CG 21) and wrote the music five years later using a beat-up old piano in Unalaska, Alaska. In describing his motivation, Captain Van Boskerck reported that for as long as he could remember, Semper Paratus had been used as a watch phrase by members of the Revenue Cutter Service and Coast Guard.

THE ANCIENTS

In career fields that are both inherently dangerous and richly rewarding, it is fitting that we take time to recognize distinguished senior leaders who epitomize the very best. The tradition of naming a Coast Guard community's "Ancient" dates back to 1966, when aviators created the Ancient Albatross, a title that recognizes the longest-serving pilot in the Coast Guard. In 1978 the cutter community followed suit by naming the Gold and Silver Ancient Mariner to honor the officer and enlisted member with the earliest qualifications as a permanent cutterman. In 1988 aviators established the Enlisted Ancient Albatross. In 2003 the Ancient Keeper was developed to honor the member with the most experience in the small-boat community. And finally in 2016 the Gold and Silver Ancient Trident was created to honor those longest-serving members in the Marine Safety community.

Earning the title as a Coast Guard Ancient is more than an honorary position; it's a sign of respect for achieving mastery of craft, for reaching the peak of a specialty. Recipients are charged with keeping a close watch to ensure that the traditions and time-honored reputation of the Coast Guard and the professionalism of the people within the community are maintained. In their capacity as an Ancient, recipients attend or take part in official ceremonies that involve their respective community, advise headquarters program offices on issues related to their specialty, and serve as mentors to junior members to develop proficiency and pride in the execution of their missions. As the respected Ancients retire from active duty, often after more than thirty years of service, they pass on the torch in official ceremonies.

Enlisted Ancient Albatross

The Enlisted Ancient Albatross was first given in 1988 to honor the enlisted member on active duty with the earliest graduation date from "A" school in a Coast Guard aviation rating. It has been held by senior ranking enlisted members such as MCPO Eric Trent, who served as MCPOCG. Why the Albatross? It is a short, fat, stubby bird with long thin wings that, despite its initial appearance, is known for its endurance and ability to fly regardless of age or weather conditions. These traits made it a natural choice to be the face of a tradition that recognizes the Coast Guard's most senior aviators. Honorees perform the award ceremony wearing vintage "Red Baron" style flying leathers, goggles, scarves, and caps.

Silver Ancient Mariner

Cuttermen honor their senior-most members with the title Ancient Mariner. The USCG officer still on duty with the longest linage of sea time is referred to as the Gold Ancient Mariner, while the veteran enlisted member is the Silver Ancient Mariner. As their badge of office, they carry nautical spyglasses and authentic relic headgear of the Revenue Cutter Service and Lighthouse Service, respectively, when carrying out Mariner duties. Written in the Ancient Mariner's Log, which is passed down from one Ancient to another, is a note from seventeenth commandant of the Coast Guard, James Gracey, which reads, "The spirit of centuries of our fellow Coast Guardsmen and their forebears, who have met the sea and conquered it, or have gone to their Maker while trying, is keeping watch with you, the Ancient Mariner. Keep a sharp eye out to ensure their tradition is continued and the time-honored reputation of the Coast Guard is kept intact."

Ancient Keeper

The Joshua James Ancient Keeper Award is a given to an active-duty member, historically a master chief or warrant officer, to recognize their longevity and outstanding performance in Coast Guard boat operations. The award's namesake, Joshua James, is the most celebrated lifesaver in Coast Guard history, with 626 lives saved during his lifetime. As a badge of their office, the Ancient Keeper wears a replica of the those worn by U.S. Life-Saving Service Station Keepers and Surfmen. Describing the award during a 2012 Change of Watch ceremony in which MC James Clemens relieved CWO Thomas Guthlein, the twenty-fourth commandant, Robert Papp, a Gold Ancient Mariner himself, said, "Holding the title of Ancient Keeper means you represent all keepers who continue to live by the creed of those who have served before them; those who continue to go to the sea's rocky realm at the peril of their own lives, in pursuit of the safety and security of others."

Silver Ancient Trident

Established in 2016, the Ancient Trident Award honors the heritage and progression of Marine Safety from the Steamboat Inspection Service, to the Bureau of Marine Inspection and Navigation, and finally to the Coast Guard. Since the transfer of the Marine Safety mission to the Coast Guard, it has been modified and expanded to include vessel and port safety and security, the protection of the marine environment, and facilitation of maritime transportation. The enlisted Ancient Trident wears a silver trimmed hat with an anchor and ring buoy insignia.

MEMORIALS

Hundreds of memorials have been dedicated to honor and remember the service and sacrifice of Coast Guardsmen, whether in defense of the nation or that others might live. Although unable to list all here, several notable examples are described. If you ever need to feel inspired, want to thank those responsible for our freedoms, or need a reminder that you are a part of an organization dedicated to higher causes, you're encouraged to visit one of these sites.

Coast Guard Monument, Arlington National Cemetery, Section 4, Arlington, VA

Dedicated 23 May 1928 as a tribute to the Coast Guardsmen who lost their lives in World War I, this monument is made of white marble and pyramid-shaped. It stands twelve feet high. The foundation and pyramid are suggestive of rocks standing in the sea along the coast or marks of danger to navigation, and they represent the service ideals of steadfastness and endurance. The front of the monument has the service emblem on it and a seagull, symbolic of the Coast Guard's watchful and untiring spirit. The names of those in the Coast Guard who lost their lives in World War I are inscribed on the monument. The northwest side is dedicated to CGC *Seneca*, which lost eleven Coast Guardsmen while endeavoring to salvage the torpedoed British steamer *Wellington* in the Bay of Biscay on 17 September 1918. The southeast side is dedicated to CGC *Tampa*, sunk by an enemy submarine in Bristol Channel on 26 September 1918. All 115 on board were lost. The architect was George Howe, and the sculptor was Gaston Lachaise. Carved on the foundation: Thy Way Is In The Sea.

USS *Serpens* Memorial, Arlington National Cemetery, Section 34, Arlington, VA

When the Coast Guard–manned USS *Serpens* (AKA 97) exploded and sank on 29 January 1945 at Lunga Beach, Guadalcanal, Solomon Islands, it marked the largest single disaster suffered by the Coast Guard in World War II. Only 2 survived the blast, while 250, including 193 Coast Guardsmen, were lost. The Serpens Monument is octagon-shaped and has inscribed upon it an alphabetical listing of the deceased servicemen's names, ranks, and branches of service. It marks the second-largest mass grave at Arlington.

Vietnam Veterans Memorial, Washington, DC

Among the 58,196 names etched on the Vietnam Veterans Memorial (Wall) in our nation's capital are seven Coast Guardsman. The wall, dedicated in 1982, is located adjacent to the Lincoln Memorial.

Coast Guard Academy, New London, Connecticut

Bertholf Plaza

Completed in 1992, the plaza is named for Ellsworth P. Bertholf, the first commandant of the modern-day Coast Guard. The plaza is the site of several plaques commemorating Coast Guard personnel who served in wartime.

Hall of Heroes Memorial

Located in Chase Hall barracks, the Hall of Heroes was established in April 2005 by the Class of 1959 to commemorate heroic alumni of the academy. The Hall of Heroes includes the Wall of Remembrance, which honors academy graduates who perished while carrying out an operational mission, and the Wall of Gallantry, which honors academy alumni who have been formally recognized for acts of heroic service.

CAPT Hopley Yeaton Memorial

The tomb of the first commissioned officer of the Revenue Marine, Hopley Yeaton, now lies on the academy's grounds. He was originally buried in Lubec, Maine, but in 1975 his burial site was threatened by modernization. The Corps of Cadets sailed the *Barque Eagle* to Lubec where his remains were exhumed and then laid to rest at the academy.

Coast Guard Enlisted Memorial, Training Center Cape May, New Jersey

This memorial is dedicated to enlisted men and women who have died in the line of duty and is located on the training center's parade field. It consists of three granite walls etched with the names of over 1,500 service members, centered around a pyramid topped with a bronze flame.

Coast Guard Aviation Memorial, Elizabeth City, North Carolina

This cenotaph is dedicated to more than two hundred airmen who did not return. It is located at the Wings of Faith Memorial Chapel.

Coast Guard at Normandy Memorial, Utah Beach, Normandy, France

On 6 June 1994, the Coast Guard Combat Veterans Association dedicated a plaque to the Coast Guard veterans who served at the invasion of Normandy. The plaque's inscription reads: "Dedicated this 6th day of June, 1994, to the members of the United States Coast Guard who participated in the initial invasion of Normandy, especially to those who gave their lives here, and to all United States Coast Guard forces who served worldwide on land, sea and air during WWII. The nations of the world shall long remember Normandy; the United States armed forces, their allies and the cost of freedom at this place. The United States Coast Guard motto is, as always, 'Semper Paratus' Always Ready."

U.S. Coast Guard Cutter *Ingham* (WHEC 35) Maritime Museum and National Historic Landmark, Key West, Florida

In accordance with a directive from the Commandant's Office, *Ingham* is the official memorial site to Coast Guardsmen killed in action in World War II and Vietnam. The list of casualties is kept on the quarterdeck. It is the only Coast Guard cutter afloat today to receive two Presidential Unit Citations for extraordinary heroism in action against an armed enemy.

Coast Guard Bicentennial Monument, Newburyport, Massachusetts

This monument was dedicated by the City of Newburyport on 4 August 1989, in anticipation of the celebration of the two hundredth birthday of the Coast Guard.

Newburyport is the birthplace of the Coast Guard. The first revenue cutter, *Massachusetts*, was launched upriver at MacKay Shipyard, not far from where this monument stands on the waterfront behind the maritime museum.

Coast Guard Governors Island Monuments, Governors Island, New York

To commemorate the Coast Guard's thirty-one-year stay on the New York harbor island, the Coast Guard placed four monuments at each compass point on the island. The west monument is a tribute to area lighthouses. The east monument is a remembrance of service during World War I and World War II. The south monument honors those from New York who fought for our nation's freedom. The north monument, facing Lower Manhattan, is a tribute to the "cradle of Coast Guard history," remembering Alexander Hamilton and our nation's first Congress, then convened in New York City, which passed legislation establishing the Revenue Marine, on 4 August 1790.

CGC *Cuyahoga* Memorial, Yorktown, Virginia

When CGC *Cuyahoga* collided with the Argentine motor vessel *Santa Cruz* near the mouth of the Potomac River in Virginia on 20 October 1978, ten Coast Guardsman and one Indonesian naval officer died. This monument honors them.

CGC *Blackthorn* (WLB 391) Monuments, St. Petersburg, Florida, and Galveston, Texas

On the evening of 28 January 1980, CGC *Blackthorn* (WLB 391) collided with the tanker *Capricorn* in Tampa Bay, Florida, near the Sunshine Skyway Bridge. Homeported in Galveston, *Blackthorn* had just completed an extensive yard period in the Tampa area and had many crew members who had never sailed on board the vessel. The incident claimed the lives of twenty-three Coast Guardsmen. A monument commemorating the sinking is located at the Sunshine Skyway Bridge north base rest area and points out the location where *Blackthorn* and *Capricorn* collided. The Florida legislature named the wayside parks at each end of the Skyway Bridge "Blackthorn Memorial Park." At Base Galveston, a buoy contains a commemorative plaque that is lit permanently. It looks out over Galveston Bay.

White Alder Memorial Park, New Orleans, Louisiana

CGC *White Alder* collided with the freighter *Helena* on 7 December 1968. The memorial is dedicated to seventeen Coast Guardsmen who were killed in the collision.

Columbia River Bar Memorial, Astoria, Oregon

The Columbia River Bar is one of the most treacherous in the nation, if not the world. The monument in nearby Astoria honors nine Coast Guardsmen who paid the ultimate sacrifice, including five from the MLB *Triumph* in 1961, three from UTB 41332 in 1977, and MK1 Charles W. Sexton in 1991.

Douglas Munro Monuments

Several monuments have been erected to Munro, including a bronze statue at the Coast Guard Training Center in Cape May, a larger-than-life bust at the Douglas A. Munro Coast Guard Headquarters complex, a black marble pillar etched with Munro's visage at the Coast Guard Academy, a memorial marker at the Point Cruz Yacht Club in Honiara in the Solomon Islands, and a monument in Crystal River, Florida. In 2006 Munro was added to the Wall of Heroes of the National Museum of the Marine Corps, becoming the only non-Marine represented on it.

3

COAST GUARD MISSIONS AND ORGANIZATION

The Coast Guard serves three primary roles and maintains a persistent maritime presence in order to perform them at all times. These roles are to

- Protect those on the sea (maritime safety);
- Protect the nation from threats delivered by sea (maritime security); and
- Protect the sea itself (maritime stewardship).

The full spectrum of Coast Guard operations are executed through the Prevention–Response operating concept to prevent, protect against, respond to, and recover from maritime incidents. Coast Guard missions are described in the United States Code (U.S.C.), the Code of Federal Regulations (CFR), executive orders, and international law.

STATUTORY MISSIONS

The Coast Guard is responsible for carrying out eleven statutory missions, meaning missions or duties assigned to the service by federal law. By performing these assigned missions, the Coast Guard fulfills its intended roles in maritime safety, security, and stewardship.

Maritime Safety

One of the most basic responsibilities of the U.S. government is to protect the lives and safety of Americans. There are two different missions that fall under maritime safety.

Organization

Search and Rescue (SAR). Minimizing the loss of life, injury to persons, and property damage or loss by rendering aid to those in distress at sea has always been a Coast Guard priority. Coast Guard SAR response involves multimission boat stations, cutters, and aircraft linked by communications networks. The National SAR Plan divides the U.S. area of SAR responsibility into internationally recognized inland and maritime SAR regions. The Coast Guard is the Maritime SAR Coordinator. To meet this responsibility, the Coast Guard maintains SAR facilities on the East, West, and Gulf Coasts; in Alaska, Hawaii, Guam, and Puerto Rico; and on the Great Lakes and inland U.S. waterways. The Coast Guard also operates the automated mutual-assistance vessel rescue (AMVER) system, which keeps updated positions of vessels that can render assistance to any vessel in distress far out to sea. The Coast Guard is recognized worldwide as a leader in the field of search and rescue.

Marine Safety. The Coast Guard performs its marine safety mission by conducting marine inspection, marine investigation, waterways management, port safety, and merchant mariner credentialing activities. Its personnel inspect commercial vessels for safety and seaworthiness, respond to pollution releases, investigate accidents and deaths on board vessels, approve and monitor marine events, and test and evaluate prospective merchant mariners.

Maritime Security

Law enforcement is the oldest of the Coast Guard's missions, dating back to our founding as a system of cutters in 1790. Today the service is the lead federal agency for maritime law enforcement, and its commissioned, warrant, and petty officers possess authorities to prevent, detect, and suppress violations of federal law on, under, and over the seas. There are four primary missions that fall under maritime security.

Port Waterways and Coastal Security (PWCS). Since 1917 the Coast Guard has been responsible for the security of our ports and waterways. The PWCS mission involves the prevention and disruption of terrorist attacks, sabotage, espionage, or other subversive acts and the response to and recovery from those that do occur.

Defense Readiness. Under Title 14 U.S.C., the Coast Guard is "at all times an armed force of the United States." As part of the Joint Force, the Coast Guard maintains its readiness to carry out military operations and is a vital component of the U.S. National Fleet. The service maintains a high state of readiness at all times in order to operate alongside the Navy and Marine Corps. At any time, its personnel and assets may be called upon to execute essential military operations during national crisis, armed conflict, or war. The Coast Guard supports U.S. combatant commanders by

- Deploying law enforcement teams on board U.S. Navy and allied ships to enforce embargos;
- Training maritime security forces to secure ports and offshore platforms;
- Training foreign nations in maritime law enforcement, port security, and maritime search and rescue;
- Conducting alert, intercept, communication, surveillance, and escort activities for National Air Defense; and
- Functioning as a service under the Navy in time of war or when directed by the president.

Drug Interdiction. The Coast Guard is the lead federal agency for maritime drug interdiction and shares lead responsibility for air interdiction with U.S. Customs and Border Protection. The Coast Guard's mission is to reduce the supply of drugs from the source by denying smugglers the use of air and maritime routes in the transit zone, a 6 million square mile area that includes the Caribbean, Gulf of Mexico, and Eastern Pacific. In meeting the challenge of patrolling this vast area, the Coast Guard coordinates closely with other federal agencies and countries within the region.

Migrant Interdiction. The Coast Guard is tasked with enforcing immigration law and interdicting undocumented migrants at sea, denying them entry via maritime routes to the United States, its territories, and its possessions. Thousands of people try to enter this country illegally every year using maritime routes, many via smuggling operations. Interdicting migrants at sea means they can be quickly returned to their countries of origin without the costly processes required if they successfully enter the United States. When successful, illegal immigration can potentially cost U.S. taxpayers billions of dollars each year in social services. In addition to relieving this financial burden on our citizens, the Coast Guard's efforts help to support legal migration systems. Primarily, the Coast Guard maintains its humanitarian responsibility to prevent the loss of life at sea since most migrant vessels are dangerously overloaded, unseaworthy, or otherwise unsafe.

Maritime Stewardship

America's waters and their ecosystems are vital to the health, well-being, and economy of our nation. Our marine environment is among the most valuable and productive natural resources on Earth, containing one-fifth of the world's fishery resources. There are five primary missions performed within the realm of Maritime Stewardship.

Other Law Enforcement. Preventing illegal foreign fishing vessels from encroaching in our exclusive economic zone (EEZ) is a mission with high national priority. This protects the integrity of our maritime borders and ensures the health of our fisheries. The Coast Guard focuses on three high-threat areas for illegal foreign incursions: the U.S.–Mexico border in the Gulf of Mexico, the Russian Maritime Boundary Line in the Bering Sea, and the eight noncontiguous EEZ zones in the Western/Central Pacific.

Living Marine Resources. The Coast Guard is the primary at-sea enforcement arm for living marine resource laws administered by the National Oceanic and Atmospheric Administration (NOAA), notably the Magnuson-Stevens Fisheries Conservation and Management Act (MSFCMA), the Marine Mammal Protection Act (MMPA), and the Endangered Species Act (ESA).

Marine Environmental Protection and Prevention. Protecting the marine environment includes planning and operations to prevent, enforce, investigate, respond to, and mitigate the threat, frequency, and consequences of oil discharges and hazardous substance releases in U.S. waters.

Aids to Navigation. Aids to navigation (ATON) are used by mariners to navigate U.S. waterways in and around the continental United States, Alaska, Hawaii, and U.S. territories such as those in the Caribbean and the western Pacific. The Coast Guard establishes and maintains these aids, to include buoys, fixed structures, and the virtual aids used by today's electronic navigation systems.

Ice Operations. The Coast Guard conducts icebreaking to assist vessels and communities in emergencies and to facilitate essential commercial maritime activities in the Great Lakes and Northeast regions. Beyond domestic operations, the Coast Guard operates the only U.S.-flagged heavy icebreakers capable of providing year-round access to the polar regions. The International Ice Patrol facilitates commerce by broadcasting position information on icebergs crossing south of 48° N latitude.

THE MARITIME DOMAIN

The maritime environment is dangerous, complex, and difficult to govern. Its geography, weather, temperatures, and sea states are dynamic and present significant challenges.

Operational Areas

The United States claims sovereignty over 3.4 million square nautical miles of maritime territory, which includes 95,000 miles of coastline and 361 ports that compose our Maritime Transportation System (MTS). The MTS also includes 12,000 miles of interconnected navigable rivers, the Intracoastal Waterway (ICW), and the Great Lakes, which connect inland metropolitan areas, industrial complexes, and the agricultural heartland of the country with one another and with the oceans. The Coast Guard has defined three geographic operational areas.

Inland. The inland operational area includes all waters shoreward of the U.S. baseline that are subject to the jurisdiction of the United States, including the Western Rivers System.

Coastal. The coastal operational area includes waters extending from the U.S. baseline seaward to a distance of fifty miles offshore, including the territorial sea, contiguous zone, and that part of the U.S. EEZ extending seaward to fifty miles from the baseline. This area includes the waters seaward of the baseline on the U.S. side of the international boundary on the Great Lakes.

Offshore. The offshore operational area extends seaward from fifty nautical miles of the U.S. baseline to those waters claimed by foreign nations. This area includes the polar regions, defined as north of 60° N latitude (Arctic) and south of 60° S latitude (Antarctic).

Jurisdictional Complexity

Coast Guard operational areas overlap with recognized U.S. and international geographic regimes. These regimes of ocean and airspace directly affect maritime operations by determining the degree of jurisdiction a coastal state may exercise.

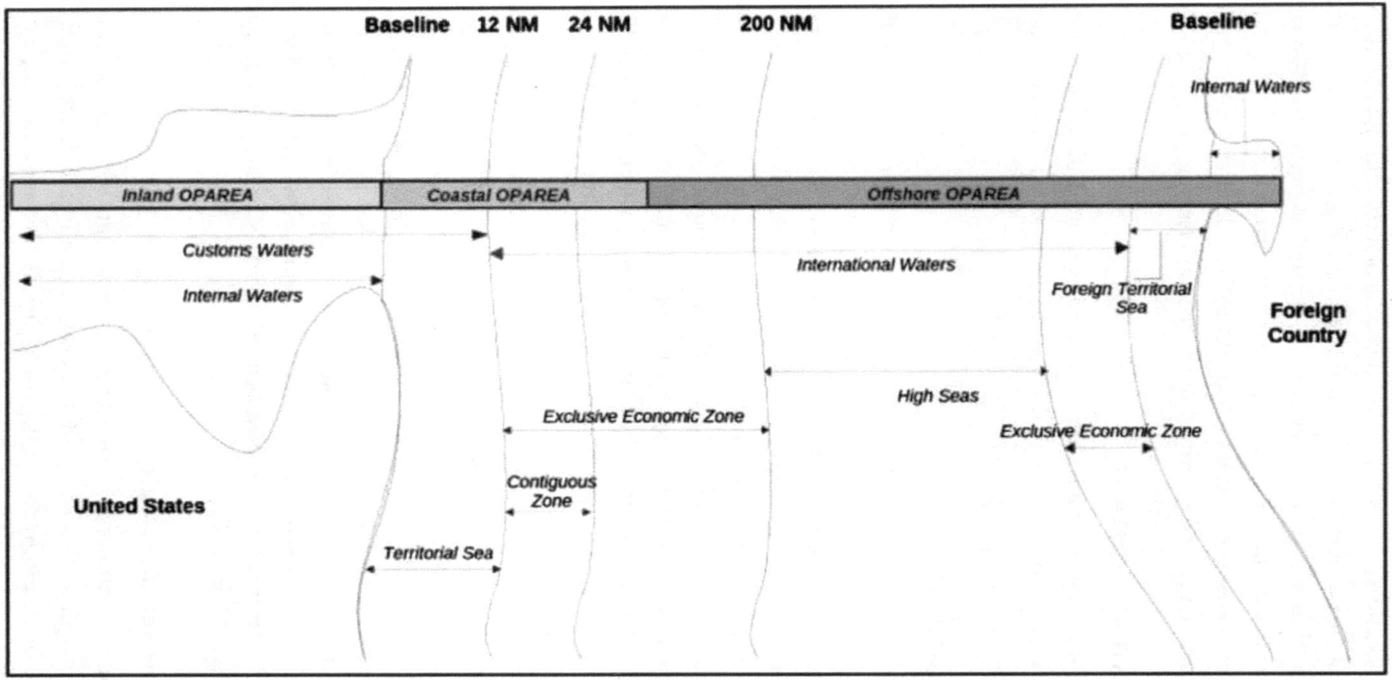

Operational areas

National waters include internal waters and territorial seas of a coastal state. International waters include waters seaward of the territorial seas of any state.

Internal Waters. All U.S. waters shoreward of the baseline, including all waters on the U.S. side of the international boundary of the Great Lakes.

Territorial Sea. The waters within the belt that is twelve miles wide and adjacent to the U.S. coast measured seaward from the baseline. For the purpose of enforcing some domestic U.S. laws, the territorial sea extends only three miles seaward of the baseline.

Contiguous Zone. The waters within the belt adjacent to and seaward of the territorial sea and extending to twenty-four miles from the baseline.

Customs Waters. Generally defined as the waters shoreward of a line drawn twelve miles from the baseline, including territorial sea and inland waters with ready access to the sea.

Exclusive Economic Zone (EEZ). The zone of waters beyond and adjacent to the territorial sea not extending beyond two hundred miles from the baseline.

International Waters. Waters seaward of the outer limit of the territorial sea of any nation but including the high seas, EEZ, and contiguous zones.

High Seas. All parts of the ocean seaward of the EEZ. If a nation has not proclaimed an EEZ, the high seas begin at the seaward edge of the territorial sea.

Foreign Territorial Sea. The United States recognizes territorial sea claims of other nations up to a maximum distance of twelve miles from the baseline.

ORGANIZATION

Since most Coast Guard operations are domestic, the service has always been organized along regional lines. The Coast Guard's shore establishment divides the continental United States and its territories into two area commands. These areas are divided into districts, and each district is further divided into sectors. Operational units report to various levels in this organizational structure. For example, small boat stations report to sector commanders while air stations report to district commanders. Unlike the ships of the U.S. Navy, Coast Guard cutters are not organized into standing fleets and instead individual cutters report to commanders at various levels, with patrol boats reporting to districts while larger national security cutters report to areas.

Although focused primarily on domestic operations, the Coast Guard has three major operational commands located outside the United States to facilitate its high seas missions and to deliver support to the Department of Defense in support of national security objectives.

Senior Leaders

President of the United States (POTUS)

The head of state and head of government of the United States of America. Directs the executive branch of the federal government and is the commander-in-chief of the U.S. Armed Forces.

Secretary of Homeland Security (DHS)

The head of the Department of Homeland Security, the federal department tasked with ensuring public safety in the United States. Agencies within the department include the Coast Guard, the Federal Protective Service, Customs and Border Protection, Immigration and Customs Enforcement, the Secret Service, and the Federal Emergency Management Agency. The commandant of the Coast Guard reports to the secretary of homeland security. When operating under the Department of the Navy, the commandant reports to the secretary of the navy.

Coast Guard Headquarters

The Douglas Munro Coast Guard Headquarters building (above) is a 1.2 million square foot office complex in the National Capital Region that opened in 2013 and houses approximately 3,700 military personnel and civilian employees. Headquarters serves as the home of the senior-most organizational level for the service consisting of the commandant, the vice commandant, deputy commandants, assistant commandants, headquarters directorates, and supporting staffs.

Commandant of the Coast Guard (COMDT/CCG)

The service chief and its highest-ranking member, the commandant of the Coast Guard is an admiral, appointed for a four-year term by the president of the United States upon confirmation by the U.S. Senate. Unlike the chiefs of other services, the commandant maintains operational command over the Coast Guard. Reporting directly to the commandant are the vice commandant, master chief petty officer of the Coast Guard, chaplain of the Coast Guard, Civil Rights Directorate, and the chief administrative law judge.

Vice Commandant of the Coast Guard

Second in command, the vice commandant oversees the Coast Guard's senior operational and mission-support commanders and its headquarters staff and serves as service chief in the absence of the commandant. Reporting directly to the vice commandant are the judge advocate general and chief counsel, the director of governmental and public affairs (CG-092), the assistant commandant for resources/chief financial officer (CG-8), the deputy commandant for operations, and the deputy commandant for mission support, Atlantic Area, and Pacific Area.

Master Chief Petty Officer of the Coast Guard (MCPOCG)

The MCPOCG is the senior enlisted member of the U.S. Coast Guard and the principal adviser to the commandant of the Coast Guard on all enlisted personnel matters. The MCPOCG is appointed by the commandant of the Coast Guard.

Chaplain of the Coast Guard

The chaplain of the Coast Guard advises the commandant on religious matters, promotes the well-being of personnel, and manages members of the Navy Chaplain Corps assigned to the Coast Guard to ensure the Coast Guard family receives the full continuum of pastoral care.

Deputy Commandant for Mission Support (DCMS)

The DCMS enables Coast Guard missions through support of people, platforms, and systems. The DCMS leads the following assistant commandants:

- Human resources (CG-1)
- Engineering and logistics (CG-4)
- Assistant commandant for C4&IT (CG-6)
- Acquisitions (CG-9)

Deputy Commandant for Operations (DCO)

The DCO develops and oversees the execution of operational planning, policy, and international engagement at the strategic level. The DCO leads the following assistant commandants:

- Coast Guard intelligence (CG-2)
- Response policy (CG-5R)
- Prevention policy (CG-5P)
- Capability (CG-7)
- Director of international affairs and foreign policy (CG-DCO-I)

National Command Center

The Coast Guard's National Command Center (NCC) is located at Coast Guard Headquarters in Washington, D.C., and operates twenty-four hours a day, every day of the year, to maintain situational awareness of Coast Guard operations worldwide and of all significant external events or activities with potential Coast Guard involvement. The NCC serves as the command center for the commandant, coordinating information flow and providing support to national-level decision makers; Coast Guard forces; federal, state, local, and tribal authorities; and industry.

Headquarters Units

Coast Guard Headquarters units may be geographically distant from Washington, D.C., but they operate independently of area and district commanders. These units were established to provide certain support services for the Coast Guard as a whole or to satisfy requirements in a specific geographic area. Headquarters units include the following:

Academy (CGA), New London, Connecticut (CG-1)
Air Station, Washington, D.C. (CG-7)
Asset Project Office (APO), Baltimore, Maryland (CG-9)
Aviation Logistics Center (ALC) Elizabeth City, North Carolina (CG-4)
Command, Control, Communications, Computers, and Information Technology Service Center (C4ITSC), Alexandria, Virginia (CG-6)
Counterintelligence Service (CG-CI), Washington, D.C. (CG-2)
Cryptologic Group (CG-CG), Fort Meade, Maryland (CG-2)
Cyber Command (CGCYBERCOM), Arlington, Virginia (CG-2)
Director of Operational Logistics (DOL), Norfolk, Virginia (DCMS)
Finance Center (FINCEN), Chesapeake, Virginia (CG-8)

Health, Safety, and Work-Life Service Center (HSWLSC), Norfolk, Virginia (CG-1)
Hearing Office (CGHO), Arlington, Virginia (CG-094)
Intelligence Coordination Center (ICC), Washington, D.C. (CG-2)
Investigative Services (CGIS), Arlington, Virginia (CG-2)
Legal Services Command (LSC), Norfolk, Virginia (CG-094)
Marine Safety Center (MSC), Washington, D.C. (CG-5)
Marine Safety Laboratories, Groton, Connecticut (CG-5)
National Data Buoy Center, Bay St. Louis, Mississippi (CG-5)
National Maritime Center (NMC), Martinsburg, West Virginia (CG-5)
National Pollution Funds Center (NPFC), Arlington, Virginia (CG-8)
National Vessel Documentation Center (NVDC), Falling Waters, West Virginia (CG-5)
Navigation Center (NAVCEN), Alexandria, Virginia (CG-5)
Personnel Service Center (PSC), Arlington, Virginia (CG-1)
Research & Development Center (RDC), New London, Connecticut (CG-9)
Surface Forces Logistics Center (SFLC), Baltimore, Maryland (CG-4)
Shore Infrastructure Logistics Center (SILC), Norfolk, Virginia (CG-4)
Uniform Distribution Center (UDC), Woodbine, New Jersey (CG-1)

Area Commands

The Coast Guard is divided into two theater commands: Atlantic Area (LANTAREA), headquartered in Portsmouth, Virginia, and Pacific Area (PACAREA), headquartered in Alameda, California. Area commanders are responsible for control of forces including polar icebreakers, major cutters, long-range-surveillance fixed-wing aircraft, and deployable specialized forces (DSF).

- LANTAREA's area of responsibility spans five Coast Guard districts and forty states, extending east from the Rocky Mountains and including the Atlantic Ocean and Gulf of Mexico, the Great Lakes, the Western Rivers, Europe, Africa, and the Arabian Gulf.
- PACAREA's area of responsibility extends from the Pacific coast to Asia, and from the Arctic to Antarctica.

District Commands

Nine Coast Guard districts are under the command of district commanders, rear admirals who are responsible for control of sectors, buoy tenders, patrol boats, icebreaking tugs, and air stations. The Coast Guard presently has nine districts:

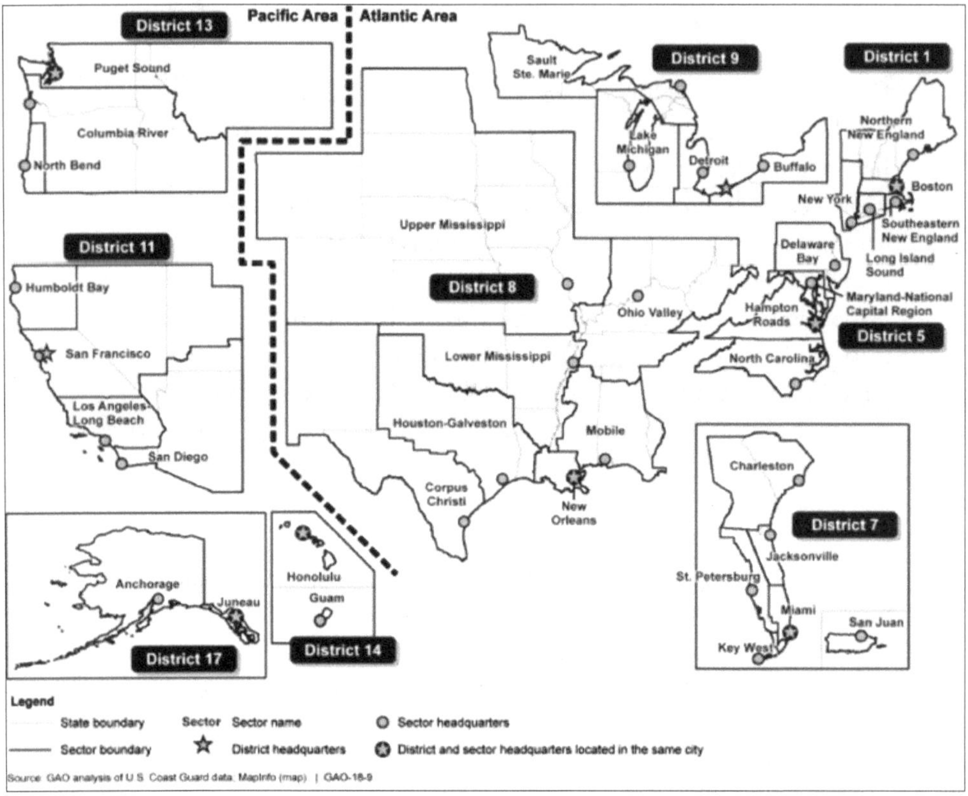

The nine Coast Guard districts

- 1st Coast Guard District (D1). Headquartered in Boston, Massachusetts, D1 covers northern New Jersey, New York, Connecticut, Rhode Island, Massachusetts, New Hampshire, Vermont, and Maine.
- 5th Coast Guard District (D5). Headquartered in Portsmouth, Virginia, D5 covers the mid-Atlantic region from central New Jersey through North Carolina.
- 7th Coast Guard District (D7). Headquartered in Miami, Florida, D7 covers the southeast United States and the Caribbean basin including Florida, Georgia, South Carolina, Puerto Rico, and the U.S. Virgin Islands.
- 8th Coast Guard District (D8). Headquartered in New Orleans, D8 covers all or part of twenty-six states throughout the Gulf Coast and heartland of America. It stretches from the Appalachian Mountains and Chattahoochee River in the east to the Rocky Mountains in the west, and from the border

between the United States and Mexico and the Gulf of Mexico to the Canadian border in North Dakota.

- 9th Coast Guard District (D9). Headquartered in Cleveland, Ohio, D9 covers the five Great Lakes, the Saint Lawrence Seaway, and parts of the surrounding states.
- 11th Coast Guard District (D11). Headquartered in Alameda, California, D11 covers the states of California, Arizona, Nevada, and Utah, and the Pacific coast of Mexico, Central America, and South America.
- 13th Coast Guard District (D13). Headquartered in Seattle, Washington, D13 covers Washington, Oregon, Idaho, and Montana as well as the Pacific Northwest coast.
- 14th Coast Guard District (D14). Headquartered in Honolulu, Hawaii, D14 covers the Hawaiian Islands, Guam, and American Samoa.
- 17th Coast Guard District (D17). Headquartered in Juneau, Alaska, D17 covers the entire state of Alaska and portions of the North Pacific Ocean, Arctic Ocean, and Bering Sea.

Overseas Commands

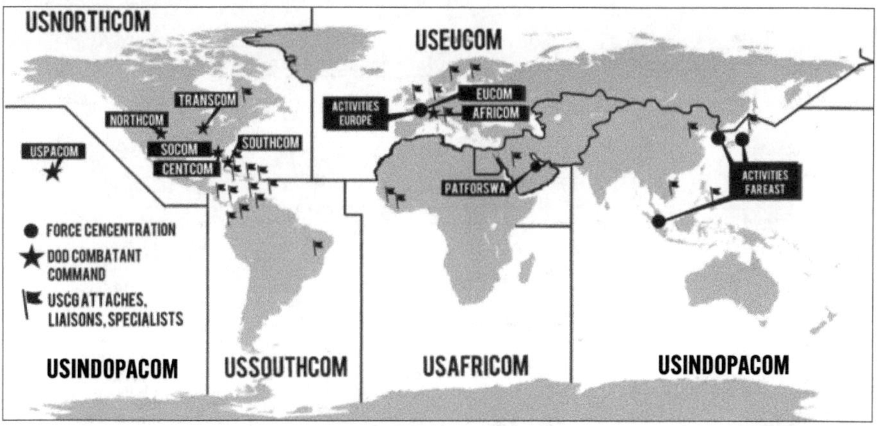

- USCG Far East Activities (FEACT) is located at Yokota Air Base, Japan, and provides maritime safety, security, training and support to U.S. Naval Forces in the region. FEACT notably commands port security units deploying to South Korea in support of U.S. Naval Forces Korea.
- USCG Activities Europe (ACTEUR) is located in Schinnen, Netherlands, and represents the U.S. Coast Guard throughout Europe, the Middle East, and Africa.

- Patrol Forces Southwest Asia (PATFORSWA) is based out of Manama, Bahrain. Established in 2002, the mission of PATFORSWA is to train, organize, equip, support, and deploy combat-ready Coast Guard forces in support of CENTCOM and national security objectives.

Trident of Forces

Coast Guard operations are conducted by the maritime trident of forces: shore-based forces, maritime patrol forces, and deployable specialized forces.

Shore-Based

Comprised of sector commands and specific subordinate units that operate in ports, waterways, and coastal regions of the United States and its territories, shore-based sector commands include a command-and-control element and staff with organic mission-support and intelligence functions as well as prevention and response elements. Sector commanders assume tactical control of boats, cutters, aircraft, and teams to execute operations.

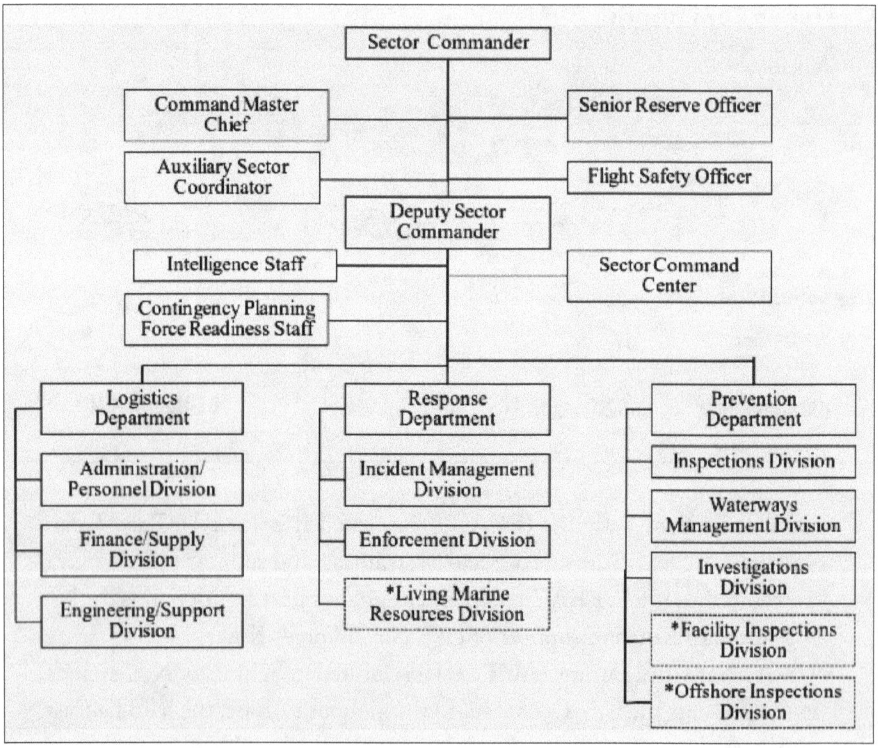

Shore-based sector command organization

Sector commanders have specific legal authorities that enable operations, including the following:

- Captain of the Port (COTP) authorities to enforce port safety, security, and marine environmental protection regulations;
- Officer in Charge of Marine Inspection (OCMI) authority over vessel standards compliance and mariner competency;
- Federal On-Scene Coordinator (FOSC) authority over oil and hazardous material spill response and preparedness;
- Federal Maritime Security Coordinator (FMSC) authority to establish and appoint members to the Area Maritime Security (AMS) Committee and implement and exercise the AMS Plan in coordination with the AMS Committee; and
- Search-and-Rescue (SAR) Mission Coordinator (SMC) authority to direct SAR operations, including rendering decisions to suspend active searches.

There are thirty-five sectors across the nine Coast Guard districts:

Sector	Location of Sector Command
Sector Northern New England	South Portland, Maine
Sector Boston	Boston, Massachusetts
Sector Southeast New England	Woods Hole, Massachusetts
Sector Long Island Sound	New Haven, Connecticut
Sector Delaware Bay	Philadelphia, Pennsylvania
Sector Baltimore	Baltimore, Maryland
Sector Hampton Roads	Portsmouth, Virginia
Sector North Carolina	Atlantic Beach, North Carolina
Sector Charleston	Charleston, South Carolina
Sector Jacksonville	Jacksonville, Florida
Sector Miami	Miami, Florida
Sector Key West	Key West, Florida
Sector St. Petersburg	St. Petersburg, Florida
Sector San Juan	San Juan, Puerto Rico
Sector Ohio Valley	Louisville, Kentucky
Sector Upper Mississippi	St. Louis, Missouri
Sector Lower Mississippi	Memphis, Tennessee
Sector New Orleans	New Orleans, Louisiana

(continued)

Sector	Location of Sector Command
Sector Houston-Galveston	Houston, Texas
Sector Corpus Christi	Corpus Christi, Texas
Sector Buffalo	Buffalo, New York
Sector Detroit	Detroit, Michigan
Sector Lake Michigan	Milwaukee, Wisconsin
Sector Sault Ste. Marie	Sault Ste. Marie, Michigan
Sector San Francisco	San Francisco, California
Sector Humboldt Bay	McKinleyville, California
Sector Los Angeles–Long Beach	San Pedro, California
Sector San Diego	San Diego, California
Sector Puget Sound	Seattle, Washington
Sector Columbia River	Astoria, Oregon
Sector North Bend	North Bend, Oregon
Sector Honolulu	Honolulu, Hawaii
Sector Guam	Santa Rita, Guam
Sector Anchorage	Anchorage, Alaska
Sector Juneau	Juneau, Alaska

Shore-based forces include the following units:

Boat Stations. Boat stations conduct operations in inland and coastal areas using assigned motor life boats, response boats, and special purpose craft.

Aids-to-Navigation Team (ANT). ANTs conduct operations in inland and coastal areas using assigned ATON boats.

Marine Safety Unit (MSU) and Marine Safety Detachment (MSD). MSUs and MSDs perform a range of prevention activities such as inspections, investigations, and waterways management.

Vessel Traffic Services (VTS). VTSs facilitate the safe and efficient transit of commercial vessel traffic along high-density routes in inland and coastal areas. They coordinate commercial vessel movements through specified waterways and support Coast Guard operations by providing domain awareness.

River, Construction, and Inland Buoy Tenders. These buoy tenders maintain ATON in inland waterways. The are available to conduct other response activities (e.g., SAR, maritime security, law enforcement).

Harbor and Icebreaking Tugs. These tugs break ice in inland waterways, including the Great Lakes. They are available to conduct other response activities (e.g., SAR, maritime security, law enforcement).

Maritime Patrol Forces

Composed of Coast Guard cutters, aircraft, and their crews, these assets deploy primarily in coastal and offshore areas but can operate in inland areas when required. Cutters provide armed, persistent presence and command-and-control capabilities throughout the maritime domain. Maritime patrol aircraft conduct intelligence, surveillance, and reconnaissance (ISR). These forces provide unique capabilities to the Department of Defense for joint operations, including warfighting when under operational control of a combatant commander. Maritime patrol forces include the following:

Major Cutters. These large oceangoing cutters conduct the full range of Coast Guard operations primarily in the offshore operational area. Superior endurance and seakeeping capabilities provide the ability to maintain a persistent presence in a range of environmental conditions. Major cutters include national security cutters (NSC), high- and medium-endurance cutters (WHEC and WMEC, respectively), the planned offshore patrol cutters (OPC), and their assigned small boats. Major cutters routinely deploy with embarked rotary-wing aircraft.

Patrol Boats. Patrol boats conduct the full range of Coast Guard operations primarily in the coastal operational area. Patrol boats include fast-response cutters, patrol boats (WPB), and coastal patrol boats (CPB). They are designed for rapid response, patrol, and interdiction and may deploy overseas in support of combatant commanders for foreign coastal interdiction operations in the offshore operational area.

Great Lakes Icebreakers. These vessels ensure access to U.S. ports and navigational channels in the Great Lakes. They are purpose-built with reinforced and tailored hulls for open-water icebreaking.

Polar Icebreakers. Polar icebreakers protect U.S. sovereignty in the polar regions, support science and research, supply remote stations, and perform other operations across the Coast Guard mission spectrum. They are the only means of ensuring surface access in support of polar maritime security, national defense, and sea control requirements.

Oceangoing and Coastal Buoy Tenders. These cutters maintain ATON in inland and coastal operational areas. They also conduct operations across the Coast Guard mission spectrum. Certain oceangoing buoy tenders may carry an organic dive team as part of the crew to support aids-to-navigation operations.

Fixed-Wing Aircraft. These all-weather aircraft deploy from land-based Coast Guard Air Stations to conduct airborne operations and consist of long-range surveillance (LRS) and medium-range surveillance aircraft. These aircraft conduct operations supporting all Coast Guard missions and perform mission-support activities such as logistics and transport flights.

Rotary-Wing Aircraft. These aircraft deploy primarily from land-based Coast Guard Air Stations or underway cutters to conduct airborne operations supporting all Coast Guard missions. They consist of medium-range response and short-range response helicopters. These aircraft conduct operations supporting all Coast Guard missions and perform mission-support activities such as logistics and transport flights.

Deployable Specialized Forces

Deployable specialized forces (DSF) are teams of ready and globally deployable personnel and assets with specialized capabilities. DSF conduct operations across a range of Coast Guard missions where their unique capabilities are required. DSF includes the following:

Maritime Security Response Teams (MSRT). These advanced tactical teams are assigned counterterrorism and high-risk law enforcement missions. MSRTs have specialized assets and capabilities to allow chemical, biological, radiological, nuclear, and high-yield explosive (CBRNE) detection and response. Two MSRTs have been established, designated as East and West, with MSRT-East based in Chesapeake, Virginia, and MSRT-West based in San Diego, California.

Tactical Law Enforcement Teams (TACLET). These units consist of deployable law enforcement detachments (LEDET) that primarily operate from U.S. Navy or allied vessels to conduct law enforcement operations through vessel interdiction and boarding activities. LEDETs operate in the offshore areas. Two TACLETs have been established and designated South and Pacific, with TACLET South based in Opa-locka, Florida, and the PACTACLET in San Diego, California.

Maritime Safety and Security Teams (MSST). These deployable teams are equipped with boats to conduct waterborne and limited shoreside security operations across Coast Guard mission areas, primarily antiterrorism and law enforcement. Ten MSSTs have been established and operate in the inland operational area.

Port Security Units (PSU). These reserve units conduct expeditionary missions in support of combatant commander requirements. Eight PSUs have been established and consist of a command element, security forces, boat forces, and organic mission-support capabilities.

The National Strike Force (NSF). The National Strike Force includes the National Strike Force Coordination Center and three national strike teams that provide high-end pollution and hazardous materials response in all operational areas. The NSF has specialized capabilities to detect and respond to CBRNE incidents.

Regional Dive Lockers (RDL). The regional dive lockers conduct military diving operations in support of PWCS, ATON, and polar operations missions.

Unit-Level Organization

Cutters, boat stations, and air stations have standard organization charts that define their internal chain of command and division of work. Units are authorized to make additions and deletions of functions and duties. Such changes can be found in each unit's organization manual.

Cutters

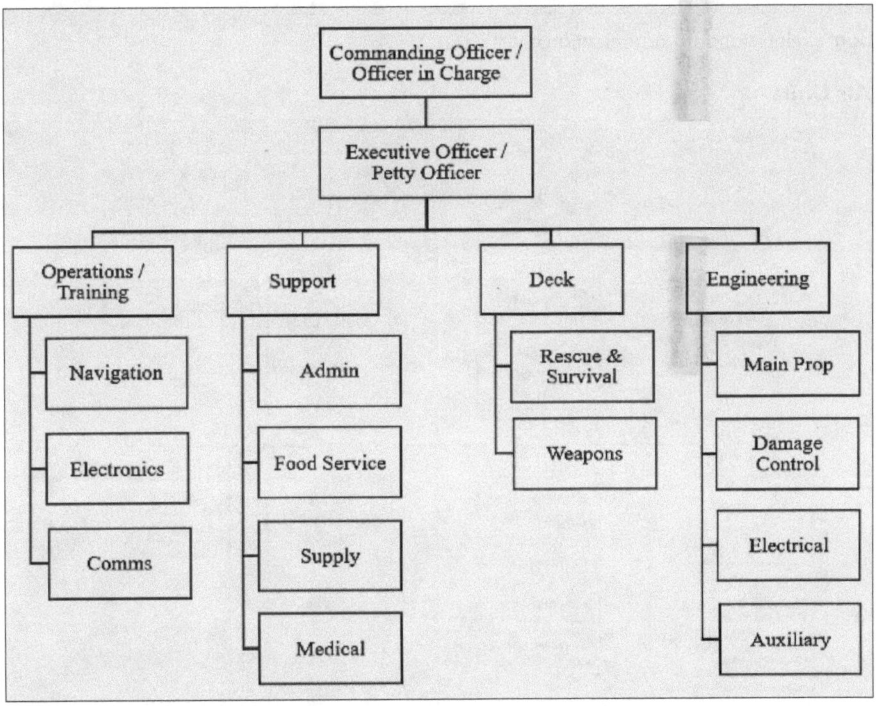

Cutter standard organization chart

Boat Units

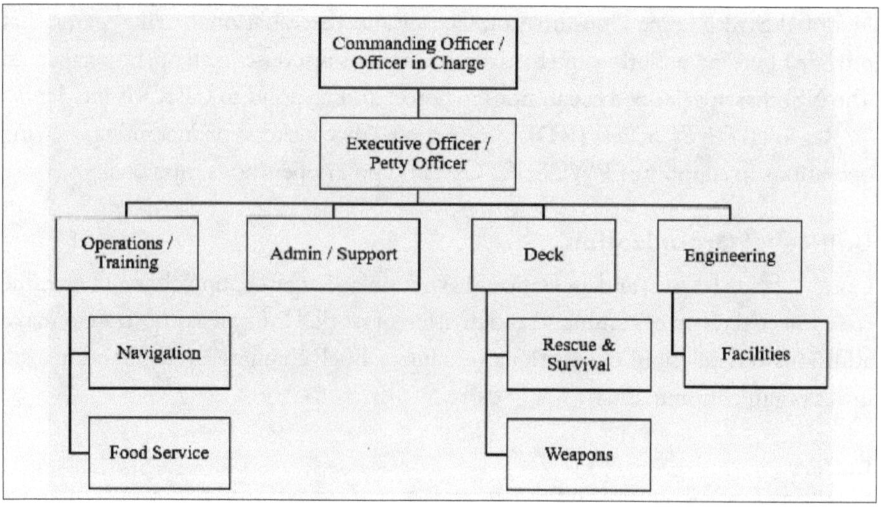

Boat station standard organization chart

Air Units

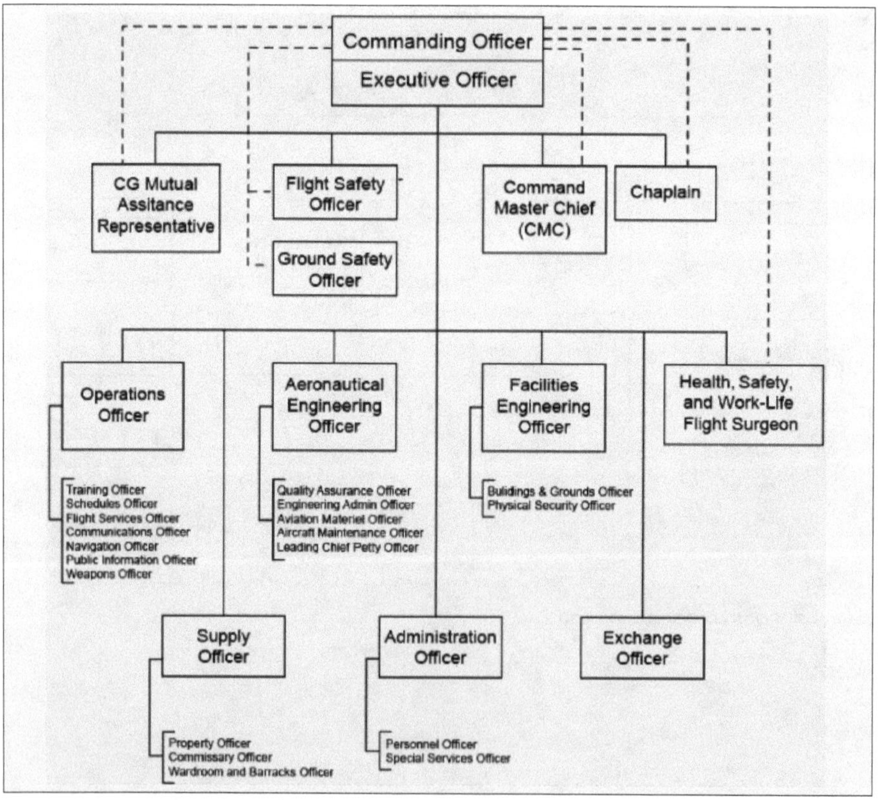

Air station standard organization chart

PERSONNEL

As a military, multimission, maritime service within the Department of Homeland Security, the Coast Guard is composed of nearly 48,000 active-duty and reserve personnel, 8,500 civilian employees, and over 30,000 members of the Coast Guard Auxiliary.

Active Duty

Coast Guard active-duty personnel are on duty twenty-four hours a day, seven days a week conducting search and rescue; halting the flow of illegal drugs, aliens, and contraband; preventing illegal fishing; and responding to environmental threats. The active-duty component comprises approximately 32,428 enlisted, 6,714 commissioned officers, and 1,725 warrant officers.

Reserve

The U.S. Coast Guard Reserve provides trained and qualified uniformed military personnel available for active duty in time of war or other national security emergencies, and for augmentation of regular Coast Guard forces, particularly during a serious natural or man-made disaster, accident, or catastrophe. The Coast Guard Reserve has 6,293 men and women in service, many assigned to DSF and others integrated directly with regular Coast Guard units. PSUs and Navy Expeditionary Combat Command units have the highest priority for ensuring full complements. The assistant commandant for Reserve (CG-R), a rear admiral, establishes policies of the Coast Guard Reserve program. Under U.S. Code Title 14, Recall Authority, the secretary of homeland security may involuntarily recall reservists to serve in domestic emergencies, in which case the local district commander determines the specialties and number of personnel to recall.

The Ready Reserve consists of reserve personnel who are available for immediate recall to active duty. All Ready Reservists are considered to be in an active status. In addition to the Ready Reserve, the service maintains an active status list (ASL) and an inactive status list (ISL) for Standby Reserve and Retired Reserve. Individual Ready Reserve (IRR) members desiring to return to the Selected Reserve (SELRES) must successfully complete a series of medical, dental, and administrative tasks before they are eligible to transfer to the SELRES. The Ready Reserve includes the following:

Selected Reserve (SELRES). The SELRES is individuals within the Ready Reserve designated as so essential to initial contingency requirements that they have priority over all other Reserve elements. They are assigned to Coast Guard or selected Joint Service units and are required to train for mobilization. Coast Guard SELRES members are generally authorized forty-eight paid inactive-duty training (IDT) drills and twelve paid active-duty training (ADT) days per fiscal year.

Individual Ready Reserve (IRR). The IRR is a manpower pool principally consisting of individuals who have had training and have previously served in the active forces or in the SELRES. The IRR consists of individuals who must fulfill their military service obligation or have fulfilled it and voluntarily remain in the IRR. Members are not required to meet the same training requirements as Selected Reservists.

UNIFORMED PAY GRADES AND TITLES

Pay grade is defined as the level established by law for pay and allowances. There are a total of twenty-three pay grades: ten enlisted, three chief warrant officer, and ten commissioned officer. Enlisted pay grades start with the letter "E," chief warrant officers with the letter "W," and officers with the letter "O."

Enlisted Rates	Pay Grade
Master chief petty officer of the Coast Guard (MCPOCG)	E-10
Master chief petty officer (MCPO)	E-9
Senior chief petty officer (SCPO)	E-8
Chief petty officer (CPO)	E-7
First class petty officer (PO1)	E-6
Second class petty officer (PO2)	E-5
Third class petty officer (PO3)	E-4
Seaman/fireman/airman (SN/FN/AN)	E-3
Seaman/fireman/airman apprentice (SA/FA/AA)	E-2
Seaman recruit (SR)	E-1
Officer Ranks	**Pay Grade**
Admiral (COMDT)	O-10
Vice admiral (ADM)	O-9
Rear admiral upper half (ADM)	O-8
Read admiral lower half (ADM)	O-7
Captain (CAPT)	O-6
Commander (CDR)	O-5
Lieutenant commander (LCDR)	O-4
Lieutenant (LT)	O-3
Lieutenant junior grade (LTJG)	O-2
Ensign (ENS)	O-1
Chief warrant officer (CWO)	W-4 W-3 W-2

Both enlisted personnel and officers are divided into three distinct groups based on pay grade:

Enlisted

Nonrated personnel (E-1 to E-3)
Petty officers (E-4 to E-6)
Chief petty officers (E-7 to E-9)

Officer

Chief warrant officers (W-2 to W-4)
Junior officers (O-1 to O-4)
Senior officers (O-5 to O-6)
Flag officers (O-7 to O-9)

Rating is defined as the plain language name for an enlisted person's occupational specialty. Only enlisted personnel have a rating. Normally, only enlisted members in pay grades E-4 through E-9 have a rating. Personnel in pay grades E-1, E-2, and E-3 have no occupational specialty and are considered nonrated. Chief warrant officers are prior enlisted personnel who receive a commission but maintain their occupational specialty. All other commissioned officers do not have an occupational specialty.

Rate is defined as the plain language name which identifies the enlisted person by occupation and pay grade. While officers have rank, enlisted have rate. Examples of rates are chief boatswain's mate (BMC) and storekeeper third class (SK3).

AUXILIARY

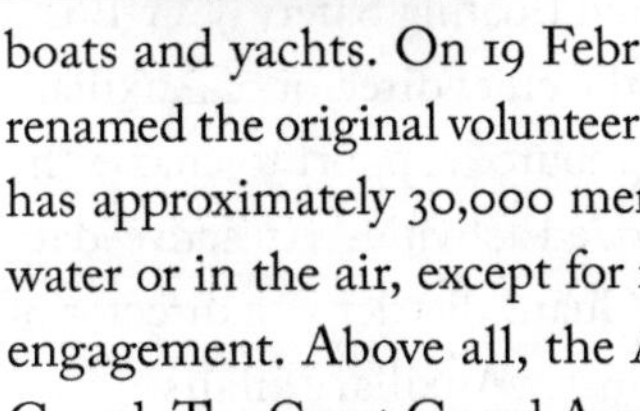

The Coast Guard Auxiliary (USCGA, USCGAUX, CGAux, or USCG Aux) is the civilian uniformed volunteer component of the Coast Guard. Congress established the organization as the Coast Guard Reserve on 23 June 1939, when it was composed of unpaid, volunteer U.S. citizens who owned motor boats and yachts. On 19 February 1941 Congress created a military Reserve and renamed the original volunteer Reserve as the Coast Guard Auxiliary. The Auxiliary has approximately 30,000 members and exists to support USCG missions on the water or in the air, except for roles that require direct law enforcement or military engagement. Above all, the Auxiliary serves as a force multiplier for the Coast Guard. The Coast Guard Auxiliary maintains the following three watchwords:

Relevance. We will continually evaluate our dedication to purpose with an ongoing gap analysis effort. Beyond being well-trained and committed to the

Auxiliary missions and vision, we have many members with valuable life/work experience that can be utilized to help foster successful Coast Guard initiatives. We must strive to remain a necessary force multiplier.

Flexibility. We will remain flexible in our training efforts and readiness posture to ensure timely responses to changing backfill and surge requests from the active duty. Day-to-day, our flexibility and our adaptability underscore our relevance.

Professionalism. We are a volunteer service, but we are an integral part of the U.S. Coast Guard. Consequently, our every effort, while performing our requested missions, must serve to bring positive credit to the U.S. Coast Guard. This includes our appearance, our demeanor, and our devotion to assigned missions.

Organization

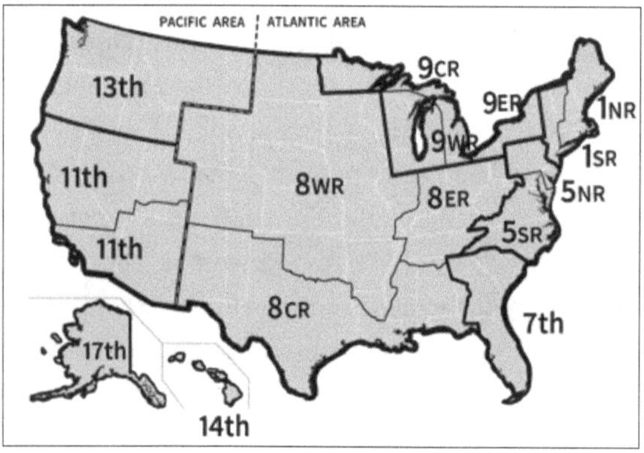

Auxiliary divisions

The Auxiliary has units in all fifty states, Puerto Rico, the Virgin Islands, American Samoa, and Guam. Membership is open to men and women seventeen years or older who are U.S. citizens. Ownership of a "facility"—whether a radio station, boat, or aircraft—is desirable but not mandatory for membership. Leadership of this component is provided by both active-duty Coast Guard personnel and by civilian Coast Guard Auxiliary as follows:

Military. The chief of the Office of the Auxiliary and Boating Safety (CG-BSX) is an active-duty Coast Guard captain who serves as the chief director of Auxiliary and provides the administrative, management, and resource support to enable the Auxiliary to carry out its assigned missions and authorized activities. An active-duty Coast Guard commander is assigned in each Coast Guard district as a director of Auxiliary (DIRAUX) to actively promote and administer Auxiliary affairs.

Civilian. The national commodore (NACO) is the senior member of the Auxiliary and reports to the commandant of the Coast Guard. Assisting the NACO is a vice national commodore and the deputy commodores for operations, mission support, information technology and planning, and recreational boating safety. District commodores (DCO) report to and represent the Auxiliary to Coast Guard district and area commanders. At least four flotillas form an Auxiliary division, and several divisions may exist within a single Coast Guard district, all headed by a division commander (DCDR) who reports to the district commodore. Divisions provide leadership, direction, and staff assistance to the flotillas. A flotilla is the smallest organizational level of the Auxiliary and consists of a minimum of ten members headed by a flotilla commander (FC) who reports to the division commander. Most of the day-to-day work of the Auxiliary is performed at the flotilla level, and all members join the Auxiliary through a flotilla. Visitors and prospective members are usually welcome to attend monthly flotilla meetings.

CIVILIANS

Civilians employed by the Coast Guard serve in many roles and provide valuable continuity as military personnel transfer away from assignments. Civilians often work behind the scenes, away from the spotlight, but without them, cutters wouldn't sail, aircraft wouldn't fly, and our stations would fall into disrepair.

4

CUTTERS, BOATS, AND AIRCRAFT

As the nation's maritime governance force, the Coast Guard possesses unique authorities, capabilities, and partnerships. Coupled with capable cutters, boats, and aircraft operated by highly proficient personnel, we maximize these authorities and capabilities to execute layered security throughout the entire maritime domain.

CUTTERS (USCGC)

A Coast Guard cutter is any vessel 65 feet or longer. Cutters provide armed, persistent presence and command-and-control capabilities throughout the maritime domain and deploy primarily in coastal and offshore areas to conduct prevention and response operations through patrol, presence, and at-sea operations. In addition to conducting Coast Guard operations, cutters project U.S. presence and protect U.S. sovereignty. Cutters also provide unique capabilities to the Department of Defense for joint operations, including warfighting under the combatant commander with operational control.

The Coast Guard uses the same ship classification system as the U.S. Navy. This is a two- or three-letter abbreviation based on the type of ship. All Coast Guard cutters are given the additional prefix "W." For example, a national security cutter has a hull abbreviation of MSL. So a Coast Guard national security cutter would be classified as WMSL. Hull numbers are marked on all Coast Guard vessels forward of the Coast Guard Stripe.

Barque *Eagle* (WIX 295)

The 295-foot-tall ship *Eagle* is a three-masted sailing barque with 21,350 square feet of sail, homeported at the Coast Guard Academy, New London, Connecticut. It is the only active (operational) commissioned sailing vessel in the U.S. maritime services. Seventh in a line of cutters to bear its name, the CGC *Eagle* was built in 1936 by Blohm and Voss in Hamburg, Germany, as a training vessel for German naval cadets.

It was taken as a war prize in 1946, commissioned into Coast Guard service as the *Eagle*, and sailed from Bremerhaven, Germany, to New London, Connecticut. The *Eagle* serves as a seagoing classroom for approximately 175 cadets and instructors from the academy. On the *Eagle*, cadets apply the navigation, engineering, and other skills they develop in classes at the academy.

Eagle's hull is built of steel, four-tenths of an inch thick. It has two full-length steel decks with a platform deck below and a raised forecastle and quarterdeck. The weather decks are three-inch-thick teak over steel. When at home, the *Eagle* is moored at the Fort Trumbull State Park on the Thames River.

Length: 295 feet
Beam: 39 feet
Displacement: 1,824 tons
Propulsion: 1,000 horsepower diesel, 21,350-square-foot sail area
Speed: 10 knots under power; 16 knots under sail
Range: 5,450 nautical miles under power

Major Cutters

These large oceangoing cutters conduct the full range of Coast Guard operations primarily in the offshore operational area. Superior endurance and seakeeping capabilities provide the ability to maintain persistent presence in a range of environmental conditions. Major cutters include national security cutters (NSC), medium-endurance cutters (WMEC), and the planned offshore patrol cutters (OPC). Major cutters carry cutter boats and routinely deploy with embarked rotary-wing aircraft.

National Security Cutter (WMSL)

The 418-foot Legend-class national security cutter, or maritime security cutter, large (WMSL), is the largest active patrol cutter class of the U.S. Coast Guard. Entering into service in 2008, eleven hulls are planned. The WMSLs are the Coast Guard's primary platform for conducting the service's full range of offshore and open ocean missions and its primary deployer in support of U.S. combatant commanders. A helicopter deck and hangar accommodate two H-65s or one H-65 or H-60 and two vertically launched unmanned aerial vehicles. Davits allow for launching and recovering 26-foot CB-OTH-IV cutter boats, and a stern ramp can accommodate both the CB-OTH-IV and the 35-foot CB-LRI-II, with capacity for a total of three cutter boats to be carried on board. The cutter's weapon and defense systems provide antisurface and limited air-defense capability as well as the capability to provide naval gunfire support. Each NSC is capable of deploying independently or as part of task groups and serving as a mobile command-and-control platform for surge operations. The first cutter, *Bertholf* (pictured above), was commissioned 4 August 2008.

Displacement: 4,500 long tons
Length: 418 feet
Beam: 54 feet
Draft: 22.5 feet
Speed: 28+ knots
Range: 12,000 nautical miles
Endurance: 60–90-day cycles
Complement: 113 (14 officers, 99 enlisted)
Boats & landing: 2 × over-the-horizon cutter boats, 1 × 35 ft long-range interceptor
Aviation: Flight deck and hangar, one MH-65, 2 × short-range unmanned aerial systems
Armament:
1 × MK 110 57-mm gun
1 × 20 mm Phalanx close-in weapons system
4 × crew-served .50-caliber Browning M2 machine guns
2 × crew-served M240B 7.62-mm machine guns

Cutters in this class:

Bertholf (WMSL 750), Alameda, California
Waesche (WMSL 751), Alameda, California
Stratton (WMSL 752), Alameda, California
Hamilton (WMSL 753), Charleston, South Carolina
James (WMSL 754), Charleston, South Carolina
Munro (WMSL 755), Alameda, California
Kimball (WMSL 756), Honolulu, Hawaii
Midgett (WMSL 757), Honolulu, Hawaii
Stone (WMSL 758), Charleston, South Carolina
Calhoun (WMSL 759), Charleston, South Carolina
Friedman (WMSL 760), Charleston, South Carolina

Offshore Patrol Cutter (WMSM)

Construction began on the 360-foot Heritage-class OPC, or maritime security cutters, medium (WMSM), in January 2019, with up to twenty-five hulls planned for the class. OPCs will provide the majority of offshore presence for the Coast Guard's cutter fleet across its full range of missions, bridging the capabilities of the

418-foot national security cutters, which patrol the open ocean, and the 154-foot fast-response cutters, which serve closer to shore.

OPCs are outfitted with three davit-launched cutter boats and a flight deck and hangar to support two H-65s and one H-60 or one HH-65 and two vertical take-off and landing unmanned aerial vehicles.. The cutter's weapon and defense systems provide antisurface and limited air-defense capability as well as the capability to provide naval gunfire support. Each OPC will be capable of deploying independently or as part of task groups and serving as a mobile command-and-control platform for surge operations. The cutters will also support Arctic objectives by helping regulate and protect emerging commerce and energy exploration in Alaska.

Length: 360 feet
Beam: 53 feet
Draft: 16 feet
Propulsion: 2 × 9,900 horsepower MTU diesels, 1 × 30,000 horsepower gas turbine
Speed: 24.5 knots
Range: 10,200 nm
Endurance: 60 days
Complement: 126
Boats & landing: 3 over-the-horizon cutter boats
Aviation: Flight deck and hangar, one H-60 or H-65, short-range unmanned aerial system
Armament:
1 × MK 110 57-mm gun
1 × Mk38 Mod 3 25-mm gun with 7.62-mm coaxial gun
2 × M2 .50-caliber machine guns mounted on a Mk50 stabilized small-arms mount
4 × M2 .50-caliber machine guns

Planned cutters in this class:

Argus (WMSM 915), San Pedro, California
Chase (WMSM 916), San Pedro, California
Ingham (WMSM 917), Kodiak, Alaska
Rush (WMSM 918), Kodiak, Alaska
Pickering (WMSM 919), Newport, Rhode Island
Icarus (WMSM 920), Newport, Rhode Island

Active (WMSM 921), TBD
Diligence (WMSM 922), TBD
Alert (WMSM 923), TBD
Vigilant (WMSM 924), TBD
Reliance (WMSM 925), TBD

282-Foot Alex Haley–Class Medium-Endurance Cutter (WMEC 39)

The *Alex Haley* is a one-of-a-kind Coast Guard cutter, named for the service's first chief journalist, who later wrote the novel *Roots* and was awarded a Pulitzer Prize. Commissioned in 1971 as the Navy salvage and rescue ship USS *Edenton* (ATS 1), the vessel was transferred to the Coast Guard in November 1997 for conversion. The cutter's primary missions are law enforcement, domestic fisheries enforcement, and search and rescue in Alaskan waters. Outfitted with two davit-launched cutter boats and a flight deck and hangar to support on H-65s or one H-60. The cutter's weapon and defense systems provide antisurface and limited air-defense capability as well as the capability to provide naval gunfire support.

Length: 282 feet
Beam: 50 feet
Draft: 17 feet
Propulsion: 4 × Caterpillar diesels
Speed: 18 knots
Range: 10,000 nm
Complement: 10 officers, 90 enlisted, 4 aircrew
Boats & landing: 3 over-the-horizon cutter boats

Aviation: Flight deck and hangar, one MH-60 or MH-65
Armament:
2 × Mk38 25-mm cannons
2 × M2 .50-caliber machine guns

270-Foot Famous-Class Medium-Endurance Cutters (WMEC)

The first of thirteen Famous-class medium-endurance cutters, the *Bear* (WMEC 901) entered service in 1983 with a mission profile covering the full range of the Coast Guard's offshore missions, with emphasis on law enforcement, particularly patrolling the newly established two hundred nautical mile exclusive economic zone. Outfitted with two davit-launched cutter boats and a flight deck and hangar to support an H-65, the cutter's weapon and defense systems provide antisurface and limited air-defense and naval gunfire support capability.

Length: 270 feet
Beam: 38 feet
Draft: 14.4 feet
Propulsion: 2 × turbo-charged ALCO V-18 diesels
Speed: 20 knots
Range: 3,800 nm
Complement: 100 (14 officers, 86 enlisted)
Boats & landing: 2 cutter boats
Aviation: Flight deck and hangar, one H-65

Armament:

1 × Mk75 76-mm gun
2 × M2 .50-caliber machine guns

Cutters in this class:

Bear (WMEC 901), Portsmouth, Virginia
Tampa (WMEC 902), Portsmouth, Virginia
Harriet Lane (WMEC 903), Portsmouth, Virginia
Northland (WMEC 904), Portsmouth, Virginia
Spencer (WMEC 905), Boston, Massachusetts
Seneca (WMEC 906) (shown on p. 70), Boston, Massachusetts
Escanaba (WMEC 907), Boston, Massachusetts
Tahoma (WMEC 908), Kittery, Maine
Campbell (WMEC 909), Kittery, Maine
Thetis (WMEC 910), Key West, Florida
Forward (WMEC 911), Portsmouth, Virginia
Legare (WMEC 912), Portsmouth, Virginia
Mohawk (WMEC 913), Key West, Florida

210-Foot *Reliance*-Class Medium-Endurance Cutters (WMEC)

Entering service between 1964 and 1969, the *Reliance*-class cutters were the first true post–World War II Coast Guard cutters to reflect evolving Coast Guard operations during the latter part of the twentieth century. Although principally employed for general law enforcement operations at sea, in geographic choke points, and within the littoral waters, these cutters are equipped and manned to perform the full range of service of service missions. Outfitted with two davit-launched cutter boats and a flight deck to support an H-65, the cutter's weapon and defense systems provide antisurface and limited air-defense and naval gunfire support capability.

Length: 210 feet
Beam: 34 feet
Draft: 10.7 feet
Propulsion: 2 × 2,500 horsepower diesels
Speed: 18 knots
Range: 6,100 nm
Complement: 77
Boats & landing: 2 cutter boats
Aviation: Flight deck, one H-65
Armament:
- 1 × Mk38 25-mm autocannon
- 2 × M2 .50-caliber machine guns

Cutters in this class:

Reliance (WMEC 615), Kittery, Maine
Diligence (WMEC 616), Wilmington, North Carolina
Vigilant (WMEC 617), Patrick Air Force Base, Florida
Active (WMEC 618), Port Angeles, Washington
Confidence (WMEC 619), Port Canaveral, Florida
Resolute (WMEC 620), St. Petersburg, Florida
Valiant (WMEC 621), Miami Beach, Florida
Steadfast (WMEC 623), Warrenton, Oregon

Patrol Boats

These cutters conduct the full range of Coast Guard operations primarily in the coastal operational area. Patrol boats include fast-response cutters, patrol boats (WPB), and coastal patrol boats (CPB). They are designed for rapid response, patrol, and interdiction. They may also deploy overseas in support of combatant commanders for foreign coastal interdiction operations.

154-Foot Sentinel-Class Fast-Response Cutter (WPC)

The first cutter in this class, *Bernard C. Webber*, was delivered in February 2012 with a level of capability exceeding that of any legacy patrol boat. Sentinel-class cutters are equipped and manned to perform the full range of service missions in the coastal environment and are outfitted with a stern-launched cutter boat. The cutter's weapon and defense systems provide antisurface and limited air-defense and naval gunfire support capability. To honor past Coast Guard members, each cutter is named for an enlisted individual who served heroically in the Coast Guard or one of its predecessor agencies.

Length: 154 feet
Beam: 25 feet
Draft: 9.5 feet
Propulsion: 2 × MTU diesel
Speed: 28 knots
Range: 2,500 nm
Endurance: 5 days
Complement: 24 (4 officers, 20 crew)
Boats & landing: 1 cutter boat
Armament:
  1 × Mk38 25-mm autocannon
  2 × M2 .50-caliber machine guns

Cutters in this class:

Bernard C. Webber (WPC 1101), Miami Beach, Florida
Richard Etheridge (WPC 1102), Miami Beach, Florida
William Flores (WPC 1103), Miami Beach, Florida
Robert Yered (WPC 1104), Miami Beach, Florida
Margaret Norvell (WPC 1105), Miami Beach, Florida
Paul Clark (WPC 1106), Miami Beach, Florida
Charles David Jr. (WPC 1107), Key West, Florida
Charles W. Sexton (WPC 1108), Key West, Florida
Kathleen Moore (WPC 1109), Key West, Florida
Raymond Evans (WPC 1110), Key West, Florida
William Trump (WPC 1111), Key West, Florida
Isaac Mayo (WPC 1112) Key West, Florida
Richard Dixon (WPC 1113) San Juan, Puerto Rico
Heriberto Hernandez (WPC 1114) San Juan, Puerto Rico
Joseph Napier (WPC 1115), San Juan, Puerto Rico
Winslow Griesser (WPC 1116), San Juan, Puerto Rico
Donald Horsley (WPC 1117), San Juan, Puerto Rico
Joseph Tezanos (WPC 1118), San Juan, Puerto Rico
Rollin A. Fritch (WPC 1119), Cape May, New Jersey
Lawrence O. Lawson (WPC 1120), Cape May, New Jersey
John F. McCormick (WPC 1121), Ketchikan, Alaska
Bailey T. Barco (WPC 1122), Ketchikan, Alaska
Benjamin B. Dailey (WPC 1123), Pascagoula, Mississippi
Oliver F. Berry (WPC 1124), Honolulu, Hawaii
Jacob Poroo (WPC 1125), Pascagoula, Mississippi
Joseph Gerczak (WPC 1126), Honolulu, Hawaii
Richard Snyder (WPC 1127), Atlantic Beach, North Carolina
Nathan Bruckenthal (WPC 1128), Atlantic Beach, North Carolina
Forrest Rednour (WPC 1129), San Pedro, California
Robert Ward (WPC1130) (shown on p. 73), San Pedro, California
Terrell Horne (WPC 1131), San Pedro, California
Benjamin Bottoms (WPC 1132), San Pedro, California
Joseph Doyle (WPC 1133), San Juan, Puerto Rico
William Hart (WPC 1134), Honolulu, Hawaii
Angela McShan (WPC 1135), Cape May, New Jersey

Daniel Tarr (WPC 1136), Galveston, Texas
Edgar Culbertson (WPC 1137), Galveston, Texas
Harold Miller (WPC 1138), Galveston, Texas
Myrtle Hazard (WPC 1139), Santa Rita, Guam
Oliver Henry (WPC 1140), Santa Rita, Guam
Charles Moulthrope (WPC 1141), Manama, Bahrain
Robert Goldman (WPC 1142), Manama, Bahrain
Frederick Hatch (WPC 1143), Santa Rita, Guam
Glen Harris (WPC 1144), Manama, Bahrain
Emlen Tunnell (WPC 1145), Manama, Bahrain
John Scheuerman (WPC 1146), Manama, Bahrain
Clarence Sutphin (WPC 1147), Manama, Bahrain
Pablo Valent (WPC 1148), St. Petersburg, Florida
Douglas Denman (WPC 1149), Ketchikan, Alaska
William Chadwick (WPC 1150), Boston, Massachusetts
Warren Deyampert (WPC 1151), Boston, Massachusetts
Maurice Jester (WPC 1152), Boston, Massachusetts
John Patterson (WPC 1153), Boston, Massachusetts
William Sparling (WPC 1154), Boston, Massachusetts
Melvin Bell (WPC 1155), TBD
David Duren (WPC 1156), TBD
Florence Finch (WPC 1157), TBD
John G. Witherspoon (WPC 1158), TBD
Earl Cunningham (WPC 1159), TBD
Frederick Mann (WPC 1160), TBD
Olivia Hooker (WPC 1161), TBD
Vincent Danz (WPC 1162), TBD
Jeffrey Palazzo (WPC 1163), TBD
Marvin Perrett (WPC 1164), TBD

110-Foot Island-Class Patrol Boat (WPB)

The Coast Guard 110-foot Island-class patrol boats are modified versions of a well-regarded British-designed patrol boat. They are quipped and manned to perform the full range of service missions in the coastal environment. Built and in service since the late 1980s, all WPBs have been or are scheduled to be decommissioned and replaced by 154-foot Sentinel-class fast-response cutters.

Length: 110 feet
Beam: 21 feet
Draft: 7.3 feet
Propulsion: 2 × Paxman Valenta or Caterpillar diesels
Speed: 29.5 knots
Range: 2,900 nm
Endurance: 5 days
Complement: 16 (2 officers, 14 crew)
Boats & landing: 1 cutter boat
Armament:
 1 × Mk38 25-mm chain gun
 2 × M2 .50-caliber machine guns

87-Foot Marine Protector–Class Coastal Patrol Boat (CPB)

Named after marine predators, the first cutter in this class, *Barracuda*, was delivered in 1998, and 74 were commissioned. They are equipped and manned to perform the full range of service missions in the coastal environment. The Protector class is the first outfitted with a stern-launched cutter boat. Four cutters in the class were purpose built to protect Navy ballistic missile submarines while they are in transit in and out of Kings Bay, Georgia, and Bangor, Washington.

Length: 87 feet
Beam: 19 feet
Draft: 5.6 feet
Propulsion: 2 × MTU diesels
Speed: 25 knots

Range: 900 nm
Endurance: 3 days
Complement: 11
Boats & landing: 1 cutter boat
Armament:
2 × M2 .50-caliber machine guns

Polar Icebreakers

The Coast Guard operates two oceangoing icebreakers, the newest of which, the CGC *Healy* (WAGB 20), commissioned in July 2000, is the service's largest ship. Icebreakers are painted with an "icebreaker red" hull to make them noticeable in ice-covered waters. The Coast Guard and U.S. Navy, through an integrated program, awarded a contract in 2019 for a new class of polar icebreaker to replace the aging WAGBs. Design of the first of three planned polar security cutters (PSC), CGC *Polar Sentinel*, began in 2021 with delivery expected in 2025. The new PSCs will be national assets that will ensure access to both polar regions and will be capable of executing the full range of Coast Guard missions worldwide in the most extreme environmental conditions found in the polar, tropical, and temperate regions.

420-Foot Healy*-Class Polar Icebreaker (WAGB)*

The Coast Guard's largest ship, the CGC *Healy*, was launched in 1997 and commissioned in 2000, joining the two Polar-class icebreakers in their homeport of Seattle, Washington. The *Healy* is designed to conduct a wide range of research activities, providing more than 4,200 square feet of scientific laboratory space, numerous electronic sensor systems, oceanographic winches, and accommodations for up to fifty scientists. *Healy* is capable of breaking 4.5 feet of ice continuously at three knots and can operate in temperatures as low as minus 50 degrees F. As a Coast Guard cutter, the *Healy* is also a capable platform for supporting other potential missions in the polar regions and is capable of accommodating two H-65 Dolphin helicopters or one Dolphin and one H-60 Jayhawk helicopter. On 5 September 2015 *Healy* became the first unaccompanied United States surface vessel to reach the North Pole.

Length: 420 feet
Beam: 82 feet
Displacement: 16,000 tons
Power plant: 4 × 46,350 horsepower diesels
Speed: 17 knots
Range: 16,000 nautical miles
Complement: 19 officers, 12 CPO, 54 enlisted, 51 scientists
Boats & landing: 2 × cutter boats; 1 arctic survey boat
Aviation: Flight deck and hangar, 2 × H-65
Armament: Small arms

399-Foot Polar-Class Icebreaker (WAGB)

The Polar-class icebreakers, built in the 1970s, were designed for open-water ice breaking and have reinforced hulls, special icebreaking bows, and a system that allows rapid shifting of ballast to increase the effectiveness of their ice breaking. Their curved bow allows the cutters to ride up on ice and use the ship's weight to break the ice. These cutters are capable of continuous progress through ice six feet thick at a speed of up to three knots. They are built to serve in the Arctic and Antarctic, supporting science and research as well as providing resupply to remote stations, but their capabilities also enable them to perform search and rescue, ship escort, environmental protection, and law enforcement in places most cutters cannot reach.

Length: 399 feet
Beam: 83.5 feet
Draft: 28 feet
Displacement: 13,194 tons
Propulsion: 6 × 3,000 horsepower Alco diesels, 3 × 25,000 gas turbine
Speed: 18 knots
Range: 28,275 nautical miles
Complement: 15 officers, 127 enlisted, 33 scientists, 12 aircrew
Boats & landing: 3 × cutter boats
Aviation: Flight deck and hangars, 2 × H-60 or H-65
Armament: Small arms

Cutters in this class:

Polar Star (WAGB 10) (shown on p. 78), Seattle, Washington
Polar Sea (WAGB 11), deactivated, Seattle, Washington

Great Lakes Icebreakers

Great Lakes icebreakers and icebreaking tugs ensure access to U.S. ports and navigational channels in the Great Lakes. They are specifically designed with reinforced and tailored hulls for open-water icebreaking. The Coast Guard operates one dedicated red-hull icebreaker and several black-hull icebreaking tugs on the Great Lakes.

240-Foot Great Lakes–Class Icebreaker (WLBB)

The CGC *Mackinaw* (WLBB 30), like its predecessor of the same name, was designed specifically for the Great Lakes, where its mission has been to keep the shipping lanes open through as much of the winter as possible. Like the former *Mackinaw* (WAGB 83), the ship is homeported in Cheboygan, Michigan, and remains the only U.S. heavy icebreaking resource assigned to the Great Lakes. The cutter is designed to break through thirty-two inches of ice at three knots. It also has a twenty-ton crane for servicing aids to navigation and an oil spill recovery system on board.

240-foot Great Lakes–class icebreaker (WLBB)

Length: 240 feet
Beam: 58.5 feet
Draft: 16 feet
Displacement: 3,500 tons
Power plant: 2 × 3,350-kw azipod
Speed: 15 knots
Range: 4,000 nm
Complement: 9 officers, 46 enlisted
Boats & landing: 2 × cutter boats
Armament:
    2 × M2 .50-caliber machine guns
    Various small arms

140-Foot Bay-Class Icebreaking Tugs (WTGB)

Named after American bays, Bay-class icebreaking tugs are small, multimission cutters specially configured for icebreaking and aids to navigation on the Great Lakes and in coastal waters and rivers of the northeastern United States. These tugs can break ice up to three feet using a low-pressure air hull lubrication or bubbler system that forces air and water between the hull and ice. This system improves icebreaking capabilities by reducing resistance against the hull, thereby reducing horsepower requirements. A 120-foot aids-to-navigation (ATON) barge augments the CGCs *Bristol Bay* and *Mobile Bay*.

140-foot Bay-class icebreaking tug (WTGB)

Length: 140 feet
Beam: 37.5 feet
Displacement: 662 tons
Power plant: 2 × Fairbanks-Morse diesel engines, 2,500 horsepower
Speed: 14.7 knots
Range: 4,000 nautical miles
Complement: 3 officers, 14–24 enlisted
Endurance: 2–10 days

Cutters in this class:

Katmai Bay (WTGB 101) (above), Sault Ste. Marie, Michigan
Bristol Bay (WTGB 102), Detroit, Michigan
Mobile Bay (WTGB 103), Sturgeon Bay, Wisconsin
Biscayne Bay (WTGB 104), St. Ignace, Michigan
Neah Bay (WTGB 105), Cleveland, Ohio
Morro Bay (WTGB 106), New London, Connecticut
Penobscot Bay (WTGB 107), Bayonne, New Jersey
Thunder Bay (WTGB 108), Rockland, Maine
Sturgeon Bay (WTGB 109), Bayonne, New Jersey

Buoy Tenders

Buoy tenders are Coast Guard cutters responsible for maintaining ATONs such as navigational buoys and fixed structures including lighthouses. Although built primarily to service ATONs, buoy tenders remain multimission cutters and are capable of performing search and rescue, maritime law enforcement, homeland security, and defense missions. They range in size from 65 feet to 225 feet in length, and their hulls are painted black. Although not classified as icebreakers, buoy tenders do provide light icebreaking capability. There are three classes: seagoing buoy tenders, coastal buoy tenders, and river tenders.

225-Foot Juniper*-Class Seagoing Buoy Tender (WLB)*

Juniper-class cutters tend coastal and offshore buoys in exposed and open ocean locations. Equipped with a dynamic positioning system to enable precision positioning of floating aids, the cutters are able to maintain position within a ten-meter circle in winds of up to thirty knots and waves of up to eight feet.

Length: 225 feet
Beam: 46 feet
Draft: 13 feet
Displacement: 2,000 tons
Buoy deck area: 2,875 square feet
Power plant: 2 × Caterpillar 3608 diesels, 6,200 base horsepower
Speed: 15 knots

Range: 6,000 nautical miles
Complement: 8 officers, 40 enlisted
Boats & landing: 3 × cutter boats
Armament:
  2 × M2 .50-caliber machine guns
  Various small arms

Cutters in this class:

Juniper (WLB 201), Honolulu, Hawaii
Willow (WLB 202), Charleston, South Carolina
Kukui (WLB 203), Sitka, Alaska
Elm (WLB 204), Astoria, Oregon
Walnut (WLB 205), Pensacola, Florida
Spar (WLB 206), Duluth, Minnesota
Maple (WLB 207), Atlantic Beach, North Carolina
Aspen (WLB 208), San Francisco, California
Sycamore (WLB 209), Newport, Rhode Island
Cypress (WLB 210), Pensacola, Florida
Oak (WLB 211) (shown on p. 82), Newport, Rhode Island
Hickory (WLB 212) Homer, Alaska
Fir (WLB 213), Astoria, Oregon
Hollyhock (WLB 214), Port Huron, Michigan
Sequoia (WLB 215), Apra Harbor, Guam
Alder (WLB 216), Duluth, Minnesota

175-Foot Keeper-Class Coastal Buoy Tender (WLM)

Keeper-class buoy tenders are named after famous American lighthouse keepers and service aids to navigation in coastal environments. They are equipped with Z-drive propulsion units instead of the standard propeller and rudder configuration in order to independently rotate 360 degrees using a thruster in the bow to support precision positioning of floating aids.

Length: 175 feet
Beam: 36 feet
Displacement: 845 tons
Propulsion: 2 × Caterpillar diesels; 2 Ulstein Z-drive, 2,040 base horsepower

175-foot Keeper-class coastal buoy tender (WLM)

Speed: 12 knots
Range: 2,000 nautical miles
Complement: 25 (2 officers, 23 enlisted)
Endurance: 1–4 days

Cutters in this class:

Ida Lewis (WLM 551), Newport, Rhode Island
Katherine Walker (WLM 552), Bayonne, New Jersey
Abbie Burgess (WLM 553), Rockland, Maine
Marcus Hanna (WLM 554), South Portland, Maine
James Rankin (WLM 555), Baltimore, Maryland
Joshua Appleby (WLM 556) (shown above), St. Petersburg, Florida
Frank Drew (WLM 557), Portsmouth, Virginia
Anthony Petit (WLM 558), Ketchikan, Alaska
Barbara Mabrity (WLM 559), Mobile, Alabama
William Tate (WLM 560), Philadelphia, Pennsylvania
Harry Claiborne (WLM 561), Galveston, Texas
Maria Bray (WLM 562), Atlantic Beach, Florida
Henry Blake (WLM 563), Everett, Washington
George Cobb (WLM 564), San Pedro, California

100- and 65-Foot Inland Buoy Tender (WLI)

Inland buoy tenders service short-range ATON in inland waterways. These vessels maintain buoys that are beyond the capabilities of the nearest ATON team and that

65-foot inland buoy tender (WLI)

are located in areas either too shallow or otherwise too restricted for larger tenders to reach. There are two subclasses, 65 foot and 100 foot.

100-Foot Inland Buoy Tender

Length: 100 feet
Beam: 24 feet
Displacement: 174 tons
Propulsion: 2 × diesels, 600–660 horsepower
Speed: 10.5 knots
Range: 2,000–2,700 nautical miles
Complement: 1 officer, 22 enlisted
Endurance: 2–4 days

Cutters in this class:

Bluebell (WLI 313), Portland, Oregon
Buckthorn (WLI 642), Sault Ste. Marie, Michigan

65-Foot Inland Buoy Tender

Length: 65 feet
Beam: 17 feet

Displacement: 71 tons
Power plant: 2 × diesels, 400 horsepower (WLI 65401); 1 × diesel, 300 horsepower
Speed: 9 knots; 11.3 knots
Range: 1,700 nautical miles
Complement: 1 officer, 13 enlisted
Endurance: 2–4 days

Cutters in this class:

Bayberry (WLI 65400), Long Beach, North Carolina
Elderberry (WLI 65401) (shown on p. 85), Petersburg, Alaska

River Tenders (WLR)

The Coast Guard operates eighteen tenders of 75-foot and 65-foot lengths to service short-range ATON on the Western Rivers. They set, relocate, and recover buoys to mark the navigable channel in the rivers as the water level changes. They also establish and maintain fixed aids, lights, and day beacons within their areas of responsibility. WLRs push barges equipped with cranes to work ATON. Some are equipped with "jetting" devices used to set and anchor buoys in rivers with sandy or muddy bottoms. The barges are an integral part of the ATON mission, and they range in length from 90 to 130 feet.

75-Foot *Kankakee*-Class River Tender
Length: 75 feet
Beam: 22 feet
Displacement: 175 tons
Propulsion: 2 × Caterpillar diesels, 1,024 horsepower

Speed: 10 knots
Range: 600 nautical miles

Cutters in this class:

Kankakee (WLR 75500), Memphis, Tennessee
Greenbrier (WLR 75501), Natchez, Mississippi

75-Foot *Gasconade*-Class River Tender
Length: 75 feet
Beam: 22 feet
Displacement: 140 tons
Propulsion: 2 × Caterpillar diesels, 660–750 horsepower
Speed: 10 knots
Range: 3,100 nautical miles

Cutters in this class:

Wedge (WLR 75307), Demopolis, Alabama
Gasconade (WLR 75401), Omaha, Nebraska
Muskingum (WLR 75402), Sallisaw, Oklahoma
Wyaconda (WLR 75403), Dubuque, Iowa
Chippewa (WLR 75404) [shown on p. 85], Buchanan, Tennessee
Cheyenne (WLR 75405), St. Louis, Missouri
Kickapoo (WLR 75406), Vicksburg, Mississippi
Kanawha (WLR 75407), Pine Bluff, Arkansas
Patoka (WLR 75408), Greenville, Mississippi
Chena (WLR 75409), Hickman, Kentucky

65-Foot River Tenders
Length: 65 feet
Beam: 21 feet
Displacement: 145 tons
Propulsion: 2 × Caterpillar diesels, 660–725 horsepower
Speed: 10 knots
Range: 3,500 nautical miles

Cutters in this class:

Ouachita (WLR 65501), Chattanooga, Tennessee
Cimarron (WLR 65502), Buchanan, Tennessee
Obion (WLR 65503), Owensboro, Kentucky

Scioto (WLR 65504), Keokuk, Iowa
Osage (WLR 65505), Sewickley, Pennsylvania
Sangamon (WLR 65506), East Peoria, Illinois

Inland Construction Tenders (WLIC)

First commissioned in the 1940s, the Coast Guard's inland construction tenders are broken into three classes, all designed for the construction, repair, and maintenance of fixed ATON within inland waterways. The WLIC is the only Coast Guard platform with the capability to drive and remove piles, erect towers, and effect major structural repairs. There are three subclasses of construction tenders: 75 foot, 100 foot, and 160 foot. The 75-foot WLIC construction tender pushes 68-foot and 84-foot construction barges. The one 100-foot WLIC pushes a 70-foot construction barge. The 160-foot WLICs are single units without barges.

160-Foot Inland Construction Tender

Length: 160 feet
Beam: 30 feet
Displacement: 411 tons
Propulsion: 2 × Caterpillar diesels, 1,000 base horsepower
Speed: 11 knots
Range: 1,205 nautical miles

Cutters in this class:

Pamlico (WLIC 800), New Orleans, Louisiana
Hudson (WLIC 801), Miami Beach, Florida

Kennebec (WLIC 802), Portsmouth, Virginia
Saginaw (WLIC 803), Mobile, Alabama

100-Foot Inland Construction Tender

Length: 100 feet
Beam: 24 feet
Displacement: 178 tons
Propulsion: 2 × Caterpillar diesels, 1,250 horsepower
Speed: 10 knots
Range: 1,200 nautical miles

Cutters in this class:

Smilax (WLIC 315, oldest commissioned cutter), Atlantic Beach, North Carolina

75-Foot Inland Construction Tender

Length: 75 feet
Beam: 22 feet
Displacement: 145 tons
Propulsion: 2 × Caterpillar diesels, 750 horsepower or 1,350 horsepower
Speed: 10 knots
Range: 1,050–2,500 nautical miles

Cutters in this class:

Anvil (WLIC 75301), Charleston, South Carolina
Hammer (WLIC 75302), Mayport, Florida
Sledge (WLIC 75303), Baltimore, Maryland
Mallet (WLIC 75304), Corpus Christi, Texas
Vise (WLIC 75305) (shown on p. 88), St. Petersburg, Florida
Clamp (WLIC 75306), Galveston, Texas
Hatchet (WLIC 75309), Galveston, Texas
Axe (WLIC 75310), Morgan City, Louisiana

Waterways Commerce Cutter (WCC)

The Coast Guard established the waterways commerce cutter program to replace the capability provided by its inland tender fleet. Sixteen river buoy tenders, eleven inland construction tenders, and three inland buoy tenders will be procured under the program. The new tenders will have greater endurance, speed, and deck load capacity than their predecessors and will feature improved habitability to accommodate mixed-gender crews. All three variants will be monohull ships, meaning self-propelled cutters instead of tug and barge configurations. The river buoy tender and inland construction tender variants are expected to be common except for hull length, working deck layouts, and deck equipment, including the crane. A contract for the design and construction of this new class was awarded in October 2022 with twenty-seven hulls expected to enter service over the next decade.

Harbor Tugs (WYTL)

Built between 1962 and 1967, 65-foot harbor tugs (WYTL) are multimission cutters that have the distinction of being used only on the East Coast, from Maine to Virginia. With a crew of six, their primary missions are domestic ice breaking and aids to navigation in rivers and littoral areas. They are capable of breaking ice up to twelve inches thick.

Length: 65 feet
Beam: 16 feet
Displacement: 72 tons
Propulsion: 1 × diesel, 500 base horsepower
Speed: 10 knots
Range: 2,700 nautical miles

Cutters in this class:

Capstan (WYTL 65601), Philadelphia, Pennsylvania
Chock (WYTL 65602), Baltimore, Maryland

Tackle (WYTL 65604) (shown on p. 90), Rockland, Maine
Bridle (WYTL 65607), Southwest Harbor, Maine
Pendant (WYTL 65608), Boston, Massachusetts
Shackle (WYTL 65609), South Portland, Maine
Hawser (WYTL 65610), Bayonne, New Jersey
Line (WYTL 65611), Bayonne, New Jersey
Wire (WYTL 65612), Saugerties, New York
Bollard (WYTL 65614), New Haven, Connecticut
Cleat (WYTL 65615), Philadelphia, Pennsylvania

BOATS

Coast Guard vessels under 65 feet in length are classified as boats. The service has about 1,680 boats altogether. These craft include cutter-based boats, surf-capable motor lifeboats, response boats, special-purpose craft, and ATON boats. Sizes range from 64 feet in length down to 12 feet. When powered by gasoline outboards, the standard crew for most open boat types is two, a coxswain and a crew member. For boats powered by a diesel engine, the standard crew is three, a coxswain, an engineer, and a crew member. The equipment supplied with each boat varies with the conditions under which it must operate and the missions assigned, but a scalable integrated navigation system (SINS II) with radar, global positioning, automatic identification system, along with an electronic chart plotter, VHF-FM radios, and basic rescue equipment, are standard. Vessels capable of high speeds, defined by the service as greater than thirty knots, are equipped with shock-absorbing seats to protect crews from musculoskeletal injuries. Most boat types have installed mounts able to accommodate crew-served weapons such as the M240B for security and defense missions. In fiscal year 2021 the Coast Guard's fleet of boats were responsible for 44 percent of all Coast Guard resource hours, 65 percent of search-and-rescue missions, 71 percent of lives saved, 76 percent of property saved, and 80 percent of maritime law enforcement boardings. All Coast Guard small boats, whether assigned to a cutter or not, have service numbers of five digits. The first two digits give the length of the boat and the last three its individual number.

Cutter Boats

Cutter boats enable their parent cutters to perform assigned missions, primarily the interdiction and boarding of suspect vessels, the maintenance and servicing of ATONs, and the ferrying of personnel and supplies to and from shore and other ships. Cutter boats propelled by outdrives or outboard engines are launched via single- or dual-point davits installed along the beam or at the stern of cutters. Jet-propelled boats may be recovered via davits but are designed primarily for launch and recovery via stern ramp, a method requiring fewer personnel on deck.

35-Foot Long-Range Interceptor (LRI)

The 35-foot long-range interceptor is an aluminum boat purpose built to deploy on board national security cutters that launch and recover the boat from a stern ramp. Constructed with a deep-v hull and propelled by water jets powered by dual diesel engines, the LRI II can travel at speeds in excess of forty knots. A semi-enclosed cabin and shock-mitigating seats provide for crew protection. It is designed primarily for intercepting and delivering boarding teams to suspect vessels, although it is capable of supporting the full range of offshore missions. Twelve hulls are currently in service.

26-Foot Cutter Boat—Over the Horizon (CB-OTH)

The 26-foot cutter boat—over the horizon is an aluminum boat built to deploy from a variety of cutters, including national security cutters, offshore patrol cutters, Famous- and *Reliance*-class medium-endurance cutters, and Sentinel-class cutters. Constructed with a deep-v hull and propelled by either a water jet (CB-OTH-IV/J) or an outdrive powered by a single diesel engine (CB-OTH-III), the CB-OTH can travel at speeds in excess of forty knots. The jet version of the design, the CB-OTH-J, can be launched and recovered from the stern ramp of cutters with the technology. There are 101 hulls currently in service.

With an estimated production of close to 200 boats between 2021 and 2029, the OTH-V will be the primary cutter boat for the Coast Guard's major cutters, and the program foresees two boats for each national security cutter, one for each fast-response cutter, three for each offshore patrol cutter, two for Polar-class security cutters, and one for each medium-endurance cutter. OTH-V characteristics include a top speed of forty-two knots, a diesel engine with waterjet, and dimensions limited to 26 feet long by 8.5 feet wide.

38-Foot Arctic Survey Boat (ASB)

The Arctic survey boat is a 38-foot boat carried on board Polar-class icebreakers for use in a variety of tasks to support cutter navigation and science operations, including performing cold-climate surveys, taking depth soundings ahead of the cutter, assisting in rescue operations, ferrying personnel, and serving as a lifeboat. It is constructed of a single skin, glass-reinforced hull with a reinforced bow for minor icebreaking and a reinforced belt around the waterline for protection against ice fields. Diesel-powered, the ASB has a maximum speed of ten knots and a range of 320 miles. The normal crew is six people.

24-Foot Cutter Boat—Large (CB-L)

A 24-foot aluminum boat carried on board the *Reliance*-class medium-endurance cutters (WMEC) and *Juniper*-class seagoing buoy tenders (WLB). Constructed with a deep-v hull and propelled by an outdrive powered by a single diesel engine, the CB-L can travel at speeds in excess of thirty knots.

38-foot arctic survey boat (ASB)

24-foot cutter boat—large (CB-L)

17-Foot Cutter Boat—Medium (CB-M)

The CB-M is a 17-foot aluminum boat propelled by a waterjet powered by a single diesel engine or a gasoline outboard. It is carried on board the 87-foot Marine Protector–class and 110-foot Island-class patrol boats.

17- to 28-Foot Utility Boat—Light (UTL)

UTLs are nonstandard boats fielded on board varied cutters based on unique mission requirements, operating environment, and standard hull availability.

16- to 24-Foot Cutter Boats—Aids to Navigation (CB-ATON)

Recognized types of CB-ATONs include the 24-foot ATON—large (CB-ATON-L), the 18-foot ATON—medium (CB-ATON-M), and the 16-foot ATON—skiff or small (CB-AB-S / CB-ATON-SKF).

Shore-Based Boats

The Coast Guard operates approximately 190 multimission boat stations throughout the country along the shores of the Atlantic Ocean, Pacific Ocean, Gulf of Mexico, and Great Lakes. Stations are traditionally associated with search and rescue, but they may perform the full range of Coast Guard missions. Coast Guard personnel live and work at or near their stations so they can rapidly respond to emergencies as they arise. Station locations are established and maintained based on a response standard to arrive on the scene of any case within their area of responsibility within two hours, which includes up to thirty minutes of preparation prior to launch, and factors such as expected survival times of people in the water, location of critical infrastructure, and other resource availability. In general, coastal stations are available to respond up to fifty nautical miles from shore.

47-Foot Motor Lifeboat (MLB)

The 47-ft aluminum MLB is primarily designed as a fast-response rescue vessel for high seas, surf, and heavy-weather environments, but it can perform the full range of Coast Guard missions. It is capable of surviving winds up to sixty knots, seas up to thirty feet, and breaking surf up to twenty feet. If the boat should capsize after being knocked over by waves or surf, it self-rights in less than thirty seconds with all equipment fully functional. First entering service in 1997, there are currently 117 MLBs, most assigned to surf and heavy-weather stations. Surf stations are located in regions where breaking surf greater than eight feet occurs thirty-six or more days

47-foot motor lifeboat (MLB)

per year. Heavy-weather stations are located where seas or swells are greater than ten feet or winds reach gale force for thirty-six or more days per gear.

Surf Stations

Quillayute River, Washington
Grays Harbor, Washington
Cape Disappointment, Washington
Tillamook Bay, Oregon
Depoe Bay, Oregon
Yaquina Bay, Oregon
Siuslaw River, Oregon
Umpqua River, Oregon
Coos Bay, Oregon
Chetco River, Oregon
Humboldt Bay, California
Noyo River, California
Bodega Bay, California
Golden Gate, California
Morro Bay, California
Merrimack River, Massachusetts
Barnegat Inlet, New Jersey
Oregon Inlet, North Carolina
Hatteras Inlet, North Carolina

Heavy-Weather Stations

Neah Bay, Washington
Monterey, California
Jonesport, Maine
Southwest Harbor, Maine
Portsmouth Harbor, Maine
Boothbay Harbor, Maine
South Portland, Maine
Gloucester, Massachusetts
Provincetown, Massachusetts
Menemsha, Massachusetts
Sandy Hook, New Jersey
Montauk, New York
Sandy Hook, New Jersey
Manasquan Inlet, New Jersey
Atlantic City, New Jersey
Indian River, Delaware
Ocean City, Maryland
Chincoteague, Virginia
Fort Macon, North Carolina
Oak Island, North Carolina

45-Foot Response Boat—Medium (RB-M)

The RB-M is designed as a high-speed, self-righting, multimission boat capable of operations in gale conditions and seas up to ten feet. It is jet propelled and able to reach speeds of more than forty-five knots. It entered service in 2008.

29-Foot Response Boat—Small II (RB-S II)

The RB-S II is a high-speed multimission-capable boat powered by twin outboard gas engines. Brought into service in 2003 to replace shore-based nonstandard boats, the RB-S is now in its second generation (RB-S II). The RB-S IIs are 29 feet long and capable of operations in twenty-five knot winds and six-foot seas. Approximately 268 RB-S II boats are in service.

Special-Purpose Craft

Special-purpose craft are designed to meet specific mission requirements or provide a capable and safe asset in a unique operating environment.

64-Foot Special-Purpose Craft—Screening Vessel (SPC-SV)

Purpose-built for Maritime Force Protection Units to perform their mission of protecting U.S. Navy ballistic missile submarines while surfaced and transiting between sea and the naval submarine bases at Bangor, Washington, and Kings Bay, Georgia.

52-Foot Special-Purpose Craft—Heavy Weather (SPC-HWX)

Built by the Coast Guard Yard in Curtis Bay, Maryland, in the late 1950s and early 1960s, the steel 52-foot motor lifeboats are the oldest boats still in operation. Used to perform offshore search and rescue under the most severe sea and surf conditions, the boats are self-righting and can tow vessels as large as 750 tons in thirty-foot seas. These boats are the only Coast Guard vessels under 65 feet to receive names. All are stationed in the Pacific Northwest.

45-foot response boat—medium (RB-M)

29-foot response boat—small II (RB-S II)

64-foot special-purpose craft—screening vessel (SPC-SV)

Victory, 52312, Yaquina Bay
Invincible, 52313, Grays Harbor
Triumph, 52314, Station Cape Disappointment
Intrepid, 52315, Station Coos Bay

42-Foot Near-Shore Lifeboat

This lifeboat is a jet-propelled shallow-draft fast-response aluminum boat capable of operations in high seas and surf. It is used exclusively at Station Chatham, Massachusetts, where the depth of the Chatham Bar prohibits operation of the propeller-driven 47-foot motor lifeboat.

38-Foot Special-Purpose Craft—Training Boat (SPC-TB)

These boats are operated by Coast Guard Training Center Yorktown as a platform to teach seamanship and boat handling to boatswain's mate "A" school students. The school's fourteen training boats, combined with several standard hulls, create the largest boat fleet at any one unit in the service.

33-Foot Special-Purpose Craft—Law Enforcement (SPC-LE)

The SPC-LE is a high-speed multimission craft designed and built primarily for counterdrug and migrant interdiction missions along the U.S. maritime border. The SPC-LE was the first joint acquisition with U.S. Customs and Border Protection.

32-Foot Transportable Port Security Boat (TPSB)

TPSBs are operated exclusively by port security units (PSU) for defense readiness operations domestically and in support of combatant commanders overseas. They are propelled by diesel outdrives and capable of speeds in excess of forty knots, and

32-foot transportable port security boat (TPSB)

are heavily armed with a .50-caliber machine gun and two M60 machine guns. There are fifty-two in service.

Port Security Units (PSU), number and location:

301 Cape Cod, Massachusetts
305 Fort Eustis, Virginia
307 Clearwater, Florida
308 Stennis Space Center, Mississippi
309 Cape Perry, Ohio
311 San Pedro, California
312 Alameda, California
313 Everett, Washington

24-Foot Special-Purpose Craft—Shallow Water (SPC-SW)

The SPC-SW is an aluminum multimission boat powered by gasoline outboards fitted to a hydraulic jacking plate to reduce its draft to less than 1.5 feet. It is used primarily by stations with large bodies of shallow water and to support natural disaster response missions in flooded areas.

18–24-Foot Special-Purpose Craft—Air (SPC-Air)

The SPC-Air are fan-driven boats assigned to stations in the Ninth District to support long-haul ice rescue and flood response missions. Long haul is defined as greater than one nautical mile from shore. They are capable of operating over water, ice, and land.

Long-Haul Capable Ice Rescue Stations:

Marblehead, Ohio
Sault Saint Marie, Michigan
Saginaw River, Michigan
Sturgeon Bay, Wisconsin

15-Foot Skiff—Ice (SKF-ICE)

The SKF-ICE are inflatable boats with a removable outboard assigned to stations and cutters in the Ninth District to support short-haul ice rescue missions in enclosed ports, waterways, bays, and from Coast Guard cutters, and to respond to flooding during natural disasters.

16-Foot Flood Punts

Flood punts are aluminum and inflatable boats fitted with removable outboard engines and are used for flood response during natural disasters.

Aids-to-Navigation Boats

These boats are operated by shore-based aids-to-navigation teams whose primary mission is to maintain the nearly 50,000 navigation aids of the maritime transportation system located within inland waters, bays, sounds, and harbors. Approximately 145 ATON boats are in service.

64-Foot Aids-to-Navigation Boat (ANB)

Built in the 1990s, three ANBs are in service. They are equipped with a 4,500-pound lifting-capacity crane and a V-cut stern to allow it to maintain contact with a floating aid under wave conditions. They have a crew size of four, a maximum speed of nine knots, and an operating range of five hundred nautical miles.

55-Foot Aids-to-Navigation Boat (ANB)

These boats are designed primarily for fixed-aid maintenance but are equipped with a crane to hoist and secure various buoys for servicing with a lifting capacity up to 3,000 pounds. They have a crew size of 4, a maximum speed of 21.5 knots, and an operating range of 175 nautical miles. Built by the Coast Guard Yard between 1977 and 1988, 26 were brought into service.

49-Foot Buoy Utility Stern Loading (BUSL)

Built between the 1990s and 2001 by the Coast Guard Yard, the forty BUSLs to enter service are designed with a V-notched stern and equipped with A-frame and hoisting winches to secure buoys up to 4,500 pounds for servicing. They have a crew size of 4, a maximum speed of 10 knots, and an operating range of 300 nautical miles.

45-Foot Bow-Loading Buoy Boats (BU)

In service since the 1950s, BUs have a landing craft style design with a bow ramp. Fitted with a crane having a 4,000-pound lifting capacity for servicing bouys, they have a crew size of 4, a maximum speed of 8.5 knots, and an operating range of 550 nautical miles. BUs have been almost entirely replaced by BUSLs.

26-Foot Trailerable Aids-to-Navigation Boat (TANB)

The TANB is a 26-foot aluminum center-console boat fielded as the primary asset of most ATON teams. It is outfitted with an electric winch-operated davit and boom with a hoisting capacity of 500 pounds to service small buoys and has a T-top canopy that allows servicing fixed aids from the boat without the need to climb. TANBs have a crew size of 4, a maximum speed of 40 knots, and an operating range of 170 nautical miles.

16-Foot Aids-to-Navigation Boat—Skiff (AB-SKF)

The AB-SKF is designed to service fixed aids in shallow water and is fitted with a sixty-horsepower outboard engine on a hydraulic jacking plate to achieve a six-inch draft.

AIRCRAFT

The Coast Guard has three different types of aircraft: manned fixed-wing planes with jets or turboprops, manned rotary-wing helicopters, and unmanned aerial systems (UAS). Manned Coast Guard aircraft operate from the service's air stations, and Coast Guard helicopters routinely deploy on board flight-deck-equipped cutters. The Coast Guard operates 202 manned fixed- and rotary-wing aircraft from 21 Coast Guard Air Stations (CGAS) around the country. Unmanned systems are operated from shore and cutters at sea by a variety of units.

The Coast Guard uses the standard Department of Defense system to designate aircraft. A typical example follows:

HC-130H
- H: modified mission symbol
- C: basic mission symbol
- 130: model number
- H: modification

Some MH-designated helicopters are armed, and some are equipped with armor to protect against small-arms fire.

Fixed-Wing Aircraft

All-weather aircraft are deployed primarily from land-based CGAS to conduct airborne operations. They consist of command-and-control, long-range surveillance (LRS), and medium-range surveillance aircraft. These aircraft conduct operations supporting all Coast Guard missions and perform mission support activities such as logistics and transport flights.

C-37A Gulfstream V Command-and-Control Aircraft

The Coast Guard operates two Gulfstream V aircraft as its principal command-and-control transport for travel by the secretary of the Department of Homeland Security, the commandant of the Coast Guard, and other U.S. officials. The twin turbofan fixed-wing aircraft, C-37A, carries a crew of four and up to twelve passengers. A contract for a new C-37B was awarded by the service in October 2021 to ensure the service retains a two-aircraft fleet.

Power plant: 2 × 14,750 horsepower turbofan engines
Cruising speed: 459 knots
Service ceiling: 51,000 feet
Range: 5,500 nautical miles
Wingspan: 94.5 feet
Length: 96 feet

C-37A Air Station
Washington, D.C. (Ronald Reagan Washington National Airport)

UHC-130H Hercules and HC-130J Super Hercules, Long-Range Surveillance (LRS)

The Coast Guard's history with the "Herc" dates to 1958, when the service first ordered the HC-130B model. A design proven successful for Coast Guard operations over decades, the fleet of modern Hercs now in service numbers approximately thirty. The HC-130J is a long-range four-engine turboprop fixed-wing aircraft that provides a versatile platform capable of on-scene command-and-control and surveillance with extended loitering capabilities. The aircraft features a nose-mounted electro-optical/

HC-130J Super Hercules, long-range surveillance (LRS)

infrared camera and surface-search radar. It can carry 92 passengers or 51,000 pounds of cargo. Operated by a crew of seven, it performs various missions including maritime patrol, law enforcement, search and rescue, disaster response, and cargo and personnel transport. Life rafts, dewatering pumps, flares, or dispersant for oil pollution control can be airdropped from the ramp at the rear of the plane. HC-130Hs are being retired as HC-130Js are accepted into service.

Power plant: 4 × 5,600 horsepower turboprop engines
Cruising speed: 280 knots; maximum 362 knots
Service ceiling: 39,000 feet
Range: 5,200 nautical miles
Wingspan: 132.6 feet
Length: 99.6 feet
Endurance: 20+ hours

HC-130 Air Stations:
Sacramento, California
Clearwater, Florida
Elizabeth City, North Carolina
Kodiak, Alaska
Barbers Point, Hawaii

HC-144A Ocean Sentry "CASA" Medium-Range Maritime Patrol Aircraft (MPA)

Entering service in 2009, the HC-144 is a medium-range two-engine turboprop fixed-wing aircraft that provides a versatile platform capable of on-scene command-and-control with extended loitering capabilities. Operated by a crew of six, the eighteen HC-144s now in service perform various missions including maritime patrol, law enforcement, search and rescue, disaster response, and cargo and personnel transport. Life rafts, dewatering pumps, flares, or dispersant for oil pollution control can be airdropped from the ramp at the rear of the plane.

Power plant: 2 × 1,750 horsepower turboprop engines
Cruising speed: 230 knots; maximum 246 knots
Service ceiling: 25,000 feet
Range: 2,000 nautical miles
Wingspan: 84.7 feet
Length: 70 feet
Endurance: 10+ hours

HC-144 Air Station:
Mobile, Alabama
Cape Cod, Massachusetts
Miami, Florida
Corpus Christi, Texas
Astoria, Oregon
Borinquen, Puerto Rico

HC-27J Spartan Medium-Range Maritime Patrol Aircraft

Entering Coast Guard service in 2014, the HC-27J is a modified surveillance variant of the U.S. military's joint cargo aircraft, made possible by the transfer of fourteen C-27Js from the U.S. Air Force. Medium-range two-engine turboprop fixed-wing aircraft, Spartans perform various missions including maritime patrol, law enforcement, search and rescue, disaster response, and cargo and personnel transport. Life rafts, dewatering pumps, flares, or dispersant for oil pollution control can be airdropped from the ramp at the rear of the plane.

Power plant: 2 × 4,637 horsepower turboprop engines
Cruising speed: 315 knots; maximum 325 knots
Service ceiling: 30,000 feet
Range: 1,100 nautical miles
Wingspan: 94.1 feet
Length: 74.5 feet
Endurance: 12 hours

HC-27J Air Station:
Sacramento, California

Helicopters

These aircraft deploy primarily from land-based CGAS or underway cutters to conduct airborne operations supporting all Coast Guard missions. They consist of medium-range response and short-range response helicopters. These aircraft conduct operations supporting all Coast Guard missions and perform mission support activities such as logistics and transport flights.

MH-60T Jayhawk Medium-Range Recovery Helicopter

The Coast Guard began to operate the all-weather, medium-range H-60 in 1990 and now has forty-two MH-60Ts in service. Jayhawks are crewed by two pilots, a flight mechanic, and a rescue swimmer and can carry up to six seated survivors. The aircraft is capable of operations out to three hundred nautical miles, with a forty-five-minute on-scene time, and can be deployed from the flight decks of some cutters. In addition to a rescue hoist rated for six hundred pounds, Jayhawks are equipped with a heavy-lift external sling with a capacity of six thousand pounds. The MH-60 carries sensors and equipment for search and rescue, law enforcement, and homeland security missions. All Jayhawks employ full night-vision-device capability and are equipped for airborne use of force (AUF).

Power plant: 2 × 1,560 shaft horsepower turboshaft engines
Rotor diameter: 53.7 feet
Length: 64.7 feet
Maximum speed: 180 knots
Service ceiling, 13,000 feet
Range: 700 nautical miles
Gross weight: 21,884 pounds
Armament:
  1 × M107 .50-caliber precision rifle
  1 × M240 7.62-mm machine gun

MH-60T Units:
Mobile, Alabama
Kodiak, Alaska
Sitka, Alaska
San Diego, California
Clearwater, Florida
Cape Cod, Massachusetts
Elizabeth City, North Carolina
Astoria, Oregon

MH-65C/D/E Dolphin Short-Range Recovery Helicopter

The H-65 Dolphin is the Coast Guard's oldest helicopter, dating to the 1980s when it was selected for the short-range rescue mission after becoming the first helicopter certified with a four-axis autopilot, allowing for hands-off hover over a predetermined location. A twin-engine, single-rotor helicopter, the H-65 is certified for all-weather and nighttime operations and equipped for airborne use of force (AUF), although it is prohibited from flying under known icing conditions. It is the Coast Guard's standard shipboard-deployable aircraft and operates from all flight deck–equipped cutters.

Power plant: 2 × 853 shaft horsepower turboshaft engines
Rotor diameter: 39.1 feet
Length: 44.4 feet
Cruising speed: 120 knots; maximum 175 knots
Operational ceiling: 10,000 feet
Range: 375 nautical miles
Gross weight: 9,480 pounds
Armament:
1 × M107 .50-caliber precision rifle
1 × M240B 7.62-mm machine gun

H-65 Air Stations:

Traverse City, Michigan
Barbers Point, Hawaii
Borinquen, Puerto Rico
Atlantic City, New Jersey
Corpus Christi, Texas
Detroit, Michigan
Houston, Texas
Humboldt Bay, California
Los Angeles, California
Miami, Florida
Mobile, Alabama
New Orleans, Louisiana
North Bend, Oregon
San Francisco, California
Port Angeles, Washington
Savannah, Georgia
Kodiak, Alaska
Jacksonville, Florida

Unmanned Aerial Systems (UAS)

The Coast Guard is procuring unmanned aerial system (UAS) capability as a cost-effective approach to meeting the need for a persistent airborne surveillance capability. Use of UAS expands maritime domain awareness and disseminates actionable intelligence on maritime hazards and threats. A UAS consists of an unmanned aircraft, its mission payloads, launch and recovery equipment, ground control station, and control and data links. The service recognizes three types of UAS: long range, medium range, and short range.

Long Range

Long-range UASs are runway-dependent models that provide wide area surveillance for up to twenty-four hours of flight time beyond radio line of sight, transmitting video feeds via SATCOM. Coast Guard shore-based UAS flight crews fly the MQ-9 Guardian, a maritime-variant of the Predator UAS, jointly with Customs and Border Protection from airfields in Corpus Christi, Texas, and Jacksonville, Florida. The MQ-9 has a 950 horsepower turboprop, a maximum speed of 260 knots, a range of 1,150 miles, and an operational altitude of 50,000 feet. The technology supported its first narcotics interdiction on 15 September 2011.

Medium Range

Medium-range UASs are launched on demand to support tactical commanders with up to twelve hours of flight time beyond visual line of sight, providing video feeds via radio signals. All national security cutters are outfitted with ScanEagle UAS capability, a catapult launch system with an eighteen-hour endurance, a one hundred nautical mile range, and maximum speed of ninety-six miles per hour.

Short Range (sUAS)

Short-range UASs are personal, portable, hand-launched systems providing twenty to thirty minutes of flight time within visual line of sight and transmitting video feeds via radio signals or Wi-Fi.

5

UNIFORMS, MEDALS, AND AWARDS

Your uniform identifies you as being a professional member of a military service that is over two centuries old. It shows, at a glance, that you are a member of the U.S. Coast Guard and is distinctive visual evidence of the authority and responsibility vested in their wearer by the United States. You must present a proud, professional, and consistent appearance that will positively reflect yourself, the Coast Guard, and the United States. Because many of our missions require close contact with the public, it is important that the public have faith and confidence in, and respect for, the Coast Guard. Presenting a proper appearance is a critical but easy way of developing that attitude in the public. The Uniform Regulations Manual, COMDTINST M1020.6 (series), can answer any question you may have about uniforms, grooming, and insignia.

GENERAL UNIFORM STANDARDS

- Uniform neat, clean, pressed, proper fit, in good condition, zipped, snapped, and buttoned unless otherwise authorized. Brass polished and scratch-free.
- Headgear worn outdoors with all required uniforms. Should not be worn indoors or in no-cover areas unless under arms or for ceremonial purposes.
- Oxfords, pumps, or flats shined. Boots well blackened, edges dressed, all in good condition.

Official party conducts a personnel inspection during a change-of-command ceremony

REQUIRED SEA BAG ITEMS

All Coast Guard personnel are responsible for maintaining their uniforms in serviceable condition and are responsible for procuring sufficient quantities of uniform items and properly maintaining them as appropriate to their assigned duties. All personnel are responsible for owning, wearing, and maintaining sufficient quantities of personal items such as undergarments, socks and stockings, and so on, to comply with Coast Guard standards of personal hygiene and appearance. Active-duty, reserve, retired, and Auxiliary members may purchase uniform items from the Coast Guard Exchange System at ShopCGX.com. The following items are required for all enlisted Coast Guard personnel, E-9 and below:

REQUIRED COAST GUARD SEA BAG ITEMS

	Gender	Quantity
Bag, duffel, nylon, olive green	M & F	1
Belt, black, web (nylon)	M & F	1
Belt riggers, black, web ODU*	M & F	1
Blousing band set, trouser ODU*	M & F	2
Buckle, belt, brass	M & F	1
Cap, garrison	M & F	1
Ball cap (standard)	M & F	1
Coat, service, dress blue	M & F	1
Combination hat set	M & F	1
Foul weather parka II	M & F	1
Gloves, leather, black	M & F	1
Insignia, garrison cap	M & F	1
Insignia, combination cap	M & F	1
Nametag, plastic	M & F	2
Name / U.S. Coast Guard tape set, blue	M & F	4
Necktie, blue, four-in-hand	M	1
Necktie, CG blue tab type	F	1
Necktie, black bowtie	M	1
Necktie, black, tab type	F	1
Shield, gold	M & F	1
Shirt, dress white, long sleeve	M & F	1
Shirt, light blue, long sleeve	M & F	3
Shirt, light blue, short sleeve	M & F	3 / 2
Overblouse (Air Force style)	F	1
Shirt, ODU*	M & F	4
Windbreaker, w/liner	M & F	1
ODU* utility jacket	M & F	1

*ODU is operational dress uniform

UNIFORM OF THE DAY

The standard Coast Guard uniform of the day, regardless of season or location, is service dress blue (SDB) or tropical blue (TB) uniform. District commanders prescribe the uniform for the season, day, or special occasion for personnel within the geographical limits of their district. Operational dress uniform (ODU) or authorized organizational clothing must be worn as required by the work or duties being performed.

In addition to periods of duty, the uniform of the day is also authorized to be worn in the following situations:

- On liberty
- Traveling in U.S. government vehicles, vessels, or aircraft
- Traveling on orders in commercial conveyances
- On duty under instruction
- Participating in traditional veteran's functions, parades, ceremonies, and other events honoring the United States or U.S. armed forces

It is prohibited to wear a uniform in the following situations:

- Participating in public speeches, interviews, picket lines, marches, or rallies not specifically approved or sanctioned by the Coast Guard
- Participating in political activities, private employment, or commercial interests
- Working in an off-duty civilian capacity
- When it would discredit the Coast Guard, armed forces, or U.S. government

Operational Dress Uniform (ODU)

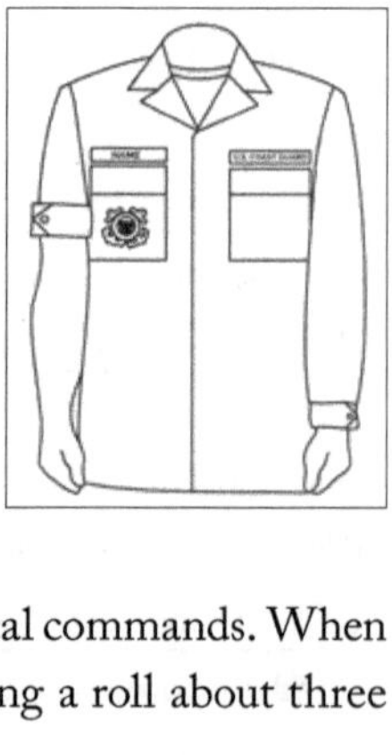

Operational dress uniform is worn year-round as required by the work or duties being performed. You must always wear and tuck in your T-shirt, but the ODU blouse may be removed when not on watch, after the standard workday, and when not engaged with the public. The ball cap or knit watch cap are authorized. The watch cap is rolled enough times to rest just above the ears. In inclement weather, it may be rolled down enough to cover the ears but not pulled straight down like a skull cap.

The sleeves of the ODU may be rolled up at the option of local commands. When authorized, sleeves will be rolled with the outside out, forming a roll about three inches wide and terminating at a point at the middle of the bicep, about two inches above the elbow. To roll the sleeves:

- Start by folding cuff over itself creating a neat fold up to the armpit.
- Fold the sleeve up again, leaving about three inches of the first fold exposed.
- Finally, pull the cuff back down over the entire fold. This results in a three-inch cuff about two inches above the elbow. Folds will vary based on the size and length of the arm.

Service Dress Blue "Bravo" (SDB)

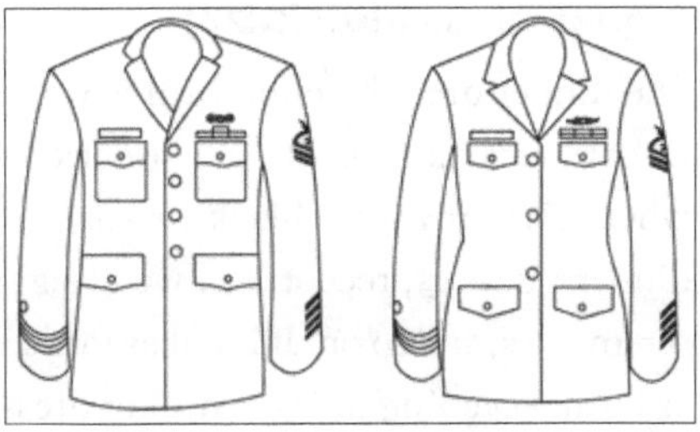

The service dress blue "Bravo" is worn year-round by all personnel for all business and social occasions where civilian equivalent is a coat and tie. This uniform requires a light-blue shirt, collar insignia for enlisted personnel, tie, tie bar/tack, name tag, ribbons, large qualification insignia and badges, and miniature devices. It may be worn with the combination or garrison cap.

Tropical Blue (TBSS)

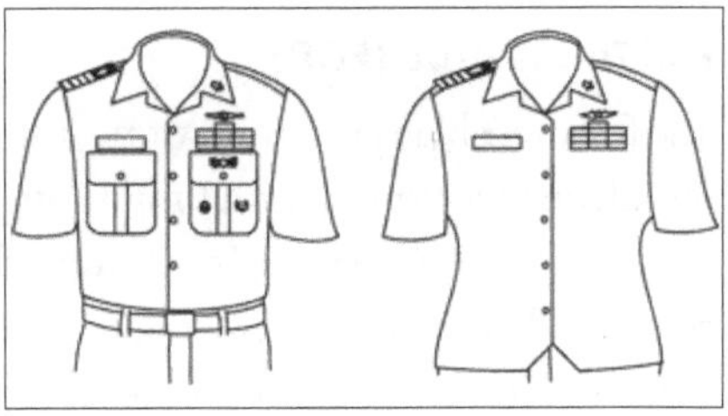

The tropical blue uniform is worn year-round at all occasions where the SDB coat is relaxed. It requires a collar insignia, name tag, tie, qualification insignia, and miniature devices and badges, and it may be worn with the combination or garrison cap.

Tropical Blue Long Sleeve (TBLS)

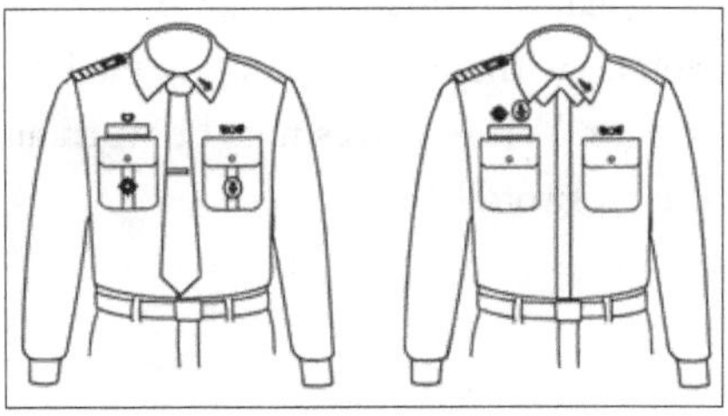

The tropical blue long-sleeve uniform is worn year-round at all occasions where a coat and tie are not required or expected. It requires a collar insignia, tie, tie bar/tack, name tag, ribbons, qualification insignia, and miniature badges, and it may be worn with the combination or garrison cap.

Winter Dress Blue (WDB) (Optional)

The optional winter dress blue uniform is worn in lieu of the SDB. It may not be worn to functions where civilians would wear a coat and tie. It requires a dark-blue dress shirt, collar insignia, tie, tie bar/tack, name tag, ribbons, qualification insignia and badges, and miniature devices, and it may be worn with the combination or garrison cap.

OTHER UNIFORMS REQUIRED FOR ALL PERSONNEL

Dinner Dress Blue (DDB)

The dinner dress blue uniform is worn in cold weather for official functions where civilians wear black tie such as dinners, dances, receptions, weddings, ceremonies, and so on. It is not as formal as formal evening dress or dinner dress jackets, and it requires a white shirt with no collar insignia, a black bow tie, small medals, and miniature qualification insignia, devices, and badges. White gloves may be worn at the prescribing officer's discretion.

Full Dress Blue (FDB)

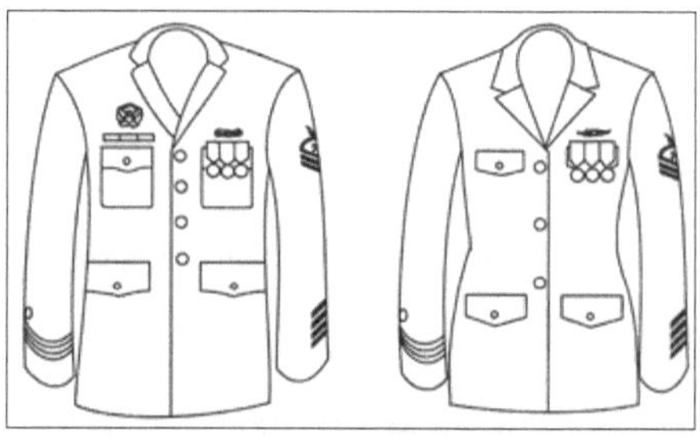

The full dress blue uniform is worn year-round when assuming or relinquishing command; participating in change of command ceremonies; or conducting official visits with honors, occasions of state, ceremonies, weddings, or solemnities. The FDB requires a white shirt with no collar insignia, a tie, large medals, ribbons without corresponding medals, and large qualification insignia, devices, and badges. White gloves may be worn at the prescribing officer's discretion. Officers wear a sword belt.

Service Dress Blue "Alpha" (SDB "A")

The service dress blue "Alpha" uniform is worn year-round by all personnel when prescribed for functions that require more formality than provided by the SDB "Bravo" uniform but less formality than the full dress or dinner dress uniforms. It requires a white shirt with no collar insignia, tie, name tag, ribbons, large qualification insignia, and miniature devices.

ADDITIONAL UNIFORMS REQUIRED FOR OFFICERS

Service Dress White (SDW)

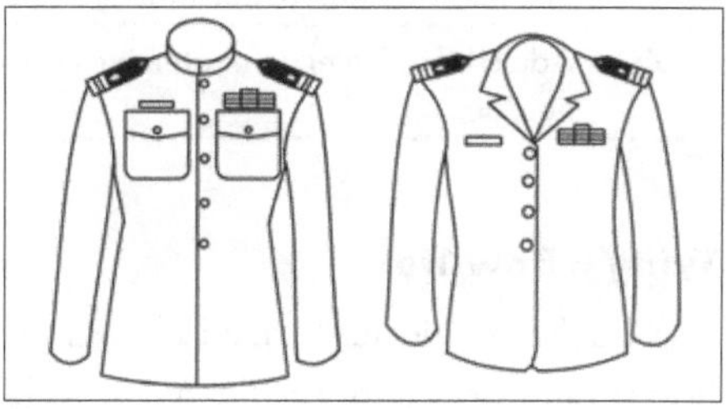

The service dress white uniform is worn in warm weather for official business, social functions, and ceremonies when senior officers wish to pay special honor to the occasion, and for parades, reviews, and evening socials.

Full Dress White (FDW)

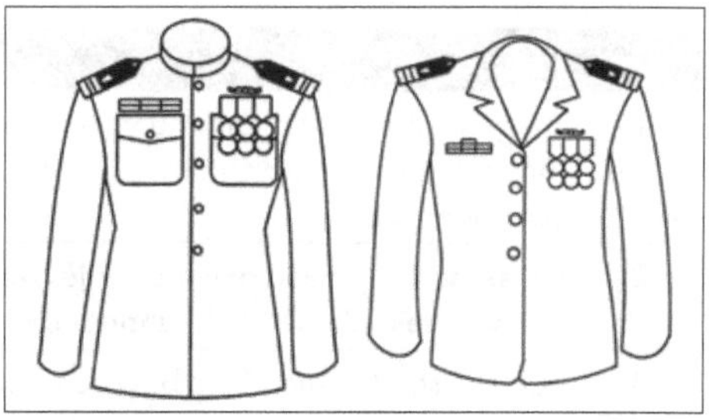

The full dress white uniform is worn in warm weather when assuming or relinquishing command, participating in change of command ceremonies, making official visits with honors, and attending occasions of state, ceremonies, weddings, or solemnities. All personnel wear the FDW.

Tying a Neck Tie

The standard necktie is CG blue, not exceeding three inches in width, and either hand-tied or clip-on. The double Windsor knot is the only authorized tie knot. To tie:

Step	Procedure
1	Fold the shirt collar up, and place the necktie around your neck. Ensure that the wide end of the tie is six to seven inches below the narrow end and cross it over the narrow end.
2	Slip the wide end of the tie through the gap between the neck and the tie cross. Keep pulling the tie all the way around until it is back behind the narrow end of the tie.
3	Still using the wide end of the tie, repeat step 2 on the other side.
4	After this, wrap the wide end of the tie and bring it back over toward the front.
5	Pull the wide end again through the gap between your neck and the knot loosely, so that a loop is created.
6	Slip the wide end of the knot through this newly created loop.
7	Adjust and tighten to neaten the knot and place your collar back down over the necktie.

Tying a Bow Tie

The black bow tie authorized for wear with dinner dress uniforms and the white bow tie authorized for the formal dress blue uniform are required to be a plain style with squared ends, approximately 4.5 inches wide, and either hand-tied or a clip-on type. To tie a bow tie:

Step	Procedure
1	Hang the bow tie flat around your neck, making sure that one end of the bow tie is longer than the other end by a few inches. The longer end of the tie should be on your right side.
2	Cross the longer end of the bow tie over the shorter end. Bring the longer end of the bow tie underneath the shorter end and out from the center of your neck
3	Fold the shorter end of the bow tie horizontally.
4	Place the longer end of the bow tie over the shorter end.
5	Fold the longer end of the bow tie horizontally.
6	Insert the longer end of the bow tie through the loop at the back of the shorter end.
7	Pull both ends gently to tighten the knot.
8	Adjust as needed.

ORGANIZATIONAL CLOTHING

Organizational clothing is Coast Guard unit-owned clothing that is issued to individuals. Personnel are accountable for all organizational clothing items, which will be entered on a Personal Clothing and Equipment Record, Form AF 538, and the clothing will be returned to the unit when the member is transferred. Types of organizational clothing include:

- Personal protective equipment
- Boat crew equipment
- Coveralls
- Flight crew clothing
- Culinary specialist uniforms
- Special command aide clothing
- Ceremonial honor guard, precision drill team, and unit-level color guard
- Hot weather clothing
- Cold weather clothing

MATERNITY UNIFORM

Maternity uniforms may be worn by women during pregnancy and up to sixty days after returning from maternity leave.

AUXILIARY UNIFORMS

Auxiliarists are authorized to wear Coast Guard uniforms with the appropriate Auxiliary insignia, with the exception of full dress uniforms. Often the public does not differentiate between Coast Guard active duty and the Auxiliary, especially when the uniform is worn. Coast Guard active-duty members are bound by strict military grooming and weight standards. Realizing that the Auxiliary is made up entirely of volunteers, the active-duty grooming and appearance standards are relaxed as they apply to Auxiliarists. Uniforms are worn as prescribed in Coast Guard Uniform Regulations, COMDTINST M1020.6 (series).

AWARDS AND DECORATIONS

The term "award" is used to describe any medal, ribbon, or attachment displayed on the uniform. "Decoration" is usually used to describe an award given to an individual for a specific act of bravery or meritorious service. Medals always have a corresponding ribbon, but some awards consist of only a ribbon and do not have a corresponding medal. If you earn the same award more than once, you will not receive the medal or ribbon again but will instead receive a metal attachment that goes on the original medal or ribbon. Attachments are also sometimes added to represent something other than repeated awards, such as an "S" on a marksmanship ribbon to indicate qualification as a sharpshooter, or the unique hurricane device authorized for the Presidential Unit Citation the Coast Guard received for its response to Hurricanes Katrina and Rita.

Medals are worn only on very formal occasions. Ribbons are worn on dress uniforms but not on working ones. When you earn your first ribbon, wear it centered a quarter of an inch above your left breast pocket. As you add ribbons, you will build them in rows of three. If you have a number not divisible by three, the uppermost row contains the lesser number, with the extra one or two ribbons centered over the row beneath. Ribbons are displayed in order from top to bottom, from center out. In other words, they are displayed as if someone else was reading lines of print on your chest.

Medals are displayed in a similar fashion. On full dress occasions, when you are wearing your medals, line them up in rows of three, side by side, or you may put five in a row if you overlap them. Any awards that only have a ribbon are worn on the right breast when full-size medals are worn. Do not wear any ribbons, however, when you are wearing miniature medals on dinner dress occasions.

Precedence of Awards

It is important that you wear your awards in the proper order. Whenever you earn a new award, you must determine its precedence, meaning where it goes in relation to the other awards you have already earned. The awards with the higher precedence are worn closer to your heart, called "farthest inboard." In general, awards are worn in the following order:

- Personal awards
- Unit awards
- Campaign and service awards
- Foreign and international awards
- Marksmanship awards

Below is a listing of awards by precedence. If you have any questions or doubts about how to arrange your ribbons, medals, and attachments, consult the Medals and Awards Manual, COMDTINST M1650.25 (series).

Personal Decorations

Medal of Honor
Coast Guard Cross
Navy Cross
Distinguished Service Cross

Air Force Cross
DHS Distinguished Service Medal
DOT Distinguished Service Medal
Defense Distinguished Service Medal
Coast Guard Distinguished Service Medal
Navy Distinguished Service Medal
Army Distinguished Service Medal
Air Force Distinguished Service Medal
DOT Guardian Medal
Silver Star Medal
Defense Superior Service Medal
Legion of Merit
Distinguished Flying Cross
Coast Guard Medal
Navy/Marine Corps Medal
Soldier's Medal
Airman's Medal
Gold Lifesaving Medal
Bronze Star Medal
Purple Heart Medal
Defense Meritorious Service Medal
Meritorious Service Medal
Air Medal
Silver Lifesaving Medal
Air Force Aerial Achievement Medal
Joint Service Commendation Medal
Coast Guard Commendation Medal
Navy and Marine Corps Commendation Medal
Army Commendation Medal
Air Force Commendation Medal
Joint Service Achievement Medal
DOT 9/11 Medal
Coast Guard Achievement Medal
Navy and Marine Corps Achievement Medal
Army Achievement Medal
Air Force Achievement Medal
Commandant's Letter of Commendation Ribbon

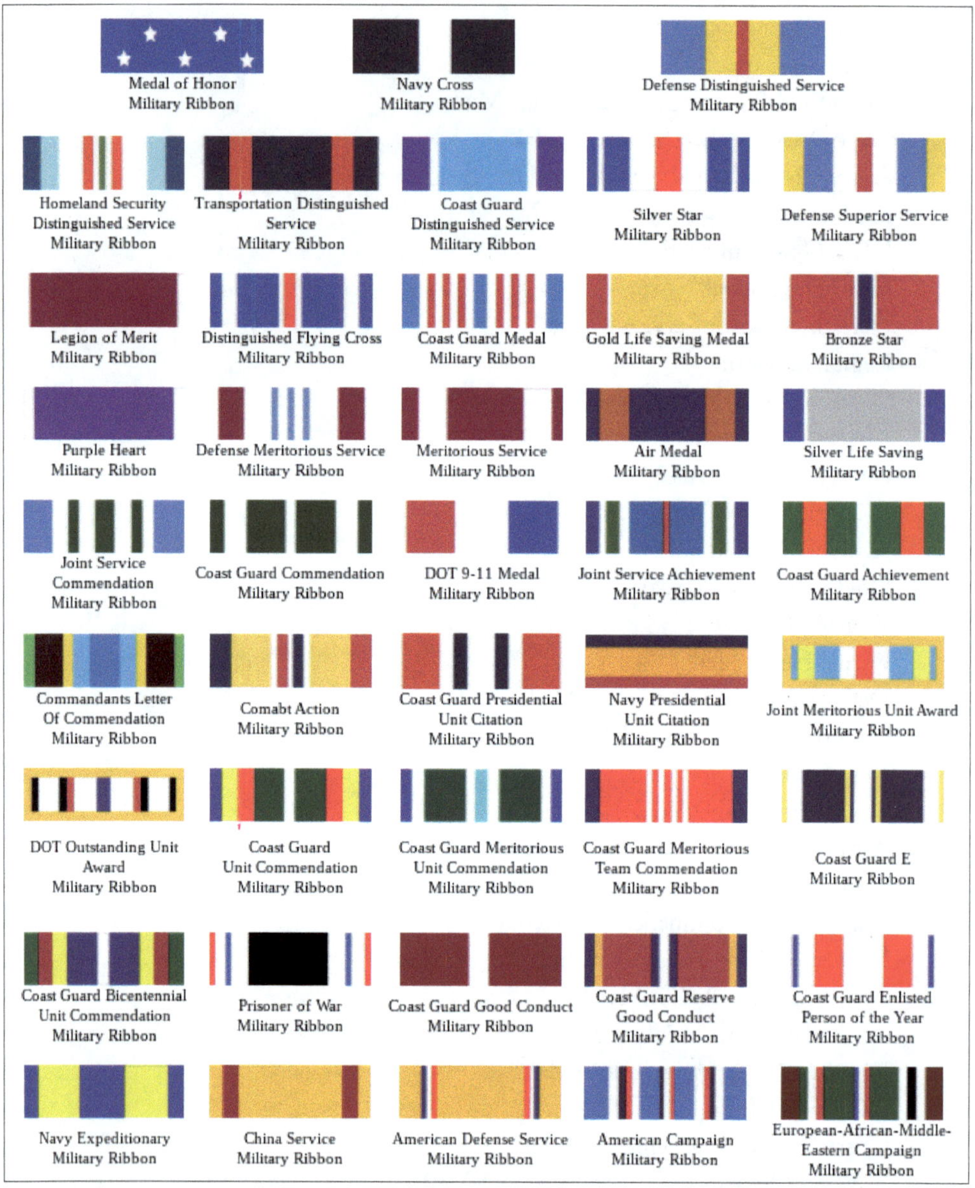

Awards by precedence (1)

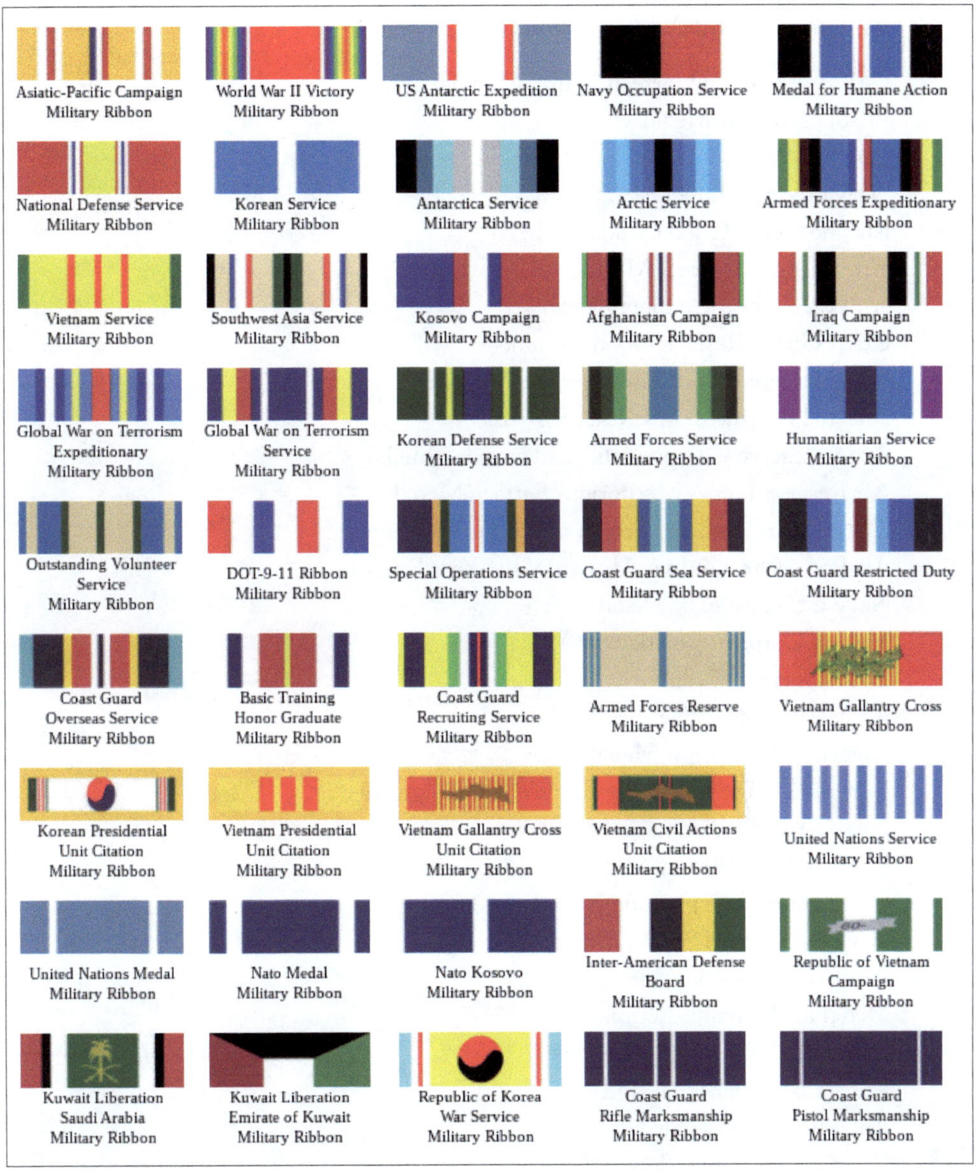

Awards by precedence (2)

Coast Guard Combat Action Ribbon
Navy Combat Action Ribbon
Air Force Combat Action Ribbon

Unit Awards (No Corresponding Medals)
Prisoner of War Medal
Coast Guard Good Conduct Medal
Navy Good Conduct Medal
Marine Corps Good Conduct Medal
Army Good Conduct Medal
Air Force Good Conduct Medal
Coast Guard Reserve Good Conduct
Naval Reserve Meritorious Service Medal
Selected Marine Corps Reserve Medal
Army Reserve Components Achievement Medal
Air Reserve Forces Meritorious Service Medal
Coast Guard Enlisted Person of the Year Ribbon
Air Force Outstanding Airman of the Year Ribbon
Navy Expeditionary Medal
Marine Corps Expeditionary Medal
China Service Medal
American Defense Service Medal
American Campaign Medal
European-African-Middle Eastern Campaign Medal
Asiatic-Pacific Campaign Medal
World War II Victory Medal
U.S. Antarctic Expedition Medal
Navy Occupation Service Medal
Army of Occupation Medal
Medal for Humane Action
National Defense Service Medal
Korean Service Medal
Antarctica Service Medal
Coast Guard Arctic Service Medal
Armed Forces Expeditionary Medal
Vietnam Service Medal
Southwest Asia Service Medal

Kosovo Campaign Medal
Afghanistan Campaign Medal
Iraq Campaign Medal
Inherent Resolve Campaign Medal
Global War on Terrorism Expeditionary Medal
Global War on Terrorism Service Medal
Korean Defense Service Medal
Armed Forces Service Medal
Humanitarian Service Medal
DOT 9/11 Ribbon
Military Outstanding Volunteer Service Medal
Navy/Marine Corps Seas Service Deployment Ribbon
Coast Guard Special Operations Service Ribbon
Navy Arctic Service Ribbon
Coast Guard Sea Service Ribbon
Naval Reserve Sea Service Ribbon
Coast Guard Restricted Duty Ribbon
Air Force Overseas Ribbon - Short
Army Sea Duty Ribbon
Coast Guard Overseas Service Ribbon
Navy/Marine Corps Overseas Service Ribbon
Army Overseas Ribbon
Air Force Expeditionary Service Ribbon
Air Force Overseas Service Ribbon (long tour)
Air Force Special Duty Ribbon
Coast Guard Basic Training Honor Graduate Ribbon
Navy Basic Military Training Honor Graduate Ribbon
Navy Ceremonial Guard Service Ribbon
Coast Guard Recruiting Ribbon
Navy Recruiting Service Ribbon
Marine Corps Recruiting Ribbon
Marine Corps Drill Instructor Ribbon
Marine Security Guard Ribbon
Armed Forces Reserve Medal
Naval Reserve Medal
Army Service Ribbon
Marine Corps Reserve Ribbon

Army Reserve Components Achievement Medal
Army Noncommissioned Officer Professional Development Ribbon
Air Force Noncommissioned Officer Professional Military Education Ribbon
Air Force Basic Military Training Honor Graduate Ribbon

U.S. Non-Military Campaign and Service Awards
Merchant Marine Defense Medal
Merchant Marine Mediterranean / Middle East War Zone Medal
Merchant Marine Pacific War Zone Medal
Merchant Marine World War II Victory Medal
Merchant Marine Combat Bar
Merchant Marine Korean Service Medal
Merchant Marine Vietnam Service Medal
Merchant Marine Expeditionary Medal
Public Health Service Hazardous Duty Ribbon
Public Health Service Foreign Duty Ribbon
Public Health Service Special Assignment Ribbon
Public Health Service Isolated Hardship Ribbon

Non-U.S. Decorations and Awards
(Foreign personal decorations are worn immediately preceding unit awards and in the order received)
Philippine Republic Presidential Unit Citation
Republic of Korea Presidential Unit Citation
Vietnam Presidential Unit Citation
Republic of Vietnam Gallantry Cross Unit Citation
Republic of Vietnam Civil Actions Unit Citation

Non-U.S. Campaign and Service Awards
Philippine Defense Medal
Philippine Liberation Medal
Philippine Independence Medal
United Nations Service Medal (Korea)
United Nations Medal
NATO Medal
Multi-National Force and Observers Medal
Inter-American Defense Board Medal
Republic of Vietnam Campaign Medal

Kuwait Liberation Medal (Saudi Arabia)
Kuwait Liberation Medal (Kuwait)
Republic of Korea War Service Medal

Marksmanship Awards
Coast Guard Expert Rifleman Medal or Sharpshooter, Marksman Ribbon
Coast Guard Expert Pistol Shot Medal or Sharpshooter, Marksman Ribbon

6

HONORS, CEREMONIES, AND DRILL

Honors are ceremonies reserved for eligible high-ranking military personnel and dignitaries whose titles and grades are specified in United States Coast Guard Regulations, M5000.3 (series). Ceremonies, in contrast, are rituals performed to commemorate a particular event, achievement, or anniversary, often without regard to the pay grade of those involved.

CALLING ATTENTION

Because both honors and ceremonies are intended as public events, there is a need to call attention to them. There are a number of commonly used means to accomplish this, the order of which is the bugle, whistle, and passing word by loudspeaker. Not more than one means should be used for a given event, and the same means should be used throughout that event; for example "Attention to port," should not be announced by a whistle then followed by the same order verbally over the cutter's loudspeaker. Subsequent commands like "Carry on" should not be announced using a different means than that used to announce "Attention." If the use of any signal could cause confusion with another ongoing operation, the use of any other means should be substituted without regard for the precedence or preference.

HONORS

In the highest tradition of the Coast Guard, arrival and departure honors are given to certain personnel in the form of varying bells and announcements. These honors are

Honors, ceremonies, and drill

reserved for visiting civilian dignitaries, commanding officers, foreign officials, and other officers. For official visits, all honors, including pipes, are planned and carried out professionally and respectfully. Official visits are also known as announced or scheduled visits. The ship's bell is used to indicate the arrival of officers or civilian officials. Side honors are used only during an official visit and may be disposed of by the senior officer present. Side boys shall NOT be paraded on Sunday or other days between sunset and 0800. The table below indicates examples of the number of bells and side boys needed for the rank listed.

Arrival and Departure

On board any Coast Guard cutter or installation, the arrival and departure of visiting commanding officers, officers O-5 and above, and high-ranking civilian officials is normally announced. Although not a recognized means of rendering honor, such customs succeed in both notifying concerned personnel of the presence and movements of high-ranking individuals and in setting a tone for the reverence and respect afforded to them.

Accompanying the arrival or departure announcement is the sounding of the bell. The number of bells sounded is the same as the number of side honors the individual rates. The bells are sounded in pairs (i.e., ding-ding) in the same manner as done to denote the passing of time.

Position	Bells	Pipe (followed by ARRIVING/DEPARTING)
President/Vice President	8	UNITED STATES
Cabinet Member	8	[Name of Department, e.g., HOMELAND SECURITY]
U.S. Senator	8	UNITED STATES SENATOR
U.S. Representative	8	UNITED STATES REPRESENTATIVE
Governor	8	[Name of State]
State Senator or Representative	8	[Name of State] SENATOR / REPRESENTATIVE
Commandant/ Vice Commandant/ Chief of Staff	8	COAST GUARD
Assistant Commandant	6	COAST GUARD [Title, e.g., Operations]
District Commander	6	FIFTH DISTRICT
Area or Headquarters staff member	4	ATLANTIC AREA [Title, e.g., Operations]
District Operations	4	FIFTH DISTRICT [Title, e.g., Operations]
Mayor	4	[City, e.g., SAN FRANCISCO]
CO/OIC USCGC*	4	[Cutter Name]
Any Other CO/OIC 4	[Name of Command],	Note: Only cutter CO/OIC gets a stinger.
Any Senior Officer (O-5 and above)	4	[Rank and Service]
Unknown Officer and Rank**	None	STAFF GANGWAY

*Once CO/OIC has crossed the bow, 1 stinger.
**Appropriate number of bells based upon rank.

The following procedure should be adhered to:

- As the person approaches the bow or begins up the accommodation ladder, the proper number of bells is sounded in pairs and the announcement that the individual is arriving or departing is passed. For example, if the captain of USCG barque *Eagle* is returning, the announcement is "Ding-Ding . . . Ding-DING. . . EAGLE (half pause) ARRIVING."

- Once the individual steps on board the cutter, they are piped "ABOARD." For example, "NOW EAGLE, ABOARD."
- Once the individual steps off the cutter, they are piped "AWAY." For example, "NOW EAGLE, AWAY."
- In the case of the cutter's commanding officer or an officer whose personal flag is flown on board the cutter, a stinger, or a final "Ding," is sounded as the person steps on board or ashore. No other voice announcement is made if tones are used. Ding . . . "NOW EAGLE, ONBOARD." Ding . . . "NOW, EAGLE, AWAY."
- The arrival of a visiting captain, commander, or officer of equivalent grade in the other services who is not an operational commander or commanding officer, or whose command identity is not known, is announced by rank and service. For example, "Colonel, U.S. Army, Arriving."
- When unable to determine the rank or position of an arriving officer or official, the phrase "STAFF GANGWAY" is announced. After the unknown officer or official boards the cutter or installation, the visitor's official title is determined and used to announce their departure.

Side Honors

In addition to announcing the arrival and departure of visiting commanding officers, officers O-5 and above, and high-ranking civilian officials on board any Coast Guard cutter, side honors are also rendered. This honor, better known as side boys, originated centuries ago in Britain's Royal Navy when officers were hoisted on board in a sort of basket. Since senior officers usually weighed more than juniors, more crew were needed on the line. When the side is piped, two to eight crew, depending on the rank of the officer or office held by a civilian dignitary, will form a passageway at the gangway. They salute on the first note of the pipe and finish together on the last note.

Side boys are selected from the crew. The number of side boys paraded varies from two to eight (always an even number), depending on the rank of the individual being saluted. They are lined up and drilled in their duties by the boatswain's mate of the watch in charge of the gangway.

Personal Flags

All flag officers and some U.S. civil officials are honored by the display of a personal flag. During an official visit to a military installation and when embarked for passage on a boat, cutter, government vehicle, or aircraft, their personal flag is flown, unless there is already a flag of an officer their senior being displayed.

Commandant of the Coast Guard flag

Vice Commandant of the Coast Guard flag

Vice Admiral flag

Rear Admiral (Upper Half) flag

Rear Admiral (Lower Half) flag

On board cutters, personal flags are flown at the after masthead day and night, or, in a mastless ship, from the loftiest and most conspicuous hoist, except when, on the occasion of an official visit, the president's personal flag is displayed from the mainmast. Personal flags are not displayed at the same masthead with a national ensign. The commission pennant, if displayed, is hauled down when a personal flag is displayed.

On board boats, the flag or pennant of the officer or official is displayed on the bow. For commanding officers not entitled to a personal flag, a commission pennant is displayed in lieu of a flag. An officer embarked for an occasion other than an official one may display a miniature of their flag.

At shore installations personal flags are flown day and night on the inboard halyard on the portside of the main flagpole.

An officer entitled to the display of a personal flag or command pennant may display a miniature of such flag or pennant in the vicinity of the coxswain's station when embarked on an occasion other than an official occasion in a boat. Sheet metal replicas of personal flags inserted in metal slots are also authorized for display on aircraft. These should be located on each side of the forward fuselage.

The Union Jack is displayed for a diplomatic representative of or above the rank of chargé d'affaires within the waters of the country to which he is accredited and for a governor-general or governor commissioned as such by the president within the area of his jurisdiction. An officer below flag rank who commands a cutter squadron or major aircraft unit flies a command or broad pennant in lieu of the normal commission pennant. A burgee is used by commanders of cutter divisions and smaller aircraft units.

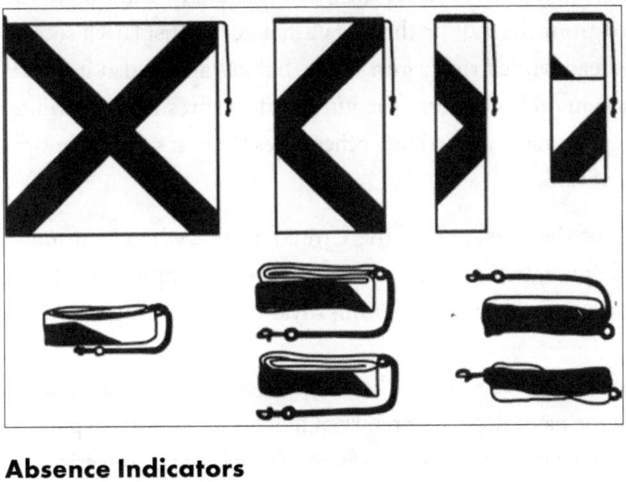

Breaking a personal flag

Absence Indicators

When the commanding officer, or any flag officer, is absent from the command for less than seventy-two hours in port, an "absentee pennant" is flown consistent with the International Code of Signals as follows:

- First Substitute (starboard yardarm): Flown when the flag officer or commanding officer whose personal flag or pennant is flying is absent.
- Second Substitute (port yardarm): Flown when the chief of staff is absent.

- Third Substitute (port yardarm): When the commanding officer is on an official absence of more than seventy-two hours, the executive officer, serving as acting commanding officer, is accorded the use of this pennant as if the executive officer were in command.
- Fourth Substitute (starboard yardarm): Flown when a civil or military official whose personal flag is flying (such as the secretary of defense) is absent.

Gun Salutes

In the days of sail it took as much as twenty minutes to load and fire a gun, so a ship that fired her guns first in honor of another ship did so as a friendly gesture, making herself powerless for the duration of the salute. Note that gun salutes are fired only by cannons or artillery on board a ship or shore battery, not by rifles or small arms. Gun salutes are prescribed by United States Coast Guard Regulations, COMDTINST M5300.3 (series). No salutes are fired between sunset and sunrise, before 0800, or on Sunday except when related to death ceremonies. In general, a cutter will fire a gun salute on Independence Day and Memorial Day, when falling in line with a ship displaying the personal flag, when present when a personal flag is broken, when entering a foreign port, or in return when saluted. In addition to the national ensign displayed from the gaff or the flagstaff, a cutter displays a second national ensign at a masthead while firing a gun salute. Salutes are fired at intervals of five seconds and always in odd numbers. The junior officer fires the first salute. When officers are of the same grade, the arriving officer fires the first salute. Multiple salutes are fired in order of rank.

- 21 guns: to the flag of the president of the United States, when a cutter enters a port of a foreign nation recognized by the U.S. government, to the president, sovereign, or member of a reigning royal family, to conform to foreign custom
- 19 guns: to the flag of the secretary of homeland security, secretary of defense, secretary of the navy, deputy secretary of homeland security, deputy secretary of homeland security, director of defense research and engineering, secretary of state when acting as a special foreign representative of the president
- 17 guns: assistant secretary of homeland security, assistant secretary of defense, general counsel, undersecretary of the navy, assistant secretary of the navy, admiral
- 15 guns: vice admiral

- 13 guns: rear admiral (upper half)
- 11 guns: rear admiral (lower half)

Boats passing U.S. or foreign men-of-war during colors must lay to, and their crews stand, face the colors, and salute.

Groups of Officers or Officials

When two or more officials or officers, each entitled to honors, make an official visit in company, only the senior shall be saluted. If they arrive or depart at different times, each shall be rendered honors to which each is individually entitled. The official visiting party may direct that honors will be rendered to more than one member of the official party. The honors prescribed for an official visit shall be rendered in opposite order upon departure.

Passing Honors

Passing honors are rendered on occasions when vessels, embarked officials, or embarked officers pass, or are passed close aboard. "Close aboard" means passing within six hundred yards for ships and four hundred yards for boats. The more common situations are between Navy ships and Coast Guard cutters, Coast Guard cutters and foreign warships, Coast Guard cutters and ships or boats with foreign dignitaries embarked, and Coast Guard cutters and merchant ships friendly to the United States. Passing honors are not normally rendered after sunset or before 0800. Passing honors consist of personnel "manning the rails," sounding "Attention" and the rendering of a hand salute by all persons on deck when the bow of one ship passes the bow or stern of the other or, if a senior is embarked in a boat, before the boat is abreast, or nearest to abreast, the quarter-deck. In order to standardize whistle signals used for passing/side honors, the following are prescribed:

- One whistle, attention to starboard, initiated by junior vessel
- Two whistles, attention to port, initiated by junior vessel
- One whistle, render salute, initiated by junior vessel when bows cross
- Two whistles, terminate salute, remain at attention, initiated by senior vessel when bridges cross
- Three whistles, carry on, when sterns cross

Passing Washington's Tomb

A special honor calling for half-masting the ensign is required of cutters passing Washington's tomb between sunrise and sunset. As the cutter comes in sight of

Mount Vernon, Virginia, a full band and guard are paraded if on board, the ship's bell is tolled, and the national ensign is half-masted at the tolling of the bell. When opposite Washington's tomb, the guard orders "Present arms," persons on deck hand salute facing the direction of the tomb, and "Taps" is sounded. The national ensign is two-blocked, and the tolling ceases at the last note of "Taps," after which the national anthem shall be played. Upon completion of the national anthem, "Carry on" is sounded.

Passing the USS *Arizona* Memorial

When a cutter is passing the USS *Arizona* Memorial in Pearl Harbor, Hawaii, between sunrise and sunset, it renders passing honors by sounding "Attention" before all personnel in view on deck and not in ranks render a hand salute.

CEREMONIES

Ceremonies are rituals performed to commemorate a particular event, achievement, or anniversary, often without regard to the pay grade of those involved. All require reverence and strict military bearing to carry out properly. Most are observable by the public and have significant meaning for all Americans.

Colors

Every Coast Guard shore command and every cutter not under way performs the ceremony of colors twice a day. This ceremony consists of paying honor to the national ensign as it is hoisted at 0800 and lowered at sunset. At shore units, the national ensign is flown from the main flagpole. On board cutters that are not under way, the national ensign is flown from the flagstaff located on the fantail or stern. If the cutter is under way, the national ensign is flown from the main mast twenty-four hours a day. On federal holidays a larger holiday flag is flown.

The Coast Guard ensign is hoisted at morning colors and lowered at evening colors in the same manner as the national ensign except it is kept slightly below the national ensign during hoisting. It is hoisted on the outboard halyard on the port side of the cross tree. The Coast Guard ensign should never rise above the national ensign. On cutters, the Coast Guard ensign is displayed on the mast twenty-four hours a day under way and in port.

Commissioned vessels not under way also hoist the Union Jack at morning colors and lower it at evening colors. It is hoisted on the jackstaff, a small flagpole at the bow of a ship. Ships that are under way do not hold morning or evening colors. They hoist the national ensign at the gaff as they get under way, but the jack is not flown.

At 0755 "First call" is sounded. At 0800 "Attention to Colors" is ordered and the national ensign is hoisted smartly to the top of the flagstaff or truck. When whistle signals or firing of a cannon are used to call attention to colors, a single blast indicates "Attention," and those not in ranks render the hand salute. When ships are nested, meaning moored alongside one another, the PREP pennant is used to initiate "Attention" and "Carry on." Only the ship with the visual guard needs to hoist the pennant, which will be observed by personnel on all ships in the nest.

While the national ensign is hoisted, the band, if there is one, plays the national anthem. If there is no band, but there is a bugler, the bugler plays "To the Colors." Everyone within sight or hearing renders honors as follows:

- If in ranks, you will be called to attention or to present arms.
- If you are in uniform but not in ranks, you stop whatever you are doing, face the colors, and salute until "Carry on" is sounded.
- If you are in a vehicle and traffic safety permits, stop and sit at attention, but do not salute. If conditions permit, the senior person in the vehicle gets out and salutes. The driver remains seated.
- If you are a passenger in a boat, remain at attention, seated or standing. The boat officer or coxswain salutes for the boat.
- If you are in civilian or athletic gear at colors, stop and face the colors at attention. If you have a hat, hold it in your right hand, over your heart. If you have no hat, salute by holding your right hand over your heart. Members of the armed forces and veterans not in uniform may render the military salute.

At the conclusion of the ceremony, "Carry on" is ordered and the PREP pennant is lowered.

At sunset a similar ceremony takes place. Five minutes before sunset "First call" is sounded. At sunset "Attention" is sounded and the band, if there is one, plays the national anthem. The flag is lowered slowly so that it reaches the bottom and is balled up within the field of blue on the last note of the music. If there is no band but there is a bugler, then the bugler plays "Retreat."

Your behavior during evening colors is the same as that for morning colors. You face the ensign at attention and hold your salute until "Carry on" is sounded.

Half-Mast

The national ensign is flown at half-mast as a tribute to the dead. On Memorial Day, the national ensign is flown at half-mast from 0800 until completion of the twenty-one-gun salute at 1200, or until 1220 if no salute is fired. During a burial at

Colors under way

sea, the national ensign is at half-mast from the beginning of the funeral service until the body is committed to the deep. After the death of a nationally known figure, all ships and stations may be directed to half-staff the ensign for a stated period.

To fly the national ensign at half-mast, first hoist the ensign to the peak of the mast. On the last note of the national anthem, lower the national ensign smartly to the half-mast position. Before lowering fully at evening colors, hoist the national ensign smartly to the peak on the first note of the music and then ceremoniously lower it.

Colors Detail

The colors detail is the detail that physically hoists and lowers the national ensign. It is a sacred responsibility and an honor to be entrusted to perform this ceremony. Ten minutes before the colors ceremony, the detail forms in a line near the flagstaff with the senior member at the rear of the formation and the folded national ensign carried at the center and any other flags carried in a position that corresponds to their halyard. After first call plays, the senior member gives the command "Detail, ATTENTION," followed by "Forward, MARCH," and the detail moves forward in a ceremonious manner, not allowing their arms to swing. The detail approaches the flagstaff and when centered the senior member commands "Mark Time, MARCH" to allow any adjustments for spacing between detail members before ordering "Detail, HALT." The senior member then calls "Right/Left, FACE" to execute a turn toward the base of the flagstaff. On the command "POST," the detail steps off as in marching and stops on the appropriate sides of the flagstaff. The senior member then commands "Ready, FACE," and the detail faces inboard toward the flagstaff. Once all members are in the correct position, the senior member commands "ATTACK THE HALYARD." Each member then approaches and detaches a halyard and attaches it to the correct eyelet so the flag is raised correctly. The detail remains at modified position of attention, holding the halyard in their left hands. When the first note of the national anthem begins or the command "ATTENTION TO COLORS" is given, the national ensign is raised to the top of the flagstaff smartly with the halyard pulled hand over hand to remain above the other flags and to reach the top of the flagstaff first. The other flags are raised with a slower hand-over-hand motion, all at the same pace so they reach the top of the flagstaff at the same time.

If the national anthem is played, the senior member salutes from the first note of the national anthem until the last note is played for morning colors and for the duration of "Retreat" at evening colors. If the national ensign is to be flown at half-mast, it will be lowered to that position on the last note of music. If no music is played, the person in charge will maintain their salute from the time "Attention to Colors" is ordered until the national ensign is fully hoisted or lowered. Upon completion of the hoist or lowering, personnel in the detail handling halyards will grasp the line in their left hands and render a hand salute. The detail does not look up at any point and instead keeps their eyes in the boat. When all flags have been hoisted, the entire detail is saluting, and the music has ended, the person in charge orders "Ready, TO." At the command "ATTACK THE HALYARD" the detail secures the halyards to the flagstaff. Once all halyards are secure, the command "POST" is given, and the detail forms a line facing the flagstaff. When in position, the senior member will order "Right/Left, FACE" then "Forward, MARCH" to return to the staging area where the detail is halted and dismissed.

For additional discussion on the commands used during the colors ceremony, see the Military Drill section later in this chapter.

Folding the national ensign

Folding the National Ensign

Folding the national ensign is a ceremony honoring the flag that requires at least two people. After the flag is completely folded and tucked in, it has the appearance of a cocked hat, ever reminding us of the soldiers who served under GEN George Washington and the sailors and Marines who served under CAP John Paul Jones to achieve the rights, privileges, and freedoms we enjoy today.

To perform the ceremony, the national ensign is held by one person at each end and folded in half the long way so the crease parallels the red and white stripes. It is folded in half again so the new crease also parallels the red and white stripes and the blue field is to the outside. The fly end, or the end away from the blue field, is folded up to the top so that the single edge lies perpendicular to the stripes. By repeatedly folding the thick triangle thus formed about the inboard edge of the triangle, the ensign is folded into the shape of a cocked hat.

Dress and Full Dress Ship

Commissioned cutters full dress on Presidents' Day and Independence Day and dress on other national holidays and to honor U.S. and, upon official invitation, foreign national anniversaries. Ships not under way dress or full dress from 0800 until sunset. Ships under way do not dress or full dress.

When dressing a cutter, the largest national ensign in the cutter's allowance is flown from the flagstaff and an additional daily-sized national ensign is flown from each masthead. Should half-masting or dipping of the national ensign be required, only the ensign at the flagstaff is half-masted or dipped. The Coast Guard ensign is flown from the outboard starboard yardarm. When a ship is full-dressed, in addition to the ensigns, a "rainbow" of signal flags is displayed from bow to stern over the mastheads or as nearly so as the construction of the cutter permits.

Rainbow

The rainbow of signal flags and pennants on a fully dressed cutter is arranged in a prescribed order. All flags and pennants in the allied flag bag are used except Screen, Subdiv, and Group. The rainbow reaches from the foot of the jackstaff to

Full dress ship

the mastheads and then to the foot of the flagstaff. Peculiarly masted or mastless ships make a display as little modified from the rainbow effect as is possible. If one set of flags does not complete the rainbow, the prescribed order is repeated starting with Flag 3, Flag 4, etc. The sequence of signal flags and pennants for the rainbow display is as follows:

(1) Flag 3	(14) Pennant 9	(27) 2nd Sub	(40) Hotel	(53) Victor
(2) Flag 4	(15) Zulu	(28) Bravo	(41) Echo	(54) Golf
(3) Pennant 1	(16) Corpen	(29) Delta	(42) Emerg	(55) Starboard
(4) Sierra	(17) Flag 8	(30) Turn	(43) Lima	(56) India
(5) 1st Sub	(18) Uniform	(31) Flag 5	(44) Pennant 7	(57) Foxtrot
(6) Alfa	(19) Flag 6	(32) Station	(45) Flag 0	(58) Quebec
(7) Prep	(20) Xray	(33) Kilo	(46) Int	(59) Pennant
(8) Charlie	(21) Negat	(34) Pennant 6	(47) Div	(60) Yankee
(9) Mike	(22) Flag 2	(35) Whiskey	(48) Pennant 4	(61) Design
(10) Speed	(23) Port	(36) Pennant 0	(49) Flag 9	(62) Flag 7
(11) Juliett	(24) November	(37) Flag 1	(50) 4th Sub	(63) Pennant
(12) Pennant 5	(25) Pennant 2	(38) Oscar	(51) Papa	(64) Squad
(13) Romeo	(26) Tango	(39) 3rd Sub	(52) Form	(65) An

Christening, Commissioning, and Decommissioning

Ceremonies held to christen and commission a ship have a long history dating back to a time when the Greeks called upon the gods to protect their ships and crew from the perilous sea. Religion has always played an important role in these ceremonies, which originated as a way to appease the gods of the natural elements. Over the years, different seagoing cultures and people changed and shaped the way these ceremonies have been performed, but the basic tenets have been sustained to the modern day.

Christening gives a ship its identity. A ship is traditionally christened or given its name at the time it is launched into the water. During the earliest ceremonies a sovereign or official would sip wine from a "standing cup," a large cup made of precious metal, then pour the remaining wine onto the deck or over the ship's bow before the cup was tossed overboard and the ship launched into the water. The naming of a woman as a ship's sponsor and the use of champagne for these ceremonies became popular at the close of the nineteenth century, and the tradition of breaking the bottle across the ship's bow was adopted. Sponsors now play an important role in the life of a new cutter, maintaining a relationship with the crew and participating in ceremonies through its decommissioning.

The actual physical process of launching a new ship from a building site to the water involves three principal methods. The oldest, most familiar, and most widely used is the "end-on" launch in which the vessel slides, usually stern first, down an inclined shipway. The "side launch," whereby the ship enters the water broadside, came into use in the nineteenth century on inland waters, rivers, and lakes. It was given major impetus by the World War II shipbuilding program. Another method involves ships built in basins or graving docks. When ready, ships constructed in this manner are floated by admitting water into the dock.

The commissioning ceremony completes the cycle from christening and launching to full status as a cutter in the U.S. Coast Guard. Unlike the christening ceremony, commissionings were not attended by the public in early American maritime history, and no accounts of them are found in contemporary newspapers. In recent years however, commissioning ceremonies have come to be public occasions. During the ceremony, the cutter is placed into service, its captain reads his or her orders and assumes command, the cutter's colors and commissioning pennant are hoisted, and the first watch is set. The setting of this first watch involves the passing of a long glass—a traditional symbol of an officer of the deck's authority in a ship of the line—from the sponsor to the cutter's first officer of the deck. At the close of the ceremony the crew mans the rail and reports for duty. Maintaining the religious nature of the early ceremonies, a chaplain delivers both an invocation and benediction.

When a ship is taken out of service, either to go into "mothballs" or to be transferred to a foreign navy, the ceremony is smaller and simpler. The commanding officer reads the decommissioning orders to officers and crew at quarters, and the ensign, jack, and commission pennant are hauled down. By tradition, the captain keeps the commission pennant.

Line Crossings and Voyages

When your cutter crosses a line such as the equator or enters a new ocean or sea, you may undergo a command-sponsored initiation and morale event that, once passed, will cause the crew to honor your achievement and commemorate it with an elaborate certificate. For example, when a ship crosses the equator, all those who have never crossed, called "pollywogs," are initiated into the "Ancient Order of the Deep" during a line-crossing ceremony in which King Neptune, Davy Jones, and a group of royal police hold court to decide who will become a "Shellback."

Events commemorated under way include:

- Shellback—crossing the equator
- The Emerald Shellback—crossing the equator at the prime meridian

- Golden Shellback—crossing the point where the equator crosses the 180th meridian
- The Order of the Blue Nose, Domain of the Polar Bear—crossing the Arctic Circle
- The Realm of the Czars—crossing into the Black Sea
- The Order of the Ditch—passing through the Panama Canal
- The Domain of the Golden Dragon—crossing the 180th meridian
- The Order of the Lakes—sailing on all five Great Lakes
- The Order of Magellan—circumnavigating the Earth
- The Magellan's Strait Jacket Club—transiting the Straits of Magellan
- The Moss Back—sailing around the tip of South America
- Persian Excursion and the Society of the Arabian Nights—serving in the Persian Gulf
- Plank owner—personnel assigned to a cutter or shore command placed in commission
- The Order of Purple Porpoises—crossing the junction of the equator and the International Date Line at the sacred hour of the vernal equinox
- The Order of the Red Nose, Domain of the Penguin—crossing the Antarctic Circle
- The Order of the Rock—transiting the Strait of Gibraltar
- The Safari to Suez—passing through the Suez Canal
- The Order of the Sand Squid (or Sand Sailor)—service ashore in the Middle East
- The Order of the Spanish Main—sailing the Caribbean
- The Order of the Sparrow—sailing on all seven seas
- The Order of the Square Rigger—serving on board USCGC *Eagle* or USS *Constitution*
- The Order of the Black Hulls—sailing on board a buoy tender

Homeward-Bound Pennant

No ship carries this flag as it is made up only when needed. It is flown by a cutter that has been overseas for nine months or more when she gets under way to return to the United States. The homeward-bound pennant flies until sunset of the day of arrival in a U.S. port. Next to the hoist the pennant is blue and has one white star for the first nine months overseas and an additional star for each six months. The rest is divided lengthwise; the upper section is red, the lower is white. The pennant is one foot long for each person on board who has been outside the United States for at least nine months but is never longer than the length of the ship.

After reaching home port and being hauled down, the pennant is ceremoniously cut into pieces, with the blue fabric going to the commanding officer and the remainder divided equally among all hands.

Advancements, Promotions, and Frockings

In the Coast Guard, officers are promoted and enlisted personnel are advanced. When a member is above the cutoff on the current advancement or promotion eligibility list, they may be considered for frocking, meaning they are authorized the insignia and title of the next higher pay grade but not the corresponding pay. Frockings are generally authorized to clearly establish the individuals' position when reporting to a new duty station.

Advancements, promotions, and frockings involve a ceremony that, at a minimum, includes reading and issuing the advancement or promotion certificate and pinning the insignia or affixing the shoulder boards commensurate with the new pay grade to the uniform. These ceremonies are held on the first day of the month or as otherwise prescribed by the Enlisted Personnel Advancement Announcement or Officer Promotion Authorization Listing. When multiple advancements or promotions are authorized for the same day, they should be sequenced in descending rank order. These ceremonies begin with the command, the individual to be advanced, and anyone selected to pin the collar devices or affix the shoulder boards assembled before the audience. The presiding official will order "ATTENTION TO ADVANCEMENT/RECOGNITION/AWARD" and then read the certificate.

Families and friends are often invited to attend these ceremonies. It is traditional for these guests to affix shoulder boards to an officer's uniform or to pin collar devices on an enlisted individual's uniform. Often collar devices worn by more senior personnel are passed on and used for the pinning, sometimes for generations. It is also traditional for those advancing to chief, senior chief, and master chief petty officer as well as those being promoted to commander or flag ranks to have their new and distinctive combination covers placed on their heads by family or friends.

A common question associated with the ceremony concerns the meaning and pronunciation of the opening line of promotion and advancement certificates which states, "To all who shall see these presents, greetings. . . ." Pronounced "pre-zents" the word means "to give or award ceremonially." It is a very formal and long-used opening intended as an announcement to those in attendance and the public at large.

The phrase in plain language means "To all who shall see this ceremony, welcome and thank you for coming."

Officers traditionally reaffirm their oath of office upon promotion, a practice increasingly customary for enlisted members who reaffirm their oath of enlistment. When taking the oath, the right hand is raised.

Oath of Office (for Officers)

> I, ___________, having been appointed an officer in the Coast Guard of the United States, do solemnly swear (or affirm) that I will support and defend the Constitution of the United States against all enemies, foreign or domestic, that I will bear true faith and allegiance to the same; that I take this obligation freely, without any mental reservation or purpose of evasion; and that I will well and faithfully discharge the duties of the office upon which I am about to enter. So help me God.

The Oath of Enlistment (for Enlisted)

> I, ___________, do solemnly swear (or affirm) that I will support and defend the Constitution of the United States against all enemies, foreign and domestic; that I will bear true faith and allegiance to the same; and that I will obey the orders of the President of the United States and the orders of the officers appointed over me, according to regulations and the Uniform Code of Military Justice. So help me God.

Following the advancement, those advanced are invited to give remarks and typically thank their families and shipmates for their support.

Change of Command

When a new commanding officer or officer in charge takes over any cutter or station, a time-honored change-of-command ceremony is held to formally restate to the crew that the continuity of command will be maintained. It is a formal ceremony conducted before the assembled company of the command and conveys to the officers, enlisted personnel, civil employees, and Auxiliarists of the Coast Guard that although the authority of command is relinquished by one person and assumed by another, it is maintained without interruption. It represents a transfer of total responsibility, authority, and accountability from one individual to another. During the ceremony, the departing commanding officer or officer in charge reads their detachment orders, and the relieving officer reads the orders to assume command. Both then inspect the crew and the cutter.

Retirements

A military retirement ceremony marks the end of a long career in a difficult occupation and a transition into civilian life. A traditional ceremony has an invocation; a presentation of colors and playing of the national anthem; a presentation of certificates of appreciation to members of the immediate family; a presentation of awards, gifts, and the retirement certificate and pin to the retiree; and speeches by the retiree, presiding official, and any guest speakers. Unique to retirement ceremonies are the presentation of a shadow box, the presentation of "Old Glory," and the reading of "The Watch."

Shadow Box

Historically, when a sailor would join a ship's crew, he would join that ship for his entire career. During the sailor's voyages to ports of call around the world, he would collect many trinkets, souvenirs, and reminders of his travels. As space on board ship was at a premium, these items tended to be small. When the sailor was piped ashore for the last time, his shipmates saw to it that a special ceremonial box, a shadow box, was constructed for him. The box would hold all the possessions that had been collected during those many voyages and would simultaneously symbolize the sailor's career and time on board the ship.

Ideally, a shadow box serves not only as a reminder of achievements and accomplishments but also as a summation, a culmination, of a career. Within the shadow box lie a sailor's most honored and cherished possessions, including the flag of the United States of America, representing a lifetime of valiant and faithful service. A shadow box should enable a stranger glancing at its contents to gain a substantial understanding of the owner's past service and achievements. It is presented so the retiree may display it proudly and remember the honor it reflects.

Old Glory

A flag ceremony called "Old Glory" is often performed at retirements to present the national ensign to the retiree. The text is divided into the number of segments needed to represent each of the pay grades held by the retiring member during his or her career. Individuals representing each of these pay grades are formed into a line at the position of attention, with all facing the lowest ranking member who holds the flag and faces toward the retiree positioned at the end of the link. Participants wear white gloves, and their movements are slow and deliberate. After each segment of "Old Glory" is read there is a pause and the tone of a bell to allow participants to pass the flag to the next higher pay grade in line, with the receiving individual saluting the flag before taking hold of it. Once passed, the flag is saluted by the junior member and the individual holding the flag performs about face. This process is repeated until the flag is presented to the retiree. The words spoken for this ceremony in honor of U.S. Coast Guard retirees is as follows:

> I am the flag of the United States of America. My name is OLD GLORY.
> I fly atop the world's tallest buildings. I stand watch in America's halls of justice.
> I fly majestically over great institutions of learning.
> I stand guard with the greatest military in the world.
> LOOK UP! and see me!
> I stand for peace, honor, truth, and justice. I stand for freedom!
> I am confident. I am arrogant. I am proud.
> When I am flown with my fellow banners, my head is held a little higher—my colors a little truer.
> I BOW TO NO ONE!
> I am recognized all over the world.
> I am worshipped—I am saluted—I am respected—I am Revered—I am loved—I am Feared!!
> For more than 200 years, I have flown from the mast of every Coast Guard vessel in both war and peace as they patrolled waters from Barrow's Shores to Ross' Island,
> from the Chesapeake to San Francisco Bays.
> Flown from cutters in action off the coast of Cape Henry, the Bay of Cardenas, the North Atlantic, the South China Sea, and the Persian Gulf.
> From lifeboats braving surf in the graveyards of both coasts,
> picket boats on Rum Row in the Roaring '20s,
> landing craft approaching contested beaches in both World Wars,
> fast boats in pursuit of smugglers in the Caribbean Sea,
> airboats saving survivors in the flooded streets of New Orleans,
> and a score of other places.
> Long forgotten by all, but those who were there with me.
> I was there.
> I led my Coast Guard men and women.
> I followed them, I watch over them, they Love Me.
> I was on a narrow beach at Point Cruz.
> I was dirty battle torn and tired but my Sailors and Marines Cheered Me!
> I WAS PROUD!

I have been soiled, burned, torn, and trampled on the streets of countries that I have helped to set free.
It does not hurt for I am invincible.
I have been soiled, burned, torn, and trampled on the streets of my own country
And when it is done by those with whom I have served in battle, it HURTS!
But I shall overcome for I am strong!
I have slipped the surly bonds of earth and, from my vantage point on the moon,
I stand watch over the new frontiers of space.
I have been the silent witness to all of America's finest hours.
But my finest hour comes when I fly at half-mast
to Honor my Coast Guardsmen
whether lost at sea as they endeavored to save others
while in pursuit of those who would break our nation's laws
or in action against or enemies.
And when I lie in the trembling arms of a grieving mother at the graveside of her fallen son or daughter
I am proud!
I am proud to have served with you during your many years of service to our country. I will be with you in retirement just as I was during your active service.
My name is "Old Glory." Long may I wave, dear God, long may I wave.
For unfailing and loyal service to your country, we present you this flag.

The Watch

At the conclusion of the retirement ceremony, the following text is read by the individual with the lowest pay grade who participates in the ceremony, usually a recent graduate of recruit training or "A" school.

And now the time has come for a sailor to retire.
Aye mates, for many years, this sailor stood the watch.
While some of us lay about our bunks at night, this sailor stood the watch.
While others of us were attending schools, this sailor stood the watch.
And yes, even before many of us were born, this shipmate stood the watch.
As our families watched the storm clouds of war brewing on the horizon of history, this shipmate stood the watch.
Though he saw his family ashore, often needing his guidance, he still stood the watch.

For 20 years, he has stood the watch so that we and our fellow countrymen could sleep soundly, in safety, knowing that a sailor would stand the watch.
Today, we are here to say, Shipmate, the watch stands relieved . . . ,
Relieved by those you have led, guided, and trained.
You stand relieved. . . . We have the watch!

Military Funeral Honors

Funeral honors are what's called a total force mission and responsibility, meaning at any time you could be selected to participate in a funeral detail charged with providing a dignified ceremony that honors our profession and qualified veterans. A qualified Coast Guard veteran is anyone who served on active duty in the Coast Guard or in the Coast Guard Reserve, was on active duty at the time of death, was retired, or had received an honorable discharge or general discharge under honorable conditions.

To conduct funeral honors requires a minimum of a two-person detail, at least one of whom is in the same service as the deceased. When reasonably available, the grade or rating of the senior member of the detail will be the same as, or higher than, that of the deceased. The appropriate uniform is service dress blue with combination cover. The ceremony will include, at a minimum, the folding of a United States flag, presentation of the flag to the family of the veteran, and the playing of "Taps." Ceremonies may also include the firing of three volleys. The typical sequence of events for an Honors Detail at a military funeral is as follows:

Muster at the ceremony site.
Stand at ease in designated places as family members arrive.
Stand at attention when deceased veteran arrives.
Chaplain remarks.
Firing of three volleys.
Playing of "Taps."
Flag-folding and presentation.
Chaplain remarks.
Dismissal at conclusion of ceremony.

Firing of Three Volleys

Anyone who is entitled to a military funeral will receive three rifle volleys, subject to availability of a firing team. The firing team can consist of any number, but is usually a team of eight, with a petty officer in charge of the firing detail. Whether the team consists of three or eight, or ten, each member fires three times. The flag detail often slips three shell casings, each representing one volley, into the folded flag before presenting the flag to the family.

These three volleys are often mistaken as a twenty-one-gun salute, although it is different. In the military, a "gun" is a large-caliber weapon. The three volleys are fired from rifles, not guns. The three volleys come from an old battlefield custom. The two warring sides would cease hostilities to clear their dead from the battlefield, and the firing of three volleys meant that the dead had been properly cared for and the side was ready to resume the battle.

"Taps"

A bugle call composed by BG Daniel Butterfield of the Union army in 1862, "Taps" was intended to replace an earlier bugle call titled "Tattoo" that was played at lights out. The call became known as "Taps" because it was often tapped out on a drum in the absence of a bugler. Officially adopted by the U.S. Army in 1874, its first use for funeral honors was during the Peninsula campaign in 1862 when a soldier was buried at a time when his unit occupied an advanced position concealed in the woods. It was unsafe to fire the customary three volleys over the grave due to the proximity of the enemy, and it occurred to the unit's commanding officer that the sounding of "Taps" would be the most appropriate ceremony that could be substituted. The custom, thus originated, was soon taken up throughout the military.

A bugler, to include a member using the ceremonial bugle with an electronic insert, sounds "Taps" from a prominent position, normally to the far right of the assembled guests. Portable stereo players should be used as a last resort and be out of sight of the funeral party.

Flag-Folding and Presentation

A U.S. flag drapes the casket of deceased veterans to honor the memory of their service to the country. The ceremonial folding and presentation of the flag is a moving tribute of lasting import to the veteran's family. The flag is place on a closed casket so the union blue field is at the head and over the left shoulder of the deceased. The flag should not be lowered into the grave or allowed to touch the ground.

After "Taps" is played, the flag is ceremoniously folded into a triangle in accordance with the folding ceremony. When folded, no red or white is to be evident, leaving only the blue field with stars. It is then presented to the next of kin or an appropriate family member as a remembrance of their loved one's service.

The flag presentation protocol is as follows:

> Stand facing the flag recipient and hold the folded flag waist high with the straight edge facing the recipient.
>
> Lean toward the flag recipient and solemnly present the flag to the recipient.
>
> "On behalf of the President of the United States, the United States Coast Guard, and a grateful Nation, please accept this flag as a symbol of our appreciation for your loved one's honorable and faithful service."

Burial at Sea

During combat or at other times when operations prevent holding a body for services ashore, the dead are buried at sea. When schedules permit, retired persons may be buried at sea. The ceremony may be very simple in combat or more elaborate when time permits but in all cases is solemn, moving, and carried out with the greatest dignity. Prior to the start, the body of the deceased is covered by the national ensign or their urn is placed atop a table holding a folded national ensign. All hands are present for the ceremony and assembled in ranks facing the deceased when the order is given to "Stop the engines" and "Dip the ensign to half-mast." The captain or chaplain will lead the ceremony, giving the names, ranks, and branches of service for each decedent before offering a prayer and reading Psalm 23. The ship's company will then be ordered to attention and salute as the captain or chaplain states, "Unto almighty God we commend the souls of our brothers/sisters departed, and we commit their bodies to the deep, in sure and certain hope of the resurrection unto eternal life." At the words "we commit their bodies," those bearing the body will remove the flag and ease it into the sea on a tilted board or will disperse the ashes from an urn. The rifle squad will then be ordered to fire three volleys. The national ensign is then folded and presented to the commanding officer before "Taps" closes the service.

Next of kin are later provided with the folded national ensign, a shell casing from the volley, a small chart showing the approximate site where the remains were committed, and a letter of condolence from the captain.

Missing Man Table

A missing man table, also known as a fallen comrade table, is a ceremony and memorial set up in military dining facilities and during official dining functions in honor of fallen, missing, or imprisoned service members. The table serves as the focal point of ceremonial remembrance. The missing man table consists of the following elements:

- A small table set for one symbolizes the isolation of the absent service member. The table is usually set close to, or within sight of, the entrance to the dining room. For large events, the table may be set for six places to represent each of the armed services. The table is round to represent the everlasting concern the survivors have for the missing.
- One or more head covers may be placed upon the table to represent the armed service of the missing persons.
- A white tablecloth symbolizes the pure intentions of the service members who responded to the country's call to arms.
- A single rose in a vase symbolizes the blood that service members have shed in sacrifice to ensure the freedom of the United States of America. This rose also represents the family and friends who keep the faith while awaiting the return of the missing service members.
- A red ribbon represents a love of country that inspired the service members to serve the country.
- A slice of lemon on the bread plate represents the bitter fate of the missing.
- Salt sprinkled on the bread plate symbolizes the tears shed by waiting families.
- An inverted glass represents the fact that the missing and fallen cannot partake.

- A lit candle symbolizes a light of hope that lives in hearts to illuminate the missing's way home.
- An empty chair represents the absence of the missing and fallen.

Ceremonial Units

Ceremonial Honor Guard

The Coast Guard Ceremonial Honor Guard represents the commandant, the military district of Washington, D.C., and the U.S. Coast Guard through ceremonial operations held before world leaders and dignitaries. Ceremonies can include parades, funerals, and White House dignitary arrivals as well as presenting colors at local and official functions. Honor Guard members participate in joint service activities as well as Coast Guard functions. The Honor Guard performs more than 1,600 ceremonies annually.

The Honor Guard is composed of seventy-three members, with a lieutenant (O-3) serving as the Honor Guard company commander, two junior officers (usually O-2) serving as operations/weapons officer and supply/training officer, a chief petty officer (E-7) as the Honor Guard chief, and four petty officers (ranging from E-4 to E-6). The remaining members of the Honor Guard are "first-tour" nonrated personnel (E-3) coming directly from Training Center Cape May. The officers and nonrates serve a two-year tour of duty in the Honor Guard, while the chief petty officer and petty officers serve four-year tours.

Coast Guard Band

Since 1945 the U.S. Coast Guard band has been one of the five premier service bands in the United States. The Coast Guard band routinely tours throughout the United States and has performed in the former Soviet Union and in England. Additionally, the band regularly furnishes programs to the National Public Radio for broadcast throughout the country. Concerts have been broadcast in Australia, Japan, and Europe.

Each year the band conducts tours with free public concert appearances. The band also fulfills a limited number of public requests for patriotic programs at national conventions and meetings of nationally recognized civic, patriotic, and veterans' organizations.

Coast Guard Pipe Band

The Coast Guard pipe band is an unofficial, volunteer organization affiliated with the U.S. Coast Guard that is organized as a nonprofit corporation. Membership in the U.S. Coast Guard pipe band is open to active-duty, reserve, and retired members of the U.S. Coast Guard, male or female, officer or enlisted, honorably discharged wartime veterans of the Coast Guard, and members of the Coast Guard Auxiliary.

The Coast Guard pipe band wears the U.S. Coast Guard tartan, which was adopted as the official tartan of the U.S. Coast Guard on 1 May 2002. The Coast Guard pipe band marches annually as a full band at the Grand Haven Coast Guard Festival and the Coast Guard National Memorial, plays at numerous Coast Guard change-of-command and retirement ceremonies, and participates in full dress military funerals.

Flyovers

Coast Guard aircraft participate in flyovers for civic-sponsored public ceremonies such as Armed Forces Day, Memorial Day, Independence Day, Veterans Day, and similar local holidays overseas and for national conventions of veterans' organizations and other public events such as airshows and fairs.

Although not part of the required funeral honors ceremony discussed in this chapter, flyovers can also be requested to support memorial services for eligible persons including active-duty aviation officers and enlisted personnel, active-duty and retired flag officers, war heroes, and dignitaries of the U.S. armed forces and federal government.

MILITARY DRILL

Close-order drill

A group of personnel in uniform and bearing arms must be under the direct control of a senior officer in order for them to move in an orderly fashion. A close-order drill will accomplish this. Such a drill also serves to teach discipline by instilling habits of automatic and precise response to orders; to prepare for military operations on land; to facilitate movements of companies from one place to another; to promote teamwork; and to make a favorable appearance in a parade or other public ceremony. The primary reference document for close-order drill is the *Marine Corps Drill and Ceremonies Manual.*

MANUAL OF ARMS

The manual of arms is performed with an issued or ceremonial rifle. The primary reference document for the manual of arms is the *Marine Corps Drill and Ceremonies Manual.*

MANUAL OF THE COLORS

The standard color guard consists of four individuals of approximately equal height. Two petty officers are the color bearers and two other members, junior to the color bearers, are the color guards. The senior color bearer carries the national colors and commands the color guard. The junior color bearer carries the organizational colors, the Coast Guard standard, which is always on the left of the national colors. The primary reference document for the manual of the colors is the *Marine Corps Drill and Ceremonies Manual.*

Manual of the colors

7

GOOD ORDER AND DISCIPLINE

Good order and discipline are vital to every branch of the armed forces because military service entails performing your duties under difficult conditions, which creates special challenges not encountered in civilian walks of life. Order is the arrangement of people and resources in relation to each other and is how we ensure we are best organized to achieve mission readiness and operational success. In the Coast Guard things are ordered and organized a certain way. This is for good reason. When operating in otherwise chaotic circumstances, an organized service makes things more predictable, resources more readily available, communication more consistent and effective, and outcomes more successful. Things work better when people and resources are where they need to be when they need to be there.

Discipline is an individual's willing adherence to the set of rules that regulate conduct and order. These rules may be the laws, regulations, policies, or customs that are known and respected by members of an organization. Parental guidance, mentorship, religious beliefs, and other values contribute to discipline in individuals. Discipline in all the military services is based on the Uniform Code of Military Justice (UCMJ). The word "discipline" comes from a Latin word that means "to teach." The Coast Guard core values, along with patriotism, unit pride, and a culture of service over self, are relied upon to create a system of discipline that ensures Coast Guardsmen will do what is expected of them even when it is different from what they may want to do. When discipline is working best within a unit, the individuals who make up the unit have a good attitude, do their work efficiently, and exhibit high morale. In a well-disciplined unit, the crew often do the right thing because they want to, not because they have to. They possess the willingness and motivation to do things without being told, often with little guidance.

Good order and discipline are indispensable to the military, and one of the most crucial responsibilities you have is to contribute to them within your team, unit, and service. Three simple rules will serve you well in meeting expectations for how Coast Guardsmen conduct themselves:

- Always be aware that you are a representative of the Coast Guard and of the United States of America.
- Always assume that someone is watching.
- Never do anything you would not want to read about in the headlines, or that you would not want to have to explain to your commanding officer or the people who raised you.

UNIFORM CODE OF MILITARY JUSTICE

The Uniform Code of Military Justice, or UCMJ, is the set of rules that define military conduct. The UCMJ became law on 5 May 1950 and is the basis for the *Manual for Courts-Martial.* Before 1950 the individual services had different disciplinary systems. The only variations that now exist among the services are those that result from regulations specific to the department under which the services operate. The basic principles of the UCMJ are the same for all. They often parallel civilian rules but include many punitive articles unique to the military that demand standards of conduct much stricter than what is expected of civilians. In the civilian workplace it is not uncommon for people to be late for work, to feign illness, to yell at their bosses, or simply walk away from a job entirely without any substantial consequences. In the military, however, these actions would be considered violations of the UCMJ and could result in severe punishments.

Persons Subject to the UCMJ

The following personnel are subject to the UCMJ at all times whether on or off duty:

- Active-duty personnel
- Cadets and midshipmen
- Reserve personnel during drills
- Retired personnel receiving pay
- Retired personnel receiving hospitalization

- Persons serving a sentence imposed by a court-martial
- Members of National Oceanic and Atmospheric Administration, Public Health Service, and other organizations when serving with the armed forces
- Prisoners of war in custody of the armed forces
- In time of war, persons serving with or accompanying armed forces in the field

Members of military auxiliaries such as the Coast Guard Auxiliary and civilians are not subject to the UCMJ.

Discipline Types

Discipline is often confused with punishment, but the two words do not mean the same thing. Although punishment may be used as a form of discipline, it is only one aspect of a wide spectrum of how good order and desired conduct are achieved and maintained. In fact, three distinct types of discipline are recognized: self, positive, and negative.

Self-Discipline

Self-discipline, or doing the right thing because you have the ability and willingness to do the right thing despite any temptation to the contrary, is the most desirable form. Self-discipline is developed through knowledge, experience, and attitude, all shaped by positive and negative discipline. When you perform well both under supervision and independently, on and off duty, you show self-discipline and avoid any negative discipline. Positive discipline may still be required, especially as advancing technology introduces new challenges for conduct never considered by past generations.

Positive Discipline

Includes the training, instruction, and mentoring you receive so that you know what doing the right thing looks like and can choose to do it in the absence of supervision. Coast Guardsmen must be taught the core values and standards for their performance and conduct in order to develop the self-discipline that will prevent disciplinary problems. Positive discipline also includes the use of rewards such as promotions, high marks on evaluations, and awards to promote desired behaviors and encourage individuals and crews to meet and exceed standards.

Negative Discipline

Negative discipline is associated with punishment and is intended to create a fear of consequences for any conduct that is counter to maintaining good order. When people demonstrate a lack of self-discipline, negative discipline may be required, especially if their actions harm others or jeopardize mission success. Negative discipline involves taking things away from you that you value such as rank, pay, liberty, or freedom. If you find yourself in a situation where you are being punished, you may feel that you have been wronged, but if you reflect on the situation and are honest with yourself, it is far more likely you will find that you lacked self-discipline and failed to act in accordance with your training.

Your leaders do not find any satisfaction in using negative discipline, but when necessary to maintain good order, they have to and will. Depending on the pay grade of your command officer or officer in charge, punishments that can be awarded can range from extra duties to referral for courts-martial, a criminal proceeding discussed later in this chapter. These actions can have severe impacts on both your military career and on your future. The punishment used by the Coast Guard, such as loss of liberty, loss of pay, and confinement, have three purposes:

- To deter you from breaking the rules;
- To encourage you to do your duty; and
- To provide an example to others.

Articles to Be Explained

Article 137 of the UCMJ requires that the UCMJ be explained to all who enter active duty and mandates that a copy be made available for personal examination. It further requires that the articles be explained again after six months of active duty and upon each reenlistment. This is positive discipline. By requiring that each of us receive repeated explanations of the code that will be used to hold us accountable, we are being trained to understand what conduct will violate its punitive articles and are advised to avoid the same. If you have questions concerning the UCMJ or are unsure of what is required of you, it is YOUR responsibility to ask. The oath you took on entering service obligates you to observe the UCMJ. As in civilian life, ignorance of the law is no excuse for breaking it.

Methods of Discipline

Although the full range of disciplinary options described in the UCMJ is available to commands, not every lapse of self-discipline warrants a court-martial. Consideration must be given to the severity of the conduct, the overall performance of the offender, and other related factors when deciding how to best address and resolve the matter. Informal, administrative, and criminal methods of negative discipline allow significant command discretion in dealing with poor conduct and maintaining good order.

Informal

Reserved for minor and first-time offenses, informal discipline is generally limited to verbal counseling and training. Incidents may be documented locally on a memorandum but are typically not referred for entry in permanent service records. Examples of minor offenses include horseplay in the berthing area, reporting late for the workday or duty, or quitting work early without permission.

Chief Petty Officer Council. This council is an informal procedure designed to guide, improve, and correct behavior of enlisted members, E-6 and below, through nonjudicial means. It is a meeting before the command master chief and a select number of other chief petty officers whose goal is to hear both sides of a story, counsel any and all involved, and provide recommendations to the command for follow-up. The intent is to intervene in a situation at an early point before permanent negative documentation is entered into a service member's record.

Administrative

Administrative actions are used for more serious offenses or for repeated minor offenses. The incidents are formally documented on administrative remarks forms and entered into the permanent service record. Extra-military instruction is an example of an administrative action.

Extra-Military Instruction. Extra-military instruction is a training device rather than a punishment and is used only to correct a deficiency in performance or knowledge or to improve the efficiency of a command; for example, when an individual fails to meet initial certification timelines required for boat crew competencies. It is assigned by supervisors during working hours and by the commanding officer or officer in charge for after hours. It is limited to two hours per day and is not assigned on the individual's Sabbath.

- Withholding of privileges such as driving on base
- Canceling liberty or leave
- Nonjudicial punishment (NJP)
- Administrative separation

Criminal

- Violations criminal in nature and serious offenses related to dereliction of military duty are referred for courts-martial.

Reserve-Specific Concerns

Although Reservists are subject to the UCMJ only while on orders for active or inactive duty, the Coast Guard may impose administrative measures regardless of status for discrediting acts such as those resulting in an arrest by law enforcement. These measures include but are not limited to negative performance evaluations, denial of advancement recommendation, documentation of substance or alcohol abuse, and administrative separation.

Unsatisfactory Participation

Unsatisfactory participation occurs when a Reservist fails to comply with any contractual obligations or program requirements. Participation is also considered unsatisfactory when Reservists acquire at least nine unexcused absences from scheduled training within a twelve-month period of duty. Failure to meet minimum participation standards will subject Reservists to military discipline and administrative penalties, up to and including separation.

Enforcing the UCMJ

All levels of your chain of command are tasked with enforcing discipline, and any Coast Guard member who knows or suspects an offense has been committed can place a person subject to the UCMJ on report. Ordinarily, an offense is reported to the supervisor, who relays the information to the commanding officer or officer in charge, who then takes the necessary action. Article 7 of the UCMJ grants authority to all commissioned officers, warrant officers, and petty officers to stop quarrels, fights, and disorders among persons subject to the UCMJ and to apprehend anyone subject to the UCMJ if they have a reasonable belief the person committed an offense.

In addition to your chain of command, at some bases Coast Guard police departments patrol the installation, and at others petty officers assigned to security watches or contracted civilian security personnel perform identical functions. Department of Defense installations have similar security forces. You must obey superiors and those personnel who have been charged with maintaining order. Any failure to do so will result in disciplinary action.

Mistaken Loyalty

A mistaken concept of loyalty to shipmates sometimes causes otherwise well-intentioned persons to refrain from reporting crimes. Those who constantly run to officers with reports of minor derelictions of duty by other crew members are a nuisance, but you should be able to distinguish between this and giving information about a serious offense. If an act or a neglect constituting a legal offense has been committed, or if you have serious reason to believe one has been committed, it is your duty to report it to the proper authority. If you do not, you too are punishable under the UCMJ. In permitting an offender to escape punishment, you weaken the disciplinary system that protects you and your shipmates. You owe loyalty to those who keep the laws, not to those who break them.

Report of Offense and Disposition

When an individual is suspected of a violation, the accused is "placed on report" by the commissioned, warrant, or petty officer who observed the conduct or who was notified that it had occurred. In the Coast Guard, individuals are placed on report using Coast Guard form CG-4910, Report of Offense and Disposition. In addition to detailing the alleged offense and identifying relevant parties and witnesses, the form follows the progression of the actions taken to address of the matter, from the assignment and recommendations of a preliminary inquiry officer to the command's decision on whether to dispose of the matter at nonjudicial punishment or refer for court-martial and the outcome of such proceedings.

Preliminary Inquiry

The commanding officer orders a preliminary investigation to develop information. The accused is informed of the charges and is permitted, but not required, to make a statement. The accused is advised that any statement they make may be used against them. After the investigating officer has developed the facts, a full report is made to the commanding officer or officer in charge. If it is determined that the offense has been committed and is within the jurisdiction of the commanding officer, a mast may be held or the matter referred for courts-martial.

Protection of the Accused

Consistent with the Constitution of the United States, the laws of military justice protect those accused by recognizing their right to be informed about the charges against them, to reply to the charges, and to confront and question their accusers. They also have the right to legal counsel of their own choosing. There are limitations on the questioning of prisoners and rules requiring humane treatment. Trials must be conducted fairly. Punishments are limited in accordance with the seriousness of the offense; cruel and inhuman punishments are prohibited. Finally, there is an elaborate system of reviews to ensure that all these provisions for protecting the rights of the accused have been observed and that the court's decision is just.

If you find yourself facing formal punishment, your first best move is to seek legal advice on the process and your rights. The only right to legal counseling related to NJP is the right to consult with a lawyer about whether to accept NJP. Commands usually have a legal officer assigned to assist you in these cases. You also have the right to retain civilian counsel at your own expense.

Dual Action

Although there are certain reservations, military personnel are also subject to civilian trial and punishment. Coast Guard personnel are not answerable to civil authorities for violations of a strictly military nature, such as unauthorized absence, desertion, or misbehavior before the enemy. These offenses are subject to trial by military authorities only. However Coast Guard personnel may be subject to dual action, meaning both civil and military for offenses such as murder, robbery, rape, or driving under the influence of drugs or alcohol. Under these circumstances you could be tried twice for the same offense. Both a state and the federal government may prosecute the same offense; however, due to the legal principle of double jeopardy, a person cannot be tried by the federal government and be subject to a court-martial for the same offense. Actions resulting in a civil court conviction bring discredit upon the Coast Guard and, except for minor traffic violations, are required to, at a minimum, reflect in the performance evaluations of both officer and enlisted members.

NONJUDICIAL PUNISHMENT (ARTICLE 15)—CAPTAIN'S MAST

An Article 15 hearing or captain's mast is a nonjudicial hearing in which the commanding officer or officer in charge handles a relatively minor offense in which lesser administrative measures are considered insufficient. It is commonly used for infractions of discipline such as being late for duty or a mandatory formation, appearing in a deficient uniform, or not promptly following the orders of a superior. Few incidents like these warrant the time and effort that goes into a full court-martial, and there is a strong incentive on the part of the officers and senior enlisted personnel in a particular command to take care of discipline in house. While it is not a "trial" in the civil justice sense, a captain's mast is held with formality and is not a pleasant experience for anyone involved. If an admiral is overseeing the mast, then the procedure is referred to as an admiral's mast or a flag mast.

In naval tradition, the proceeding takes the form of a public meeting held at the mast of a ship, and all sailors not on duty are expected to be present. A green tablecloth covers the table used by the captain presiding over the mast, a tradition dating back to the 1500s when a green woolen baize cloth was used to cover the table and decks in the quarters of Royal Navy captains.

There is no right to an attorney at NJP; rather, the offender appears without representation. Because these proceedings are nonjudicial, the offender may be punished but will not have a criminal record. Members assigned to shore units may refuse NJP and demand trial by courts-martial. Members assigned to cutters may not refuse NJP. Upon review of the case, the commanding officer or officer in charge may:

- Dismiss the case if it is not supported by evidence;
- Dismiss the case with a warning;
- Assign some other administrative discipline;
- Impose punishment;
- Impose, but suspend punishment;
- Refer the case to a superior officer for NJP; or
- Refer the case to trial by court-martial.

As the rank of the presiding individual increases, the authority to impose punishment increases. Conversely, as the rank/rate of the individual appearing before the NJP increases, the punishment that may be imposed decreases. The chart below shows the maximum punishment for enlisted personnel.

OFFICER OR WARRANT OFFICER MAXIMUM PUNISHMENTS

	PUNISHMENT		ADMONITION	
	Reprimand	Arrest in Quarters	Restriction	Forfeiture of Pay
Flag Officer	Yes	30 days	60 days	½ pay × 2 months
LCDR or Above	Yes	No	30 days	No
LT or Below	Yes	No	15 days	No

ENLISTED MAXIMUM PUNISHMENTS

	E3 AND BELOW			E6 AND BELOW		
	Admonition/ Reprimand	Correctional Custody	Extra Duties	Restriction	Forfeiture of Pay	Reduction
LCDR or Above	Yes	30 days	60 days	45 days	½ pay × 2 months	Yes
LT or Below	Yes	7 days	14 days	14 days	7 days	Yes
Enlisted OIC	No	No	14 days	14 days	3 days	No

Right to Appeal

Those who receive punishment under NJP do have the right to appeal that punishment under two circumstances. Either the punishment was unjust or it was disproportionate:

- Unjust: The evidence does not support the guilty finding or there was a substantial procedural error or illegal aspect of the NJP.
- Disproportionate: The punishment is disproportionate to the offense, too harsh, or unfair under the specific circumstances of the case.

In these cases, the accused has five working days from the date of the NJP to file an appeal in writing. The appeal must be routed by the accused via the officer who imposed NJP, who will draft an endorsement with their comments before forwarding it up the chain of command. Your legal officer can help you with this process.

Other Masts

To avoid confusion, you should also be aware that there are other forms of "mast" in the Coast Guard that have nothing to do with the UCMJ. Besides the captain's mast, if you asked to see the commanding officer for an important reason, this would be called a "request mast." On many cutters and stations, a certain time of the day is set aside by the command to hear requests from members of the crew. This practice is called request and complaint mast. Request and complaint mast provides an effective and alternate way to bring problems not resolved at lower levels to the command's attention for resolution. All requests to communicate with your commanding officer are submitted on a special request chit and forwarded through your chain of command.

COURTS-MARTIAL

If an alleged offense is too severe to dispose of by captain's mast, or if the accused exercises his right to refuse NJP, the case will go to court-martial. There are three types of court-martial: summary, special, and general.

Summary Court-Martial

If the offense is minor and if nonjudicial action has been ruled out, the command may refer the charges to trial by summary court-martial. A summary court-martial is a single-officer proceeding that looks more like nonjudicial punishment than a trial.

Commissioned officers, warrant officers, and cadets may not be tried by summary courts-martial. All Coast Guard members may refuse summary courts-martial. If refused, the matter may be referred for trial by a special or general court-martial. Members are afforded an opportunity to consult with a lawyer, military or civilian, prior to accepting or declining summary court-martial but are not entitled to military representation at the proceeding. Sentences that may be imposed at summary court-martial are similar to the punishments available at NJP, with a few key distinctions. For pay grades E-4 and below, it is possible to be reduced to the lowest enlisted pay grade and to be sentenced to up to one month's confinement in a military brig. For pay grades E-7 and above, it is possible to be reduced by one pay grade, which is not possible at NJP.

Special Court-Martial

Offenses that equate to misdemeanors under civilian law are tried by special court-martial, an intermediate level of the military court system that consists of a lone military judge or a four-person panel and a military judge. There is no ability to refuse this type of court-martial. The staff judge advocate of the convening authority, meaning the command's lawyer, will normally appoint a prosecutor, called the trial counsel, and a defense counsel. At courts-martial the jury is called a panel. Unlike a jury, panel members are not selected at random from among the accused's peers; rather, they are appointed by the commanding officer who convenes the courts-martial based on their age, education, training, experience, length of service, and judicial temperament. Also unlike a jury, unanimity is not required to convict but rather the concurrence of two-thirds of the members. If a finding of guilty is returned, the panel or military judge will hear evidence from the government and the accused concerning what the appropriate sentence should be. There are no sentencing guidelines under the UCMJ but rather maximum punishments established by statute or by the president through an executive order in his role as commander-in-chief. After hearing sentencing evidence, the panel or military judge will deliberate and deliver the sentence. Punishment may include one or a combination of the following:

- Maximum of six months' confinement
- Maximum of six months of forfeitures
- A bad-conduct discharge; this is the lowest forum where a punitive discharge may be adjudged
- Reduction to pay grade E-1

General Courts-Martial

A general court-martial is the highest court level under the UCMJ and is used to try offenses that equate to felonies under civilian law. Trial counsel and defense counsel will be appointed by flag officers. General courts-martial consist of a military judge and at least eight panel members. If the accused is enlisted, one-third of the panel may be enlisted upon request. A general court-martial differs from lesser forums primarily because any lawful punishment may be assigned depending on the offense, including the death penalty.

PUNITIVE ARTICLES

Although distinct from civilian law, many aspects of military justice are similar. Many of the offenses that are covered in the UCMJ require no special explanation. Theft is theft and arson is arson no matter what system of justice you answer to. But several articles deal with military specific offenses, like unauthorized absence and missing movement, that are probably unfamiliar to those who are new to the service. Below are several punitive articles that address conduct unique to military service.

Article 83: Malingering

It is an offense for a member of the armed forces, for the purpose of avoiding work, duty, or service, to fake illness, physical disablement, mental lapse, or derangement; or to intentionally inflict self-injury. Maximum punishment: In time of war—dishonorable discharge, forfeiture of all pay and allowances, and confinement for ten years; Other than in war—dishonorable discharge, forfeiture of all pay and allowances, and confinement for five years.

Article 85: Desertion

It is an offense for any member of the armed forces to leave his unit or remain absent without authority with the intent to remain away permanently; leave his unit with the intent to avoid hazardous duty or shirk important service; or enlist in the same or other service or foreign service without having been properly discharged. Maximum punishment: In time of war—any punishment, including death; Other than war—dishonorable discharge, forfeiture of all pay and allowances, and confinement for five years.

Article 86: Unauthorized Absence or Absence without Leave

It is an offense for any member of the armed forces to, without authority, fail to go to the appointed place of duty at the time prescribed; go from that place; or absent himself or remain absent from the appointed place at the time prescribed.

Maximum punishment: Dishonorable discharge, forfeiture of all pay and allowances, confinement for eighteen months.

Article 87: Missing Movement

Any member of the armed forces who through neglect or design, misses the movement of a ship, aircraft, or unit with which she is required in the course of duty to move is guilty of violating Article 87. Maximum punishment: Dishonorable discharge, forfeiture of all pay and allowances, confinement for two years.

Article 89: Disrespect toward a Superior Commissioned Officer; Assault of Superior Commissioned Officer

It is an offense to strike a superior commissioned officer or threaten an officer with a weapon or violence against her while in the execution of her office. The maximum punishment: In time of war—any punishment including death; Other than war—Bad-conduct discharge, forfeiture of all pay and allowances for one year, confinement for ten years.

Article 90: Willfully Disobeying Superior Commissioned Officer

It is an offense for any member of the armed forces to willfully disobey a command of his superior commissioned officer. Maximum punishment: In time of war—any punishment, including death; Other than war—Dishonorable discharge, forfeiture of all pay and allowances, confinement for five years.

Article 91: Insubordinate Conduct toward Warrant Officer, Noncommissioned Officer, or Petty Officer

It is an offense to strike or assault a warrant officer, noncommissioned officer, or petty officer while that officer is in the execution of her office; to willfully disobey the lawful order of a warrant officer, noncommissioned officer, or petty officer; or to treat with contempt or disrespect a warrant officer, noncommissioned officer, or petty officer while that officer is in the execution of her office. Maximum punishment: Dishonorable discharge, forfeiture of all pay and allowances, confinement for five years.

Article 92 (part 1): Failure to Obey Order or Regulation

In addition to obeying direct orders given by your superiors, you are required to obey service regulations and the lawful orders of those who have authority over you. You can be punished for violating a lawful general order, certain punitive regulations issued by a flag officer through a commandant instruction, or other means of wide promulgation regardless whether you are aware of the regulation or not. In other words, if a regulation meets the criteria of a lawful general order, it has the force of law and ignorance of it is not an excuse. Coast Guard lawful general orders will typically prohibit conduct that you should instinctively know is wrong. If an order or regulation does not meet the standard for a lawful general order, then the accused needs to have actually known of the order to be punished for violating it. Maximum punishment: Dishonorable discharge, forfeiture of all pay and allowances, confinement for two years.

Article 92 (part 3): Dereliction of Duty

You are required to perform all military duties, whether established by law, regulation, lawful order, standard operating procedure, or service custom. A member of the armed forces commits an offense if the member is derelict in the performance of any duty that may be properly expected or, through negligence or culpable inefficiency, fails to perform expected duties. Maximum punishment for willful cases of dereliction of duty: Dishonorable discharge, forfeiture of all pay and allowances, confinement for six months.

Sexual Assault

Sexual assault is any sexual contact that is against a person's will or without consent. This includes situations where force, violence, or weapons are used as well as situations where the victim is too intoxicated or scared to give consent. Some attackers use weapons, physical force, or the threat of force to gain the upper hand, but most often it is the use of alcohol or a position of authority used to perpetrate a sexual assault. Sexual assault includes rape, nonconsensual sodomy (oral or anal sex), and unwanted inappropriate sexual contact or fondling or any attempt to commit these acts. Sexual assault is a criminal act and a violation of UCMJ articles 120 (Rape), 125 (Forced Sodomy), and 134 (General). Perpetrators will be prosecuted to the fullest extent to the law. All convicted military sex offenders must register with the sex offender registry.

Article 134: General Article

Anyone who commits an act not otherwise mentioned in the UCMJ that is prejudicial to good order and discipline, of a nature to bring discredit to the armed forces, or a noncapital crime in violation of other (state/local) law will be punished as a court-martial may direct. Maximum punishments vary but, depending on the offense, can range from confinement for one month or up to twenty years, punitive discharge up to dishonorable discharge, and partial to total forfeiture of pay. Examples of the wide range of offenses defined under Article 134 include adultery, assault, fraternization, child endangerment, child pornography, failing to pay debts, disloyal statements, drunkenness, fraternization, jumping from a vessel, kidnapping, prostitution, and wearing an unauthorized decoration.

Offenses Related to War

Certain offenses are particular to wartime situations. For each of the offenses listed below, the death penalty may be imposed in time of war:

- Misbehavior before the enemy
- Subordinate compelling surrender
- Forcing a safeguard
- Improper handling of captured or abandoned property
- Aiding the enemy
- Misconduct as a prisoner of war
- Spying

Lesser Included Offenses

If, during a trial, a person is charged with desertion and it develops that the defendant was absent without leave but did not intend to abandon the service permanently, the court may convict that person of unauthorized absence and punish accordingly. This concept is referred to as a lesser included offense. For example, an attempt to commit an offense is a violation even though the crime is not consummated. An attempt to commit a particular crime is always a potential lesser included offense when a perpetrator is tried for the crime.

Conspiracy

When two or more persons agree to violate the UCMJ by concerted action and any one of them performs any action seeking to accomplish the agreed objective, all of the parties to the agreement are guilty of conspiracy.

COMPLAINTS OF WRONGS

Anyone who believes himself wronged by his commanding officer should address the wrong to the commanding officer. If refused redress by the commanding officer, an individual may complain to any superior commissioned officer who will address the complaint to the commanding officer's superior.

NO CONTACT AND MILITARY PROTECTIVE ORDERS

These orders provide commands with tools similar to the role of restraining orders in the civilian community designed to either prohibit a person from taking an action likely to cause irreparable harm or to maintain the status quo between two or more people. A no contact order is designed as a temporary restraining. It is limited in scope to contact between the subject of the order and the protected person(s), and it is of limited duration. Military protective orders cover a wider range of conduct, can be for longer duration, and trigger notification to civilian authorities and entry into the National Criminal Information Center restraining order file. These orders are typically issued in sexual assault cases and those involving intimate relationships where accusations of domestic violence surface.

SEXUAL ASSAULT PREVENTION, RESPONSE, AND RECOVERY

When you hear the term "sexual assault," what do you think of? If you imagine a stranger jumping out of the bushes on a dark night and attacking someone, you are only partly right because most sexual assaults are not committed by strangers but by coworkers and friends. This means that everyone must be constantly aware of their environment, for situations that may be developing, and to intervene or disrupt the chain of events that could lead to sexual assault. Alcohol plays a key role in the majority of sexual assaults, and it has been estimated half of all sexual assaults involve alcohol consumption by the perpetrator, victim, or both. Alcohol reduces inhibitions as well as the victim's ability to protect themself. Reduce your risk of being sexually assaulted by

- Staying sober;
- Stating what you want. If you do not want to be intimate, communicate that clearly;
- Traveling with a friend or stay in a group;
- Never leaving your drink unattended;
- Keeping your doors locked; and
- Not going anywhere alone with someone you do not know well

Reduce Your Risk of Offending

You can reduce your risk of becoming a sexual assault offender by communicating your expectations, avoiding excessive alcohol, and ensuring that your partner consents to sexual activity. You must have consent from your partner before you can engage in any sexual activity. Consent is words or overt acts by a competent person indicating a freely given agreement to the sexual conduct or activity. Legally, an individual cannot consent to sex if they are so impaired by alcohol or drugs of any kind that they are substantially incapacitated, which includes being incapable of understanding what is happening, incapable of declining participation, or unable to communicate their unwillingness.

Consent is the basis for every sexual encounter. Both individuals must agree to the sexual encounter and either person, at any time, may no longer consent and want to stop the activity. A person may also give nonverbal signals that they are uncomfortable. If your partner seems unsure or hesitant, you should immediately stop and verbally engage them to ensure they still consent. Remember that no means no. Even if you have been kissing or had sex before. To determine consent ask . . .

- Does the person, who I am with, want to give consent?
- Are they capable of giving consent?
- When in doubt, walk away.

Sexual Predators

It is an all-hands effort to continually know the indicators and behaviors of sexual predators and predator grooming. A sexual predator is a person who obtains or tries to obtain sexual contact with another person in a predatory or abusive manner. Grooming includes tactics a predator will use to gradually and methodically build trust with a potential victim both emotionally and psychologically to later gain access and time alone with their future victim. Predators may pretend to be a supportive friend. They establish themselves as the kind of person you would never suspect—someone who seems nice, helpful, charming, and likable. The goal of the predator is to manipulate the victim until they get whatever they are after. To reduce your chances of falling prey to a sexual predator,

- Do not share too much with someone you have just met;
- Do not accept gifts of money;
- Be wary of too much flattery;
- Don't stay if the situation becomes unhealthy/uncomfortable;
- Be cautious around someone you just met;
- Trust your instincts;
- Keep personal information private; and
- Drink responsibly.

Bystander Intervention

We must all work together to create a culture intolerant of sexual assault or behaviors that enable it. There are NO bystanders in the Coast Guard! You are responsible for taking action whenever you see something suspicious occurring. We must take responsibility for protecting our shipmates through personal responsibility, awareness, and intervention so we can eliminate sexual assault from our service. Many sexual assault offenders continued their horrible act for years because others who suspected and or even knew of their heinous actions took the role of bystander and did not do or say anything. We cannot have bystanders in our service.

If you notice a problem or suspect a problem, Take Action! Assess the situation to ensure that all parties are safe and determine whether to call the authorities. Decide if you should take action and the type of action to take:

- Direct action: Point out someone's inappropriate behavior in a way that deescalates the situation. Talk to the person who is acting inappropriately, or to the potential victim.
- Indirect action: Use creative options to distract the people involved in order to deescalate the situation. Have the bartender stop serving drinks, involve others, refocus the situation, or call the police/authorities.

Reporting Sexual Assault

The Coast Guard believes its first priority is for the victim to receive the medical treatment, care, and counseling they deserve. We understand that some may not want to initiate legal action, which is why two reporting options are available.

Unrestricted Reporting. The service member who desires medical treatment, counseling, and an official investigation of the allegation can report the matter using official reporting channels—for example, to a duty watch stander, supervisor, or their chain of command. Once a Coast Guard official receives a report that a sexual assault has occurred, the Coast Guard Investigative Service is notified immediately and an investigation is initiated, although a victim always has the choice to not participate with an investigation.

If a victim informally confides in a shipmate or friend that they have been assaulted, the shipmate or friend will not have a duty to report that sexual assault. However, this communication is not privileged; therefore, any information disclosed by a person in whom the victim confided is not protected, and they may be required to disclose the information if questioned by law enforcement.

Restricted Reporting. This option can be used by the service member who desires medical treatment and counseling without triggering the official Coast Guard investigative process and notification of the victim's command. Victims may report the matter to a sexual assault response coordinator, a victim advocate, or a health professional. Safe Helpline offers members of the armed services one-on-one expert advice and information, available 24/7 at safehelpline.org or 877-995-5247.

If You Are Sexually Assaulted

- Get to a safe place.
- Call someone who can help you.
- Don't shower, drink, eat, or change your clothes. These activities can destroy physical evidence that could be useful if you decide to pursue an unrestricted report that will hold the offender accountable.
- Go to medical or to the nearest hospital. Even if you feel embarrassed about your injuries, it is very important to receive medical assistance.

How to Help Others Who Have Been Sexually Assaulted

If someone has been sexually assaulted, they will need a great deal of support from the people around them. You should immediately report any activity that indicates a sexual assault has or may have taken place. You should

- Get assistance, but never leave the victim;
- Offer shelter;
- Listen to the victim without asking for details;
- Do not make judgments about the victim or the alleged offender;
- Encourage the victim to report the crime;
- Preserve the possibility of a restricted report by not discussing the assault with anyone; and
- Make sure the victim knows they are not to blame.

Support organizations include

- Military medical facility or clinic;
- Local hospital;
- Local police;
- National Rape Crisis Hotline 1-800-656-HOPE; and
- Employee Assistance Program Coordinator / Sexual Assault Response Coordinator, at 1-800-222-0364, which is on call 24/7.

It is important to note: Although victims may use any facility they desire for help or treatment, some agencies must call the police when a rape has been reported.

SEXUAL HARASSMENT

Sexual harassment is any unwelcome sexual advance, requests for sexual favors, or other verbal or physical conduct of a sexual nature when used as a term of employment or a basis of employment decisions, or used to create a hostile or offensive working environment. Sexual harassment includes

- Verbal harassment (offensive jokes, soliciting sexual acts);
- Physical harassment (standing too close, unwanted touching); and
- Visual harassment (sexual gestures, leering, inappropriate pictures, etc.).

Types of Harassing Behavior

Any behavior that relates to sex, is intentional or repeated, is unwelcome, and interferes with a person's ability to do their job or has an adverse effect on their working conditions can be classified as sexually harassing behavior.

Gender Harassment. Gender harassment consists of sexist statements and behaviors that convey insulting or degrading attitudes relating to sex or gender. Obscene jokes, offensive graffiti or photographs, insulting remarks or humor about sex, sexual orientation, or gender in general are examples of gender harassment.

Seductive Behavior. Seductive behavior is any unwanted, inappropriate, and offensive sexual advances. Unwelcome, persistent requests for dinner, drinks, or dates; repeated unwanted sexual invitations; and unwanted letters, phone calls, or other invitations are examples of seductive behavior.

Sexual Bribery. Sexual bribery is the solicitation of sexual activity or other sex-related behavior as a basis for a promised reward. Rewards could include a promotion or a raise.

Sexual Coercion. Sexual coercion is coercion of sexual activity or other sexually related behavior by threat of punishment, including threat of termination or demotion, withholding of promotion, or negative performance appraisal.

Sexual Imposition. Sexual imposition is an uninvited physical sexual violation or sexual assault. This includes forceful grabbing, feeling, or touching.

Preventing Sexual Harassment

Understand your stereotypes and biases. Once you recognize that your behavior could be taken as inappropriate, you will be better prepared to avoid those behaviors.

- Think about the impact before you speak.
- Recognize harassment and discrimination when it occurs.
- Never ignore harassment or discrimination.
- Use inclusionary language.
- Never immediately discount someone based on their gender.

Responding to Sexual Harassment

When a person experiences sexual harassment, they should not ignore the problem or assume it will stop. Harassment usually intensifies when it is ignored because the lack of corrective actions is seen as acceptance or encouragement. The most effective way to prevent harassment or discrimination is to tell the harasser to STOP. It is possible the person may not be aware that their behavior is inappropriate. The person might believe such behavior is acceptable.

The person experiencing the harassment should report the problem to their chain of command as soon as possible. If a supervisor is the alleged harasser, the harassment should be reported to the next higher level in the chain of command. Any individual witness to harassing behaviors should report the inappropriate conduct to their chain of command immediately so that it may be addressed. Additionally, at any point, individuals experiencing harassment or retaliation have the right to initiate an equal employment opportunity / equal opportunity (EEO/EO) complaint or seek guidance from a civil rights service provider (CRSP). It is important to document the harassment, which should include dates, times, places, and any witnesses.

All commanding officers and officers in charge will take prompt action to address reports of harassment under the antiharassment and hate incident (AHHI) policy found in the U.S. Coast Guard Civil Rights Manual, COMDTINST M5350.4E, while at all times respecting the privacy and confidentiality of individuals reporting harassment or providing information relating to harassment. Note that filing an AHHI complaint does not replace, substitute, or satisfy the separate requirements of filing an EEO/EO complaint.

Depending on the severity of the sexual harassment, different punishments can be administered. The performance evaluations of anyone found to have engaged in or condoned sexual harassment could reflect appropriate comments and marks by the supervisor and chain of command. Military members can be processed for administrative separation on the first substantiated incident of sexual harassment. Unwelcome and unauthorized physical contact of a sexual nature, if charges are in violation of the UCMJ, could result in a punitive discharge.

Acts of Reprisal

Acts of reprisal are illegal. If anyone feels retaliated against for attempting to stop harassment, he may also file a complaint regarding the retaliation. No person shall take reprisal action against a person who, in good faith, raises an allegation or provides information concerning an incident of alleged sexual harassment.

INTERPERSONAL RELATIONSHIPS

The Coast Guard recognizes that personal relationships are acceptable and in fact encouraged to enhance unit morale and esprit de corps, provided they do not, either in actuality or in appearance, jeopardize the member's impartiality, undermine the respect for authority in a member's rank or position, result in a member improperly using the relationship for personal gain or favor, or violate a punitive article of the UCMJ. Relationships can develop into romantic relationships and even lead to marriage. A relationship, including marriage, does not violate service policy unless the relationship or the member's conduct fails to meet the standards of conduct set by the UCMJ or other regulations.

Character of the Relationships

Personal. A non-intimate, unromantic association between two or more people, such as occasional attendance at recreational or entertainment events (movies, ball games, concerts, etc.) or meals, does not involve conduct that violates the UCMJ.

Romantic. A mutually consensual sexual or amorous relationship does not involve conduct that violates the UCMJ.

Marital. Marriage between an officer and enlisted member is not considered unacceptable or prohibited providing the marriage existed prior to one of the service member's accession to the office corps.

Types of Relationship

Acceptable. Acceptable relationships include, for example, two crew members going to an occasional movie, dinner, concert, or other social event or member jogging or participating in wellness or recreational activities together.

Unacceptable. Unacceptable relationships include any interaction of an unprofessional nature between prohibited parties.

Romantic. Romantic relationships are unacceptable between members when

- Members have a supervisor and subordinate relationship, including periodic supervision of duty section or watch standing personnel;

- Members are assigned to the same small shore unit numbering less than sixty personnel;
- Members are assigned to the same cutter;
- The relationship is between chief petty officers (E-7/8/9) and junior enlisted personnel (E-4 and below); or
- The relationship is manifested in the work environment in a way that disrupts the effective conduct of daily business.

Resolving Unacceptable Relationships

Avoiding unacceptable personal relationships is in the best interest of all concerned. Prompt resolution at the lowest level possible is desirable. Members may request or a command may recommend reassignment of a member involved in a questionable relationship. However, reassignment is not a preferred option. When reassignment is not an option, members may be directed to end a relationship. The Coast Guard is not obligated to reassign personnel due to a member's desire or based solely on a relationship.

Prohibited Relationships

These relationships violate the UCMJ, and resolution may be either administrative or punitive. Coast Guard policy prohibits the following relationships of conduct, regardless of rank, grade, or position of the persons involved:

- Engaging in sexually intimate behavior on board any Coast Guard vessel or in any Coast Guard controlled workplace;
- Personal and romantic relationships between instructors and students at training commands; or
- Romantic relationships outside of marriage between officers and enlisted.

This provision is a punitive general regulation, applicable to all personnel subject to the UCMJ.

Responsibility

All personnel are responsible for avoiding unacceptable or prohibited relationships. Personnel finding themselves involved in or contemplating unacceptable relationships should report the situation and seek early resolution from their supervisor, commanding officer, officer in charge, command enlisted adviser, or Coast Guard chaplain. Bringing an unacceptable relationship to early command attention will increase the opportunity for an early and positive resolution.

HAZING AND BULLYING

Hazing and bullying policies apply to all Coast Guard personnel at all times, on or off duty, at sea or ashore, on or off base. Any violation or solicitation of another to violate either of these policies may subject involved members to administrative or disciplinary action. Hazing and bullying are contrary to honor, respect, and devotion to duty. They negatively impact esprit de corps. There is no place in the Coast Guard for dehumanizing treatment, and any incident of hazing will be investigated, and appropriate disciplinary action be initiated against the perpetrators.

Hazing

Hazing is any conduct in which a military member, without proper authority, causes another military member(s) to suffer or be exposed to any cruel, abusive, humiliating, oppressive, demeaning, or harmful activity, regardless of the perpetrator's and recipient's service or rank, for the purpose of inclusion. Soliciting or coercing another to conduct such activity also constitutes hazing. Hazing does not need to involve physical contact among or between members; it can be verbal or psychological in nature. No service member may engage in hazing or consent to being hazed.

Prevention of Hazing

Hazing serves no useful purpose and is contrary to our core values. The demeaning and abusive activities associated can inhibit performance and debase our personal dignity and may result in serious injury. To prevent hazing, we must first be aware of what constitutes hazing and understand that these actions have a negative impact on the individual and our service. Hazing can include, among other things, the following activities:

- Playing abusive or mean-spirited tricks intended to ridicule, humiliate, or ostracize
- Throwing personnel over the side from a ship or pier
- Tacking on crows or other devices by forcibly applying them to a member's clothes or body
- Forcing or encouraging the consumption of substances;
- Encouraging a member to consume excessive amounts of alcohol or requiring the consumption of alcohol in any amount
- Forcibly cutting or shaving hair
- Branding, tattooing, or painting another

- Taping, tying, or otherwise restraining a member's arms, legs, or mouth
- Touching in an offensive manner

Voluntary Consent

Personnel often attempt to justify their hazing activities by voluntary participation. Even genuine voluntary participation can cause detrimental consequences. Often apparent willing participation is actually prompted by subtle compulsion or peer pressure and is not truly voluntary. Does this mean we do not have fun in the Coast Guard? No. Military customs and traditions are an essential part of the Coast Guard. We do have traditional and authorized service initiation ceremonies including crossing the equator, International Date Line, and Arctic and Antarctic Circles. However; commanding officers ensure these events do not include any degradation of character, sexual overtones, bodily harm, or harassment and comply with Coast Guard policy.

Bullying

Bullying is defined as abusive conduct by a military member or members that harms a military member or any other persons, either physically or psychologically, without a proper military or other governmental purpose and with the intent to exclude the member. Bullying is threatening, humiliating, or intimidating. Bullying can also be work interference, undermining performance, or verbal abuse. Bullying may also be described as psychological abuse, psychological harassment, status-blind harassment, and mobbing.

Prevention of Bullying

The prevention measures for the hazing section also apply to bullying.

Bullying can include, among other things, the following:

- Physically striking another in any manner or threatening to do the same
- Intimidating, teasing, taunting
- Oral or written berating of another for the purpose of belittling or humiliating
- Encouraging another to engage in illegal, harmful, demeaning, or dangerous acts
- Playing abusive or malicious tricks; branding, handcuffing, duct taping, tattooing, shaving, greasing, or painting
- Subjecting to excessive or abusive use of water (to include drinking)
- The forced consumption of food, alcohol, drugs, or any other substance
- Degrading or damaging the person or their property or reputation

Procedure if Hazing or Bullying Occurs

On or off duty, in or out of uniform; there is no place in the Coast Guard for dehumanizing treatment. Every member must ensure hazing and or bullying does not occur in any form at any level. Victims of actual or attempted hazing and witnesses to these activities must report it to the appropriate level of the chain of command.

POLITICAL ACTIVITIES

The military must remain politically neutral and divorced from partisan politics. As a member of the Coast Guard, the manner in which you exercise your political rights is somewhat limited. While you have the right and duty as an American citizen to vote and voice your opinions concerning political matters, you must be very careful your personal opinions and activities are not directly, or by implication, represented as those of the Coast Guard.

Monetary Contributions and Fundraising

You may make monetary contributions to a political organization or a political committee favoring a particular candidate or slate of candidates and may attend partisan and nonpartisan political fundraising activities as a spectator, but not in uniform and only when no inference of Coast Guard approval or endorsement can reasonably be drawn. You may NOT:

- Use your official authority or influence to solicit contributions
- Solicit or receive a campaign contribution from another member of the armed forces or from a civilian officer or employee of the United States for promoting a political objective or cause
- Sell tickets for or otherwise actively promote fundraising events
- Make campaign contributions to a partisan political candidate or to another member of the armed forces or an officer or employee of the federal government for promoting a political cause
- Solicit or otherwise engage in fundraising activities in federal offices or facilities, including military reservations, for a partisan cause or candidate

Participation in Political Groups and Events

Specific to your participation in political groups and events as a Coast Guard member, you may attend political meetings or rallies as a spectator and join a political club and attend its meetings. You may NOT:

- Participate in partisan political management, campaigns, or conventions or make public speeches in the course of such activity
- Conduct a political opinion survey under the sponsorship of a partisan political group or distribute partisan political literature
- Perform clerical or other duties for a partisan political committee during a campaign or on Election Day
- Participate in any organized effort to provide voters with transportation to the polls, attend, as an official representative of the armed forces, partisan political events, even without actively participating
- Engage in the public or organized recruitment of others to become partisan candidates for nomination or election to a civil office

Groups, Organizations, and Unions

Because of its incompatibility with military service, active participation in the following types of organizations is prohibited:

- Organizations supporting supremist or extremist causes or attempting to create illegal discrimination based on race, creed, color, sex, religion, or national origin
- Organizations advocating the use of force or violence or otherwise engaging in efforts to deprive individuals of their civil rights

Political Speech

Relative to making political speeches, as a Coast Guard member you may sign a petition for specific legislative action or a petition to place a candidate's name on an official election ballot, but only if the signing does not obligate you to engage in partisan political activity and is done as a private citizen and not as a representative of the armed forces. You may write a letter to the editor of a newspaper expressing your personal views concerning public issues if those views do not attempt to promote a partisan political cause You may also display a political sticker on your private vehicle or wear a political button; however, you may NOT

- Wear a button on your uniform or at any time while on duty
- Write a personal letter not for publication expressing preference for a specific political candidate or cause, provided it is not part of an organized letter-writing campaign on behalf of a partisan political cause or candidate
- Allow or cause to be published partisan political articles signed or authorized by you for soliciting votes for or against a partisan political party or candidate

- Speak before a partisan political gathering of any kind for promoting a partisan political party or candidate
- Participate in any radio, television, or other program or group discussion as an advocate of a partisan political party or candidate
- March or ride in a partisan political parade
- Use contemptuous words against political office holders
- Display a large political sign, banner, or poster (as distinguished from a political sticker) on the top or side of your private vehicle

Serving as an Election Official

Election officials set up voting equipment, register, and sign in voters, and record vote totals. You may serve as an election official if such service does not interfere with your military duties. You must perform your service out of uniform, and you may not be a representative of a partisan political party. You must have the prior approval of your commanding officer.

Possession or Distribution of Printed Materials

You may not be prohibited from distributing or posting materials solely on the grounds that these materials are critical of government policies or officials. However, you cannot distribute any written or printed materials at a Coast Guard installation when the commanding officer or authorized designee determines the materials:

- Are a clear danger to the loyalty, discipline, or morale of members of the armed forces
- Interfere with accomplishing the military mission

These materials may be impounded and returned to you when you leave the installation, unless the impounded materials are determined to be evidence of a crime.

Writing for Publications

As a Coast Guard member, you may NOT write for unofficial publications during duty hours. Nor can you produce an unofficial publication, such as an underground newspaper, using government or non-appropriated fund property or supplies. Any publication that contains language the utterance of which is punishable by the UCMJ or other federal laws may subject the person(s) involved in its printing, publishing, or distribution to prosecution or other disciplinary action.

Off-Limit Establishments

Certain establishments may be placed off limits under the armed forces Disciplinary Control Boards and Off-Installation Liaison and Operations, COMDTINST 1620.1 (series). An establishment runs the risk of being placed off limits if its activities include counseling service members to

- Refuse to perform their duties;
- Desert their military service; or
- Involve themselves in acts that would result in significant adverse effects on health, welfare, or morale of military members.

Demonstrations

As a member of the Coast Guard, you are prohibited from participating in demonstrations when

- On duty;
- In a foreign country;
- The demonstrators are in uniform;
- The activities constitute a breach of law and order; or
- Violence is likely to result.

Political Dissent

Political dissent refers to any expression designed to convey dissatisfaction with or opposition to the policies of a governing body. Such expression may take forms from vocal disagreement, to civil disobedience, to the use of violence. The basic mission of the Coast Guard is to safeguard the security of the United States. As a Coast Guard member, your right of expression—consistent with good order, discipline, and national security—must be preserved to the maximum extent possible.

ETHICS

There are standards of ethical conduct that Coast Guard personnel are accountable for as embodied in the Code of Federal Regulations. The purpose of these regulations for executive branch employees is to maintain public confidence in the integrity of the federal government. Since you have committed yourself by sworn oath to a term of service in the Coast Guard, this information is of vital importance to you.

Misuse of Position

There are four rules governing misuse of your government position:

- You cannot use your public office for your own personal gain or that of your family or friends.
- You cannot engage in financial transactions using information that you received through your official position that is not available to the general public.
- You must preserve and protect government property.
- You cannot use official time to do personal business.

Impartiality in Performing Official Duties

Impartiality is defined as equal treatment and fairness. You cannot conduct activities, act in the interest of outside entities, or do anything similar while conducting government business that gives the appearance that you are no longer acting impartially. Basic rules for impartiality include the following:

- You are prohibited, without specific permission, to ask a party with whom you conduct government business to develop a personal or business relationship with that party on behalf of yourself or members of your family.
- You are prohibited from engaging in activities that affect the financial interest of your spouse, minor children, and friends or in which you serve in a position as trustee or director or any organization in which you are an active participant.

Gifts

The term "gift" is defined to include nearly anything of monetary value. It is something given to someone without payment; a present. Gifts can come in many forms and may include alcohol, trips, vehicles, watches, or tickets. Gifts do not include items that are clearly not intended to be gifts, such as

- Publicly available discounts;
- Commercial loans;
- Modest items of food and refreshments offered other than as part of a meal, such as soft drinks, coffee, and donuts;
- Greeting cards and items with little intrinsic value, such as plaques, certificates, and trophies, which are intended solely for presentation;

- Rewards and prizes given to competitors in contests or events, including random drawings open to the public;
- Pensions and other benefits resulting from continued participation in an employee welfare or benefits plan maintained by a former employer;
- Anything that is paid for by the government;
- Any gift that is accepted by the government under specific statutory authority (e.g., gifts of travel expenses from non-federal sources); or
- Anything for which the employee pays market value.

Gifts between Employees

Employees are not allowed to ask for a gift from or give a gift to someone who is their supervisor or in a higher rank or pay grade than them unless there is a personal relationship between the two. Gifts must always be voluntary. There are exceptions to this general rule:

- For events like office birthday parties, or holidays, gifts valued at ten dollars or less may be given and received.
- For infrequent but significant personal occasions like marriage, birth of a child, or retirement, gifts may be given and received.

Gifts from Outside Sources

The general rule for gifts from outside the Coast Guard is that Coast Guard members are prohibited from asking for or accepting gifts from prohibited sources or gifts given because of their official position. Persons or parties with whom your agency does business, or that your agency regulates, are called prohibited sources. A prohibited source is anyone who

- Seeks official action by the Coast Guard;
- Does business or seeks to do business with the Coast Guard; or
- Conducts activities regulated by the Coast Guard.

There are exceptions to these rules:

- Unsolicited gifts with a market value of twenty dollars or less per occasion, totaling not more than fifty dollars per calendar year from a single source;
- Gifts given because of family relationship or personal friendship;
- Widely available discounts and similar benefits;
- Gifts, other than cash or investment interests, with an aggregate market value of two hundred dollars or less if such gifts are a bona fide award or

incident to a bona fide award that is given for meritorious public service or achievement;

- Gifts of ten dollars or less for events such as office birthday parties or holidays; or
- Gifts given for infrequent but significant personal occasions like marriage or birth of a child or retirement.

OUTSIDE EMPLOYMENT

Outside employment is any legitimate and ethical enterprise or employment, including self-employment, during off-duty hours. Coast Guard personnel on active duty are in a twenty-four-hour duty status, and their military duties shall at all times take precedence on their time, talents, and attention. Personnel who accept off-duty employment must realize that even though they are on leave or liberty, they are subject to recall and duty at any time. Outside employment cannot expose Coast Guard personnel to bodily injury, interfere with their military duties, bring discredit upon the Coast Guard, or involve a conflict of interest. You cannot have a part-time job outside the Coast Guard that conflicts with your Coast Guard position. Additionally, the Coast Guard requires you obtain permission from your chain of command before you can apply for a part-time job outside the Coast Guard and must include a brief description of the specific responsibilities of the position and an estimate of the number of hours per week required for employment.

Potential Conflict

If you find yourself in a potential conflict of interest, there are people available who can help. It is critical to your career that you remember there are ethics officials available to assist you. Ethics officials are usually lawyers working in Coast Guard legal offices.

Government Vehicles

While performing your official duties you may be required from time to time to operate different types of Coast Guard–owned motor vehicles. As the operator of these vehicles, it will be your responsibility to make sure you are familiar with all of the regulations pertaining to the authorized and unauthorized use of government-owned vehicles.

Authorized Use

Coast Guard employees entrusted with government-owned or -leased vehicles are responsible for the proper care, operation, maintenance, and protection of the vehicles entrusted to them. Government motor vehicles operated by active-duty personnel, Coast Guard civilians, or contracted personnel shall be used for official purposes only. Use of government vehicles for other than the direct accomplishment of Coast Guard missions may be authorized if these missions are not compromised and the risk of loss or damage are at a minimum. Examples of non-mission-related functions include the following:

- Official functions are authorized, provided the transportation begins and ends at the workplace or official duty station. Under no circumstances is it appropriate to pick up or return a passenger to any other location than his workplace.
- If the employee and the spouse travel together from the duty location to an official function in which the spouse was invited, the spouse may be transported.
- The vehicle may be used for temporary duty as long as the employee does not make any unofficial stops in route. The vehicle may not be used for any local trips except the following:
    - › Duty site
    - › Lodgings
    - › Dining facilities
    - › Places of worship
- The vehicle may be used for similar places required for the traveler's subsistence, health, or comfort—for example, barber shop, drugstore, and so on.
- The vehicle may be used for public ceremonies, provided civilian or military personnel are participating in an official public ceremony, parade, or event that promotes Coast Guard relations with the public.

Unauthorized Use

Unauthorized use of government-owned vehicles includes use for non-duty-related purposes, unauthorized side trips while on official business, and transportation of civilians not on government business, including family members of employees. Transportation of non-official passengers shall be discouraged and may result in legal claims against the government and public criticism. Unauthorized use of government-owned vehicles includes but is not limited to the following:

- Luncheons and office functions, including send-off functions
- Permissive travel
- Picking up or giving rides to hitchhikers or strangers

Military members are subject to disciplinary action for unofficial use of government motor vehicles in accordance with the provisions of the UCMJ.

License

All operators of government vehicles must have a valid state driver's license. No employee is authorized to drive a Coast Guard–owned vehicle while their state license or driving privileges have been suspended or revoked. An OF-346, U.S. Government Motor Vehicle Identification Card, is issued when required to operate a vehicle or trailer combination that requires special skills and experience. Commanding officers may revoke an employee's privilege to operate any or all government motor vehicles whenever they question the employee's ability to operate a vehicle safely.

Violations and Accidents

A Coast Guard member or employee may be held personally liable for expenses incurred through the unauthorized use of a government motor vehicle, including accident damage or damage to personal or government property. By law, an agency cannot use government funds to pay fines for traffic violations regardless of the circumstances.

You must report any motor vehicle accidents/mishaps or damage to vehicles and trailers, as well as loss of license plates, as soon as possible to motor vehicle managers and their supervisors. If you are involved in an accident, immediately take the following actions:

- Check for injured persons.
- Check for witnesses.
- Notify your supervisor.
- Call the police.
- Obtain information from other drivers.
- Secure the vehicle and its contents.

If you are the driver of a government vehicle during an accident, you must file a Motor Vehicle Accident Report, Form SP-91. In order to properly complete this form, you will need driver and vehicle information from all parties involved in the accident, such as name, address, driver's license, and vehicle information. If you can take pictures, do so as this will help you document the event. If you are the passenger in a government vehicle during an accident, you must fill out a Statement of Witness, Form SF-94.

8

SECURITY

The security of the United States in general, and of military operations in particular, depends greatly upon the success attained in the safeguarding of classified information. Such information may be printed or may be other forms of communication including radio, visual signals, mail, and knowledge of ships' movements. Security, in the full sense of the word, has to do with the safeguarding of materials and the physical safety of ships and stations. It is important that everyone understands what classified information is, its importance, how to safeguard it, how to obtain clearance to work with or have access to it, and the penalties for security violations. Refer to COMDTINST M5500.11A for more information.

CLASSIFIED NATIONAL SECURITY INFORMATION

Executive Order 13526 prescribes a uniform system for the federal government to classify, safeguard, and declassify national security information, meaning information that has been determined to require protection against unauthorized disclosure. National security information is any knowledge that can be communicated regardless of its physical form or characteristics and that is owned by, produced by or for, or is under the direct control of the U.S. government. Classification levels are assigned to this information to determine the level of protection directly related to the damage that unauthorized disclosure could cause to our national security.

Top Secret. Top Secret information is of such importance to the national security of the United States that its unauthorized disclosure or loss would result in exceptionally grave damage to the United States or its allies. Examples of exceptionally grave damage include hostilities against the United States or its allies, disruption of foreign relations affecting the national security, and the compromise of vital national defense plans or complex cryptologic and communications systems.

Secret. Information of such importance that its unauthorized disclosure could reasonably be expected to cause serious damage to national security is classified as Secret. Examples of serious damage include the disruption of foreign relations, compromise of significant military plans or intelligence operations, and compromise of significant scientific or technological developments relating to national security.

Confidential. Information of such importance that its unauthorized disclosure could reasonably be expected to cause damage to the national security is classified as Confidential. Examples of damage include the compromise of information that indicates the strength of ground, air, and naval forces in the United States and overseas; disclosure of technical information used for training, maintenance, and inspection of classified munitions of war; and performance characteristics, test data, design, and production data on munitions of war.

In addition to these classification levels are categories of information that do not meet the criteria to be protected as Classified National Security Information but still require protection from disclosure under other programs.

Sensitive Compartmented Information (SCI). SCI is highly sensitive national security information to which access is based on a strict need-to-know basis. The SCI system is a national intelligence community security program derived from intelligence sources, methods, or analytical processes that are required to be handled exclusively within formal control systems established by the director of national intelligence.

Sensitive but Unclassified (SBU) Information. SBU information includes but is not limited to, For Official Use Only (FOUO); Sensitive Security Information (SSI); Law Enforcement Sensitive (LES), Chemical-Terrorism Vulnerability Information (CVI), and Protected Critical Infrastructure Information (PCII).

Access

Access to classified national security information is granted by the commanding officer or officer in charge or a designated official. To be eligible for access to classified information, you must

- Be a U.S. citizen
- Have a favorably adjudicated personal security investigation
- Possess a valid need to know

- Be granted a final or interim security clearance by the Security Center
- Receive a Clearance Access Briefing
- Execute a Classified Information Nondisclosure Agreement, SF-312

SECURITY CLEARANCES

Before you can be granted access to classified information, you must have a valid security clearance. The clearance must be equal to, or higher than, the classification of the information to which you will have access. A security clearance is an administrative determination by competent authority that you are trustworthy and dependable and possess the integrity required for a position of trust.

Various types of security investigations are used to grant security clearance. Every person entering the Coast Guard has a National Agency Check (NAC) completed. An integral part of all background investigations, the NAC consists of searches of the FBI Identification Division's name and fingerprint files and other files or indexes, when necessary. Additional investigations are performed depending on the level of access required and the sensitivity of the position being filled.

National Agency Check and Inquiries (NACI). This is the basic and minimum investigation required on all new federal employees. It consists of a NAC with written inquiries and searches of records covering specific areas of a person's background during the past five years. Those inquiries are sent to current and past employers, schools attended, references, and local law enforcement authorities.

Access NACI (ANACI). This is a new investigation designed as the required initial investigation for federal employees who will need access to classified national security information at the Confidential or Secret level. The ANACI includes NACI and credit coverage with additional local law enforcement agency checks.

NAC with Local Agency Check and Credit (NACLC). This initial investigation is designed for contractors at the Confidential and Secret national security access levels. The NACLC is also to be used to meet the reinvestigation requirement for all individuals who have Confidential or Secret clearances.

Background Investigation (BI). This in-depth investigation is required of those going into highest risk public trust positions.

Single-Scope Background Investigation (SSBI). This is the government-wide investigation required of those who need access to Top Secret classified national security information. This background investigation covers the past seven years of the subject's activities. It includes verification of citizenship and date and place of birth as well as national agency records checks on the subject's spouse or cohabitant and interviews with selected references and former spouses.

SSBI-Periodic Reinvestigation (SSBI-PR). This is the required five-year update investigation for those who have Top Secret security clearances. It consists of personal investigative coverage of employment and residences since the previous investigation, including interviews with all former spouses divorced during the coverage period.

e-QIP

Investigations begin with your completion of an Electronic Questionnaire for Investigations Processing (e-QIP). e-QIP is a secure website that contains all forms, including the Questionnaire for National Security Positions, SF86; Questionnaire for Public Trust Positions, SF-85P; and Questionnaire for Non-Sensitive Positions. The unit command security officer will initiate the request to give you access to the site where you will complete the personnel security questionnaire online.

Your eligibility for access to classified information and suitability for service will be continuously evaluated throughout your career, and you will be required to update e-QIP at regular intervals. It is recommended you print your e-QIP submissions and retain a copy in your personal records to help with future updates.

Final Clearance

A final security clearance can only be granted by the Coast Guard Security Center. This clearance may be granted after a background investigation has been completed along with a review of your personnel and health records.

Access to classified information is granted based on a need to know. If you need to have classified information to perform your assigned duties, you will be granted access by your commanding officer or officer in charge. Having access does not mean that you will be able to look at all the classified material held by the command. You will receive only the information you need to perform your job.

Security investigations remain valid and may serve as the basis for the issuance of future clearances unless derogatory information becomes available and indicates a need for further investigation, if the individual is assigned to a particularly sensitive billet requiring a greater level of clearance, or if continuous active service in the armed forces or civilian employment government service is broken by a period longer than one year.

Derogatory Information

Commands are required to report all derogatory information on personnel to the Security Center via an electronic Personnel Security Action, CG-5588. Derogatory information and types of conduct that could jeopardize an individual's clearance include the following:

- High levels of debt
- Not reporting foreign contacts
- Pattern of security violations, meaning incidents that could be reasonably expected to result in an unauthorized disclosure of classified information.

Examples of security violations include

- Open and unattended safes
- Inappropriate classified discussions with unauthorized persons
- Classified processing on unsecured networks or printers
- Transmitting over nonsecure means
- Using a strip cutter to destroy classified material

Safeguarding Classified Information

All classified material entrusted to you will be clearly marked with its appropriate classification. There is never an excuse for the careless handling of classified material. If you accidentally come across some classified matter—whether a letter, booklet, or a device—that has been left unguarded, misplaced, or not properly secured, do not read or examine it or try to decide what to do with it. Notify the nearest officer or petty officer in charge, and then stand by to keep unauthorized personnel away until a responsible person arrives to take charge. If you become aware of a compromise of classified material, immediately notify supervisory personnel.

Anyone responsible for the loss or unauthorized disclosure of classified material, and anyone who violates security regulations, will be subject to disciplinary action under the UCMJ. Disciplinary action may include trial by court-martial.

To keep yourself from getting involved in a compromise, remember that what you do, see, and hear at work stays at work. Bragging or talking freely about what you know and do may impress your family and friends, but it could jeopardize your safety and the safety of your shipmates. There was a saying during World War II, "Loose lips sink ships." That saying is as applicable today as it was then.

Restricted Areas

Restricted areas are established at units to provide additional safeguards for matériel and equipment. These areas will be clearly marked with warning signs. Persons on watch in such areas should not hesitate to ask visitors for identification. No responsible person will object to being stopped politely, but firmly, pending identification.

Three levels of restricted areas meet the security requirements for the protection of the information stored in the space.

Exclusion Area. Access to this area constitutes, for all practical purposes, access to classified material or equipment.

Limited Area. A limited area is an area in which uncontrolled movement would permit access to classified material or equipment.

Controlled Area. Controlled areas are the areas adjacent to or surrounding exclusion or limited areas.

CRYPTOGRAPHY

Since all communication may be intercepted by an adversary, messages that contain vital information must be "disguised" so they cannot be read. This is done by putting messages in codes or ciphers. Cryptography is the science of enciphering and deciphering messages; cryptanalysis is the breaking of enemy ciphers or codes. All ciphers are designed to allow a large volume of traffic and at the same time defeat enemy cryptanalysts.

Codes are word-for-word substitutions of letter or number groups, with prearranged meanings for entire phrases, such as those found in signal books. They are used where speed is preferred to security and are changed frequently since they are subject to compromise.

Ciphers offer letter-for-letter substitutions and greater flexibility and security than codes and are easier to use in rapid communication. Most ciphers are changed periodically. Simple tactical codes and ciphers are changed frequently; those considered more secure may be used for months or years.

EMERGENCY DESTRUCTION BILL

Emergency destruction is an action, ordered by the commanding officer or officer in charge, to burn or otherwise destroy registered publications and cryptographic material to avoid their falling into enemy hands. Procedures are detailed in a bill found in the Unit Standing Orders. Personnel assigned to duties under this bill are given specific and detailed instructions. In shallow water less than one hundred fathoms, papers must be burned. In deep water more than one hundred fathoms, papers may be permitted to go down with the ship.

FREEDOM OF INFORMATION AND PRIVACY ACTS

The Freedom of Information Act, also known as the FOIA, was enacted in 1966 and is the primary means by which the public has access to records in the possession of executive branch agencies of the federal government. The FOIA is operated under the premise that the public has a right to know what the government is doing, how it is being done, and what information is being collected. However, to prevent individuals, businesses, and government from harm resulting from the release of certain information, the FOIA provides a means by which limited information can be withheld from disclosure. See COMDTINST M5260.3, chapters 8 and 14, for a discussion of the exemptions that limit disclosure of certain information. FOIA requests may be submitted by the public in writing via mail or electronically via e-mail to the headquarters FOIA coordinator.

The Privacy Act provides safeguards for the protection of records the federal government collects on U.S. citizens or lawfully admitted permanent residents. Under the Privacy Act of 1974, an individual U.S. citizen or legal permanent resident may seek to access, correct, or amend records that are retrieved by name or other personal identifier such as one's social security number contained in a system of records. Requests may be made in any written form that bears the requester's signature.

OPERATIONAL SECURITY

All personnel are responsible for ensuring sensitive information is not released publicly, whether in an official, unofficial, or personal capacity. Coast Guard personnel are expected to exercise good operational security (OPSEC) and common sense when considering the release of information. Even when off duty and on a personal site or social networking application, as a Coast Guardsman, YOU are subject to the UCMJ and are subject to administrative or legal action if you violate policy or security regulations. OPSEC will help prevent you from putting your country, your family, and yourself at risk. Some examples of OPSEC issues that shall not be released are the following:

- Rules of engagement or use of force policy
- Details of schedules, daily routine, specific locations, and courses of action for ongoing or future operations and activities
- Speculation or discussion of terrorist events, activities, and force protection posture
- Deployment plans and procedures for Coast Guard units or other U.S. forces
- Names of personnel assigned to sensitive deployments or duty stations
- Personnel recall rosters
- Family member information, home telephone numbers, and addresses
- Physical security vulnerabilities of Coast Guard or other forces
- Detailed information about unit capabilities and sensitive or controlled missions

- Maps or specific locations of sensitive operational units
- Specific movement of flag officers, senior leaders, VIPs, and key command personnel including itineraries, hotel reservations, events, agendas, and so on

Cyber Security

The Coast Guard uses computers and office equipment to store and process data on operations, personnel, communications, and many other areas. There are clear risks associated with the use of this equipment. As such, the Coast Guard has set clear policies and procedures for their use. Before you can access your standard workstation for the first time, it is important that you acknowledge the policy on personal use of government office equipment and its limitations. Office equipment, such as computers, telephones, cell phones, faxes, copiers, and so on, is intended for official purposes in support of Coast Guard operations and missions. Limited personal use is permitted, but the rules and regulations regarding personal use must be strictly upheld and enforced. This policy is a general order; applies to all military, civilian, and contract employees; and is punishable under the UCMJ. Personnel must be authorized to use government office equipment or services for official government business before such equipment may be available for limited personal use.

Personal use of government office equipment or services may be authorized for CG personnel if the use

- Involves minimal expense to the government;
- Does not interfere with official duties;
- Does not inhibit the security of information or information systems; and
- Does not cause degradation of network services.

It is important to note that managers and supervisors may further restrict personal use based on the needs of the command or office. There are specific circumstances when use of government office equipment or services is prohibited:

- Viewing, downloading, storing, displaying, transmitting, or copying any sexually explicit materials.
- Using personal e-mail sites, such as Gmail, Hotmail, and so on. These sites bypass Coast Guard firewalls and virus security measures and may compromise the systems.
- Using Coast Guard e-mail address for subscribing to anything other than official, professional, or job-related communication.
- Loading software or making configuration changes to a government computer system.

- Downloading movies, books, streaming media, stock tickers, YouTube, and so on.
- Visiting Web-based or social networking sites such as Facebook, Twitter, and Skype unless they are official Coast Guard Public Affairs sites.
- Engaging in any fundraising activity.
- Deliberately introducing or failing to report the introduction of viruses, worms, or malware.
- Accessing dating or gambling websites.
- Connecting personally owned equipment to the Coast Guard network.

The following are inappropriate uses of government office equipment and may result in adverse administrative actions:

- Making personal long-distance calls unless there is an emergency or a brief call home within the local commuting area to check on a sick child, arrange transportation, and so on.
- Creating, copying, or transmitting any communication that is illegal or offensive to fellow employees or the public; for example, hate speech.
- Using personal shopping sites.

You can only transmit classified information if you are on a system that is accredited for processing classified information, operating at the appropriate security classification level, and you occupy a position that requires you to transmit this information as part of your duties.

SBU correspondence is information that may be withheld and protected from public release. This would include e-mails that contain personnel data (i.e., social security numbers, home phone numbers, contract awarding information). E-mail messages that are SBU, FOUO, LES, and so on must be plainly marked and cannot be transmitted over the Internet. Only Coast Guard networks are authorized for transmission and posting of FOUO and SBU information. If presented with a situation where you are unsure whether you are in compliance with Coast Guard policy, always ask your supervisor for guidance.

The Coast Guard requires all users of government information systems to sign an Authorization to Access Automated Information Systems form. Individuals using Coast Guard information and computer resources are subject to having all of their activities monitored and recorded by system administration personnel. Anyone using these systems is advised that if such monitoring reveals possible evidence of criminal activity, management may authorize system administration personnel to provide the evidence of such monitoring to law enforcement officials or a Coast Guard Security representative.

Social Media

While social media is great for communication, America's adversaries also use the Internet and social media sites to gather information about Coast Guard operations and personnel. With current technologies there is potential to unintentionally disclose sensitive and or protected information to millions of people or engage in a conversation without authority.

Personal information and imagery disclosed, or private comments posted, are easily searchable online and very difficult to remove from the Internet. Members need to be acutely aware of their role in ensuring the appropriate level of OPSEC is maintained and our services operational and informational strategies are protected from our enemies. When establishing social media accounts, personnel will ensure the following:

- Accounts shall not be established with government e-mail addresses.
- Accounts shall not use government logos or protected words or symbols.
- Accounts shall not be used to conduct official business.

Coast Guard members must be aware that they lose control over content they post on the Internet and that many social media sites have policies that give these sites ownership of all content and information posted or stored on those systems. Coast Guard members should always use their best judgment and anticipate how the content of their posts will reflect upon themselves, their unit, and the Coast Guard. Coast Guard military members violating social media policies may be held accountable for violations of the UCMJ.

DOs

DO make OPSEC common sense. Start by reading the Coast Guard External Affairs Manual, COMDTINST M5700.13.

DO be aware of your surroundings at all times.

DO keep any personal electronics in your control at all times.

DO use the acronym SAPP to determine if information is releasable: Security, Accuracy, Policy, and Propriety.

DO be wary of suspicious e-mails that use your name or appear to come from inside your organization.

DO create strong passwords.

DO carefully consider who you accept as Internet friends and, if possible, validate before acceptance.

DO be extra vigilant, especially in airports and other public places.

DO use caution when connecting to services in hotels, airports, and other public places. If you are directed to a log-on page before you can connect by VPN, the risk of loading malware or a data compromise is increased.

DO use good judgment when using social media.

DON'Ts

DON'T release details of schedules, daily routine, specific locations, and courses of action for ongoing or future operations and activities.

DON'T discuss deployment plans and procedures for Coast Guard units or other U.S. forces.

DON'T release names of personnel assigned to sensitive deployments or duty stations.

DON'T leave unattended personnel recall rosters containing any combination of personally identifiable information or unique identifiers of individuals.

DON'T give out family member information, home telephone numbers, and addresses.

DON'T publicize maps or specific locations of sensitive operational units.

DON'T discuss information on security systems or hazardous materials.

DON'T discuss specific movement of flag officers, senior leaders, VIPs, and key command personnel including itineraries, hotel reservations, events, agendas.

DON'T post digital pictures of Coast Guard assets involved with embedded GPS information (some electronic pictures, especially cell phone photos, actually contain embedded GPS coordinates).

DON'T discuss Coast Guard operational information on Facebook, Twitter, and so on, or by texting.

9

SAFETY AND RISK MANAGEMENT

We do dangerous work in a perilous environment. Our heritage is based in large part on the selfless acts of courageous men and women who use their tools and their judgment under the most demanding conditions to save the lives of others and uphold the rule of law. This tradition continues as we perform duties that often place us in harm's way. We honor our heritage daily by casting off all lines or lifting off the runway in severe weather to perform any federal action required at sea, while carefully weighing the risk of losing our own lives. Safety is an essential element of our duties, and you owe it to yourself and to your shipmates to make it a part of your everyday routine. The Safety and Environmental Health Manual, COMDTINST M5100.47, contains information about the programs developed to inform you of hazards, the procedures you will use to report hazardous conditions, and the measures available to protect you and others from harm.

Risk Management, COMDTINST M35300.3 (series), provides a standardized risk assessment process applicable to all communities of work to aid in recognizing and reducing risks to you and your crew. No manual can cover every hazard you could encounter at sea and ashore, or all of the possible ways to mitigate them, but we will cover some of the major hazards you will face and the programs and tactics employed to limit their potential for mishap.

YOUR RIGHTS AND RESPONSIBILITIES

Safety is a job for all hands at all times. Nearly every operation poses potential danger. It is the responsibility of everyone to observe all safety precautions, both for themselves and for their shipmates. The "Occupational Safety and Health Rights and Responsibilities" poster is required to be displayed in all workplaces. This poster spells out your rights and responsibilities and lists the safety representatives' names and phone numbers. You have the right to

- Know what hazards exist in your workplace;
- Access copies of Coast Guard standards, procedures, and injury and illness statistics;
- Comment on proposed Coast Guard occupational safety and health standards;
- Assist safety and health officials in conducting inspections of Coast Guard facilities;
- Request an inspection of a work area alleged to possess unsafe or unhealthful conditions; and
- Appeal if you disagree with the action taken on unsafe or unhealthful conditions. If you have exhausted all means available to achieve a satisfactory resolution of unsafe conditions and you remain dissatisfied, you may write to the Office of Federal Agency Safety and Health Programs, Occupational Safety and Health Administration.

You may exercise these rights without fear of reprisal, coercion, discrimination, restraint, or interference. If you think you are being treated unfairly because you exercised your rights, report it to the most immediate safety and health official.

As a member of the Coast Guard, you have the right to a safe workplace. Your participation in the program is essential in providing such an environment.

REPORTING HAZARDS

If you come across a hazardous situation or even perceive a situation is dangerous or hazardous, say something! You must report safety and health hazards when you find them. Never leave a hazardous condition for the next person to find the "hard way." There are two ways to report a hazardous condition. The first is a simple oral report to the person responsible for the area in which the hazard exists. Most hazards reported in this way can be corrected on the spot. The second is a written report submitted on an Employee Hazard Report, Form CG-4903. Fill out section 1 at a minimum. Instructions for completing the entire form are contained in

COMDTINST M5100.47. The information from this form will be entered into a database so other people can use it to correct similar hazards or document a widespread problem. Even if you make an oral report, a CG-4903 should be submitted for record purposes.

Hazard Abatement

Abatement is the action taken to reduce, control, or eliminate risk. All reported hazards must be abated. Following safety procedures and requiring the use of protective equipment, clothing, and devices are examples of abatement methods.

Hazard Communication

Each unit will have an appointed safety officer responsible for the local hazard communication program that serves as the cornerstone of compliance with OSHA standards for a safe workplace. This program lists required training and provides a hazardous materials inventory, specific criteria for container labels and warning, and safety data sheets (SDS). It will also identify administrative and engineering controls in the workplace established to prevent exposure to chemical hazards. Administrative controls refer to written safety policies, rules, schedules, and training intended to reduce the duration, frequency, and severity of exposure to hazardous chemicals. Engineering controls reduce the exposure to a chemical or physical hazard through the use of machinery or equipment such as ventilation, personal protective equipment, permissible exposure limits, and hygiene practices.

Hazard Types

The Hazard Communication Standard identifies the two main types of hazards as physical and health. Physical hazards are characterized by flammability, explosiveness, reactivity, and the presence of compressed gas. Health hazards include inflammatory irritants, allergens, corrosives, mutagens, and carcinogens. These hazards are found in chemicals that may be in the form of powders, dusts, liquids, gaseous vapors, or fumes. Routes these chemicals may take to enter your body and cause harm include absorption through the skin or eyes, inhalation into the lungs, or ingestion by swallowing.

PICTOGRAMS

Hazard pictograms form part of the international Globally Harmonized System of Classification and Labelling of Chemicals (GHS). Two sets of pictograms are included within the GHS: one for the labeling of containers and for workplace hazard warnings and a second for use during the transport of dangerous goods. Pictograms on labels alert users of the chemical hazards to which they may be exposed.

Hazardous Materials Inventory

All workplaces must maintain an inventory that lists chemicals classified as hazardous material. This inventory will list the chemical name, the product name, the location

Globally Harmonized System of Classification and Labeling of Chemicals

of the SDS, and other pertinent information. It is used to keep an accurate account of all hazardous materials during safety inspections and in the event of an emergency.

Authorized Chemicals List

All authorized chemicals used to perform maintenance activities on Coast Guard vessels are included in the authorized chemicals list, which follows published maintenance and procedure cards. If a chemical you are using is not on the authorized chemical list, stop using it and notify your supervisor.

Container Labels

Labels identify what chemical is inside a container and how to safely handle it. All labels on incoming containers of hazardous chemicals from manufacturers and importers must have pictogram(s), a signal word, hazard and precautionary statements, the product identifier, and supplier identification. Secondary or in-house containers, such as those used when a substance is transferred from a bulk shipment, like a fifty-five-gallon drum, to a more usable size container require only the product identifier and a word, picture, or symbol that provides at least general information regarding the chemical hazards. A Department of Defense hazardous chemical warning label is used to identify hazardous materials if, for some reason, the original manufacturer's label cannot be applied or if the original label is lost, damaged, illegible, or destroyed.

If you are unsure of what is in or what was once in a container, do not use it. Mixing chemicals could lead to a fire or explosion, endangering the lives of you and your shipmates. Even a small amount of residue from one chemical could create a dangerous situation when mixed with another.

Safety Data Sheets

Although chemical containers are required to be labeled, specific hazard information may not be present. An SDS is provided with each shipment of a product to the unit. Upon the product's arrival, the safety officer will review the SDS and file a copy in the right-to-know file station.

This file station provides for twenty-four-hour access by employees. The SDS for each product is filed by the chemical name and preparation date to ensure that the most accurate information is available. To prevent improper use or handling of a chemical, you must consult the SDS, which provides a wealth of information.

Hazardous Materials Stowage

Upon discovering unidentifiable or improperly stowed hazardous or combustible material, immediately notify your chain of command, who will take required actions to store the hazardous materials in the appropriate storage area.

Products stored in the wrong area may interact with other chemicals and cause fires, release highly toxic fumes such as hydrogen chloride gas, or corrode containers of other chemicals. Unexpected chemical reactions may cause the evacuation of mission-essential functions, control centers, and watch stations. As fires and fumes spread, they may also cause equipment malfunctions.

RESPONDING TO SPILLS, UNUSUAL ODORS, OR VAPORS

The largest contributing factors to hazardous materials spills are inattentiveness to the task at hand, hurrying to complete required tasks, and fatigue. When responding to and cleaning up a hazardous spill, systems and operations in the affected areas may have to be stopped until the cleanup is finished. Crew members dressed in the proper protective equipment may operate systems on an emergency basis, as required for

Hazardous materials response

mission critical operations, but may be slowed down due to the additional burden of the equipment. If immediate actions to contain the situation are not taken, hazardous fumes or vapors may spread, causing other areas or watch stations to evacuate and operations in those areas to be shut down. Harmful vapors, if they are flammable, may cause flash fires, further impacting operations.

The first signs of a hazardous materials release or spill may be unusual odors, incapacitated crew members, visual identifiers, or an alarm. In each case, personnel should immediately evacuate the area, isolate the space, and don appropriate respiratory protection. If you observe or are notified of a hazardous material spill or detect an unusual odor or vapor in your workspace, take the following actions:

Step	Procedure
1	Stop work.
2	Don an emergency escape breathing device or other emergency breathing device, if possible, for unusual odor/vapors or alarm.
3	Egress the space (all personnel).
4	Report the casualty to the officer of the deck (OOD), engineering officer of the watch, or supervisor, detailing the location, type of spill/odor/vapor.
5	Isolate the space, as able/directed.
6	Stand by for further directions from supervisory personnel. Consult the involved material's SDS for more information.

Attempting to rescue personnel incapacitated by a hazardous material spill or unusual odors/vapors may result in further casualties if proper personal protective equipment is not donned. Do NOT attempt to render first aid until properly outfitted.

PERSONAL PROTECTIVE EQUIPMENT

Personal protective equipment (PPE) does not eliminate hazards; it only reduces the risks associated with them. The job will dictate which clothing or equipment you need. You must be trained to wear or use any PPE, and several types have substantial supporting programs, notably eye, hearing, and respiratory, which cross all communities of work.

Eye Protection

There are many processes and work situations that can cause potential hazards to the eyes. To protect against those hazards the Coast Guard has implemented a sight conservation program. For those working with potential hazards, it is imperative that they wear eye protection, be aware of the location of nearest eyewash station, and understand simple first-aid measures. Frequent minor mishaps are corneal abrasions and particles in the eye. Many nonindustrial jobs, such as dusting overhead, can also be hazardous and may require eye protection. Protective goggles or face shields must be worn when cutting, drilling, chipping, sanding, or working with hazardous chemicals. Eyewash stations are in areas where there is a frequent risk of exposure to eye damage and provide clean, potable water to spray water directly into and flush the eyes.

Hearing Protection

Many jobs we do every day present high-noise-level hazards. To protect personnel from these, the Coast Guard has a hearing conservation program that assigns you certain responsibilities, including

- Advising the immediate supervisor of any unhealthy or unsafe working conditions or practices;
- Wearing appropriate hearing protection devices;
- Ensuring that good hearing conservation practices are continued on and off duty; and
- Completing annual and any required follow-up audiometric testing if enrolled in the hearing conservation program.

The purpose of the hearing conservation program is to prevent occupational hearing loss. Environments or equipment that produce continuous noise pressure above 84 dBA or above 104 dBA peak noise pressure are considered hazardous, and protective measures must be taken to reduce exposure to personnel. Inserts or earmuff-type hearing protective devices shall be worn by all personnel when they enter or work in an area where the operations generate noise levels equal to or greater than 84 dBA. A combination of insert and earmuff-type protection must be worn in all areas where noise levels exceed 104 dBA. If the noise cannot be reduced, then restrictions must be placed on the amount of time personnel can spend in the areas. Placards with the noise levels and the amount of protection required are posted at all entrances to a space with high sound levels. Ear protection is required when working in, or making rounds in, an enclosed engineering space.

Respiratory Protection

Respiratory protection is required to prevent death or illness caused by harmful dusts, fumes, smoke, aerosols, mists, gases, vapors, or atmosphere without enough breathable oxygen. The Coast Guard strives to prevent contamination by confining the hazard, providing adequate ventilation, or substituting a less toxic material. When respirators are required, you are responsible for using them properly and reporting any malfunctions such as leaks and problems with fit and air flow. Your unit safety officer and respiratory protection coordinator will assist your unit in selecting the right respirator for any given situation. There are two basic types of respiratory protection:

- Air-purifying respirators (APR) can remove contaminants in the air that you breathe by filtering out particulates like dusts, metal fumes, mists, etc. Other APRs purify air by adsorbing gases or vapors on an adsorbing material in a cartridge or canister.
- Supplied-air respirators (SAR) supply clean air from a compressed air tank or through an air line. This air is not from the work room area. Respirators with tight-fitting respiratory inlets have half or full facepieces. Types with loose-fitting respiratory inlets can be hoods or helmets that cover the head and neck, or loose-fitting facepieces with rubber or fabric side shields. These are supplied with air through air lines. Examples of these classes of respirators include self-contained breathing apparatus (SCBA).

Respirator Wear and Maintenance

To be effective, the respirator must first be fitted to you by a certified technician, normally your unit's respiratory protection program coordinator. Prior to commencing work that requires the wearing of a respirator

Step	Procedure
1	Inspect the respirator for signs of damage and cleanliness.
2	Check to ensure that the cartridges are the proper ones for the job you will be doing and that both cartridges are identical.
3	Place the respirator over your mouth and nose.
4	Adjust the head straps so the respirator fits very snugly. Ensure you have a good seal around the outer edge of the respirator by feeling to see if air is getting around it when you breathe. If air is getting around the seal, then tighten the head straps until you get a good seal.

Clean and inspect the respirator before and after each use. The face seal, mouth, and nose areas should be wiped with an alcohol wipe or soapy water and rinsed. Take the cartridges off the respirator before you clean it. The cartridges will be replaced as required by the instructions for that cartridge. After cleaning, store the respirator in a clear plastic, resealable bag. The respirator itself can be used over and over again if it is kept in a good, clean condition. Only the cartridges need to be replaced periodically.

Immediately Dangerous to Life or Health (IDLH)

Some types of atmospheres contain concentrations of hazardous substances that places the worker in immediate danger because these concentrations would impair the ability to leave the work area and self-rescue or would potentially cause irreversible health effects, including serious injury or death in a matter of minutes. To determine IDLH level of a contaminant, consult the NIOSH Pocket Guide to Chemical Hazards.

BLOODBORNE PATHOGENS

Coast Guard helicopters, small boats, and cutters are routinely dispatched to transport persons with serious injuries or life-threatening illnesses. As a result, the interior decks and rescue or medical equipment may become contaminated with bloodborne pathogens and other biohazardous material. This material may consist of bodily fluids, bandages, needles, scalpels, ampoules, and equipment used to aid in respiration. Each unit is required to have written plans for the protection of personnel from bloodborne pathogens and for the decontamination of resources.

Universal precautions are used by all members whenever the potential for exposure to bloodborne pathogens exists to minimize risk of exposure to all blood and other body fluids, all of which are considered infectious materials. A universal precaution is an infection control approach developed by the Centers for Disease Control and Prevention that assumes every direct contact with body fluids is potentially infectious. Protective measures to eliminate or minimize exposure to infectious materials include

- Using barrier protection at all times (i.e., a CPR mask);
- Using gloves for protection when working with or around blood and body fluids and changing gloves between patients;
- Using glasses, goggles, masks, shields, and waterproof gowns/aprons to protect face from splashes;
- Washing hands if contaminated and after removing gloves;
- Decontaminating all surfaces and devices after use;
- Using the approved disinfectant (e.g., 1:10 household bleach);
- Absorbing, containing, and decontaminating spills;
- Properly disposing of contaminated materials and placing infectious waste in labeled "red bag"/designated containers; and
- Promptly seeking medical attention and counseling if exposed to contaminated materials.

LOCKOUT/TAGOUT

Serious injuries and deaths occur during the repair or servicing of equipment when accidentally energized. The Coast Guard's lockout/tags-plus system is designed to prevent these types of mishaps by providing practical and useful safety guidelines to achieve a safer work environment. You must not remove these tags. They can be removed only by the person who placed them. If your job requires you to operate a tagged-out switch, valve, or control, report it to your supervisor and do not operate it.

HEAT STRESS

It is likely that you will work in hot climates throughout your career in the Coast Guard, which will raise your risk of sunburn, heat cramps, heat exhaustion, and heat stroke. Adaptation to heat takes approximately thirty days. Look out for your shipmates. If you see signs of a heat-related illness or injury, intervene. Consider the following precautions when working in hot environments:

- Maintain a good fitness level.
- Don't take salt tablets to restore the sodium/water balance unless prescribed by a physician.
- Clothing should be loose, absorbent, and light in color.
- Drink as much water as you can tolerate comfortably before and after exercise. Take sips of water during exercise. Avoid cold beverages.
- If you are thirsty, then you are in the early stages of dehydration.
- Avoid alcohol; it will actually cause you to lose more fluid and possibly cause dehydration.
- Do not use an oil-based sunscreen or suntan oil because the oil could clog your pores and hinder sweating.

COLD STRESS

During cold weather, always practice prevention of and be aware of potential hazards involving frostbite and hypothermia. The following safety precautions must be adhered to when working or training in cold weather:

- Always be conscious of the wind chill factor.
- Eat adequately; metabolism increases during cold weather.
- Dress properly: Maintain body temperature without excessive sweating. Dressing in multiple layers of clothing will help trap air and insulate the body. Use fabrics that draw moisture away from the skin as the first layer. A wind-resistant fabric or wool may be used as an outer layer. Since 30 percent

of the body's heat is lost from the surface area of the head, a warm hat or face mask is advised. A wool blend fabric is recommended.
- Drink plenty of fluids during cold-weather activities. You can become dehydrated during cold weather just as easy as during hot weather.
- Avoid alcohol and smoking, which increases the danger of frostbite. Alcohol dehydrates the body, which will reduce the body's efficiency. Smoking restricts blood vessels, which in turn reduces the blood's ability to distribute heat.
- Avoid exposure to water. Moisture intensifies the cold.

LIFELINES, SAFETY NETS, AND LIFE JACKETS

Being washed overboard is a very real danger. The best-case scenario is your immediate recovery by a cutter, boat, or helicopter. But, if not done quickly, you can experience hypothermia or drowning. There are some safety measures in place, but when seas get rough or when jet blast is a hazard, being washed overboard without a life jacket greatly reduces your chances of ever seeing your ship, home, or family again.

The lifelines (railings) on a cutter are there to prevent people from falling overboard. You should never lean, sit, stand, or climb on any lifeline whether the ship is under way or in port. Lifelines should not be dismantled or removed without specific permission from the first lieutenant, and then only if temporary lifelines are promptly rigged.

Anytime you go topside or on deck in heavy weather, you must wear a life jacket. Life jackets are also worn at times when the nature of your work increases the chances of your falling or being knocked overboard. Life jackets are also worn when working over the side in port and at sea, whether you are suspended over the side on a stage or in a boatswain's chair. You are required to wear a life jacket when traveling in a small boat regardless of weather conditions.

LINE HANDLING

You may find yourself filling the role of a line handler. Lines are used for many purposes on board cutters and boats, and although it sounds straightforward, experienced sailors know that lines can be very dangerous. Always treat lines with respect. Several hazards exist with line handling.

The first is called a bight. Any line, rope, or wire can be coiled, intentionally or by chance. A loop, coil, or a section of loose line can injure or kill if the line suddenly tightens. The sudden tension of a line results not only in the tightening of any bights but also the rapid movement of the line toward the line of tension. This may result in any rope positioned off this line moving rapidly in that direction, lifting anyone or anything in its path. Do not, under any circumstances, stand in the bight (loop) of a line. Know where your feet are while you work and regularly check that your feet are outside any bights.

Second, lines can part under strain and instantaneously sever limbs or kill people. If that line suddenly goes taut, it can seriously injure you or drag you into danger.

Never step on a line that is taut. Do not stand in its direct line of pull when heavy loads are applied—these are called snapback zones, and you want to avoid them. Develop the mindset of treating a line as you would treat a loaded gun and approach any line with the idea that it can be very dangerous if it parts.

Third, pinch points can develop where lines meet other objects. The force of a line against a stationary object can catch, crush, or sever body parts. Loose gloves and clothing are also at risk of being caught in pinch points and getting you into a dangerous situation.

TANKS AND VOIDS

No one is permitted to enter any closed compartment, tank, void, or poorly ventilated space on board a Navy ship until the space has been ventilated and determined to be gas-free by a qualified gas-free engineer. In an emergency, if a space must be entered without gas freeing, a breathing apparatus such as a SCBA must be worn. In all cases, at least two persons must be present when such a space is occupied. One remains outside the space and acts as line tender and safety observer. The space entered should be continuously ventilated, and a reliable person must be stationed at the entrance to keep count of the number of persons inside as well as to maintain communications. Suitable fire-extinguishing equipment must be on hand, non-sparking tools are to be used, and persons entering should not carry matches or lighters or wear articles of clothing that could cause a spark.

Confined Spaces

Tanks, voids, and unventilated spaces are classified as confined spaces. Confined spaces are any areas that are large enough to enter and work but not designed for continuous human occupancy and fitted with only limited or restricted means for entry and exit. All spaces must be considered unsafe unless tested by a gas-free engineer. Within a confined space you may encounter serious hazards, such as the

- Lack of sufficient oxygen;
- Excessive oxygen (risk of fire or explosion);
- Presence of flammable or explosive material; or
- Presence of toxic gases or materials.

GOING ALOFT

Going aloft is defined as going onto upper areas of the superstructure or onto the ship's masts where there are radio, radar, or electronic antennas mounted. This can be very dangerous for a number of reasons. Electrical shock is a hazard, but you can also be seriously injured by electronic emissions, be struck by a rotating antenna, or be injured in a fall. You should never go aloft except with an approved "man-aloft chit" and when working should consider the following:

- Never go aloft without the OOD's permission.
- Ensure that electrical and electronic equipment is tagged and turned off.
- Always use an industrial safety harness with safety and working line.
- Always attach tools to the harness with preventer lines to keep them from being dropped.
- If working in the path of stack gases, always use a SCBA.
- Always have a ready-to-climb safety observer.
- Lower tools with a hand line or suitable tackle.
- Radio and radar equipment on other ships create a charge you may feel when climbing. The shock won't hurt, but it may startle you and cause you to fall.
- Do not touch antennas, aloft or otherwise. You could be burned by the energy from your own ship or from another vessel.
- Watch for rotating radar antennas. They could knock you to the deck.
- Report any unsafe rigging aloft to the OOD.

GOVERNMENT VEHICLE SAFETY

Safety must always be your first priority when operating any type of vehicle. All government vehicle operators must possess and maintain a U.S. government motor vehicle operator's identification card, Form OF346. Prior to towing any trailer or boat, you must complete unit-level job qualification requirements. The following are several important safety considerations you MUST follow before, during, and after operating a Coast Guard owned vehicle.

- Review unit driver training requirements and instructions. Request training for any vehicle you are required to operate.
- Visually inspect the vehicle for any obvious safety concerns, for example, tires, mirrors, or leaks. Report vehicles that are in an unsafe condition immediately.
- Find out how to fill the fuel tank, the type of fuel required, and how and where to make a purchase.

- Seat belts must be worn at all times by the operator and all passengers.
- Text messaging or talking on a cell phone while operating a vehicle is strictly prohibited.
- Smoking is prohibited in all government vehicles at all times.
- Upon return of the vehicle, top off the fuel and clean out any trash or refuse items to be discarded.
- Report any problems or safety concerns directly to the vehicle manager or local motor pool manager.
- When trailering, always assign spotters and use them whenever backing a trailer.
- Comply with all applicable municipal, state, and federal motor vehicle regulations.
- Ensure that vehicles and their contents are properly secured when left unattended.
- Promptly report accidents, mishaps, and any traffic violations detected by law enforcement personnel or electronic traffic surveillance equipment. Complete accident and mishap reports as required by the Motor Vehicle Manual, COMDTINST M11240.9 (series).

Backing a boat trailer

SPORTS AND RECREATIONAL

Sports and recreation come with risks. Basketball has the highest percentage of disabling injuries among team sports. Softball and football are the next-largest producers of longtime injuries. The most commonly reported injuries are to the knee, lower leg, and ankle. Fractures occur most often in football and softball, while sprains and strains occur more frequently in basketball. Pickup games result in more injuries than organized, officiated games. Sports injuries are largely attributed to four basic factors: poor conditioning, inadequate ability and skill, lack of protective clothing and equipment, and violation of recognized rules.

Recreational mishaps indicate that most injured personnel do not understand or appreciate the magnitude of certain hazards. Alcohol is also a common factor. Jogging, bicycling, and physical fitness–oriented activities produce the greatest number of reportable lost-time injuries. The type of injuries cited most often are fractures, sprains, and strains. Gun handling has a high potential for fatal injury. For example, hunters or sport shooters cleaning guns are common scenarios that lead to death. Enjoy your off-duty time, but make sure you understand the risks and hazards with the hobbies and sports you do for enjoyment.

BRIEFINGS

Briefings are required before launching an operation or beginning an evolution. Briefings help create a shared mental picture of what is expected to happen and strives to set rules for the mission. Include the mission objective, known information and risks regarding the mission, and the planned course of action. Be specific in assigning duties and responsibilities. Expectations should be understood by all. Do not make the crew second-guess what needs to be done or, in special situations, how it should be done. Establish a positive climate for teamwork by encouraging the crew to double-check each other, point out errors, speak up when they have relevant information, and ask questions when they do not understand.

RISK MANAGEMENT

Risk management (RM) is a process that assists individuals and crews in reducing or offsetting risk and making decisions that weigh risks against mission or activity benefits. RM is fully integrated into operational planning, preparation, and execution at all levels of the Coast Guard. There are two primary and interrelated levels of RM that dictate the level of effort and scope that should be undertaken when assessing risk, separated only at the point where the planning phase transitions to the execution phase.

Deliberate RM. Deliberate RM is a pre-mission/activity planning that normally involves the full application of the complete five-step RM process outlined below. In-depth RM planning is normally implemented well in advance of the mission or activity. Throughout the deliberate RM process, the experience, expertise, and knowledge of experienced personnel is included to identify known hazards/risks as well as strategies to effectively mitigate those hazards/risks.

Real-Time RM. Real-time RM is making decisions in "real time" during the "execution" or tactical phase of a mission or activity. This level of RM can be used to supplement previously conducted deliberate RM. Real-time RM is usually done "on the fly" due to short-notice taskings, weather-driven activities, emergency responses, or spontaneous off-duty activities using basic RM process steps to identify and mitigate hazards in a new or changing situation. As time is normally constrained or limited in these situations, deliberate RM is impractical.

Four Principles of Risk Management

Integrate RM into All Phases of Missions and Operations. Risk assessments are most successful when they are part of the normal preplanning of a mission or activity and not a last-minute add-on. Any amount of preplanning is better than no planning at all.

Accept No Unnecessary Risk. Take only those risks that are necessary to accomplish the mission or task. The best choices for accomplishing a mission are those that meet all mission requirements while exposing personnel and resources to the lowest risk.

Accept Necessary Risk When Benefits Outweigh Costs. Compare all identified benefits to all identified costs. Avoiding unnecessary risk is not the same thing as avoiding all risks, and even a risky activity can be worthwhile if the reward is great enough. Balancing risks and rewards may be a subjective process, and ultimately the appropriate decision authority may have to determine the balance.

Make Risk Decisions at the Appropriate Level. The appropriate level for risk decisions is the person responsible for minimizing or accepting the risk. Members must be aware of how much risk they can accept before raising RM decisions to a higher level. Those accountable for the success or failure of the mission or activity must be fully engaged in the risk decision process.

Five-Step Process

The RM process consists of five steps to make informed decisions regarding risk. These steps define the formal RM process used in deliberate RM and form the basis for real-time RM considerations.

Step 1—Identify the Hazards

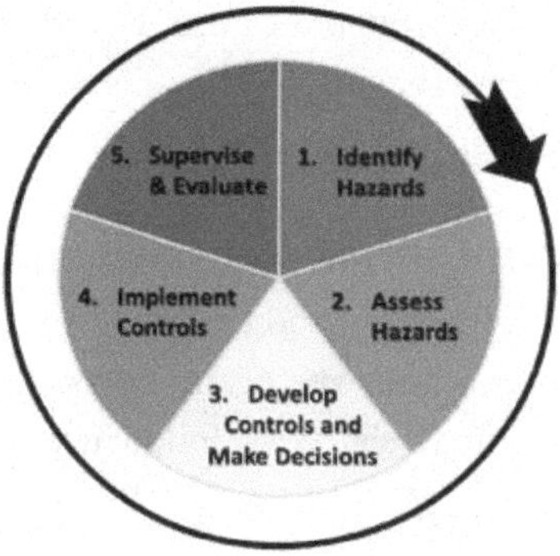

Hazards are defined as any real or potential condition that could cause mission degradation; injury, illness, or death to personnel; or damage to equipment or property. Key aspects of this step include the following:

- Mission/Task Analysis: Review all operations and tasks associated with the mission or activity.
- List Potential Hazards: Identify and list hazards and factors that could generate hazards.
- List Causes of the Potential Hazards: Identify the sources of hazards and try to identify any root cause(s).

Step 2—Assess the Hazards

After identifying the hazards, determine the probability and severity of consequences that may result. Key aspects of this step include the following:

- Assess Hazard Exposure: Evaluate the duration, proximity, volume, or frequency of exposure to a hazard.
- Assess Hazard Severity: Determine the severity of the impact on personnel, equipment, or mission/activity should a hazard result in a mishap.
- Assess Probability: Determine the probability that a mishap will occur.

Step 3—Develop Controls and Make Decisions

Next evaluate strategies and controls to reduce or eliminate some or all of the three components (probability, severity, or exposure) of risk. Key aspects of this step include the following:

- Identify Control Options: Identify as many options as possible to reduce risks to an acceptable level.
- Determine Control Effects: Reassess each hazard by repeating Step 2 while taking into consideration the effects the control options will have.
- Prioritize Risk Controls: For each hazard, rank the available controls with respect to how well they reduce the risk to an acceptable level.
- Select Risk Controls: For each hazard, select the risk controls that best reduce the risk to an acceptable level while remaining consistent with mission/activity objectives and optimum use of available resources. If the risk level

of a hazard cannot be reduced to acceptable levels, inform the next level in the chain of command that the mission's risk exceeds its benefits, and it should be modified.

Step 4—Implement Controls

Next develop and carry out an implementation strategy that identifies the who, what, when, where, and cost(s) associated with the control measure. Key aspects of this step include the following:

- Make Implementation Clear: Provide a roadmap for implementation and a vision of the end state, and describe successful implementation.
- Establish Roles and Responsibilities: While every member has the responsibility to contribute to the RM process and no single person is responsible for the success of control measures, someone must be accountable for approving control measures and ensuring their implementation.
- Provide Support: For control measures to be successful, command/leadership must provide the personnel and resources necessary to implement them.

Step 5—Supervise and Evaluate

Once controls are in place, the process must be periodically evaluated and reviewed to ensure controls remain effective and mission-supportive over time. Key aspects of this step include the following:

- Monitor: Monitor the operation/activity to ensure the controls remain in place and that risk and controls are reevaluated when the personnel, equipment, environment, or mission/activity changes.
- Evaluate: Determine if the implementation of controls was effective and performed according to RM principles, and identify ways to improve the controls in the future.
- Feedback: Feedback is required to ensure that any potential hazards or controls identified during the mission/activity are analyzed and appropriate actions are taken.

RM Models

When used in tandem, PEACE and STAAR generate the necessary information to establish a shared mental model up and down the chain of command, assist in decision-making, and encourage robust risk management.

PEACE

PEACE is a memory aid for elements considered essential in the safe and effective execution of any operation, activity, or mission. Deficits or inefficiencies in these elements increases mission risk exposure. The PEACE model is an effective tool to conduct risk assessments and ensures that all critical mission elements are considered in the analysis. Each PEACE element is described below. For any mission, review tasks and activities to determine whether or how each of the PEACE elements increase risk exposure. A risk assessment sheet using the PEACE elements is provided as appendix G of this manual.

P Planning: Mission plans often have a shelf life and can be out of date for any number of reasons shortly after getting under way. Always anticipate possible deviations, especially if you suspect your information is incomplete. Questions about planning prompt you to consider problems that may come up as to the quality of the mission context-related information. Consider what could go wrong with equipment, personnel, the environment, or mission if the team has incorrect or insufficient information; if the team has not clarified methods of performing key tasks; or if roles are left unclear or unassigned.

E Event Complexity: Event complexity depends on the amount of data, number of participants, and number of steps that must be performed with little margin for error. Each unit defines their own level of comfort with an evolution, typically based on their capabilities and recent experience. A breakdown in ability to process data or execute a series of activities correctly at the right time can lead to mishaps. Consider what could go wrong with equipment, personnel, or the environment if coordination with other agencies, assets, or units breaks down; the crew performs a series of finely tuned activities incorrectly; or the crew is unable to continually monitor multiple dynamic data streams.

A Assets: Specific assets may be associated with specific hazards for a given evolution. Assets include equipment, event platform performance tolerance thresholds, and personnel details such as experience and confidence. Considering assets provides individuals the opportunity to apply their own subjective assessment. Are they stressed? Are they alert, confident? Consider what could go wrong with equipment or personnel if a platform is used in its current condition for the evolution; if the operational experience, fitness, and confidence of the crew is inadequate; or if the fitness level (e.g., rest, hydration, nutrition) of the crew is not satisfactory.

C Communication and Supervision: Poor communication and supervision can impair the crew's ability to maintain situational awareness and receive feedback about decisions, including making risk decisions at the appropriate level. Consider what could go wrong with the equipment, personnel, environment, or mission if the crew cannot communicate with the command center or if there are communications problems among the crew.

E Environment: How will weather, geographic influences, physical barriers, workplace climate, and available light affect the event? Consider what could go wrong with equipment, personnel, the environment, or mission given the weather, illumination, congested waters, or airspace conflicts.

STAAR

The STAAR model outlines potential strategies to mitigate or control risk.

S Spread Out: Move forces, equipment, or tasks to other areas to avoid risk to the entire mission. For example, placing assets in a single area can lead to catastrophic losses if an explosion or fire breaks out. Spreading your resources can mitigate this potential risk by reducing the exposure of these resources in a single, combined area.

T Transfer: Transfer some or all of the mission or task to another individual, unit, or platform that is better positioned, more survivable, or more expendable. Transfer does not decrease the probability or severity of the risk to the unit but reduces risk to the total force.

A Avoid: Avoid specific risks by "going around" them or doing the mission or task in a different way. For example, risks associated with a night mission or task may be avoided by planning for daytime. This might present other hazards that would need to be identified and assessed.

A Accept: Accept risk when the benefits clearly outweigh the costs, but only as much as necessary to accomplish the mission or task. For example, position PPE or administrative people on deck for shorter periods of time.

R Reduce: Reduce the number of individuals, equipment, or resources exposure to a particular risk is a simple way of mitigating overall risk. Although this strategy may reduce risk, it must be weighed carefully against potential rewards. Reduction can sometimes have the negative consequence of not having back-up options available when you need them.

Warranted Risk

Coast Guard operations are inherently complex, dynamic, and dangerous; by nature, they involve the acceptance of some level of risk. However, the decision to accept risk must balance the potential gains of conducting the activity or mission with the inherent risks of the operation. Coast Guard leaders must constantly carry out this evaluation. With any activity or mission, persons in charge must operate their assigned asset to minimize the inherent risk involved to the greatest degree possible.

Boat Operations

- Damage to or sacrifice of the boat is an acceptable risk in the defense of the United States, its citizens, or its installations.
- The probability of saving human life warrants a maximum effort. When no suitable alternative exists and the mission has a reasonable chance of success, the risk of damage to or abuse of the boat is acceptable, even though such damage or abuse may render the boat unrecoverable.
- The possibility of saving human life or the probability of preventing or relieving intense pain or suffering warrants the risk of damage to or abuse of the boat if recovering the boat can reasonably be expected.
- The probability of saving property of the United States or its citizens warrants the risk of damage to the boat if the value of the property to be saved is unquestionably greater than the cost of boat damage and the boat is fully expected to be recoverable.
- The possibility of recovering evidence and interdicting or apprehending alleged violators of federal law does not warrant probable damage to or abuse of the boat, unless in defense of self or others.

Aviation Operations

Warranted Risk for National Defense. Risk of damage to or sacrifice of the aircraft and crew is acceptable if the gain is the defense of the United States, its citizens, or its installations. Accepting this risk for national defense applies to flight activities performed during certain operational defense-readiness missions and certain operational ports, waterways, and coastal security missions. In all cases, the appropriate rules of engagement or use-of-force policy shall take precedence.

Saving Human Life. If a mission is likely to save human life, it warrants a maximum effort. When no suitable alternatives exist and the mission has a reasonable chance of success, the risk of damage to or abuse of the aircraft is acceptable, even though such damage or abuse may render the aircraft unrecoverable. Probable loss of the aircrew is not an acceptable risk.

Preventing or Relieving Pain or Suffering. If a mission is likely to prevent or relieve intense pain or suffering, or if it may result in the possibility of saving human life, it warrants the risk of damage to or abuse of the aircraft if recovering the aircraft can be reasonably expected.

Saving Property. If a mission is likely to save property of the United States or its citizens, it warrants the risk of damage to the aircraft if the value of the property to be saved is unquestionably greater than the cost of aircraft damage and the aircraft is fully expected to be recoverable.

Warranted Risk for Missions Involving Law Enforcement and Evidence Recovery. The possibility of recovering evidence and interdicting or apprehending alleged violators of federal law does not warrant probable damage to or abuse of the aircraft. Airborne use-of-force missions are not considered routine in nature and may warrant the risk of damage to or abuse of the aircraft following the issuance of a statement of no objection. The aircraft is fully expected to be recoverable, and probable loss of the aircrew is not an acceptable risk.

Warranted Risk for Logistics and Other Missions. Logistics or other missions having little or no urgency shall not be prosecuted if they expose the aircraft to hazards greater than those encountered during the course of routine training missions.

Debriefings

Crew debriefings should be performed after most missions. The debriefing is the best opportunity to evaluate performance and recognize individual and team accomplishments. When correctly performed, the debriefing can serve as a valuable tool for continuous improvement. It can show the way from just "doing things right" to knowing how to do "right things right." Recap major events of the mission (e.g., preparations, transit, on-scene operations). Determine level of performance within key events. Key events include the initial brief, critical navigation segments of the transit, bar crossings, crossings, approaches to vessels, personnel transfers, and hazardous parts of the assigned mission.

The crew and, when possible, the command discuss what human behavior or risk decisions affected the outcome in these events. This discussion is for professional growth and learning. Determine if the goal to improve one or two weak coordination areas has been met. Set, change, or affirm a specific goal for improving one or two weak areas in crew coordination. Goals are set or changed with the knowledge and guidance of the command.

10

MAINTENANCE

Every Coast Guard asset, system, and piece of equipment, no matter how simple, must be inspected, cleaned, monitored, and serviced to be ready for use in operations and emergencies. You may think tasks like cleaning and painting are not important, but they are. Poorly maintained equipment eventually fails, and that failure could come at a time when your life, the life of a shipmate, or someone you're called to rescue depends upon it.

MAINTENANCE ELEMENTS

Managing maintenance of boats, cutters, aircraft, lighthouses, and shore facilities involves all of the elements discussed below.

Preventive Maintenance

Preventive maintenance consists of routine and scheduled inspections, cleaning, testing, monitoring, lubricating, servicing, and preservation to maintain standards and to prevent failures that might diminish operational capability, safety, or expected service life. Such tasks are normally scheduled weekly, monthly, quarterly, semiannually, annually, or post-use. Common preventive maintenance items include routine painting of vessels and fixed structures and post-use inspections of rescue equipment.

Corrective Maintenance

Corrective maintenance is also called "unscheduled maintenance." It consists of actions to troubleshoot, repair, and restore equipment, systems, software, hull, and structures in response to failures or casualties. The amount and severity of corrective maintenance required may be minimized considerably by preventive maintenance.

Levels of Maintenance

Most formal maintenance programs in the Coast Guard are bilevel and consist of the following.

Organizational. Organizational-level maintenance is normally performed by an operating unit on a day-to-day basis in support of its own operations. The organizational-level maintenance mission is to maintain assigned assets, equipment, and facilities in a full mission-capable status while continually improving the process. Operating units are responsible for completing all organizational-level planned and corrective maintenance. Organizational maintenance encompasses a number of categories, such as inspections, servicing, handling, preventive maintenance, and corrective maintenance. Maintenance tasks falling under this category include all planned maintenance except that requiring tools or other resources not held within the asset or requiring technical skills of personnel beyond those available in the asset's operating crew.

Depot. Depot-level maintenance involves processes beyond the organizational-level capability such as those needed to repair, overhaul, upgrade, rebuild, manufacture, modify, or analyze. Many depot-level tasks require industrial or commercial facilities and personnel of high technical skills. Typical tasks falling under this category involve major hull repairs, replacement of major components such as main engines and marine gear, or new construction.

Maintenance and Procedure Cards

Many maintenance tasks are supported by maintenance and procedure cards (MPC) that identify each step of the maintenance process, document how to physically perform the steps, and identify the resources needed to complete the task, including the number and specialty of people assigned, the time, and any tools or parts required.

Tool-Control Program

All units have a tool-control program to provide an instant tool inventory using containers with each tool positioned in an individually tailored location. The primary objective of the program is to eliminate accidents and equipment damage attributed to uncontrolled tools and minimize tool replacement costs. Unit instructions and procedures will define the specific program and preclude the completion of any maintenance action without accounting for all tools. Tool control is the responsibility of all maintenance personnel and all levels of the chain of command.

Asset Logistics and Maintenance Information System (ALMIS)

ALMIS is an asset logistics system that provides maintenance tracking, parts ordering/inventory, and mission information for aviation and surface assets. Its electronic asset logbook allows total asset visibility to support mission planning and support through the use of maintenance records and mission records.

Operational units report the status of assets in the electronic asset maintenance records with the assigned code identifying the asset's capability to meet all assigned missions from an engineering standpoint. Statuses include the following:

- Fully mission capable
- Partially mission capable
- Not mission capable. Reasons for "not mission capable" may be maintenance, supply, depot, or lay-up.

Discrepancies

Asset conditions that result in a rating of other than fully mission capable are called discrepancies and must be documented in a maintenance record that describes the nature of the discrepancy, an evaluation of whether the asset is safe to use, and the work or services being performed to correct it.

Maintenance Due List

Scheduled maintenance is listed on the maintenance due list (MDL), which provides a work list for on-duty personnel. Each item on the list will have an associated MPC explaining how to perform the necessary maintenance. If assigned a task from the MDL, follow the MPC and report its completion to your supervisor for entry into ALMIS. Remember, if it's not in ALMIS, it never happened.

CLEANING

Maintaining clean conditions on board ship and ashore has always been an important part of Coast Guard activity. Cleaning is a job that involves practically every member, from the compartment cleaner to the inspecting officer. The close surroundings of Coast Guard life require each of us to have a personal interest in our living and working areas, not only for the sake of appearance but for our health and safety as well.

The Cleaning Bill

Each area of a ship or station is assigned to one of its various departments for upkeep. The cleaning bill describes these areas and outlines which departments are responsible for each space. Departments detail their personnel to spaces and assign specific tasking to individuals to ensure all items on associated cleanup lists are completed. The cleaning bill is carefully planned to ensure that all interior and exterior areas of a ship or station are assigned and that no areas overlap or are left out.

Compartment Cleaning

If you are assigned to clean the living or berthing compartments or spaces such as passageways and heads, you will be responsible for keeping your spaces exceptionally clean and in good order. Newly assigned personnel are closely supervised to make sure they understand what to clean and how to clean it. Items that may be unfamiliar to you, such as electrical and mechanical devices, are in almost every space on board ship. With this in mind, caution must be observed at all times. Ask your supervisor to point out hazardous items in your compartment, and observe all special cleaning instructions.

To reduce waste and provide for bulk storage, cleaning gear is stocked in and issued from a supply locker. Each division is periodically issued cleaning gear and is responsible for its proper storage and care. Because cleaning compounds and solvents are usually either flammable, toxic, or both, they must never be left unattended or stored improperly. Always read warning labels, follow their directions carefully, and if you have questions refer to the safety data sheet whose number is written on the container. Gear such as brooms and swabs must be cleaned after each use and placed in their storage racks. Gear adrift, such as rags, clothing, or personal gear, must be "policed up" immediately. Any items, if left adrift, could cause serious damage to the ship.

Sweepers

"Sweepers" is piped shortly after reveille, at the end of the regular working day, and at other times as scheduled. At these times, all persons assigned as sweepers draw their gear and sweep and swab down their assigned areas. All trash and dirt should be picked up in a dustpan and placed in a trash receptacle. If dirt is swept over the side, the wind may blow it back on board or the dirt may stick to the side of the ship. In either case, additional work will be necessary to clean the ship. At this time, you should empty all trash receptacles and cigarette butt kits, ensuring that none are still burning. Never dump trash or garbage over the side of the ship without first obtaining permission. There are various times when all trash must be kept in a safe area on board the ship until it can be properly disposed of.

The Cleaning Process

Dirt, soil, and contamination all describe the same thing—a foreign material on a surface where it is not wanted. Soil includes grease, oil, tarnish, rust, food residue, and stains. Most exposed surfaces that have been soiled may be cleaned by the proper use of cleaning agents.

Detergents are materials that have the ability to remove contamination and soil. There are other ways of cleaning besides the use of detergents or cleaning compounds. These include purely mechanical processes, such as removing rust from steel by sandblasting or cleaning decks by sweeping. For many cleaning problems, chipping, sweeping, sanding, or brushing may be needed. When detergent compounds are coupled with the mechanical action, however, a cleaner surface is usually produced with less time and labor.

The steps used in most detergent cleaning operations are as follows:

- Wetting: The soil and the surface of the object being cleaned must be wetted. If the surface is not wetted properly, the cleaning results will be poor. Contrary to popular belief, water has very poor wetting properties. Its wetting ability, and therefore its cleaning ability, is improved greatly by the addition of materials such as soap or synthetic detergents, which cause the water to flow into tiny crevices and around small particles of soil.
- Scrubbing: Dirt is loosened by the mechanical action of rubbing or scrubbing. Oil droplets, for example, are emulsified; that is, they are coated with a thin film of soap. The coating prevents them from recombining, and they rise to the surface. In a somewhat similar manner, solid particles are suspended in solution.
- Rinsing: The rinse is very important. It removes loosened dirt from the surface along with the cleaning material.

Zone Inspection

Frequent inspections ensure that all spaces, machinery, and equipment are in a satisfactory state of operation, preservation, and cleanliness. One type of inspection, the zone inspection, divides the ship or station into zones. Each zone is assigned to an inspection party or team. Usually the commanding officer will head one team, while an officer or chief petty officer will head each of the remaining teams. If you are assigned to present a compartment, you present the space to the inspecting officer by saluting and greeting the inspector in the following manner: "Good morning/

afternoon, sir/ma'am; (your rank and name) standing by compartment (name or number) for your inspection, sir/ma'am." You will stay with the inspecting officer during the inspection of your spaces to answer questions and provide assistance. Such things as storage cabinets, lockers, and drawers should be unlocked prior to the inspection for easy access. Usually the inspecting officer will give an overall grade to the space; for example, a grade of "outstanding" would indicate that no new discrepancies were noted and all previous discrepancies have been corrected. This is a grade of which you can be proud.

Field Day

Field days, or cleaning days, are held periodically. All hands "turn to" and thoroughly clean the ship inside and out, usually in preparation for an inspection. Fixtures and areas that sometimes are neglected during regular sweepers such as overhead cables, piping, corners, and spaces behind and under equipment are cleaned. Bulkheads, decks, ladders, and all other accessible areas are scrubbed. Knife edges and door gaskets are checked and any paint, oil, or other substances are removed. All brightwork is shined. Clean linen is placed on each bunk. Field days improve the appearance and sanitary condition of the ship, aid in the preservation of the ship by extending paint life, and reduce the dirt taken into operating equipment. Dirt intake must be held to a minimum to prevent the overheating of electrical equipment and abrasive action in rotating machinery.

Deck coverings on board ship receive more wear than any other material. Unless proper care is given to them, costly replacement is required. There are several materials used for covering decks, but resilient and nonskid types are most common. Resilient deck coverings include the vinyl tile and linoleum that cover interior decks. These deck coverings do not need painting, but daily sweeping and wiping away of spills as soon as possible is necessary. This type of deck covering is frequently cleaned with a damp swab, allowed to dry, and then buffed with an electric buffing machine. For more thorough cleaning, when the deck is unusually dirty, apply a solution of warm water and detergent with a stiff bristle brush or circular scrubbing machine and rinse with clean water to remove residual detergent. Stubborn dirt and black marks left by shoes can be removed by rubbing lightly with a scouring pad, fine steel wool, or a rag moistened with mineral spirits. After the deck covering is washed and dried, it can be polished with or without waxing with a buffer, or it may be given a coat of self-polishing wax and allowed to dry without buffing. Resilient deck coverings can be buffed several times before rewaxing. Nonskid deck coverings contain pumice, which helps to provide better footing. To clean a nonskid painted deck on the exterior, use only water to prevent the introduction of chemicals into the sea. Nonskid deck coverings should not be waxed or painted because to do so will reduce their nonskid properties.

Because of weather conditions, there will be many days at sea when the ship's topside areas cannot be cleaned. At the first opportunity, all topside surfaces should be cleaned with fresh water and an inspection made for signs of rust and corrosion. If such signs are discovered, you should tend to the area immediately. A little work at that time will save you a lot of work later.

PAINTING

The Coast Guard uses paint primarily for the preservation of surfaces. It seals the pores of steel and other materials, prevents decay, and arrests rust and corrosion. Paint also serves several other purposes. It is valuable as an aid to cleanliness and sanitation because of its antiseptic properties and because it provides a smooth, washable surface. Paint also is used to reflect, absorb, or redistribute light. For example, light-colored paint on a ship's interior distributes natural and artificial light to its best advantage.

Learning to paint properly includes knowing how to select suitable paints for the surfaces to be covered, how to properly prepare the surfaces before painting, and how to correctly apply paint. Although the selection of suitable paints will not concern you at this time, you should know how to prepare the surface and how to apply paint with a brush and roller. Improper surface preparation and paint application, in that order, are the most common reasons for paint failure. This section covers the basic procedures and safety precautions you should observe while painting. More detailed information can be found in the Coatings and Colors Manual, COMDTINST M10360.3D.

Types of Paint

- Anticorrosive paint is used on underwater hull areas to prevent salt water from acting on metal, and on the boot topping. It is always used with antifouling paint and must go on first. It contains heavy pigments and must be stirred frequently. It dries very rapidly, so it must be applied in short, quick strokes.
- Antifouling paint is applied to a ship's bottom to prevent marine growth such as barnacles and plants. It contains copper oxide, which will pit steel plating wherever it touches, so it must always be used over anticorrosive paint.
- Boot topping is a special paint applied to the hull at the waterline, covering the area between light-load and full-load drafts. Pretreatment coating is applied to all bare metal surfaces, if possible, before the primer coat is applied. It dries fast—within thirty minutes—and must be discarded if not used within eight hours. It is highly flammable; observe all safety precautions.

- Primers are base coats applied to wood and metal to make a smooth surface for final coats. Always apply two coats of primer and a third at all outside edges and corners.
- Exterior topside paints consist of white or spar for vertical surfaces and deck gray for horizontal surfaces; spar is used on certain other areas.
- Deck paints are green for officers' quarters and wardrooms, gray for other living and working spaces, and red for machinery spaces and shops.
- Interior bulkhead and overhead paint is usually fire retardant. White is commonly used on both bulkheads and overheads. Other colors are green for offices, radio rooms, pilothouses, and medical spaces; gray for electronic offices; acid-resisting black for compartments in which acids are used; and various pastel shades in crew's living and messing compartments.
- Machinery paint is usually gray enamel.

Surface Preparation

For paint to stick to a surface, all salt, dirt, oil, grease, rust, and loose paint must be removed completely, and the surface must be thoroughly dry. Salt and most dirt can be removed with soap or detergent and fresh water. Firmly embedded dirt may require scrubbing with scouring powder. When oil and grease fail to yield to scrubbing, they may be removed with paint thinner or other approved solvents. After scrubbing or scouring, always rinse the surface with fresh water.

The removal of rust, scale, and loose paint will likely require the use of hand tools or power tools and paint and varnish removers. Hand tools usually are used for cleaning small areas, and power tools are for larger areas and for completely cleaning decks, bulkheads, and overheads covered with too many coats of paint. Paint and varnish removers are used to remove paint from wood.

Hand Tools

The most commonly used hand tools are sandpaper, wire brushes, and hand scrapers. Sandpaper is used to clean corners and to feather paint, meaning to taper the edges of chipped areas down to the cleaned, bare surface so that no rough edges remain. Paint will bond best to a clean surface that has been lightly sanded.

- Sandpaper is graded according to the size of the abrasive grit on its surface. For example, it is graded from 12 to 600, with the coarsest being 12 and the finest, 600 grit. Very fine emery paper is sometimes used to polish unpainted surfaces.

- A wire brush is a handy tool for light work on rust or on light coats of paint. Such a brush also is used for brushing weld spots and cleaning pitted surfaces.
- Hand scrapers are made of tool steel. The most common type is L-shaped, with each end tapered to a cutting edge like a wood chisel. They are most useful for removing rust and paint from small areas and from plating less than one-fourth of an inch thick when it is impractical or impossible to use power tools.
- Occasionally it is necessary to use a chipping or sealing hammer, but care must be taken to exert only enough force to remove the paint. Too much force dents the metal, resulting in high and low areas. The paint is naturally thinner on the high areas. Consequently, thin paint wears off quickly, leaving spots where rust will form and, in time, spread under the good paint.

Aluminum Surfaces

Aluminum surfaces present a special problem because if they are not treated properly, considerable corrosion will result. Corrosion is greater when dissimilar metals such as aluminum and steel are in contact with each other and are exposed to an electrolyte like seawater, which causes an electric current to flow between them. This current results in galvanic corrosion of the aluminum. The first sign of aluminum corrosion is a white, powdery residue in the area where the two dissimilar metals make contact, later by pitting and scarring of the aluminum surface, and finally by the complete deterioration of the aluminum in the area. Holes in aluminum plating enlarge, and the screws, bolts, or rivets pull out or may even disintegrate. When painted aluminum surfaces are prepared for repainting, power sanders, if used, must be used with great care. It is best to use hand scrapers, hand and power wire brushes, or sandpaper of a very fine grit. Never use scaling hammers.

Steel Surfaces

The most important task in painting a steel surface is the preparation of the surface to be painted. Steel surfaces must be completely free of rust, loose paint, dirt, scale, oil, grease, salt deposits, and moisture. Old paint in good condition is an excellent base for repainting. When a surface is to be repainted and the old paint is not to be removed, the surface must be smoothed and thoroughly cleaned and dried before new paint is applied. In touch-up painting, when only small areas or spots need repainting, it is essential that the removal of old paint be carried back around the edges of the spot or area until an area of completely intact paint, with no rust or blisters underneath, is reached. The edges of the remaining paint should be feathered. When an old painted surface is to be completely redone, the old paint should be taken down to bare metal and a primer put on before painting. Never leave a bare metal surface exposed overnight. Always put on a primer coat before you quit for the day to prevent oxidation from starting.

Paint and Varnish Removers

Paint and varnish removers are used mostly on wood surfaces but may be used on metal surfaces that are too thin to be chipped or wire-brushed. The three types of removers in general use are flammable, nonflammable, and water-based alkali. All three are hazardous, and their safety precautions must be strictly observed. The removers should be used only in well-ventilated spaces. The alkali type should not be used on aluminum or zinc because of its corrosive properties. Paint and varnish remover must never be used around an open flame, because some types are highly flammable. Removers must not be used in confined spaces because of their dangerous anesthetic or toxic properties that can kill or cause injury if you are exposed to them for long periods. Moreover, paint and varnish removers should not be used by persons having open cuts or sores on their hands unless rubber gloves are worn. Avoid letting the remover touch your skin and watch out particularly for your face, eyes, and mouth. If paint or varnish remover should come in contact with the skin, wash it off immediately with cold water; if it gets into your eyes or mouth, seek medical attention as soon as possible. Read the product label and safety data sheet for full details.

What Not to Paint

The following items must not be painted:

- Knife edges, rubber gaskets, dogs, drop bolts, wedges, and the operating gear of watertight doors, hatches, and scuttles
- Start-stop mechanisms of electrical safety devices and control switchboards on machinery elevators
- Sprinkler piping within magazines
- Heat-exchange surfaces of heating or cooling equipment
- Identification plates
- Lubricating gear, such as oil holes, oil or grease cups, grease fittings, lubricators, and surfaces in contact with lubricating oil
- Metal lagging

- Rods, gears, universal joints, and couplings of valve-operating gear
- Springs
- Threaded parts
- Zincs
- Electrical contact points and insulators
- Decorative plastic, such as tabletops

Painting Safety Precautions

Before paint, primers, or paint applicators are issued, your supervisor will inspect the area to be painted to ensure that all preparations have been properly made. For example, are all items not to be painted properly identified or masked? Are all safety precautions understood and properly observed? Is the surface ready to be painted? Painting can be dangerous if one is careless. Many paints are highly flammable, others are poisonous, and some are both flammable and poisonous. To avoid injury observe the following precautions:

- Keep paint off your skin as much as possible. Wash your hands, arms, and face with soap and warm water before eating. Do not put your fingers, cigarettes, or food in your mouth if they are contaminated with paint.
- Be sure you have adequate ventilation, and wear an approved paint/spray respirator whenever there is reason to believe the ventilation is inadequate in the place you are painting.
- Do not smoke, use an open flame, or use spark-producing tools in the vicinity of painting operations.
- Do not carry matches or cigarette lighters, or wear steel buckles or metal shoe plates. Too often one forgets and strikes a match or lights a cigarette lighter in areas filled with explosive fumes. Steel buttons, buckles, and tabs can strike sparks that are invisible to your eyes, perhaps, but wholly capable of igniting paint vapors.
- Never paint during electrical storms.
- Keep food and drink away from areas being painted.
- Do not use gasoline, turpentine, mineral spirits, or other solvents to remove paint from the skin, as the skin will absorb them.
- If you spill paint, you are responsible for cleaning it up. At the end of working hours, all paint and brushes must be returned to the paint locker. Paint must be stored in its proper container, and all brushes and rollers cleaned.
- Follow carefully the instructions of your supervisor.

Painting Methods

Three means of applying paint are used in the Coast Guard—brush, roller, and spray. The majority of personnel have no occasion to use paint sprayers. This section concentrates on brushes and rollers, with which everyone in the Coast Guard should be familiar.

Paint Application by Brush

Smooth and even painting depends as much on good brushwork as on good paint. There is a brush for almost every purpose, so use the proper brush and keep it in the best condition. The two most useful brushes are the flat brush and the sash brush. With a flat brush, a skillful painter can paint almost anything on board a ship. Flat brushes are wide and thick, carry a large quantity of paint, and provide a maximum of brushing action. Sash brushes are handy for painting small items, for cutting in at corners, and for hard-to-get-at spaces. The fitch brush is also useful for small surfaces. The painter's dusting brush is used for cleaning surfaces.

The following are some general hints to help you use a paintbrush properly:

- Grip the brush firmly, but lightly. Do not put your fingers on the bristles below the metal band or ferrule. The grip should permit easy wrist and arm motion. To hold it otherwise restricts your movements and causes undue fatigue.
- When using a flat brush, do not paint with the narrow edge. This practice wears down the corners and spoils the shape and efficiency of the brush. When using an oval brush, if you revolve it too much, it soon wears to a pointed shape and becomes useless. Do not poke oversized brushes into corners and around moldings. Such a practice bends the bristles, eventually ruining a good brush. Use a smaller brush that fits into such odd spots.
- Dip the brush into the paint, but not over halfway up the bristles. Remove the excess paint by patting the brush on the inside of the pot. Avoid overfilling the brush, otherwise paint will drip on the deck or other surfaces and run down the handle.
- Hold the brush at right angles to the surface being painted, with the ends of the bristles just touching the surface. Lift the brush clear of the surface when starting the return stroke. If the brush is not held correctly and is not lifted, the painted surface will be uneven, showing laps and spots and a daubed appearance. Also, a brush that is held at any angle other than a right angle will soon wear away at the ends.

- For complete and even coverage, first lay on, then lay off. Laying on means applying the paint first in long strokes in one direction. Laying off means crossing your first strokes. Using the recommended method and crossing your strokes distributes the paint evenly over the surface, covering it completely and using a minimum amount of paint. Where sections of the overhead contain many pipes running parallel with the beams, it is often difficult to lay off the paint in a fore-and-aft direction. In such situations, better results are obtained by laying off the paint parallel with the beams.
- Always paint the overhead first, working from the corner that is farthest from the entrance to the compartment. By painting the overhead first, you can wipe drippings off the bulkhead without smearing the bulkhead paint.

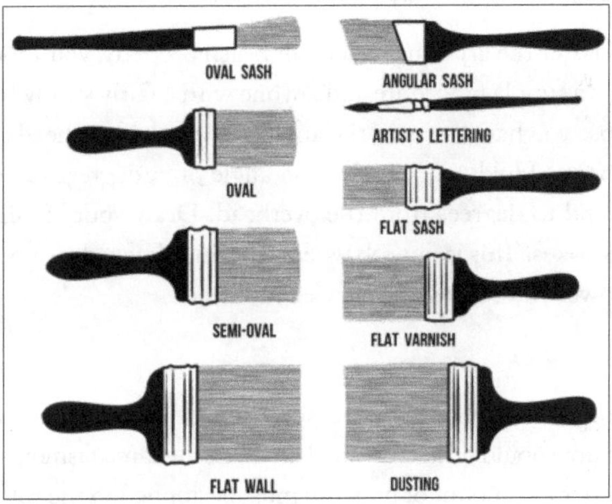

Paint brush types

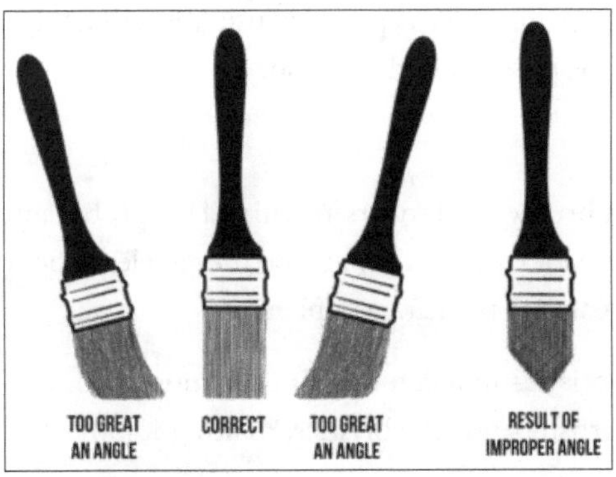

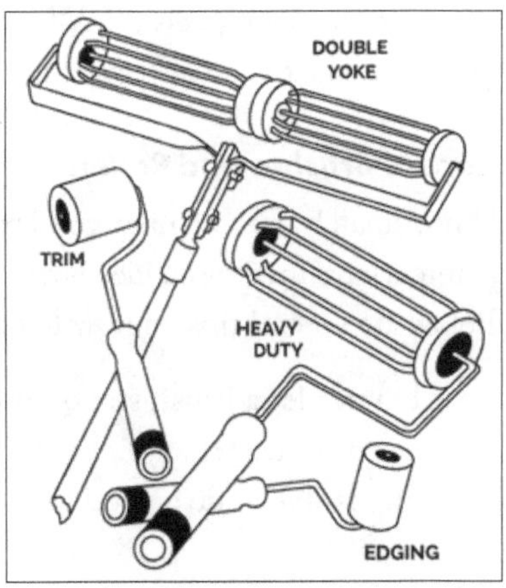

Brush technique (*above*)
Paint roller types (*right*)

- When overhead surfaces are being painted, sections should normally be painted in a fore-and-aft direction in an athwartship direction.
- To avoid brush marks when finishing up an area you have painted, use strokes directed toward the last section finished, gradually lifting the brush near the end of the stroke while the brush still is in motion. Every time the brush touches the painted surface at the start of a stroke it leaves a mark. For this reason, never finish a section by brushing toward the unpainted area. Instead, always end up by brushing back toward the area already painted.
- When painting pipes, stanchions, narrow straps, beams, and angles, lay the paint on diagonally. Lay off along the long dimension.
- Always carry a rag for wiping up dripped or smeared paint. Carefully remove loose bristles sticking to the painted surface.

After you master the art of using a paint brush properly, you should learn to cut in. Cutting in is a simple procedure and anyone with a fairly steady hand can learn it quickly. Suppose you have to cut in the angle between an overhead and a bulkhead. Start at one corner. Hold your brush at an angle of 75 degrees to 80 degrees from the bulkhead and 10 degrees from the overhead. Draw your brush along in fairly long, smooth strokes. This is one job where working slowly does not produce better results. The slower you stroke, the curvier your line.

Rollers

Large areas, such as cutter decks and sides can be painted quickly by the roller method. The paint should be laid on and laid off in the same fashion as when brushes are used. A moderate amount of pressure must be applied to the roller to make sure that the paint is worked into the surface. If pressure is not applied, the paint does not stick and soon peels off. When the paint roller is properly used, it will apply a more even coat and use less paint than is possible with a brush.

Care of Brushes and Rollers

Unfortunately, far too many good paint brushes and rollers are ruined simply because painters have little or no idea how to care for them or they are too lazy to clean them. Below are several tips on how to best care for painting supplies:

- Do not let a brush stand on its bristles in a can of paint for more than a few minutes. The weight of the brush bends the bristles, making it almost impossible to do a good job.

- Never allow paint to dry on a brush. If you intend to leave a paint-filled brush for long periods, fold wax paper or other heavy paper around the bristles and ferrule in such a way that air is kept away from the bristles. Twist the paper around the handle and secure it with rope yarn or sail twine.
- Cover your can of paint, and place both it and the brush in a safe place.
- Before starting to paint again, stir paint thoroughly with a paddle, not with the brush.
- At the end of the work, before turning in your paint and brush to the paint locker, clean as much paint from the brush as possible by wiping it across the edge of the paint can or mixing paddle.
- Only brushes that can be cleaned with soap and water are to be cleaned. Others, unless they are to be used again right away, should be allowed to dry and then thrown away. The cost of disposing of thinners as hazardous waste precludes the reuse of these brushes.
- Paint lockers usually have containers with divided compartments for stowing different types of brushes for short periods of time. These containers normally have tight covers and are equipped for hanging brushes so that the entire length of the bristles and the lower part of the ferrule are covered by the solvent or cleaning oil kept in the container. Brushes are suspended so that the bristles do not touch the bottom, thus preventing them from becoming permanently misshapen.
- Brushes to be used the following day should be cleaned in the proper cleaner and placed in the proper compartment of the container. Those not to be used again soon should be cleaned, washed in soap or detergent and water, and hung to dry. After drying, they should be wrapped in heavy paper and stowed flat.
- Do not leave a brush soaking in water. Water causes the bristles to separate into bunches, flare, and become bushy.

The proper cleaners for paint applicators are as follows:

Paint/Finish	**Solvent/Cleaner**
Natural and synthetic oil-base	Turpentine or mineral spirits paints and varnishes; chlorinated alkyd resin paints
Latex emulsion paints	Water
Chlorinated rubber paints	Synthetic enamel thinner xylene
Shellac	Alcohol (denatured)
Lacquer	Lacquer thinner

Paint rollers are cleaned in a different fashion. After use, the fabric cylinder should be stripped from the frame, washed in the cleaner recommended for the paint used, washed in soap and water, rinsed thoroughly in fresh water, and replaced on the frame to dry. Combing the pile of the fabric while it is damp prevents matting.

CORROSION

Corrosion is the wearing away of materials due to a chemical reaction, the most common example being rust. Coast Guard cutters, boats, and aircraft are operated and maintained in an extremely corrosive environment. A Department of Defense study estimates that 22 percent of maintenance costs for Coast Guard surface fleet assets are incurred due to corrosion. As cutters and boats age, the cost of corrosion continues to increase, as does the impact that corrosion has on safety and mission readiness. When corrosion is not properly managed, repair costs escalate quickly and asset availability is severely impacted. Corrosion prevention programs are established at all units to minimize costs associated with material deterioration.

The following are some specific operating environment corrosion concerns.

- Atmosphere Salt: Normal sea breezes can carry from ten to one hundred pounds of salt per cubic mile of air. Although the salt-laden air may travel inland on sea breezes for a distance of up to twelve miles, the major amount of salt fallout occurs within the first half mile of the beach. In the northern, cooler latitudes, the salt content of air is much less of a problem than in temperate and equatorial regions.
- Soot: Soot, generated from normal engine operation, is carbon, including a variety of combustion byproducts and sulfur oxides, depending on what has been burned. Soot is both corrosive and hygroscopic. It embeds itself into painted surfaces and is very difficult to clean off. Severe corrosion will result wherever paint has been chipped on an aluminum structure.
- Solar Radiation: Solar radiation (sunlight) causes chalking of paint, hydrolysis of chlorinated organics, and degrades exposed plastics and elastomers. Degradation of these materials allows an electrolyte, usually in the form of moisture and its corrosive constituents, free access to the underlying metallic surfaces.
- Chemicals: Maintenance chemicals, such as cleaners, acids, paint strippers, solvents, and so on, can present as many different problems as there are chemicals being used. Paint strippers, solvents, and some cleaning agents can,

when improperly used, deteriorate paint, plastics, and elastomers. Some paint strippers, some cleaners, and most acids are very corrosive to ship structures.

Common forms of corrosion include

- Galvanic corrosion, which results from contact between dissimilar metals and alloys.
- Pitting, which has the appearance of pin holes or cavities. This form of corrosion is very destructive since it can cause failure with only a small percent weight loss of the actual structure. Stainless steels are more susceptible to this form of corrosion than any other class of metals or alloys.
- Crevice corrosion, which occurs when a liquid corrosive is trapped in a gap between two components, in which at least one is sensitive to corrosion.
- Stress corrosion cracking, which is induced by loads applied by mounting bolts, in-service conditions, or even manufacturing processes such as welding.

Corrosion Prevention

- Ensure that assets receive a thorough fresh water washdown and, when possible, a full flush of raw water systems to remove any salt residue at the end of each operational period.
- Personnel performing work must ensure proper clean up of the work area, wiping down the area, and paying special attention to the bilges/voids, picking up any metal shavings, bolts, wire clippings, or any other metal hardware upon completion.
- Only authorized chemicals from the Authorized Chemical List and Authorized Use List may be used on an asset.
- Make note of any corrosion found on assets and immediately notify supervisors of significant corrosion or if corrosion is affecting equipment operation.

11

MARINE TRAFFIC

To fulfill the many missions of the Coast Guard, our personnel have always had to deal directly with other mariners and the marine industry. Although this relationship has not always been one of affection and admiration, it has usually been marked with a mutual respect. A general understanding of the types of vessels, and the personnel who operate them, will help acquaint new Coast Guard personnel with the marine industry and with the terms and legal definitions used by the service in its regulatory and law enforcement missions. Vessels are characterized as fitting into one of three general categories: commercial, noncommercial, and public or government vessels and warships. U.S.-flagged vessels can be further classified as documented or undocumented and inspected or uninspected.

VESSEL NUMBERING

All vessels are required to be numbered. All U.S.-built vessels measuring greater than five net tons are eligible for issuance of a certificate of documentation, and all vessels greater than five net tons that engage in coastwise trade or commercial fishing within U.S. waters are required to hold one. Foreign-built vessels are ineligible for U.S. documentation and, if over five net tons, prohibited from engaging in coastwise trade or commercial fishing in U.S. waters. Determining the net tonnage of a vessel requires either evidence of formal measurement by a surveyor from a Coast Guard–authorized measurement organization, required for vessels over 79 feet, or a boarding officer will assign a net tonnage under a simplified system. The certificate

Marine traffic

of documentation is internationally recognized and provides the vessel an official number and an endorsement for its intended use. Undocumented U.S. vessels must be issued a certificate of number, often called a registration, by the state in whose waters they are principally used. Foreign vessels are issued a number by their flag state.

COMMERCIAL VESSELS

The basic purpose of commercial vessels is to make money. Commercial vessels are conceived, designed, and built with this one ultimate goal in mind. A few terms and phrases related to commercial operations include the following:

- Jones Act: The Merchant Marine Act of 1920 provides for the promotion and maintenance of the American merchant marine and domestic yards to sustain a domestic maritime capability for national defense. It requires that all goods transported by water between U.S. ports be carried on ships that have been constructed in the United States and that fly the U.S. flag, are owned by U.S. citizens, and are crewed by U.S. citizens and U.S. permanent residents.
- Certificate of Inspection: U.S.-flagged commercial vessels are inspected by the Coast Guard on a regular basis to ensure their safety. Inspected vessels are issued a certificate of inspection that specifies conditions for

their operation such as the required number of licensed crew; maximum number of passengers; required safety, survival, and firefighting equipment; and any restrictions on their route. An uninspected vessel is a commercial vessel that is not subject to Coast Guard inspection, a classification that applies to fishing vessels, workboats, and certain small passenger vessels.
- Coastwise Endorsement: A U.S.-documented commercial vessel with a coastwise endorsement is entitled to be employed in unrestricted transportation of passengers or merchandise, dredging, towing, and any other employment for which a registry or fishery endorsement is not required while on domestic commercial voyages in which the vessel proceeds from one port or place in the United States to another port or place in the United States, normally on a route within twenty nautical miles of shore.
- Fisheries Endorsement: A fishery endorsement entitles a U.S.-built and -documented vessel to employment in the fisheries and to land its catch, wherever caught, in the United States, subject to federal and state laws regulating the fisheries. Foreign-built vessels greater than five net tons are ineligible for participation in the fisheries.
- Registry Endorsement: Entitles a U.S.-documented vessel to employment in foreign trade; trade with Guam, American Samoa, Wake, Midway, or Kingman Reef; and any other employment for which a coastwise or fishery endorsement is not required.
- Consideration: Consideration is an economic benefit, including monetary payment, to an individual, person, or entity. It does not include voluntary sharing of voyage expenses by monetary contribution or donation of fuel, food, beverage, or other supplies.
- Freight for Hire: Freight for hire is the carriage of any goods, wares, or merchandise or any other freight for a consideration, whether directly or indirectly flowing to the owner, charterer, operator, agent, or any other person interested in the vessel.
- Passenger for Hire: Passenger for hire is the carriage of any passenger for whom consideration is contributed as a condition of carriage on the vessel, whether directly or indirectly flowing to the owner, charterer, operator, agent, or any other person having an interest in the vessel.

For passenger vessels, nautical school vessels, and small passenger vessels, the inspection must occur at least once per year after the initial inspection, whereas all other vessels can go as long as five years between inspections.

Freight Vessels

Freight vessels are designed to carry a variety of dry cargoes and are generally classified as either bulk, break-bulk, or container vessels.

Bulk Carriers

Certain types of dry cargo do not lend themselves to packaging. Materials such as metal ores, grain, and minerals are more efficiently carried in bulk. Bulk carriers are commonly employed on the Great Lakes to carry iron ore from the mines to ports in the area of the steel mills. Other bulk carriers carry grain from the United States to foreign countries. The cargo holds in such vessels are designed to facilitate off-loading through the use of slightly sloping sides so that cargo settles to the middle of the hold. The cargo is usually off-loaded using cranes with large scoops aided by bulldozers lowered into the holds. Bulk carriers carrying either dry-bulk cargo or liquid-bulk cargo usually transport cargo on only one leg of a voyage.

Break-Bulk Vessels

This type of cargo vessel corresponds with the general concept of a traditional "freighter." These vessels are sometimes called "general cargo" vessels. Cargo for this type of vessel is usually in bales, bags, or small crates that are sometimes palletized to speed loading and unloading operations in port. Additionally, these vessels can carry cars, trucks, or heavy machinery. Because of their ability to carry a wide range of cargoes, and because of their self-contained cargo-handling capability, these vessels usually service ports with limited shore-side pier facilities.

Refrigerator Ships

Refrigerator, or reefer, ships are outfitted with refrigerating equipment to preserve and transport perishable cargo that requires temperature-controlled handling, such as fruits, meat, vegetables, and dairy products. The number of "reefer" spaces varies on a given vessel depending upon its trade. Usually reefer ships will carry other general cargo as well.

Container Vessels

Containerization of cargo was started in the late 1950s and early 1960s to assist in the intermodal transportation of cargoes. Cargo is placed in a container at the point of manufacture; the container is placed on a truck, transferred to a rail car, transferred to another truck, and finally placed on board a container vessel. Thus, the cargo itself is touched only two times, once during initial container loading and once at the destination where the container is unloaded. The containers themselves are constructed of fiberglass, aluminum, or steel and are in standard lengths of either twenty or forty feet.

Roll On / Roll Off (RO/RO) Vessels

RO/ROs are designed to carry cargo on multiple decks, usually in trailers, or as individual vehicles that are themselves cargo. Access from one deck to another is by way of ramps or elevators. The "cargo" is driven on board through a stern ramp or side ramp through watertight closures in the side of the vessel. The advantage of this type of vessel is that cargo can be quickly loaded or unloaded at a port without relying on elaborate shore equipment.

Tankships

Tankships, or tankers, are floating cargo tanks constructed to carry oil or hazardous material in bulk as cargo. Over the years tank ships have increased in size and capacity in the interest of fuel economy and growing consumer need for oil and chemical products, with ultra-large crude carriers measuring over 1,000 feet in length and capable of carrying more than half a million barrels of crude oil.

Liquefied Gas Carriers

Liquefied gas carriers are tankers with four to six specialized tanks located along the center line of the vessel that carry cooled or pressurized liquefied natural gas / liquefied petroleum gas (LNG/LPG). In the event tank integrity is compromised and the natural gas contained within ignites, there is a significant concern for the threat of explosion.

Passenger Vessels

Passenger vessels, as the name implies, are those designed solely to carry passengers.

Cruise Ships / Ocean Liners

Cruise ships are large oceangoing ships designed to carry large numbers of passengers for leisure or transport between ports.

Ferries

Ferries are designed to carry both passengers and vehicles on a regular schedule, normally between locations that are not more than three hundred nautical miles apart.

Charter Boat

Charter boats carry passengers for hire on short voyages associated with sightseeing tours, dinner cruises, dive trips, and fishing.

Commercial Fishing Industry Vessel

Commercial fishing industry vessels include fishing vessels, fish tenders, and fish processing or factory vessels. Fishing vessels commercially engage in the catching, taking, or harvesting of fish. Fish tenders supply, store, refrigerate, or transport fish, fish products, or materials. Fish processing vessels prepare fish or fish products for further transport to market.

Towing Vessels

There are a number of types of commercial towing vessels common to U.S. waters, all similar in that they are designed to push, pull, or haul alongside vessels and other floating objects. Typical towing vessels include harbor assist tugs that aid other vessels in maneuvers to moor or get under way, large seagoing tugs that tow vessels and barges over long distances, and towboats or "pushers" that push barges on the Western Rivers and Intracoastal Waterway.

Barges

Barges are distinguished from other vessels in that they are not self-propelled. Barges comprise a very large group of vessels that carry many types of cargo and serve a variety of functions, including lightering of cargo and the fueling or bunkering of larger vessels. In fact, there are just about as many types of barges as there are cargoes. Most barges are unstaffed, but some carry a small crew. The more common barge types are tank barges, hopper barges, LNG/LPG barges, and chemical carriers.

Dredges

Dredges are vessels equipped with excavation tools used for removing sand and other types of deposits from the sea or riverbed. Dredgers are used for several purposes, such as for making shallow coastal areas navigational, for deep-sea mining, and so on.

Workboats

A workboat is a vessel operated only over short distances for moving equipment in support of dredging, construction, maintenance, or repair work, or in support of a shipyard. Workboats are not subject to inspection or many of the requirements of towing vessels.

Mobile Offshore Drilling Units

Mobile offshore drilling units (MODU) are very special types of vessels designed for drilling operations that explore or exploit subsea resources. They are really movable platforms from which oil wells can be drilled. They are vessels only when they are moving from one site to another; at other times they are considered offshore facilities. There are three basic types of MODUs:

- Semisubmersibles are used in deep-water drilling. Upon arrival on site they are held in place using various multiple anchor systems.
- Jack-up rigs are self-elevating units used in shallow-water drilling. They are towed into place, the legs are lowered to the ocean floor, and the rig is "jacked-up" above the water level.
- Drill ships, built in the shape of regular ships but with the addition of a derrick, move to sites under their own power and are anchored in place; they are typically used for exploratory drilling.

Offshore Supply Vessels

MODUs and fixed offshore facilities are serviced by special-purpose offshore supply vessels designed to regularly carry goods, supplies, individuals in addition to the crew, or equipment in support of exploration, exploitation, or production of offshore mineral or energy resources.

U.S. MERCHANT MARINERS

The Coast Guard is responsible for specifying the minimum staff of licensed mariners required on board U.S. merchant vessels. In developing these standards the Coast Guard is concerned primarily with the safety of life at sea, meaning its goal is to ensure the vessel has a sufficient number of qualified personnel to be capable of coping with both normal operations and any emergencies that may arise during a voyage. All mariners employed on board inspected U.S. merchant vessels are required to hold a valid U.S. Merchant Mariner Credential, which serves as the mariner's qualification document, certificate of identification, and certificate of service and allows them to work in the shipboard merchant marine industry. With a Merchant Mariner Credential, an individual may work in the deck, engineering, or steward's department of a ship depending on the type of rating they hold. The deck department is concerned mainly with the navigation and operation of a ship, while the engineering department is concerned with the propulsion and maintenance of the machinery, and the steward's department with supplies and food services. At the entry level, mariners hold no rating other than ordinary seaman, wiper, or steward, but after time at sea, training, and testing they can apply for higher licenses, endorsements, and ratings based on vessel tonnage, service, propulsion type, and area of operation. Endorsements may be National-Domestic or STCW-International. STCW stands for Standards of Training, Certification, and Watchkeeping and are international standards of training applicable to all seafarers with requirements based on assigned duties. All who seek employment on an oceangoing commercial vessel making an international voyage are required to complete a STCW basic safety training course. Tonnage ratings range from twenty-five tons to unlimited, with unlimited authorizing service on vessels of any tonnage or any propulsion power. Authorized areas of operation may include the Great Lakes, inland waters, near coastal waters out to two hundred nautical miles, and oceans.

Deck Officer Endorsements and Department Ratings

Captains, mates, operators, and pilots supervise ship operations on domestic waterways and the high seas. The following are officer endorsements in the deck department:

- Master: The officer having command of a vessel.
- Mate: A qualified officer in the deck department other than the master. Includes the first, second, and third mates.
- First Class Pilot: Accompanies vessels while they enter or leave port to ensure safe navigation of entrance and departure.

Able seamen and ordinary seamen operate the vessel and its deck equipment under officer supervision. On tankers, mariners are designated to hook up hoses, operate pumps, and clean tanks. The following are ratings in the deck department:

- Ordinary Seaman: An "entry-level" unlicensed member of the deck department.
- Able Seaman: Qualified member of the deck department who may work as a watchstander.

Engine Officer Endorsements and Department Ratings

Ship's engineers operate, maintain, and repair engines, boilers, generators, pumps, and other machinery. The following are officer endorsements in the engine department:

- Chief Engineer: Head of the engineering department.
- Assistant Engineer: Qualified officer in the engine department. Includes the first, second, and third assistant engineers.

Qualified members of the engine department maintain proper running order in the engine spaces below deck under the direction of engineering officers. Wipers are entry-level workers who clean, paint the engine room and its equipment, and assist in maintenance and repairs. The following are ratings in the engine department:

- Wiper: An entry-level unlicensed member of the engine department.

Steward's Department

Stewards and cooks see to the daily needs of passengers and crew on board ship. They prepare and serve meals and maintain living quarters. A designated chief steward supervises the preparation of meals. Stewards are classified as nonqualified ratings and do not necessarily require any formal training or examination by the Coast Guard. There are exceptions to this rule though; for instance, if their ship makes international voyages, stewards will be required to satisfy certain STCW training requirements.

NONCOMMERCIAL VESSELS

Noncommercial vessels include recreational or pleasure boats, yachts, personal watercraft, and nongovernment exploration vessels, research craft, and training vessels that don't carry cargo or passengers for hire or fish commercially. A U.S.-documented vessel with a recreational endorsement is entitled to pleasure use only.

Yacht

A yacht is a sailing or power vessel used for pleasure, cruising, or racing. There is no standard definition, so the term applies to such vessels that have a cabin with amenities that accommodate overnight use. To be termed a yacht, as opposed to a boat, such a pleasure vessel is likely to be at least 33 feet in length with aesthetic qualities.

Sailboats

Sailboats are powered by wind and are commonly used for recreation and for competition. They can vary and be classified by their hull and sail or mast configurations. Hull types include monohulls, twin-hulled catamarans, and multihull crafts. Sail rigs and mast configuration types include sloop, cutter, ketch, and schooner.

The sloop is the most common mast type, where a single mast supports two sails called the foresail and the mainsail. Cutters have a single mast located further after than a sloop and supporting three sails consisting of a mainsail and two headsails, known as the staysail and the jib.

A ketch has two masts, with the aft mast known as the mizzen mast and slightly shorter than the main mast. A schooner is another configuration similar to the ketch, but the aft mast is taller than the foremast, and such vessels may have multiple masts.

Powerboats

Powerboats are propelled by machinery including outboards and stern drives that turn propellers or engines that power waterjets. They are commonly used for recreation and for competition. General construction types are cabin and open.

Convertibles. Convertibles are 35 feet or greater with large cabins, a galley, and berths for pleasure cruises and offshore fishing. A flybridge with an elevated helm helps to spot fish.

Cruisers. Cruisers are from 21 to 45 feet with a cabin designed for overnight stay. They are typically large enough for a small galley, several berths, and an enclosed head.

High Performance. Performance powerboats are built for speed, featuring narrow beam, steep deadrise, and high power to weight ratios ideal for racing or fast cruising.

Center Console. Center console boats are so named because their helm is on a console in the center of the boat. Most use outboard motors for propulsion.

Rigid-Hull Inflatable Boats. These boats have a fiberglass or aluminum hull attached to inflatable outer tubes. Outboard motors are used on the transom for power.

Runabout. A runabout is a small powerboat usually powered by an outboard or stern-drive engine. They are multipurpose and suitable for water sports, cruising, and fishing. They may have a small cuddy cabin at the bow for passenger seating or storage or may be a bowrider type with additional open-air seating at the bow.

Deck Boat. Deck boats are used for swimming and water sports; they have an open deck with seating for groups.

Flats Boat. Flats boats are shallow-draft vessels ranging from 14 feet to 18 feet and are designed to navigate shallow waters. When available depths prohibit the use of an outboard, the engine is raised and a push pole is used.

Ski Boat. Ski boats have removable, elevated tow bars and ski lockers to allow waterskiing, tubing, and wakeboarding.

Pontoon Boat. Pontoon boats have two or three aluminum tubes that support a broad platform. They are used on inland lakes and rivers and other small bodies of water.

Bass Boat. These boats are specialized for bass fishing on inland lakes and rivers.

Jon Boat. Jon boats are small utility craft primarily used for boating in shallow water.

Personal Watercraft / Jet Ski. Personal watercraft or jet skis are jet-driven vessels built for one or two people. Large models may accommodate up four people. They are used primarily for recreation and tow in surfing.

Houseboat. Floating houses are built on a barge-like hull with full kitchens, bedrooms, and living and dining areas. They are used on protected bodies of water.

Dinghy. Dinghies are companion boats to more significant vessels and are used to ferry passengers to shore.

PROFILE SUSPECT VESSELS

Vessels that match certain known characteristics, or profiles, for drug smuggling, human trafficking, and illegal migration are common targets of interest for patrolling Coast Guard assets.

Go-Fast. A go-fast boat is a small, fast boat designed with a long narrow planing hull and multiple outboard engines to enable it to reach speeds over eighty knots.

Migrant Craft. Migrants seeking to flee the conditions of their home nation often resort to constructing unsafe vessels powered by rudimentary sails or engines not designed for marine use such as lawnmower engines. During the 1994 *balsero* crisis in Cuba, the most commonly observed raft was constructed of two doors atop large truck-tire inner tubes, with the doors connected by two-foot by four-foot wooden beams. A rudimentary nine-foot mast was improvised to support a small white cloth as a flag or banner that would increase the raft's visibility to vessels traveling nearby.

Narco Sub. A narco semi- or fully submersible self-propelled vessel, also called a drug sub or narco sub, is purpose built for smuggling. These craft are difficult to detect visually, by radar, sonar, or infrared systems and typically carry several tons of cargo. In 2015 the largest-known cargo of 7.7 tons was seized on a semisubmersible.

PUBLIC VESSELS AND WARSHIPS

The term "public vessel" generally means any vessel owned and operated by a department or agency of the federal government or of a state or a subdivision thereof, which is engaged exclusively in official, noncommercial operations. This term also applies to foreign vessels owned and operated by a foreign government recognized by the United States. Public vessels include vessels of the Coast Guard, Navy, Army, Air Force, National Oceanic and Atmospheric Administration, and Army Corps of Engineers. Warships include large combatants like carriers, cruisers, and destroyers, amphibious ships, frigates and larger patrol craft, small patrol craft, submarines, and various support vessels. Foreign navies, both allied and non-allied silhouettes, are available in Jane's *Warship Recognition Guide*.

12

COMMUNICATIONS

Communication is defined as the exchange of information. We communicate effectively when we clearly and accurately send and acknowledge information, instructions, and commands and provide useful feedback. Studies have shown that words account for 7 percent of the information communicated. Vocal tone accounts for 55 percent, and body language accounts for 38 percent. To be an effective communicator you must be aware of the methods available to communicate, how to use them effectively, and how to overcome barriers to the communications process.

EFFECTIVE COMMUNICATIONS

An effective communicator is successful in establishing an active two-way link with another individual or group. Because effective information processing is critical for teams, it is essential that we learn to communicate effectively. Effective communications follows a clear process:

- Sender conceives an idea and wants to pass it onto someone else.
- Sender encodes or decides how to send the idea, then sends the idea.
- Receiver gets the message.
- Receiver decodes or interprets the information from their understanding of the message.

Sender

Communication is a two-way process that starts with the sender. The sender should convey information necessary for mission accomplishment. The sender must be proactive in making the receiver understand the message. Too often what is said is not always what is heard. To prevent this from happening, do the following:

- State one idea at a time.
- State ideas simply.
- Explain when appropriate.
- Repeat if appropriate.
- Encourage feedback.

Message

The message is simply the information the sender wants to communicate to the receiver. The message is usually verbal, but it can also be nonverbal. To reduce potential problems, the sender should do the following:

- Use correct terminology.
- Speak clearly.
- Time the message to be sent when the receiver is able to listen.
- Use appropriate vocal tone and volume. Commands must be heard and perceived as an imperative, not a question or comment. When breaking error chains or returning others to situational awareness, express a tone of urgency and importance.
- Be relevant to the receiver, not a distraction. The message needs to be inclusive and informative. Inclusive means that it contains everything necessary for the receiver to understand the sender's point. Informational means it is something that the receiver needs to know.

Receiver

The receiver needs information to accomplish the task. The effectiveness of the team often rests on its members' ability to listen. Unfortunately, most people find listening difficult. The receiver must exert control over the communication process. For receivers to have control, it is important that they ensure that the senders understand what the receivers want and why they want it.

Active Listening

Active listening is a process used by the receiver to facilitate communication and enhance performance. It requires the receiver to be active in the communications process. To actively listen, the receiver needs to do the following:

- Focus attention on the message, giving it momentary priority. If possible, look at the sender.
- Listen and look for the indirect nonverbal message content as well as hear the words. Observing nonverbal cues provides information regarding what the sender wants to convey. Your perception of the message and the sender's intent for the message may be different. Word choice, tone of voice, body position, gestures, and eye movements reflect the feelings behind the spoken word.
- Keep an open mind and suspend judgment.
- Verify what was heard. Don't assume that your perception of the message agrees with the sender's intent. Provide the sender feedback.

Message Feedback

Effective receivers verify their understanding of the message with the sender. They consider words, tone, and body language when they give feedback. Forms of feedback include the following:

- Acknowledgment: "Rogering" a message is common courtesy. It demonstrates that the receiver has heard the message. However, for critical information or complicated ideas, acknowledgment normally is insufficient to ensure understanding.
- Parroting: Repeating back verbatim the words of the speaker confirms to the speaker that the words transmitted were the words received. It is preferred in verifying receipt of standard commands. Like acknowledgment, it does not ensure the receiver understood the message.
- Paraphrasing: Rephrasing, in your own words, the content of the sender's message to the sender's satisfaction clarifies the message for both you and the sender. Paraphrasing allows you to check your understanding of the message and shows the sender that you listened accurately and that the content and intent was correctly understood. If you listened inaccurately, the sender has an opportunity to correct the communication error.

Barriers

Barriers are influencing factors that impede or break down the continuous communications loop. They block, distort, or alter the information. By identifying the barriers and applying countermeasures, team members can effectively communicate. Barriers include the following:

- Nonassertive behavior
- Inappropriate priorities
- Task preoccupation
- Organizational structure
- Anger or frustration
- Distractions
- Personal bias
- Tunnel vision
- Diversity
- Interruption

PHONETIC ALPHABET AND NUMBERS

In all communication, a phonetic alphabet is used to avoid confusion between letters that sound alike, such as C, D, and T. Each letter is designated by a word. Knowing how to pronounce the phonetic alphabet and numbers increases the chance that all voice communications are successful. To use the phonetic alphabet to spell out difficult words within a message, the actual spelling should always be preceded with the procedure words (prowords) "I SPELL."

Letter	Name	Spoken
A	ALFA	AL-FAH
B	BRAVO	BRAH-VOH
C	CHARLIE	CHAR-LEE
D	DELTA	DEL-TAH
E	ECHO	ECK-OH
F	FOXTROT	FOKS-TROT
G	GOLF	GOLF
H	HOTEL	HOH-TEL
I	INDIA	IN-DEE-AH
J	JULIET	JEW-LEE-ETT
K	KILO	KEY-LOH
L	LIMA	LEE-MAH

(continued)

Letter	Name	Spoken
M	MIKE	MIKE
N	NOVEMBER	NO-VEM-BER
O	OSCAR	OSS-CAH
P	PAPA	PAH-PAH
Q	QUEBEC	KAY-BECK
R	ROMEO	ROW-ME-OH
S	SIERRA	SEE-AIR-RAH
T	TANGO	TANG-GO
U	UNIFORM	YOU-NEE-FORM
V	VICTOR	VIK-TAH
W	WHISKEY	WISS-KEY
X	XRAY	ECKS-RAY
Y	YANKEE	YANG-KEY
Z	ZULU	ZOO-LOO

Numbers should always be preceded with the proword FIGURES. Pronounce numbers as follows:

Number	Pronunciation	Number	Pronunciation
0	Ze-ro	5	Fife
1	Wun	6	Six
2	Too	7	Sev-en
3	Thu-ree	8	Ait
4	Fow-er	9	Nin-er

Decimals should be included in a spoken number by saying the word decimal (DAY-SEE-MAL) in the proper location. When a number consists of more than one numeral or digit, one numeral at a time is pronounced with a short pause between numerals. Examples:

Figure	Spoken As
44	FOW-ER FOW-ER
500	FIFE ZE-ROH ZE-ROH
7,000	SEV-UN ZE-ROH ZE-ROH ZE-ROH
16,000	WUN SIX ZE-ROH ZE-ROH ZE-ROH
14,899	WUN FOW-ER AIT NINE-ER NINE-ER

If the letter/number combination starts with a number, use the proword "FIGURES": 12A9B, Spoken as: "FIGURES WUN TOO AL-FAH NINE-ER BRAH-VOH."

If the combination starts with a letter, use the proword "I SPELL": TS67R, Spoken as: "I SPELL TANG-GO SEE-AIR-RAH SIX SEV-UN ROW-ME-OH."

PROWORDS AND ABBREVIATIONS

Procedure words, or prowords, are words and phrases used to speed up radio communication. Commonly used prowords and abbreviations include the following:

AFFIRMATIVE: Yes.

BREAK: I hereby indicate the separation of text from other portions of the message.

CORRECT: You are correct, or what you have transmitted is correct.

CORRECTION: An error has been made in this transmission. Transmission will continue with the last word correctly sent. The correct version is . . .

ETA: Estimated time of arrival.

ETD: Estimated time of departure.

ETR: Estimated time of return or repair.

FIGURES: Indicates numbers or numerals to follow. Used when numbers occur in the text of a message.

I SPELL: I shall spell the next word phonetically.

OPS NORMAL: Used to say the patrol is normal in all respects, "operations normal."

OUT: Used following the last line of the message transmitted, signifying the end of the transmission and nothing follows. No reply is required or expected.

OVER: Used following a transmission when a response from the other station is necessary. It is an invitation to the other station to transmit.

NEGATIVE: No.

ROGER: I have received your transmission satisfactorily.

SAY AGAIN: Repeat your last message.

SILENCE (spoken three times and pronounced "SEE-LONCE"): Cease all transmissions immediately. Silence will be maintained until lifted. Used to clear routine transmissions from a channel only when an emergency is in progress.

SILENCE FINISHED (pronounced "SEE-LONCE FEE-NEE"): Silence is lifted. Indicates the end of an emergency and resumption of normal traffic.

UNKNOWN STATION: The identity of the station that you are trying to establish communications with is unknown.

WAIT: Use for pauses that require only a few seconds.

WAIT OUT: Use for pauses that require more than a few seconds.

WILCO: Will comply with your last order or request.

WRONG: Your last transmission was not correct. The correct version is . . .

MORSE CODE

Morse code is a method used in communication to encode text characters as standardized sequences of two different signal durations, called dots and dashes. Morse code is named after Samuel Morse, one of the inventors of the telegraph. International Morse code encodes the letters A through Z and numerals 0 thru 9. Each Morse code symbol is formed by a sequence of dots and dashes. The dot duration is the basic unit of time measurement in Morse code. The duration of a dash is three times the duration of a dot. Each dot or dash within an encoded character is followed by a period of signal absence, called a space, equal to the dot duration. The letters of a word are separated by a space of duration equal to three dots, and words are separated by a space equal to seven dots.

Morse code can be memorized and sent via sound waves or visible light and is most commonly transmitted by on-off keying a light or radio.

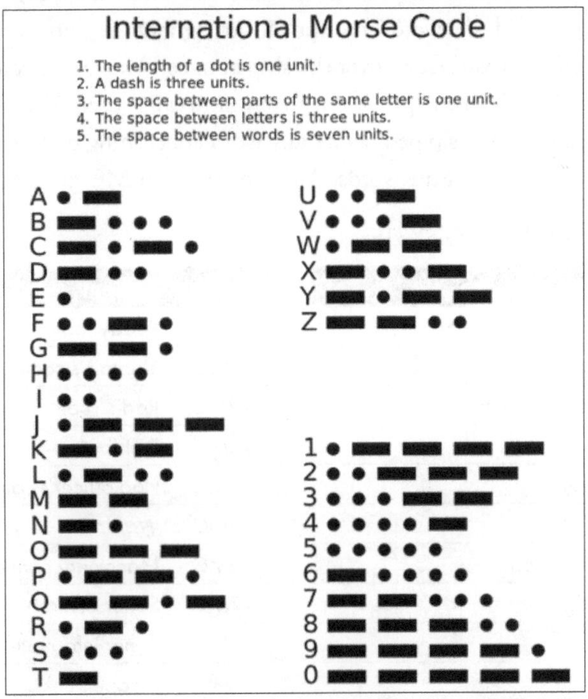

INTERNAL COMMUNICATIONS

Internal communications are those that take place within a ship or a shore-based unit. These include everything from "passing the word" over the general announcing system, the 1MC, to messengers delivering reports in person. Internal communications also include printed or written material such as the plan of the day and the "On Board–Ashore" display for the crew.

General Announcing System

The ship's general announcing system (1MC) is a one-way transmitting system heard throughout the cutter and used to pass information to the crew. Depending on the size and type of cutter, various 1MC microphone control stations, or "squawk boxes," may be available, typically on the bridge, in the damage-control center (DCC), and in the engine control room. Announcing speakers are located throughout the ship. Refer to the ship's information book for the exact locations on the cutter. Listen to announcements over the 1MC regarding the status of missions, operations, response activities, and other important information. The 1MC is also the means by which the general, chemical, and collision alarms are sounded.

Before such public address systems were developed, information or orders for the crew were passed by a boatswain's mate, who first sounded the appropriate call on the pipe and then repeated "the word" fore and aft throughout the ship. Now, on the 1MC system, the word is broadcast all over the ship at once. An announcement is usually preceded with "Now hear this," unless a boatswain's call is used. When a boatswain's pipe is used, "All hands" is piped before any word concerning drills or emergencies, and "Attention" before routine words. A complete list of MC circuits follows.

CIRCUIT ANNOUNCING SYSTEM

One-Way System	
1MC	Battle and general
2MC	Engineers
4MC	Damage control
6MC	Boat control
11MC	Turret
16MC	Turret
17MC	Antiaircraft
18MC	Bridge

Two-Way System	
19MC	Ready room
21MC	Captain's command
22MC	Radio room
23MC	Distribution control
24MC	Flag officer's command
25MC	Wardroom
26MC	Machinery control
27MC	Sonar control
29MC	Sonar information
31MC	Escape trunk

Sound-Powered Telephones

The sound-powered telephone provides the principal means of communications for damage-control operations. The handset circuits are sound-powered circuits and the transmitter of a sound-powered telephone transforms sound waves into electrical energy. The receiver transforms this electrical energy back into corresponding sound waves. Sound-powered headsets consist of a pair of earphones and a transmitter. On the bridge and in other areas, the headset telephone is stored in storage boxes. On some cutters, hooks are provided belowdecks to hang the phones. The individual ship's information book outlines the phones' location. Listed below are examples of basic circuits you must be familiar with.

- JA Circuit: Captain's battle circuit providing communication between the bridge, combat information center, weapons control center, DCC, and the plotting station
- 1JV Circuit: Maneuvering and docking circuit providing communication to coordinate ship operations such as changes in engine speed; steering orders in the event of steering system failure; checking gyro and standard compasses; reports of steering gear system derangement; and orders to line handling and transfer-at-sea stations

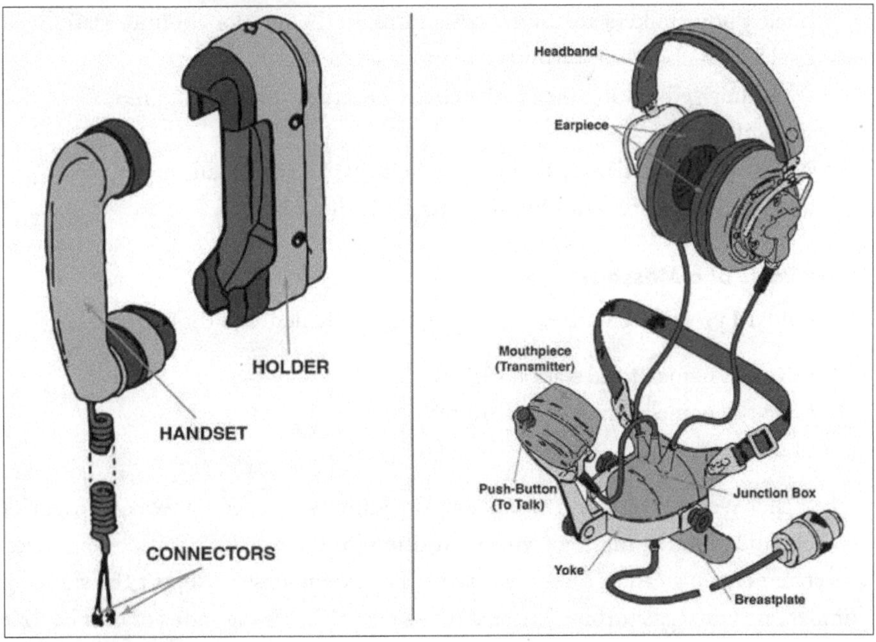

Sound-powered phone

- 2JZ Circuit: Damage and stability control circuit providing vital communication between DCC, engine rooms, repair stations, weapons control center, and other critical stations
- X4OJ Circuit: Casualty control communication circuit providing portable emergency communications between the main belowdecks stations after casualties have occurred to the primary circuits.

Rules for Circuit Discipline

Because everyone can talk and listen at once on the sound-powered telephone system, talkers must maintain strict circuit discipline. Following are the rules for circuit discipline:

- Transmit official messages only.
- Depress the transmitter button only when transmitting.
- Use only standard words and phrases.
- Do not use slang or profanity on the phone.
- Use correct nautical terms.
- Never interrupt another station unless you need to pass an emergency message.
- If you must interrupt a station on the circuit, use the phrase, "Silence on the line." Phone talkers are an important link in the interior communications chain; that chain is no stronger than its weakest link.
- Unauthorized talking means the chain has at least two weak links.
- Be professional.
- Never show impatience, anger, or excitement. You must talk slowly, clearly, and precisely. Circuit discipline means self-discipline.

Three Parts of a Message

The standard procedure for relaying a message is divided into three parts:

- Give the name of the station called.
- Give the name of the station calling.
- State the message.

In other words, you call the station for whom you have a message, identify yourself, and send the message without waiting for the receiving station to answer. Use statements instead of direct questions. For example, say "Report the status of number one bravo gas turbine" instead of asking "What is the status of number one bravo gas turbine?" When you receive a message, acknowledge receipt as soon as you understand the message. Repeat it to the sender word for word. For example, you might receive the message "Main Control—Bridge. Shift control of number one bravo gas turbine to the bridge." Your response should be "Shift control of number one bravo gas turbine to the bridge. Main Control, aye."

Do not use "aye" to answer "yes" to a question; instead use "affirmative" or some other appropriate reply. Likewise, use "negative" to answer "no."

Transmit through a Sound-Powered Earpiece

In some cases where the sound-powered phone transmitter is nonfunctional or has been damaged, the earpiece can be used as a transmitter by following the below steps:

- Hold the earpiece about one-half-inch from the mouth when talking.
- Do not use the talk button when talking.
- Do not shout or whisper into the telephone.
- Enunciate words distinctly.

Procedure for Securing and Storage

Always get permission before securing from the circuit. The station requests permission to secure from that circuit control station, whether the bridge, main control, or DCC. After securing the station, the talker must secure the phone using the correct procedure based on phone type. Handset holders are installed in enclosed spaces, and handset stowage cabinets are installed at stations exposed to weather. Stow headset on hooks in enclosed spaces. In machinery spaces and on weather decks, stow the headsets in stowage boxes, which are designed to stow from one to six sets. Never allow a loose cable to hang out of the box.

Messenger

There may be times when the ship's phone systems are out or they are overburdened and an oral or written message must be used. Messengers are a reliable way to communicate and involve the assignment of a crewman to physically transit between stations and personnel to orally communicate messages or deliver written notes and orders or preprinted message blocks. These are the general steps for transmitting information via messenger:

- Use correct terminology and standard commands for consistency and clarity of the information.
- If reporting orally, speak clearly and slowly to give the message receiver time to write down the information.

- If presenting a written report, stand by while the recipient reads it to answer any questions they might have.
- Once the receiver has accepted the message, ensure they indicate that the message is understood. Acknowledge their confirmation of the information or provide corrections.
- Return to duty station.

EXTERNAL COMMUNICATIONS

External communications are all those between a boat, cutter, aircraft, or command center and another vessel, aircraft, boarding team or unit, and the public, whether afloat or ashore.

Visual Signals

Visual signals began centuries ago, when flags and pennants were given special meanings. Columbus hoisted a lantern in the rigging to let his ships know land had been sighted. Even with dozens of radios on a ship, visual signals are still used extensively, especially in fleet operations, for two reasons—speed and security. Radio messages may be intercepted by anyone with a receiver, perhaps hundreds of miles away. Visual signals are limited to line-of-sight transmissions, meaning the receiver and sender have to be where they can see each other, and do not require complicated equipment. The three main systems of visual signals are flashing light, semaphore, and flag hoist.

Flashing Light Signaling

This system uses short and long flashes of light to spell out messages encoded in Morse code. The transmitting signalman sends one word at a time, with a slight pause between each letter. The receiving signalman flashes a dash after each word, meaning it was received and the receiver is ready for another. Methods for transmitting these signals include the following.

Directional Method. In this system the sender aims the light directly at the receiver. Only people near the receiver can see the light. The signalman controls the length of the flash by a shutter on the face of the light. Other types of directional gear are the blinker tube, or blinker gun, and the multipurpose lamp, both battery-operated with trigger switches to control the light flashes.

Nondirectional Method. Another term for this type of signaling is "all-around." Most of it is done by yardarm blinkers, lights mounted high on a mast and controlled by a signal key on the signal bridge. This method is best for sending a message to several ships at once.

Nancy. Whereas other signal systems use "white" light, Nancy uses invisible infrared light that can be seen only by those who have a special Nancy receiver that gathers the infrared rays and converts them to visible light. Nancy has a range of from 10,000 to 15,000 yards and can be used only at night. It is a very secure method of communication.

Semaphore

Semaphore is a daytime method of visual communication that employs handheld flags to send single letters that create words or coded signals. The flags are generally fifteen to eighteen inches square of a design similar to Oscar or Papa and are attached to staffs about twenty-two inches long. There are thirty flag positions.

Procedure Signs. Procedure signs, or prosigns, are a form of communication shorthand. They provide, in brief form, orders, requests, instructions, and other types of information used often in communication. Shown below are some of the more common prosigns that may be sent by semaphore or flashing light.

Prosign	Meaning
AA	Unknown station
AR	Out (end of transmission)
AS	Wait
B	More to follow
BT	Break
DE	This is (from)
EEEEEEEE	Error
F	Do not answer
HM HM HM	Emergency silence
IMI AB	Repeat all before (word)
IMI AA	Repeat all after (word)
IMI WA	Repeat word after (word)
IX	Execute to follow
K	Over (invitation to transmit)
NR	Number
R	Received (also Routine)
T	Relay to (Unit)

Flag Hoist

Flag hoists are the most rapid system of visual signaling but, like semaphore, can be used only in the daytime. There is a signal flag for each letter of the alphabet, one for each numeral from 0 through 9, and others with special uses. A complete set of signal flags has sixty-nine flags and pennants; with them, thousands of signals can be sent.

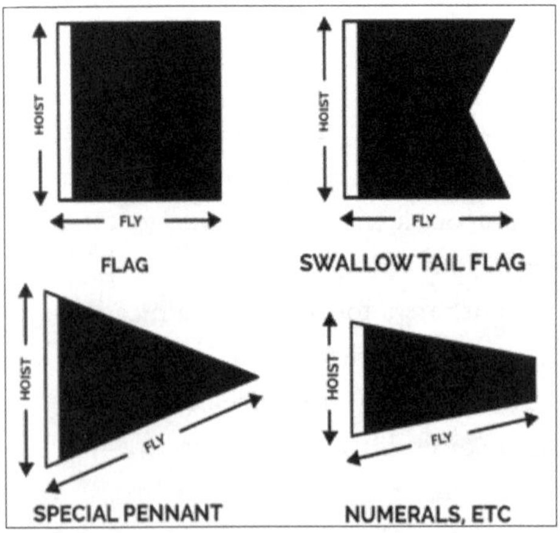

Flags and pennants

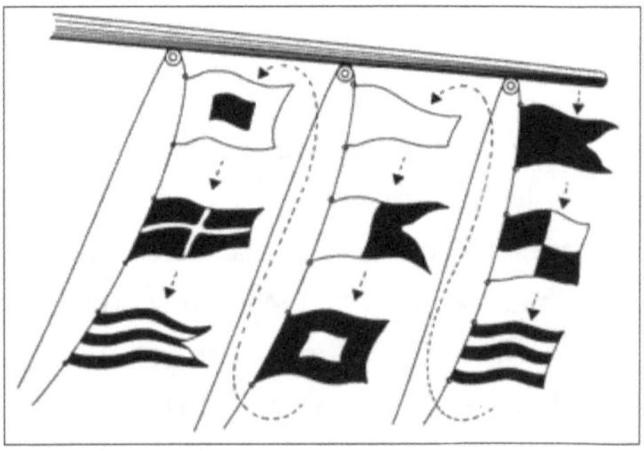
Reading flags and pennants

Most ships carry only two or three complete sets of flags, and substitutes are used when particular flags are already flying: the first substitute repeats the first flag or pennant in the same hoist, the second substitute repeats the second flag or pennant, and so on.

The following flags and their meanings should be known to every crew member, whether they stand bridge watch or not:

- Code Pennant: A red-and-white striped pennant is used by a warship to signal to another vessel that it wants to communicate. The code pennant is hoisted in a conspicuous place and kept flying during the whole time that signals are being sent and received.
- Oscar: A flag with yellow and red triangles means man overboard.
- Bravo: A red flag with a swallowtail means there is a transfer of hazardous fuel or explosives.
- Papa: A blue-and-white flag signals that all personnel are to return to the cutter.
- Five: Flag number five is yellow with a blue X and means the vessel has suffered a breakdown.
- Kilo: A yellow-and-blue flag means person working aloft.
- SOPA: Senior officer present afloat, usually referred to as SOPA, is the pennant used to indicate the vessel carrying the most senior officer.

Many occasions benefit from rigging signal flags to "break," meaning they are hoisted in a folded condition and unfurl when their halyard is pulled. A common example of when breaking is appropriate is the arrival of a flag officer.

Hand Signals

Multiple evolutions benefit from the use of hand signals, particularly in situations where noise is a factor. Hand signals are authorized and standardized for helicopter hoists, davit and crane operations, boat launch and recoveries, and surface swimmer deployments. Approved signals for each can be found throughout this manual in those sections that discuss the specific evolution.

Sound Signals

Sound signaling devices include a ship's bell, whistle, gong, and siren. Depending on the size of the vessel, the Navigation Rules, which prescribes various signals for maneuvering, navigating in fog, anchoring, grounding, and distress, may require one or several of these to be carried on board. These rules are covered in chapter 15 of this manual. Whistles, sirens, and bells may also be used by ships for transmitting emergency warning signals such as air raid alerts and, in wartime, for communication while in convoy. Procedures for sounding signals to message and maneuver convoys can be found in the International Code of Signals.

Electronic Communication

The development of the electric telegraph in 1844 marked the beginning of electronic communication that enabled real-time communications beyond visual range. Since the telegraph, the options to communicate using electronic signals have greatly expanded and today include methods that allow for communication between units regardless of their location on Earth.

Radiotelephone

Most marine communications are accomplished via radiotelephone, or voice radio. Because of its directness, convenience, and ease of operation, radiotelephone is used almost exclusively between ships and aircraft for short-range tactical communication up to twenty-five miles. Radio transmissions on very high frequencies (VHF) and ultra-high frequencies (UHF) are good for line-of-sight communication. Medium frequencies (MF) and high frequencies (HF) are used for longer-range communication. Although most radio transmissions are unsecure or made in the clear, equipment is installed on Coast Guard radios to allow secure or encrypted voice transmissions between the service's assets and those of other government agencies.

Frequencies/Channels

Just like the radio you listen to at home or in your car, VHF-FM has assigned frequencies. To simplify identification, those frequencies are assigned a channel number. For example, 157.100 MHz is the frequency assigned to Channel 22. Channel 22 is an easier number to work with. In accordance with the Navigation Rules and the Radio Telephone Act, vessels over twenty meters are required to monitor channels 13 and 16 and be capable of receiving 22A. If in a vessel traffic service area, monitoring of additional frequencies that can be found in the *Coast Pilot* will be required. Commonly monitored frequencies used by the Coast Guard include the following:

- Channel 13: Bridge-to-bridge communications in the VHF-FM band. Used between vessels to aid in the prevention of collisions.
- Channel 16: International hailing and distress frequency in the VHF-FM band. Used to request maritime assistance; to transmit distress, urgent, and safety messages; to announce marine information bulletins; and to

initially establish communications before shifting to an appropriate working frequency.
- Channel 21, 81, and 83: Intra–Coast Guard working frequency.
- Channel 22: Primary VHF-FM liaison frequency for communications between Coast Guard units and civilian stations. Also used for transmitting Coast Guard marine information and marine assistance request broadcasts.

Basic Procedure

Because the use of radios is so prevalent, it is important to learn basic procedures and ways for properly using them so messages are sent and received in the most effective and professional manner.

- Check settings to ensure the radio is set on the proper frequency.
- Check volume. It should be loud enough to hear clearly, but not so load it drowns out other radios or conversations.
- Adjust squelch control. Setting the squelch control adjusts the receiver so only signals strong enough to pass the level selected will be heard, which reduces the amount of static noise on the speaker.
- Keep the microphone about one to two inches from lips. When transmitting, shield the microphone by keeping your head and body between noise generating sources such as engines, wind, helicopter rotors, and so on, and the microphone.

Radio Checks

Radio checks test the signal strength and readability of transmitted radio signals. Checks are a simple way to determine whether the radio used to send and receive messages is working properly. This is accomplished by transmitting a request for a radio check and receiving a response from another station that provides a standardized description of the strength and readability of a transmitted signal (i.e., “Loud and Clear” or “Broken and Unreadable”). If ever in doubt about the readability of the signal being sent, a radio check should be initiated to confirm the strength and readability of the signal.

Transmitting Techniques

Because voice radio is used so widely, everyone must understand the basics of circuit discipline and transmitting techniques. Unless encrypted, voice radio can be heard by anyone with the proper equipment. The following techniques will ensure effective radio communication:

- Before beginning a transmission, listen for a few seconds to avoid interrupting other communications that are already in progress.
- Do not interrupt others.
- Before keying the transmitter, know how to say what it is you are attempting to communicate. Keep all transmissions short and to the point. Never small talk or make unnecessary transmissions on any frequency.
- Speak clearly, concisely, and in a normal tone of voice, maintaining a natural speaking rhythm.
- Use the phonetic alphabet to spell out a word or a group of letters.
- Using prowords speeds the handling of radio messages by abbreviating a single word or phrase to replace common words, phrases, sentences, and even paragraphs. Knowing and using prowords helps to reduce radio traffic by performing radio transmissions efficiently.
- Use proper prowords, ending each transmission with "over" and the last with "out." Never say "over and out."
- In cases where a pause for a few seconds between transmissions is necessary, use the proword "wait." If the pause is to be longer than a few seconds, use prowords "wait, out." Do not use "wait one" or "stand by."
- When transmitting, use caution; the microphone may pick up the conversations of people talking nearby.

The following is a list of things NOT to do while verbally communicating using the radio. Items on this list either are not protocol, they are illegal, or they cause misunderstandings of messages.

- Do not use profane or obscene language.
- Do not use unauthorized prowords, abbreviations, and procedures.
- Do not speak with extremes of voice pitch. This will cause distortion.
- Do not slur syllables or clip speech. They are hard to understand.
- Do not use phrases such as "would you believe," "be informed," or "be advised." They are unprofessional and not correct procedure.
- Do not key the microphone unless ready to transmit. Keying the microphone also transmits a signal, causing interference on that frequency.
- Do not use "10 Codes" such as those used by many law enforcement agencies.

Call Signs

Call signs are used to identify the unit type that is calling or being called over voice radio. You identify the unit you are calling (Call Sign), and then identify yourself

(Call Sign). The following summarizes the different Coast Guard facilities and how to state each call sign that may be encountered in the field.

Type	Transmit	Followed By
Boat	Coast Guard	Boat Number
Cutter	Coast Guard Cutter	Cutter Name
Shore Radio Facility (i.e., Sector)	Coast Guard	Facility Name
Aircraft	Coast Guard	Aircraft number
Aircraft SAR	Coast Guard / Rescue	Aircraft Number

- Coast Guard Small Boat example:
  "COAST GUARD TOO FIFE FIFE FO-WER FO-WER"
- Coast Guard Cutter example:
  "COAST GUARD CUTTER AMBERJACK"
- Shore Radio Facilities example:
  "COAST GUARD SECTOR SAN FRANCISCO"
- Aircraft example: "COAST GUARD SIX FIFE SIX ZE-ROH"
- Aircraft involved in SAR example:
  "RESCUE SIX FIFE ZE-ROH NINE-ER"
- Once communications are established you can cease saying
  "COAST GUARD."

Maritime Mobile Service Identification

Digital Selective Calling (DSC) is a required feature of modern VHF-FM marine radios and enables direct calling of vessels using registered Maritime Mobile Service Identification (MMSI) numbers without the need for a hail and reply. Similar to a traditional phone number, vessels are assigned a unique sixteen-digit MMSI that can be entered into a DSC radio along with the desired working channel then called with the push of a button. When called, the recipient's DSC radio will provide audible and visual alerts. Once the call is acknowledged, both the sending and receiving radios will automatically shift to the selected working frequency and can begin their transmissions. DSC can also be used to make position requests in which the receiving vessel's current GPS position is transmitted with the push of a button.

Global Maritime Distress and Safety System

The International Maritime Organization has developed the Global Maritime Distress and Safety System, usually referred to simply as GMDSS, to use satellites and digital selective calling technology to communicate distress alerts, coordinate

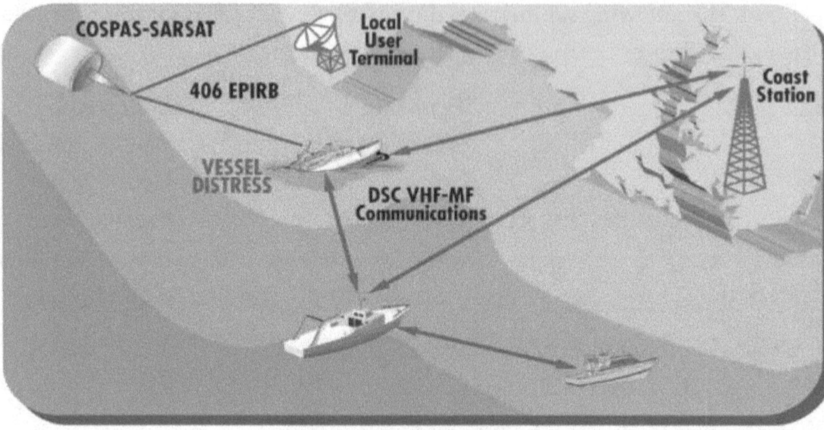

GMDSS BASIC CONCEPT

search and rescue, locate distressed mariners, and facilitate general communications. One of the primary components of this system is the VHF-DSC marine radio, which can be very useful in both routine and emergency communications. To further enhance the GMDSS system, the U.S. Coast Guard has updated its communication equipment under a program known as Rescue 21, which offers not only improved voice communications but also a direction-finding capability. Another very useful piece of GMDSS equipment, particularly for vessels operating offshore, is the 406 MHz Emergency Position Indicating Radio Beacon, or EPIRB, a small, buoyant beacon that is designed to float free from a sinking vessel. These beacons operate on the international distress frequency 406 MHz and, when activated, automatically transmit a distress message and position digitally to orbiting satellites.

Coast Guard Command Centers are manned 24/7 to monitor VHF-FM Channel 16 for voice notifications of emergencies and Channel 70 for DSC alerts. Boats, cutters, and aircraft also monitor these at all times when under way.

If voice communication is not possible or not effective, other means of communication will have to be used. These may include pyrotechnics, flag hoist signals, hand signals, or a flashing light S-O-S. Accepted distress signals are defined in the Navigation Rules and Regulations handbook as shown on the next page.

The list of accepted distress signals is not all inclusive, and any unusual signal or action could be a sign that a craft is in trouble. For example, a U.S. flag flown upside down from a vessel under way is a well-documented method of signaling distress. Investigate any peculiar or suspicious signals.

International distress signals

Emergency Situation Prowords

When an emergency occurs, prowords help to define a situation and its degree of urgency. The meaning of each proword is outlined as follows.

MAYDAY. A distress call of the highest priority. Spoken three times, it shows that a person, boat, or aircraft is threatened by grave or imminent danger and requires immediate assistance. A MAYDAY call has absolute priority over all other transmissions and is meant for all stations in range. All units hearing a MAYDAY should immediately cease transmissions that may interfere with the distress traffic and continue to listen on the distress message's frequency. When working a distress situation on channel 16, do not attempt to change or shift to a working channel until enough information is obtained to handle the distress in case communications are lost during the act of shifting. The minimal information needed is position, problem (nature of distress), and the number of people on board.

PAN-PAN (PAHN-PAHN). A signal used with an Urgent Marine Information Broadcast. Spoken three times to notify traffic of a very urgent situation concerning the safety of a ship or aircraft and to solicit the assistance of those capable of responding.

SECURITÉ (SEE-CURE-IT-TAY). A safety signal spoken three times to indicate a message concerning the safety of navigation or important weather warnings.

Digital Selective Calling

A DSC radio allows mariners with DSC-equipped and registered radios to transmit, at the push of a button, their exact GPS position and vital vessel information to the Coast Guard and other DSC-equipped vessels. When properly connected to a GPS receiver and set up with the identifying information for the vessel on which it is installed, DSC radios can send an automated message containing the vessel's MMSI number, position, and nature of distress. As a result, the Coast Guard is able to respond more quickly and to more accurately identify the location of the distress call.

Rescue 21

Rescue 21 is the Coast Guard's advanced command, control, and direction-finding communications system with coverage along the entire Atlantic, Pacific, and Gulf coasts of the continental United States and along the shores of the Great Lakes, Puerto Rico, Hawaii, and Guam. The system is used to monitor multiple VHF frequencies and features a direction-finding capability that provides lines of bearing to vessels in distress and those transmitting radio signals, digital selective calling support, geographic display, and audio recording and playback.

Alert Warning System

The Alert Warning System (AWS) is the Coast Guard's only authorized mass notification system. The objective of AWS is to disseminate alerts, threat warnings, and other critical information to Coast Guard personnel, maritime port partners, and key stakeholders in support of maritime security-level changes, Marine Transportation System recovery, small-vessel security awareness and outreach, and industry notification and outreach. Examples of AWS messages include information pertaining to weather emergencies, evacuation notices, government operating status, and other critical administrative messages.

Automated Mutual-Assistance Vessel Rescue System

AMVER is a worldwide voluntary computer-based reporting system sponsored by the U.S. Coast Guard and used worldwide by search-and-rescue authorities to arrange for assistance to persons in distress at sea. With AMVER, rescue coordinators can identify participating ships in the area of distress and divert the best-suited ship or ships to respond. Participating ships notify AMVER of their position by e-mails sent via satellite or HF radio, by compressed messages sent via satellite, HF radiotelex, or telefax.

Satellite Telecommunications and Broadband

Several commercial satellite networks provide worldwide broadband services to Coast Guard units, particularly underway cutters.

INMARSAT / KA-Band

INMARSAT provides KA-band satellite internet, phone, short messaging service (SMS) texting, and integrated services digital network (ISDN) to users worldwide via fourteen geostationary telecommunications satellites. The latest service to be fielded on board Coast Guard cutters is FleetBroadband, which is designed specifically for oceangoing vessels using portable domed terminal antennas. These antennas, and corresponding indoor controllers, are used to connect phones and computers on board the cutters with the rest of the world. All FleetBroadband antennas require line-of-sight to one of three geosynchronous orbit satellites. Legacy INMARSAT services still being used include FLEET77, which provides low- and medium-speed voice, data, and ISDN capability, and INMARSAT-C, a data only, store and forward service that does not provide real-time communications or encryption capability.

KVH / Ku-Band

KVH provides Ku-band satellite broadband and voice communications using compact TracFones and VSAT systems connected through a network of eleven satellite transponders and eight terrestrial stations. It provides primary access to the Coast Guard network, e-mail, SEAWATCH, SIPRNET, and SIPRCHAT. Service coverage is reduced from that available on INMARSAT / KA-band.

Iridium / L-Band

Iridium phones and messaging devices provide L-band satellite voice, messaging, and data services and are commonly used as a backup means of communication aboard cutters and as a primary means of communication for boats operating beyond the range of radiotelephones. The Iridium network constellation consists of sixty-six active satellites.

Networks

Information systems and networks necessary for Coast Guard business are used by USCG personnel stationed on land as well as individuals at sea on its cutters.

Coast Guard Network, CGOne

The CGOne network supports sensitive, operational, and administrative information systems as well as unclassified e-mail transmission and delivery for Coast Guard forces. This network exists as an enclave within the Department of Defense Information Network, a globally interconnected, end-to-end set of electronic information capabilities for collecting, processing, storing, disseminating, and managing digital information. This network supports all Coast Guard districts and major commands with infrastructure extending across the continental United States and Alaska and Hawaii. Coast Guard cutters in port are provided pier-side connectivity by their respective supporting stations, via a T1 capable link. When at sea, minimal connectivity is provided through a commercial satellite link.

Regular access to a CG Workstation or Virtual Desktop is required to maintain an active profile on CGOne, which requires a login every forty-five days. To reactivate a suspended account or resolve other CG Workstation issues, a help ticket must be submitted via CG FIXIT.

The Secret Internet Protocol Router Network, SIPRNet

SIPRNet is a system of interconnected computer networks used by the U.S. Department of Defense and the U.S. Department of State to transmit classified information, up to and including information classified SECRET, by packet switching. Pronounced "sipper net," it also provides services such as hypertext document access and electronic mail. In plain terms, SIPRNet is the DoD's classified version of the civilian Internet.

Record Message Systems

Record message systems are used to exchange unclassified official information such as reports, announcements, and orders regarding operations and administration and for emergency warnings and mass notifications. There are specific requirements for how messages are formatted on these systems, and personnel must follow the appropriate protocol, keep messages brief, and maintain professionalism. Information exchanges now used are accessible via a web application or through a shared mailbox in either

Microsoft Outlook or Outlook Web app. Each unit that is receiving messages is assigned an organizational administrator to assist unit personnel with establishing accounts, assigning message release authority, and accessing the shared mailbox.

Command and Control Official Information Exchange

This information exchange is used to exchange operational information, defined as any record message related to specific actions to execute and directly support strategic and tactical USCG missions.

Admin OIX

Admin OIX is used to exchange official administrative information, defined as information concerning service organization, personnel management, unit logistics, individual and unit training, discipline, contingency response, program changes, and other matters not specifically operational.

Administrative Information on CG Portal

General messages are posted on CG Portal's General Messages site and include the following:

- ALCOAST: Messages applicable to all Coast Guard commands, with blanket distribution to all Coast Guard units
- ALCGOFF: Messages applicable to all officers
- ALCGENL: Messages applicable to all enlisted
- ALCGCIV: Messages applicable to all CG civilians
- ALCPSC: Messages applicable to all Coast Guard members that originate from the Personnel Services Command

E-Mail

The Coast Guard uses e-mail to communicate both formally and informally between individuals and groups. In the course of your official Coast Guard duties, you will be required to send and receive e-mail. Therefore, it is important for you to understand how to send, receive, and write e-mail messages. If e-mail is used for conducting Coast Guard business, it must meet the same standards required for all other official correspondence. In addition, there are limitations when it comes to sending e-mails that may deal with classified, For Official Use Only (FOUO), and Sensitive but Unclassified (SBU) information.

Before sending an e-mail you must first determine who the information is intended for. Are you sending it to an individual or group? When drafting an e-mail, the primary recipient of your message should be listed in the box of the message. You can click on the global address list icon to look up the names and contact information for everyone in the Coast Guard and the Department of Homeland Security. Type their last name in first. You should send someone a copy (Cc) if you want to inform them of the contents of the e-mail but require no action or response from them; always include a copy to your supervisor. Use blind copy (Bcc) if sending to multiple recipients and you don't want them to be able identify one another. The subject line of your e-mail is very important and should be relevant to your message, telling the recipient what the e-mail is about. Be clear, since you and the recipients may use this line to refer to the e-mail in the future.

You should always begin your e-mail by addressing the recipient properly. For example, be sure to use the proper salutation, rank/title, and greeting. The purpose of e-mail is to get your message across clearly. Your message body should be clear, concise, and to the point. Use short paragraphs with blank lines between them, and be sure to use proper spelling. Don't use all uppercase as this is the same as yelling, and don't write when you are angry or when in doubt about the receipt or content of an e-mail.

Within the Coast Guard and other military services, two complimentary closings are customary. The word "respectfully" should be used by a senior when addressing a junior member and "very respectfully" should be used by a junior member when addressing a member of higher rate or rank. You may abbreviate by using "V/R" in less formal e-mail correspondence.

The last item you will include in your e-mail is your signature. This will let people know who you are and how to contact you. Omit sayings or quotes with your signature when the e-mail is official correspondence. Your signature should include

- Rate, rank/title/name
- Job title
- Unit/address
- Phone number(s)
- E-mail address

Upon receipt of an e-mail, you should send a reply back to the sender as soon as possible, especially if you are listed in the "To" line. When replying to an e-mail, you have two options: reply to the sender by clicking on the "Reply button" or reply to the sender and everyone else who also received the e-mail by clicking on "Reply All." Remember that every e-mail you send leaves an official record. You should always assume when on a standard workstation that someone is watching you.

13

WATCHSTANDING

A day's work on board a cutter or ashore involves rating duties, but it also requires crew on duty and standing the watch. Even after the workday, the watch continues. The Coast Guard is fully operational and on watch twenty-four hours a day, every day. When on duty or standing the watch, you must maintain a high state of readiness because you will be the first to detect and respond to threats and emergencies that threaten your unit. You may also be the first to receive and respond to reports of distress, violations of law, pollution discharges or hazardous materials releases, and other emergencies. You must remain always ready to carry out the Coast Guard's missions.

The reasons for establishing a live watch are safety, security, and readiness. Your watch may require you to make periodic rounds of a cutter, building, or base, remaining alert and vigilant for anything that threatens safety or security. You may also be tasked with answering phones or radio calls and making notifications or launching assets to respond to a time-critical emergency. Regardless of the nature of your duty, when standing watch you are entrusted with great responsibilities and will be held accountable for carrying them out.

WATCHES

In many naval disasters, personnel were held at fault for failing to take proper action because they were not aware of their duties and responsibilities or did not think they had the authority to act. All personnel will receive indoctrination, training, and examinations to get the prerequisite qualifications to perform the duties and functions of the assigned watch before ever being assigned as a watchstander. If you are assigned a watch and your duties are unclear, immediately notify the senior watchstander or officer of the deck (OOD). Designated responsibilities, situational awareness, and teamwork are essential to the efficient and safe operation of the ship. Confusion or conflict among watch personnel concerning responsibilities or authority could result in collision, grounding, loss of the ship, or loss of life. In many cases serious damage or loss of life might have been averted by the timely action of watchstanders working together as a coordinated team.

Departments are responsible for specified watches, and department heads are responsible for carrying out supervisory duties Traditionally, a watch lasts four hours; however, based on the conditions under which the watch will be stood, a watch can last up to six hours. Studies show that, depending on the watch rotation, alternative watch lengths can be more effective for ensuring adequate crew rest over long stretches (days or weeks) of watchstanding. Following are the standard watches stood in Coast Guard cutters both at sea and in port:

0000–0400	Midwatch
0400–0800	Morning watch
0800–1200	Forenoon watch
1200–1600	Afternoon watch
1600–1800	First dogwatch
1800–2000	Second dogwatch
2000–2400	First watch

The two "dogwatches," from 1600 to 1800 and from 1800 to 2000, serve to alternate the daily watch routine so that personnel who have the midwatch one night will not have it again on the following night.

Watch Section

Each crew member is assigned to a section of the watch. These sections are numbered; small cutters usually have three sections, and large cutters may have four. Thus, when the word is passed that the first section, the second section, or third has the watch, each person in that section immediately reports to his or her assigned watch station. Crew members in a three-section watch are spoken of as having a one-in-three watch, while those in a four-section watch have a one-in-four watch, and so on. Thus, a crew member standing a one-in-three is on watch for four hours, off for eight hours, and then back on watch for four more. In all, that person stands eight hours' watch in twenty-four hours.

Late Sleepers

Crew members on watch from 0000 to 0400—the midwatch—have a difficult time getting enough sleep. For this reason, they are permitted to sleep in after "all hands" or "turn to ship's work" is piped. They are the "late sleepers" referred to in daily routines.

General Duties of Watch Officers and Watchstanders

To ensure that proper watches are maintained, all watchstanders shall adhere to the following basic principles:

- Be present and alert at the watch station. Remain on station until properly relieved or secured by appropriate authority.
- While on watch, maintain smart, military conduct. Wear a clean and proper uniform in the correct manner.
- Use the appropriate terminology, customary to the service, when issuing orders and making reports.
- Obey all lawful orders and regulations, and require the same of all supervised persons on watch.
- Before relieving the watch, become thoroughly acquainted with all matters necessary for the proper performance of duties while on watch. The relief of the watch may be declined if, in the watch officer's opinion, such action is justified.
- Devote full attention to the proper execution of the duties and responsibilities of the assigned watch station.
- Do not be assigned or assume any duties that may distract attention from the proper execution of the duties at the watch station.
- Demand formality in all relationships while on watch.
- Make all required inspections and any such additional ones as are permitted and considered necessary to ensure that the duties of the watch are properly performed.
- When under arms, be properly qualified and instructed; exercise all of the safety rules for the carriage of weapons; and be aware of the customs and courtesies unique to the wearing of firearms, including the orders to sentries.

- While on watch or on duty as a member of a patrol or as a sentry, do not fail to report a violation of the cutter's regulations, directives, safety regulations or other breach of good order and discipline. Do not fail to act or exert with proper authority to suppress such violations, or fail to report any known violations and other malpractices that may endanger the safety or security of the cutter and its personnel.

Relieving the Watch

The watch relief process is a controlled, formal, and precise function. Accordingly, the relieving watch adheres to the following:

- The relieving watch shall inspect all spaces and equipment as required before relieving the watch.
- The relieving watch shall examine all applicable equipment log readings at the watch station since the last watch, noting any variations from normal operations. Check that the preceding watch has filled in and signed the log sheets as required. Variations shall be discussed and all questions shall be resolved prior to watch relief.
- The relieving watch shall read the remarks sections of applicable logs from the last watch, carefully noting and discussing unusual conditions, deviations from normal, or other matters of importance.
- Both the relieved watch and the relieving watch are responsible for seeing that the relieving watch is completely aware of all unusual conditions that exist. These include tactical situations, equipment out of commission or being worked on, outstanding orders, deviations from normal plant or equipment "lineup," forthcoming evolutions, and any other matters pertinent to the watch.

The relief is exercised smartly in accordance with the following guidelines:

- Relief reports, "Ready to relieve."
- Person being relieved gives a status report to the oncoming watch section.
- Relief makes a tour of the watch station.
- Person being relieved completes briefing of relief, including unexecuted orders and anticipated evolutions and answers any questions.
- Permission is obtained from the appropriate watch supervisor to relieve the watch.
- Relief, when fully satisfied and completely informed regarding the watch, relieves the watch by saying, "I offer my relief" or "I relieve you."

- At this time, responsibility for the watch station shifts to the oncoming watchstander, and the individual being relieved will state, "I stand relieved."
- The person relieved from the watch reports the relief to the same person from whom permission was requested to relieve.
- On watch stations on which a log is maintained, the log will be completed and signed by the off-going watchstander before leaving the watch station.

The general pattern of relief:

- One-half hour before the hour—Oncoming watch on station, make inspection, read logs and turnover sheets, obtain information from relief.
- Fifteen minutes before the hour—Watch members relieved.
- After watch members have relieved and watch supervisors feel the oncoming watch has situational awareness, watch supervisors relieve OOD, engineer of the watch, combat information center watch supervisor, and so on.

WATCH OFFICERS

A watch officer is placed in charge of a watch or a portion of a watch. The commanding officer or officer in charge assigns as a watch officer any commissioned, warrant, or petty officer who is certified in the position.

Officer of the Deck / Officer of the Day. The OOD is the commissioned, warrant, or petty officer on watch who is in charge of the cutter (deck) and is responsible for the safety of the unit and the performance of duties prescribed by the commanding officer and Coast Guard regulations. Every person on board is subject to the orders of the commanding officer, except the executive officer, who is subordinate to the OOD.

Engineer of the Watch. The engineer of the watch is the officer or petty officer in direct charge of the machinery and engineering department and responsible for the proper operation, care, and preservation of all machinery in the engineering department and for the leadership of its personnel.

Tactical Action Officer. The tactical action officer is the commanding officer's representative in all matters concerning the tactical employment and defense of the cutter. They are responsible for the safe and proper operation of the combat systems, the engagement of hostile forces, and other tactical actions required to fight or defend the cutter.

Combat Information Center Watch Supervisor. The CIC watch supervisor directs and supervises the personnel on watch in the CIC, ensuring that operators are detecting and reporting air and surface contacts within the capabilities of their equipment.

WATCHSTANDERS

Following is a list of the most common watchstanders with a brief description of their duties.

The Boatswain's Mate of the Watch (BMOW). When assigned, the BMOW is constantly roving throughout the cutter, delivering messages and carrying out such duties as the OOD may direct. In the absence of an assigned BMOW, the duties are shared by other members of the assigned watch, primarily the quartermaster of the watch.

Machinery Watchstander. The machinery watchstander starts, secures, and monitors the operation of main propulsion and auxiliary equipment, recording equipment-operating parameters and maintaining the material condition of engineering spaces.

Damage-Control Assistant. The damage-control assistant is responsible, under the engineering officer, for the operation, maintenance, repair, and training in the ship's survivability systems, including the control of damage, stability, list, trim, fire fighting, restoration from damage, and chemical, biological, and radiation countermeasures.

Quartermaster of the Watch. The quartermaster of the watch is stationed on the bridge to ensure the safe navigation of the cutter.

Lookouts—Sky, Surface, and Horizon. Every cutter will have at least one lookout. On larger cutters, additional lookouts may be stationed, when circumstances demand, on the flying bridge, on the fantail and bow, or elsewhere. Bridge lookouts may report in person to the OOD; others will report via radio or telephone. Although it may look to an outsider as if all a lookout has to do is stand around and enjoy the view, it takes considerable skill to stand a good lookout watch, and the safety of the cutter may depend on just how well the lookout attends to their duties.

The first duty of the lookout is to sight an object and report it. The second duty is to identify it. The report of an object sighted should not be delayed until identifying the object. After reporting, the lookout must be certain that the OOD acknowledges the report.

- Bridge lookouts: These lookouts, sometimes port and starboard, stand watch in the wings of the bridge or on the flying bridge. They report ships, aircraft, land, or any object sighted, and at night they make half-hourly reports on the ship's running lights.
- Forecastle lookout: These lookouts may be stationed during periods of reduced visibility—fog or snow—and while entering harbor.

Lookouts

- Fog lookout: This lookout may be the forecastle lookout, or may stationed on the wings of the bridge, to keep watch astern. Their primary job is to listen carefully.
- Sky and surface lookouts: On larger cutters, when enemy operations are expected, additional lookouts are assigned certain sectors of the sky and surface all around the ship.

Low-Visibility Lookouts. These lookouts are stationed during periods of reduced visibility in those locations where approaching ships can best be seen or heard in order to detect their presence.

Helmsman. The helmsman steers courses prescribed by the OOD or conning officer.

Lee Helmsman. The lee helmsman stands watch at the engine order telegraph and the ship's throttles on the bridge and rings orders to the engines to ensure that all bells are correctly answered.

Aft Steering Watch. The aft steering watch is stationed at the secondary steering stations. Duties include lining up and operating the steering equipment as directed by the OOD and taking over steering control in the event of a steering casualty.

Cold Iron Watch. This watch inspects main machinery spaces that are secured and have no regular watch posted; they report hourly to the OOD.

Anchor Watch. Crew members assigned to the anchor watch assist the OOD in such tasks as reporting signs of dragging the anchor, veering chain, or other emergency work. The number of people assigned and their specific duties vary from cutter to cutter.

Messenger. The crew member assigned to the OOD, either on the bridge when under way or on the quarterdeck in port, is commonly called the OOD messenger; they pass messages to the various personnel throughout the cutter.

Phone Talker. There may be one or more phone talkers, depending on the cutter and the operation. Phone talkers order to the stations as directed and inform the OOD of all information received.

Operations Specialists of the Watch. The operations specialist maintains the required communications in the main radio room, receiving, transmitting, and routing traffic as required.

Sounding and Security Patrol. The sounding and security patrol is a member of the engineering department who maintains a continuous patrol of unmanned spaces below decks. The patrol takes periodic soundings of designated tanks and spaces, checks damage-control closures, and is alert for evidence of sabotage, theft, fire, and fire hazards. In addition to making an hourly report to the engineer of the watch or OOD, the patrol frequently relays the report of the cold iron watch. Both of these watches are usually manned during nonworking hours at sea and in port.

Security Watch. A security watch patrols topside and below decks, whether at sea or in port. At sea, the BMOW is normally assigned this responsibility. In port, it usually falls upon the messenger, who will make these rounds between "Taps" and reveille.

LOGS

Originally a log referred to a book for recording readings from a chip log used to estimate a ship's speed through the water. Today the term has come to refer to a wide variety of official records that document everything from the positions of a cutter during an underway patrol to weather observations to the transmissions made and received over a radio. Official logbooks are legal records and are regularly used as evidence to support investigations and prosecutions for cases in which the Coast Guard played a role. Logs are generally arranged by date and marked with the time of an event or action of significance. Whenever possible, logs are maintained electronically.

Smooth Log / Deck Log. The smooth log or deck log is a complete daily record, by watches, in which is described every circumstance and occurrence of importance or interest that concerns the crew and the operation and safety of the cutter or that may be of historical value.

Weather Observations and Operational Summary Sheet (CG-4380B). Cutters log the state of the weather hourly while at sea and every fourth hour while in port.

Navigation Data Sheet. The navigation data sheet is a complete record of all direct readings of the steering compass maintained on the Log-Navigation Data Sheet, CG-4380C (current revision). It also serves as a record of gyrocompass error. When under way, steering compass comparisons are made at least every half hour and upon changing course.

BRIDGE RESOURCE MANAGEMENT

It is easy for those new to maritime operations and those who have become complacent after years of service to underestimate the dangers of poor bridge teamwork procedures. Bridge Resource Management is a formal program that outlines how the navigation teams on board cutters interact, divide responsibilities, navigate, and react to changing conditions. Bridge Resource Management principles include the following:

- Allocation, assignment, and prioritization of resources
- Effective communication
- Assertiveness and leadership
- Obtaining and maintaining situational awareness
- Consideration of team experience

Safe Navigational Watch

While under way, the cutter must keep a safe navigational watch. To do that, the crew must be aware of hazards to navigation, navigational landmarks, the vessel's location relative to the intended track, maritime traffic, weather, sea state, location and duties of watch partners, limitations in propulsion and steering systems, and maintaining appropriate communications. To maintain a safe navigational watch, the crew will

- Monitor the voyage plan closely and continuously;
- Maintain proper lookout by all available means;
- Maintain a safe speed;
- Check position, course, and speed at frequent intervals;
- Verify that the steering mode selected is appropriate;
- Maintain a suitable under-keel clearance for the draft of the vessel at all times;
- Make course changes in accordance with the voyage plan;
- Verify that the vessel's position is fixed and plotted at intervals appropriate for the vessel's speed and the area being transited;
- Ensure that aids to navigation are identified;

- Check, frequently, that the radio equipment is functioning properly;
- Take appropriate action to avoid collision with approaching vessels;
- Verify that rudder and engine orders are executed as ordered;
- Verify that gyro input to all navigation equipment is correct;
- Verify that magnetic compass and gyro errors are determined and logged;
- Verify that magnetic variation and compass deviation are correctly applied to courses and bearings;
- Verify that the person steering is competent to stand watch;
- Determine tide and current conditions;
- Verify that set and drift are determined and applied;
- Verify that weather conditions are correctly and timely recorded and reported as required;
- Verify that the running lights are working;
- Notify the commanding officer or officer in charge, when appropriate, in accordance with the commanding officer's or officer in charge's standing orders;
- Ensure that protected marine habitats, areas, and sanctuaries are properly identified; and
- Ensure that appropriate log entries are made.

Bridge Resource Management Conditions

Whenever the navigational situation demands more resources than are immediately available on the bridge, a dangerous condition exists. This can be dealt with by adding more resources (i.e., adding an additional watch officer) or lowering the navigational demands (i.e., reducing speed). Some conditions that increase demands on the bridge team include fog, heavy traffic, entering a harbor or restricted area, or an emergency. Standards known as Watch Conditions help define the personnel and procedures to be used under various situations.

Watch Condition I. Indicates unrestricted maneuverability, weather clear, little or no traffic, and all systems operating normally. In this condition, depending on the size and type of vessel and its mission, often a single person can handle the bridge watch.

Watch Condition II. Applies to situations where visibility is somewhat restricted, and maneuverability is constrained by hydrography and other traffic. This condition can be dealt with by augmenting the at-sea watch.

Watch Condition III. Reflects a condition where navigation is seriously constrained by poor visibility or close quarters, as in bays, sounds, or approach channels, and heavy traffic.

Watch Condition IV. When visibility is poor, maneuvering is tightly constrained as in channels and inner harbors, and traffic is heavy.

Watch conditions can change almost immediately due to planned or unforeseen events. Under each watch condition, OODs must manage their resources and call in extra help when necessary, assigning personnel as needed to jobs for which they are qualified and ready to perform.

SUSPICIOUS ACTIVITY

Suspicious activity refers to activity regarding vessels, buildings, facilities, persons, vehicles, or cargo in which a reasonably prudent person would consider out of the ordinary or unusual based on the facts and circumstances for that situation. You may encounter numerous people and situations while standing watch throughout your Coast Guard career. During these watches, you may see something that may indicate a terrorist activity is being planned, violations of the Uniform Code of Military Justice, theft, sabotage, or numerous other unusual circumstances.

To recognize suspicious activities, you need to know the difference between normal and abnormal behavior and activity. Watchstanders need to qualify for watch, complete break-in watches, and observe to understand what is normal and what is unusual for their area of responsibility. Know the status of personnel, spaces, and activities before you relieve the watch. If you recognize anything that may be considered suspicious, collect information quickly and report to the next senior watchstander. Some examples of suspicious activities include, but are not limited to

- Vehicles, aircraft, vessels, or persons behaving in an unusual or threatening manner;
- Unauthorized photography of base or nearby infrastructure;
- Person(s) running away or fleeing;
- Person(s) engaged in surveillance; and
- Person(s) engaged in asking unusual questions (i.e., When does your ship pull out? How many crew members are on your ship? Where is your ship going? How many security personnel work on the base?).

Noticing unusual or suspicious activity is important. Equally important is being able to communicate details of what you have witnessed. Details of unusual or suspicious activity include, but are not limited to

- People: Color of hair, eyes, skin; age; race, ethnicity; clothing; identifying marks or features such as facial hair, scars, and tattoos;

- Boats: Hull type; name/number; flag; color of hull and superstructure; make/model; length; and
- Vehicles: License plate number and state; color; make/model/year; identifying features such as stickers, decals, rims, dents, rust.

Force Protection Conditions

Typically attacks are carefully planned out and require significant time conducting surveillance on intended targets. Force protection measures help deter, detect, and interrupt such plans and are organized in a system called Force Protection Conditions, or FPCON (pronounced F-Pea-Con). As the threat changes, the FPCON changes to protect service and family members, and civilians. Unit commands design protective measures appropriate for the local situation and can apply additional measures or move to a higher FPCON as needed to respond to intelligence or provide an additional deterrent. As the FPCON progresses, all measures of the current and lower conditions are implemented.

FPCON NORMAL. Routine security measures at a specific installation are called FPCON NORMAL. Although an attack could occur at any time, the best information available indicates only routine security measures are needed.

FPCON ALPHA. A general threat exists of possible terrorist activity, directed against units and personnel, but the nature and extent are unpredictable. The measures in this FPCON must be capable of being maintained indefinitely. Examples include

- Secure buildings, rooms, and storage areas not in regular use; and
- Limit access points for vehicles and personnel to maintain a reasonable flow of traffic.

FPCON BRAVO. An increased and more predictable threat of terrorist activity exists. The measures in this FPCON must be capable of being maintained for weeks without causing undue hardship, affecting operational capability, or aggravating relations with local authorities. Examples include

- Move cars and objects at least twenty-five meters from buildings;
- Frequently inspect the interior and exterior of buildings in regular use for suspicious packages; and
- Physically inspect visitors and randomly inspect their suitcases, parcels, and other containers.

FPCON CHARLIE. Applies when an incident occurs or when intelligence is received indicating that some form of terrorist action against units and personnel is imminent. Implementation of this measure for more than a short period will probably create hardship and will affect peacetime activities of the unit and its personnel. Examples include

- Limit access points to the absolute minimum;
- Randomly search vehicles;
- Issue weapons to guards;
- Increase patrolling of installation; and
- Increase barriers.

FPCON DELTA. This applies in the immediate area where a terrorist attack has occurred or when intelligence has been received that terrorist action against a specific location is likely. Normally this FPCON is declared as a localized warning. In FPCON DELTA, all nonessential activities and movements are suspended, and resources are focused on defense against attack. Examples of security measures include

- Search all vehicles and their contents before allowing entrance to the installation;
- Control access and implement positive identification of all personnel no exceptions; and
- Search all suitcases, briefcases, and packages brought into the installation.

National Terrorism Advisory System

The Department of Homeland Security established the National Terrorism Advisory System (NTAS) to communicate information about terrorist threats to the public. Alerts issued under this system provide a summary of the threat, information on actions taken to ensure the safety of the public, and recommended steps that individuals, communities, businesses, and governments can take to help prevent, mitigate, or respond to the threat. Each alert includes a clear statement of the threat:

- ELEVATED THREAT: Warns of a credible terrorist threat against the United States.
- IMMINENT THREAT: Warns of a credible, specific, and impending terrorist threat.

NTAS alerts contain a sunset provision indicating a specific date when the alert expires. There will not be a constant NTAS alert. If the threat information changes, the secretary of homeland security may announce an updated NTAS alert.

Marine Security System

The three-tiered Maritime Security (MARSEC) System was established by the Coast Guard to easily communicate threats to the maritime sector of transportation. The commandant of the Coast Guard adjusts MARSEC level based on any NTAS alert issued by the Department of Homeland Security. The system's three levels are

- Level 1: Corresponds to no NTAS threat
- Level 2: Elevated NTAS alert
- Level 3: Imminent NTAS alert

Challenging Personnel

Whether you are assigned to a shore or an afloat command, you may be called to stand security or sentry duty, particularly during a heightened FPCON. You are responsible for knowing your eleven general orders. General Order #11 states, "To be especially watchful at night, and during the time for challenging, to challenge all persons on or near my post, and to allow no one to pass without proper authority."

For certain watches, you will be given instructions and a list of personnel who are authorized access to a military installation or secure area of operation. To ascertain the identity of personnel, you challenge them. When a person approaches a post during challenging hours, usually midwatches, challenge sharply by saying, "HALT! WHO GOES THERE?"

If the person's name doesn't appear on the authorized access list, a standard reply might be, "I'm sorry. Sir/Ma'am, your name does not appear on the authorized access list. Please stand by while I contact the officer of the day." You may also be required to apprehend and turn over to proper authority all suspicious persons involved in a disorder occurring on or near your post and all parties who try to enter your post at any time without authority or proper identification. If it is warranted, detain the offender(s) and call the OOD or the next senior watchstander for further instructions.

While assigned to patrol or sentry duty, crew members shall not

- Fail to report a violation of the unit's regulations, directives, or other breaches of good order and discipline;
- Fail to report violations of established safety procedures, precautions, and regulations; or
- Fail to exert themselves to prevent such violations.

Remember, YOU will be held accountable for your watch!

Bomb Threats

Bomb threats can be made in the form of a telephone message, written notes or letters, or e-mail, or by suspicious packages received through the mail or found at the unit. Although most bomb threats are a hoax, each is treated seriously, and action is taken immediately to prevent personal injury. Bombs can be constructed to look like almost anything and can be placed or delivered in any number of ways. The probability of finding a bomb that looks like the stereotypical bomb is almost nonexistent. The only common denominator that exists among bombs is that they are designed or intended to explode. You need to stay vigilant and report anything out of the ordinary. As a watchstander you need to know your watch and its normal surroundings. Let the trained bomb technician determine what is or is not a bomb.

Most bombs threats are received by phone, and it is your responsibility to gather as much information as possible. It is standard for commands to have a Telephonic Bomb Threat Report near phones to document any specifics communicated by a bomb threat. Act quickly, but remain calm and obtain as much information as you can. In all cases notify your chain of command and law enforcement. Follow unit procedures and orders to isolate and evacuate threatened facilities and areas.

Suspicious Packages

Bomb threats may come in the form of a suspicious package. If you detect a suspicious package, DO NOT touch or move it, or use two-way radios or a cellular phone signals nearby. Evacuate the area until police arrive and evaluate the threat, or activate the fire alarm.

Active Shooter Event

An active shooter is an individual who is engaged in killing or attempting to kill people in a confined and populated area. Victims are selected at random. The event is unpredictable and evolves very quickly. If an active shooter is in your vicinity, you must be prepared both mentally and physically to deal with the situation. You have three options.

RUN. Have an escape route and plan in mind. Leave your belongings behind and evacuate regardless of whether others agree to follow. Help others escape, if possible, but do not attempt to move the severely wounded. Prevent others from entering an area where the active shooter may be. Keep hands visible. Call 911 when you are safe.

HIDE. Lock door or block entry to your hiding place. Silence your cell phone and remain quiet.

FIGHT. Fight as a last resort and only when your life is in imminent danger. Attempt to incapacitate the shooter, acting with as much physical aggression as possible. Commit to your actions . . . your life depends on it.

14

SEAMANSHIP

Seamanship refers to the skills, techniques, practice, and art of handling a ship or boat at sea. Regardless of your rating or specialty, as a member of a seagoing service, you must know or at least be familiar with certain basic seamanship skills, many introduced to you during recruit training, in Officer Candidate School, or at the Coast Guard Academy. All boat types and cutter classes use knots and lines to secure themselves to piers and anchors to hold them in place where there are no piers. Cutters use weight-handling equipment to launch and recover boats, and much of the work on board buoy tenders relies on moving large, heavy loads about the deck. All at sea are subject to wind, waves, and currents that must be understood in order to maneuver safely between ports and over the horizon. It is important to learn the unique terminology associated with seamanship, primarily because you want to avoid confusion but also because there are certain measures of professionalism in the Coast Guard that have no official status and have nothing to do with getting you promoted but size you up as a true mariner rather than a landlubber. If you want to be recognized as a true Coast Guard professional—to earn the respect of those more experienced than you—you should think, act, and speak like a mariner. This means you should use twenty-four-hour time, call a deck a "deck" and not a "floor," and know the difference between rope and line. As a member of the world's finest coast guard, you must strive to become a professional mariner and develop the seamanship required to perform all federal actions expected of the service at sea.

VESSELS

The front end of a vessel is the bow. Moving toward the bow is going forward; when the vessel moves forward, it is going ahead. When facing the bow, the front right side is the starboard bow, and the front left side is the port bow. The central or middle area of a boat is amidships. The right center side is the starboard beam, and the left center side is the port beam. The rear of a vessel is the stern. Moving toward the stern is going aft; when the vessel moves backward, it is going astern. Standing at the stern looking forward, the right rear section is the starboard quarter, and the left rear section is the port quarter. Starboard is the entire right side of a boat, from bow to stern. Port is the entire left side of a vessel, from bow to stern. A line, or anything else, running parallel to the centerline of a vessel is fore and aft. A line or anything else running from side to side is athwartships. Outboard is from the centerline of the vessel toward either port or starboard side. Inboard is from either side of the vessel toward the centerline. The weather deck is the deck exposed to the elements. The pilothouse and cabin are covered. Windward is the direction from which the wind is blowing. Leeward is the opposite point from which the wind is blowing. Going topside is moving from a lower deck to a weather deck or upper deck. Going below is moving from an upper deck to a lower deck. Going aloft is going up into the boat's rigging.

A hold is a compartment belowdecks used exclusively for the stowage of cargo. Accommodation spaces include the galley, mess deck, lounge, quarters, restroom or head, berthing spaces, and similar spaces open to individuals. Machinery spaces are any that contain an installed engine, auxiliary machinery, or associated systems.

Directions to objects outside of the vessel's hull are given in relation to the vessel. An object in front of the vessel is ahead, while one behind is astern. Abeam refers to an object off either side. Alongside is used when lying stationary next to another vessel or to a dock wall. When two vessels are moving on the same course and are level with each other, they are abreast.

Vessel Construction

The three basic types of hull forms are displacement, planing, and semi-displacement. Displacement is the weight of a boat and is measured in long tons (2,240 lb.) or pounds. A displacement hull settles into the water and always forces, or displaces, water around it. At rest, a planing hull also displaces water, but at speed, the shape of its hull allows external forces to lift it so it skims, or planes, along the surface of the water. Once on plane, even small power increases result in large increases in speed. Semi-displacement hulls are a combination of the other hull types but are unable to achieve the same power to speed ratios of a planing hull.

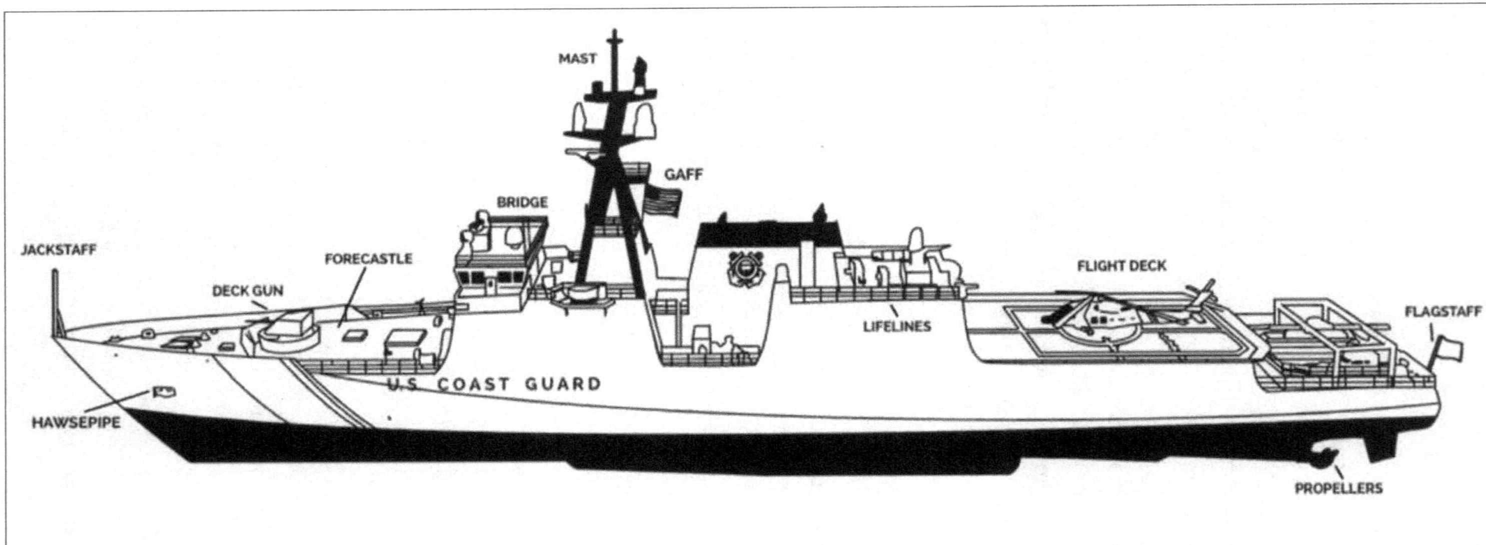

Vessel nomenclature

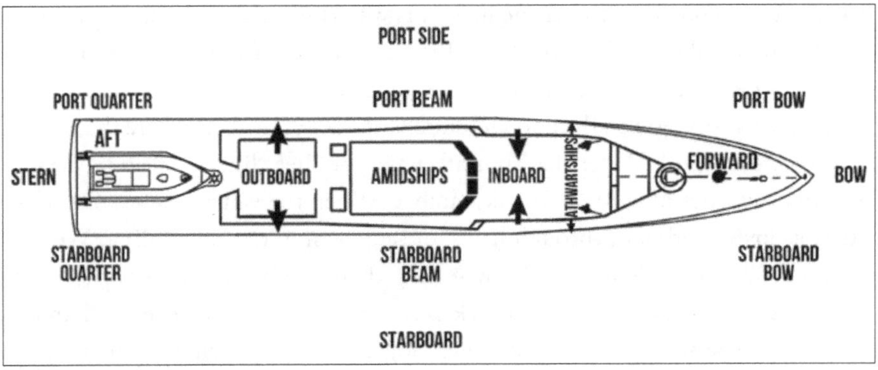

Vessel directions

The hull is the main body of a vessel. It consists of a structural framework and a skin or shell plating. The hull may be constructed of many different materials, the most common being metal or fiberglass. The keel is literally the backbone of the vessel and runs fore and aft along its center bottom. Transverse frames, or ribs, are perpendicular to the keel and give the hull its distinctive shape. These ribs are numbered from bow to stern to help quickly identify a particular location in the interior. Longitudinal frames are parallel to the keel and provide a skeletal structure over which deck plating is laid. All framing gives the hull its strength and is where the skin is attached.

The draft of the vessel is the depth of water from the surface to the keel. The waterline is the line along which the surface of the water contacts the vessel's hull and is usually painted on. Freeboard is the height of the vessel's sides from the waterline to the main deck. A deck is a seagoing floor and provides strength to the hull by reinforcing the transverse frames. The top deck is called the weather deck

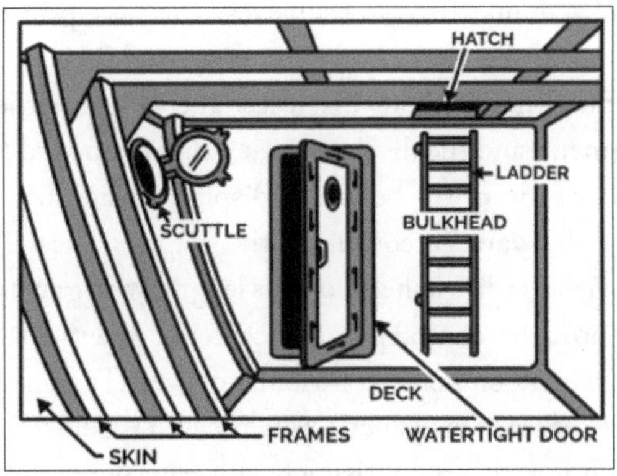

Vessel construction

because it is exposed to the elements and is watertight. In general, decks have a slight downward slope from the bow. The slope makes any water taken on board run aft. A deck also has a rounded, athwartship curve called camber. The two low points of this curve are on the port and starboard sides of the boat where the weather deck meets the hull. Water that runs aft down the sheer line is forced to the port or starboard sides of the boat by the camber. When the water reaches one of the sides, it flows overboard through holes, or scuppers, in the side railing. Weather deck hatches are made watertight by sealing them into raised frameworks called coamings. Hatches operate with quick-acting devices such as wheels or handles, or they may be secured with individual dogs. Scuttles are smaller openings in the deck fitted with a gasket and dogs to create a watertight seal. Any watertight doors are designed to resist as much pressure as the bulkheads through which they provide access. Some doors have dogs that must be individually closed and opened, while others have hand wheels or a handle that operate all dogs at once.

The interior of a boat is compartmentalized by bulkheads, decks, and hatches. With the doors and hatches closed, the space between them becomes watertight and is called a watertight compartment. These watertight compartments are extremely important. Without them, the boat has no watertight integrity and a hole anywhere in the hull will cause it to sink. By dividing the hull into several watertight compartments, the watertight integrity of the boat is significantly increased. One or more of these compartments may flood without causing the boat to sink.

Stability

The tendency of a vessel to remain upright is its stability. A vessel remains upright because of the way its weight and buoyancy interact. The greater the tendency to remain upright, and the stronger the force required to heel the vessel over in any direction, the more stability the vessel achieves. The basic principle of buoyancy is that the upward buoyant force on a body immersed in fluid is equal and opposite to the weight of the fluid that the body displaces. The weight of the fluid displaced is known as displacement, and the displaced water has an upward force, or buoyancy, that is equal to the weight of the boat. The displaced water has a central point, or center of buoyancy, that varies according to the shape of a vessel's hull.

The weight of a vessel is distributed along its length, with gravity pulling the entire vessel downward through a central point, or center of gravity. All the structure and the distribution of weight on board contribute to a vessel's center of gravity. When a boat is at rest, the center of buoyancy acting upward is below the center of gravity acting downward. It is at this point that a boat is considered to be in equilibrium.

Equilibrium is affected by static forces associated with the placement of weight within the hull that move the center of gravity or center of buoyancy or by dynamic outside forces, such as wind and waves. To keep a vessel stable in the water and prevent it from capsizing requires the center of gravity to be low and benefits from a bow-into-the-seas approach. A vessel is usually much longer than it is wide, making the longitudinal plane extending fore and aft more stable than its transverse plane or beam. Generally, the broader a vessel's beam, the more stable that vessel will be.

When a vessel rolls, the force of the center of gravity will move in the same direction as the roll. When heeling, the center of buoyancy will move toward the part of the hull that is more deeply immersed. Listing occurs if the center of gravity is not on the centerline of the vessel. These movements create a righting arm, an angle the vessel reaches where the opposing forces are able to come back to equilibrium. There comes a point at which the righting arm reaches a maximum value, normally at an angle of around 60 degrees. As the boat heels past this point, its righting arm decreases, then becomes negative, leading to a capsize.

Several boat classes in service are purpose-built for heavy weather and are self-righting, returning to the upright position quickly in the event of a knockdown.

Propulsion and Steering

The key to powered vessel movement is the effective transfer of energy from the source of the power, an internal combustion engine, to the water through a mechanism that turns the engine's power into thrust. This thrust moves the boat. There must also be an element of directional control, both fore and aft, and from side to side. Applying thrust has no use if the vessel's direction cannot be controlled, and often the device providing the propulsion also provides the steering. There are three common methods to transfer power and provide directional control:

- A rotating shaft and propeller with a separate rudder
- A movable (steerable) combination, such as an outboard motor or stern drive
- An engine-driven pump mechanism with directional control, called a waterjet

All three arrangements have their advantages and disadvantages from the standpoint of mechanical efficiency, ease of maintenance, and vessel control. Using one type of propulsion instead of another is often a matter of vessel design and use parameters, operating area limitations, and life-cycle cost. There is no single "best choice" for all applications. Regardless of which type you use, become familiar with how each operates and how the differences in operation affect vessel movement.

Vessel Motion

The following describes the different types of ship motion.

- Yaw: Rotation about the vertical axis of the cutter.
- Pitch: Rotation about the lateral axis of the cutter causing the bow to raise and lower in opposition to the stern.
- Roll: Rotation about the longitudinal axis of the cutter.
- Heave: Up and down motion of the entire cutter along its vertical axis.
- Surge: Forward and backward motion of the entire cutter along its longitudinal axis.
- Sway: Side-to-side motion of the entire cutter along its lateral axis.

WEATHER AND OCEANOGRAPHY

Weather and sea conditions can dramatically affect an underway vessel's behavior and handling. Oceanography is the broad field of study focused on winds, waves, tide, and current. It includes the biology and chemistry of the oceans and the geological formations that affect the water. Professional mariners must have an appreciation of all these factors to safely operate in the ever-changing at sea environment, one that presents the greatest challenges to seamanship and survival.

Winds

Air, like water, seeks its own level and tends to equalize its pressure by flowing from areas of high pressure to those with low pressure. Land warms quickly and radiates heat to the overlying air, warming it and causing it to rise and reduce atmospheric pressure. The air offshore over the ocean is cool, dense, and heavy, and flows inland in an attempt to equalize the pressure difference caused by the warm air rising over the land. This flow produces wind, which typically reaches its highest speeds during midafternoon, when maximum air temperatures for the day are reached.

Weather Fronts

A weather front represents a boundary between two different air masses, such as warm and cold air. If cold air is advancing into warm air, a cold front is present. If a cold air mass is retreating and warm air is advancing, a warm front exists. If the cold air is neither advancing nor retreating from the warm air mass, a stationary front is present.

In a cold front, the boundary between the cold and warm air masses is steep and causes the warm air in front of it to rise rapidly. This rising air creates energetic, billowing cumulonimbus clouds leading to showers and thunderstorms. After the

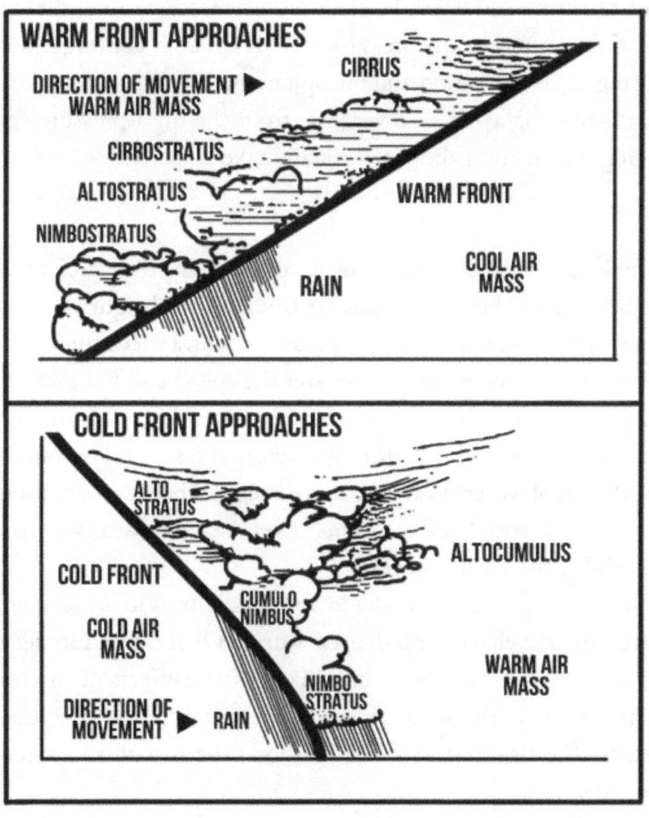

cold front passes, skies clear rapidly and temperatures cool due to the advancing cool air. With a warm front, the boundary between warm and cold air is more gradual, allowing the warm air to slowly rise and clouds to spread out into gloomy, overcast stratus clouds. Precipitation ahead of a warm front typically forms into a large shield of steady rain or snow. After the warm front passes, fair and milder weather is typical; however, a cold front is likely not far behind.

A front's strength can be assessed by the difference in temperatures between the two air masses. In general, the larger the temperature difference, the stronger the front. A stronger front has greater potential for precipitation, while a weak front may only bring a few clouds, a decrease in humidity, or a shift in winds.

Weather fronts are also associated with air pressure. It is well established that high pressure is generally associated with nice weather, while low pressure is associated with cloudy, rainy, or snowy weather. To understand how air pressure affects weather conditions, you must think vertically. Air cools as it rises, which can cause water vapor to condense into liquid droplets that form clouds and precipitation. Sinking air, in contrast, warms and dries. In areas of low pressure, air is directed slightly inward toward the center of the pressure system where it converges and is forced upward, leading to condensation and precipitation. The opposite occurs with high pressure. Air is moving away, or diverging, from the high-pressure center at the surface, causing the air from above to sink and take its place.

Waves

Wind generates waves by moving over the water's surface. The top of a wave is called the crest, and the valleys between waves are troughs. The height and time between crests of waves, or the period, is a direct result of wind speed and the amount of time the wind has been blowing. The greater the speed and the greater the fetch, or the distance wind blows unobstructed over water, the larger the waves. After forming, waves travel out from their origin across the sea and become swells that develop into a series of waves. Wave series are irregular because of constant shifting of wind direction and speed and as storms at sea create masses of waves that build up in groups higher than other waves.

Choppy waves are found in the shallow waters of bays and lakes, while broad, rounded waves are associated with deeper water. When formed in deep water far from shore, irregular waves become organized by the effects of the contact with the bottom and move in the same direction equidistant apart at similar and near constant speeds. The Beaufort Wind Scale shows the size of waves in open water for a given wind speed.

Beaufort Wind Scale

Developed in 1805 by Rear Admiral Sir Francis Beaufort of the Royal Navy, the Beaufort Wind Scale is a system of estimating and reporting wind speed. Estimation of the wind's speed and a constant awareness of the state of the sea are important responsibilities of a mariner. Sea state can have a significant influence upon a ship. Course changes, leeway, dead reckoning, engine speeds, and even rudder angles are affected by wind and sea.

Force	Wind (Knots)	WMO* Classification	Appearance of Wind Effects on the Water
0	Less than 1	Calm	Sea surface smooth and mirror-like
1	1–3	Light Air	Scaly ripples, no foam crests
2	4–6	Light Breeze	Small wavelets, crests glassy appearance, no breaking
3	7–10	Gentle Breeze	Large wavelets, crests begin to break, scattered whitecaps
4	11–16	Moderate Breeze	Small waves 1–4 ft. becoming longer, numerous whitecaps
5	17–21	Fresh Breeze	Moderate waves 4–8 ft. taking longer form, many whitecaps, some spray
6	22–27	Strong Breeze	Larger waves 8–13 ft. form, whitecaps common, more spray
7	28–33	Near Gale	Sea heaps up, waves 13–19 ft., white foam streaks off breakers
8	34–40	Gale	Moderately high (18–25 ft.) waves of greater length, edges of crests begin to break into spindrift, foam blown in streaks
9	41–47	Strong Gale	High waves (23–32 ft.), sea begins to roll, dense streaks of foam, spray may reduce visibility
10	48–55	Storm	Very high waves (29–41 ft.) with over-hanging crests, sea white with densely blown foam, heavy rolling, lowered visibility
11	56–63	Violent Storm	Exceptionally high (37–52 ft.) waves, foam patches cover sea, visibility more reduced
12	64+	Hurricane	Air filled with foam, waves over 45 ft., sea completely white with driving spray, visibility greatly reduced

Source: http://www.spc.noaa.gov/faq/tornado/beaufort.html.

*WMO is the World Meteorological Organization.

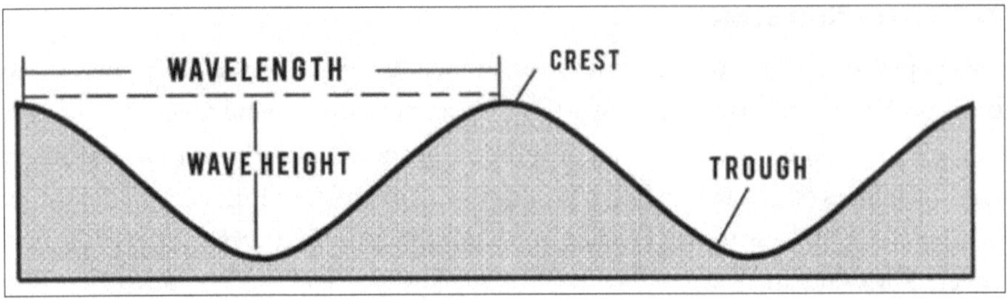

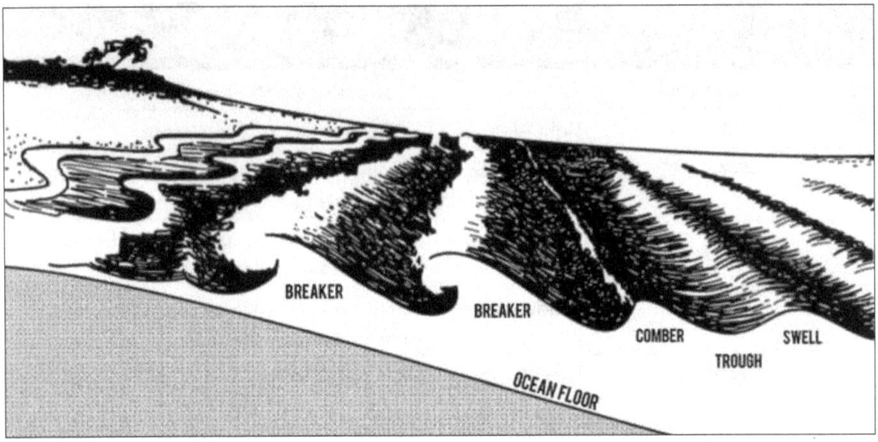

Surf

Waves dissipate as wind speed decreases, as the distance from wind increases, or upon impact with the shore. Near shore, as the depth of water becomes shallow, the waves lose speed and become closer together and steeper as they contact the bottom. This change typically happens at a point where the water is approximately one-half as deep as the wave's length.

As a wave steepens, its momentum will cause it to fall forward or to curl and break as its crest tumbles forward. The greater the size of the wave, the greater its momentum and force of the break. Breakers fall into the trough ahead, usually as a mass of foaming white water. The zone where waves give up this energy is the surf. Any last remaining energy becomes a wash rushing up the beach.

Breaking waves are the most dangerous kind of wave. How dangerous the wave is depends on the ratio of wave height to length and on wave frequency. As a wave steepens, its momentum will cause it to fall forward or curl. It is this momentum that gives a curl of breakers its tremendous force. There are three main types of breaking waves:

- Plunging waves result when there is a sudden lack of water ahead of the wave, such as in a steep rise of the ocean floor. This situation prevents the wave from traveling along, and causes the crest to be hurled ahead of the front of the wave and break with tremendous force.
- Spilling breakers result from waves of low steepness moving over a gentle sloping ocean floor. They normally have a small crest of white water spreading evenly down the wave, and they break slowly without violence.
- A surging breaker occurs on very steep beaches. The wave builds very quickly and expends its energy on the beach. It is very unlikely that a surging breaker will be encountered while on board a boat unless it is beaching on a very steep beach.

Tide and Current

Tide is the vertical rise and fall of the ocean water level caused by the gravitational attraction of the sun and moon. A tidal current is the horizontal motion of water resulting from the change in the tide. It is different from ocean currents, river currents, or those created by the wind. Flood current is the horizontal motion of water toward the land caused by a rising tide. Ebb current is the horizontal motion away from the land caused by a falling tide.

Slack water is the period that occurs while the current is changing direction and has no horizontal motion. An outgoing or ebb current running across a bar builds up a more intense sea than the incoming or flood current. The intense sea results because the rush of water out against the incoming swell slows the wave speed and steepens waves prematurely.

Three basic tidal patterns occur along the Earth's major shorelines. In general, most areas have two high tides and two low tides each day. When the two highs and the two lows are about the same height, the pattern is called a semi-daily or semidiurnal tide. If the high and low tides differ in height, the pattern is called a mixed semidiurnal tide. Some areas, such as the Gulf of Mexico, have only one high and one low tide each day. This is called a diurnal tide. The U.S. West Coast tends to have mixed semidiurnal tides, whereas a semidiurnal pattern is more typical of the East Coast.

Current direction is the compass heading toward which the water moves. The change of direction of the tidal current always lags behind the turning of the tide. This difference occurs by a time period that varies according to the physical characteristics of the land around the body of water as well as the bottom topography. For instance, with a straight coast and only shallow indentations, there is little difference between

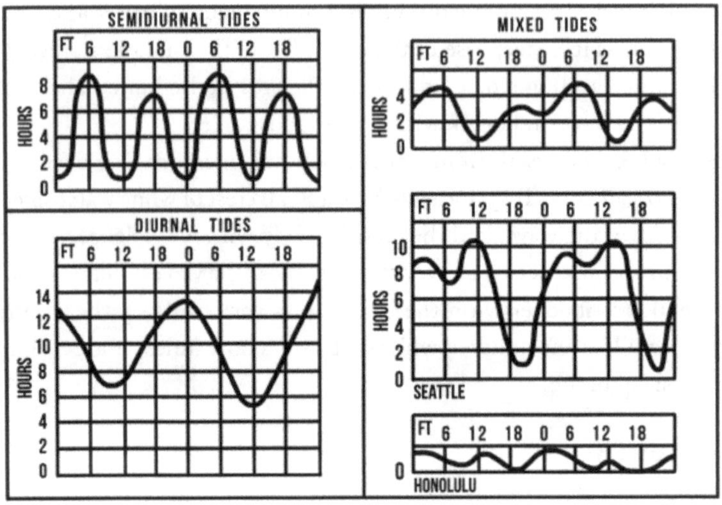

Tide types

the time of high or low tide and the time of slack water. However, where a large body of water connects with the ocean through a narrow channel, the tide and the current may be out of phase by as much as several hours. In a situation such as this, the current in the channel may be running at its greatest velocity when it is high or low water outside.

Wind affects the speed of currents. Sustained wind in the same direction as the current increases the speed of the current by a small amount. Wind in the opposite direction slows it down and may create a chop. A very strong wind blowing directly into the mouth of an inlet or bay can produce an unusually high tide by piling up the water. Similarly, a very strong wind blowing out of a bay can cause an unusually low tide and change the time of the high or low tide.

The cause of currents is not limited only to tidal shifts. Longshore currents run parallel to the shore and inside the breakers. They are the result of the water transported to the beach by the waves. Eddy currents occur at channel bends, near points of land, and at places where the bottom is uneven.

It is important for crews operating in tidal waters to know the set (direction toward) and drift (speed in knots) of the tidal currents in the area. This information can be found on many electronic charts, typically accessed by clicking on a "T" icon near your vessel's displayed position or in the area you plan to transit, or in the Tidal Current Tables published by the National Oceanic and Atmospheric Administration (NOAA).

Storms

When layers of warm, moist air rise in a large, swift updraft to cooler regions of the atmosphere, a thunderstorm results. These are violent, short-lived weather disturbances associated with lightning; thunder; dense clouds; heavy rain or hail; strong, gusty winds; and rough, confused seas. Thunderstorms are dangerous not only because of the strong winds they produce but because of lightning.

Fog

Fog consists of a multitude of minute water droplets suspended in the atmosphere that are dense enough to scatter light rays and reduce visibility. Generally formed at night or in the early morning with the cooling of the Earth's surface, fog clears slowly and is burned off by the warming air or by surface winds.

Ice

Fresh water freezes at 0° C (32° F), and seawater at -2° C (28° F). Shallow bodies of low-salinity water freeze more rapidly than deeper basins because a lesser volume must be cooled. Once an initial cover of ice has formed on the water's surface, the ice will thicken.

Marine Watches, Warnings, and Advisories

NOAA's National Weather Service Marine Weather Service Program offers a broad range of marine forecast and warning products. All are available in graphical and text format at http://www.weather.gov and via VHF-FM on NOAA Weather Radio. Some coastal stations may display flags, pennants, and colored lights to warn mariners of hazardous conditions. Below are common watches, warnings, and advisories defined by the National Weather Service Coastal Warning Program.

- Small Craft Advisory (SCA): Thresholds governing the issuance of small craft advisories are specific to geographic areas, but are generally issued when sustained winds or frequent gusts are greater than twenty-five knots and seas are greater than five feet. An SCA may also be issued when sea or lake ice exists that could be hazardous to small boats. On shore, the signal is a red pennant. At night, it is red over white lights.
- Gale Warning: A warning of sustained surface winds, or frequent gusts, in the range of thirty-four knots to forty-seven knots, either predicted or occurring, and not directly associated with a tropical cyclone. On shore, the signal is two red pennants in a vertical line. At night, it is white over red light.

- Storm Warning: A warning of sustained surface winds, or frequent gusts, in the range of forty-eight knots to sixty-three knots, either predicted or occurring, and not directly associated with a tropical cyclone. On shore, the signal is a red flag with black center. At night, it is red over red light.
- Hurricane Warning: An announcement that sustained winds of seventy-four miles per hour or higher are expected somewhere within the specified coastal area. Because hurricane preparedness activities become difficult once winds reach tropical storm force, the hurricane warning is issued thirty-six hours in advance of the anticipated onset of tropical-storm-force winds. On shore, the signal is two red flags with black centers in a vertical line. At night, it is red over white over red lights.

MARLINESPIKE SEAMANSHIP

The art of working with line or rope is called marlinespike seamanship. The name comes from a handheld, spike-like tool used in working with wire rope, which is called a "marlinespike."

Ropes, Lines, and Wires

The term "rope" refers to both fiber and metal wire. Fiber ropes include those made of such natural materials as manila and hemp and those made of synthetic materials like nylon, Kevlar, and polypropylene. Wire rope is made of metal. Here is the tricky part: Fiber rope is called "rope" only as long as it is still in its original coil. Once a piece has been cut to be used for some purpose, such as mooring or anchoring, it is called a "line." When placed into use, each line takes its name from the purpose it serves, that is, mooring line, towline. If you want to be considered a novice, call a line a rope. Rope made of metal wire is called simply "wire," even if it has been cut from its original coil and is being used for some specific purpose. You are probably safe if you forget about the word "rope" and use the words "line" and "wire." Fiber line is measured by circumference or distance around the line. Depending on its size, the line is categorized as small stuff (up to 1.5 in. circumference), line (1.5 in. to 5 in. circumference), or hawser (everything over 5 in. circumference). Wire, in contrast, is measured by diameter.

Physical properties of rope consist of the breaking strength and working load limit. The breaking strength of a rope is set by the manufacturer and is the load that will part the rope when new. Working load limit is determined by dividing the breaking strength by a design or safety factor prescribed by the Coast Guard for a particular function, meaning it must be capable of supporting a weight equal to the intended load multiplied by the safety factor. For example, the Coast Guard requires rope used for personnel lifting to have a minimum factor of 7.

Construction

Construction of a fiber line starts with small fibers that are twisted until they form larger pieces called "yarns." Rope is constructed from yarns in two basic ways, termed "twisting" and "braiding." Twisted, or laid, lines are made by twisting together yarns in the opposite direction of that used to form them, creating "strands."

Three-stand lines have been used for thousands of years, twisted first by hand and now by machine. The direction these strands are twisted, either left or right, determines the lay of the line. Most three-strand lines are right laid, and if you hold a length of it at arm's length, you will see the wrap of the line twisting to the right. It is important to know this because you should always coil a three-strand line in the direction of its lay to avoid kinking. Natural fiber lines are typically used for chafing gear and lead lines, for small projects where line strength is not a major concern, and for damage control.

Braided lines are made by weaving yarns together, meaning they are crossed over and under each other rather than twisted. These take on a rounded form and are generally smooth. Several different ways of braiding provide different qualities.

Double-braided line is the most common for Coast Guard applications and is made of two hollow-braided ropes, one inside the other. The core is made of large single yarns in a slack braid. The cover is also made of large single yarns but in a tight braid that compresses and holds the core. This line is manufactured only from

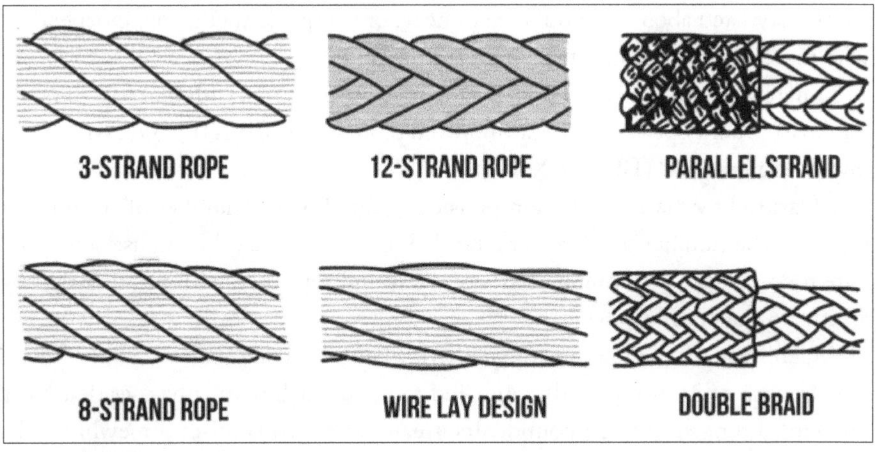

Fiber line types

synthetics, and about 50 percent of its strength is in the core. Plaited line is also common and is made by braiding eight twisted strands, four right-twisted and four left-twisted, with the resulting strands paired and worked like a four-strand braid.

Advances in construction to increase strength, ease splicing, and prolong service life have led to the adoption of a mix of synthetic line types on board Coast Guard cutters and boats. Nylon line, which has almost entirely replaced natural fiber on board Coast Guard assets, is nearly three times as strong as manila of the same size. Nylon differs from other fibers in its elasticity, meaning it stretches or elongates under load yet recovers to its normal size when tension is removed. Depending on the type of nylon used, it can be stretched up to 40 percent of its original length before it reaches a point where it will part. Nylon can stand repeated stretching with no serious effect, but as it ages the risk of parting increases. Double-braided nylon lines are the most common on board cutters and boats where they are used for mooring and alongside lines, towlines, and towing bridles. Kevlar is increasingly replacing nylon for towing applications on board Coast Guard assets for its greater strength and lower elasticity.

Polyethylene is only half as strong as nylon but is lighter, will float in water, and is usually brightly colored. It is used mainly for heaving lines and trail lines on life-ring buoys carried standard on board boats and cutters.

Wire Rope

Wire rope is used extensively for heavy lifting, hauling, hoisting, and standing rigging. It is a machine-made rope formed by individual wires made of steel or other metal helically laid together in a geometric pattern to create strands. The strands are then twisted about a central core or heart of hemp, synthetic line, independent wire rope, or a steel center strand to form the rope.

Different grades of steel are used in wire rope construction, with the most common for Coast Guard operations extra-improved plow steel (EIP/XIP) and extra-extra-improved plow steel (EEIP/XXIP).

Measured by diameter, wire rope is designated by the number of strands per rope and the number of wires per strand. Most Coast Guard units use six-strand wire rope. The most common classes of six-strand rope used with running rigging is either 6X19 or 6X36.

Much stronger than natural or synthetic fiber line, a one-inch wire has a breaking strength of 84,000 pounds compared to a four-inch manila hawser that has a breaking strength of 8,100 pounds. Its greater strength is offset somewhat by its

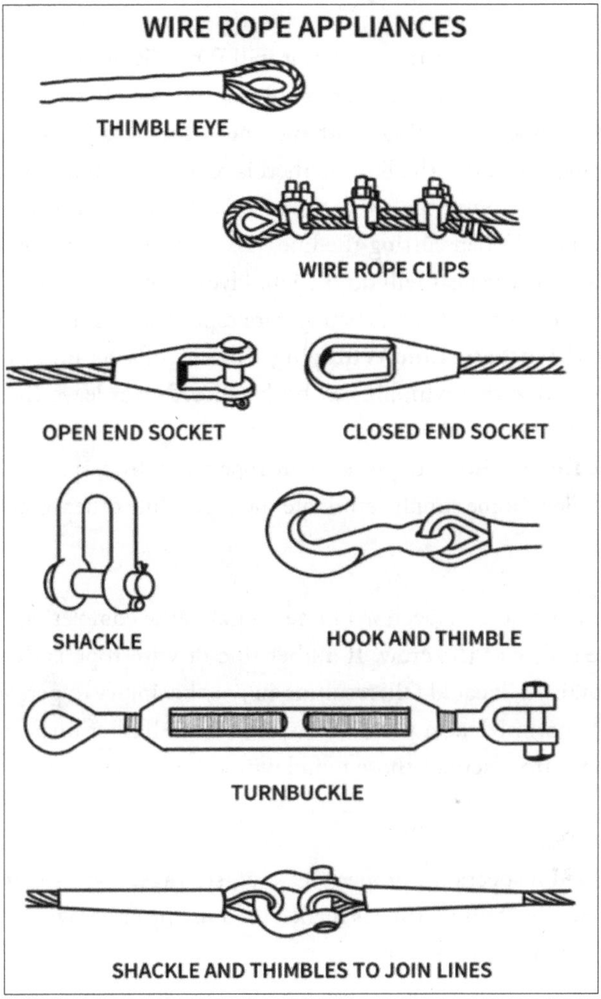

Wire rope fittings

lesser flexibility and its tendency to rust if conditions are not right. The number of wires in a strand will determine the wire rope's flexibility. The more wires, the more flexible the rope. It is very important to always wear heavy gloves when handling wire rope, and hard helmets when using it to lift. More detailed information on the care and handling of wire rope is contained in Naval Ship's Technical Manual, chapter 613, Naval Engineering Manual, COMDTINST M9000.6 (series), Wire Rope User's Manual.

Cutting

Remember, when a piece of rope is cut, it will fray. Always finish the end of the line whether before or immediately after cutting the line. The use of a hot knife is the preferred method for cutting synthetic line. Using a hot knife eliminates the need for burning the ends. The best method is to work from the outside in. First, an incision is made around the circumference of the line, and then a cut is made through the center. When cutting the line, the blade or saw should not be forced through the line, as the heat will do the job. Hydraulic, mechanical, or guillotine rope cutters are recommended for cutting wire rope. Torches should never be used.

Natural and Synthetic Line Whipping. The end of a cut line will unravel and fray if not secured with a whipping or back splice. Never leave the end of a line dangling loose without a whipping.

Wire Rope End Fittings. To prevent wire rope stands from spreading, end fittings should be installed. Some simply terminate the rope while others create useful eyes.

Inspection

Proper maintenance and inspection of line is vital to the completion of the mission as well as the safety of the crew. If a fiber line or wire rope is damaged or not properly maintained, it could fail, resulting in possible injury to personnel, damage to property, or both. Make a periodic inspection of all fiber lines and wire ropes used and remove from service those found with discrepancies.

Care of Line

Along with good inspections, proper breakout, stowage, and care of line will significantly extend its lifetime. Line is made up for use by either winding on a reel or in one of the following three ways:

- Coiling down: This simply means laying it down in circles, roughly one on top of the other. When a line has been coiled down, one end is ready to run off. This is the end that went down last and that is now on top. If you try to walk away with the bottom end, a foul-up will result. If for some reason the bottom end must go out first, turn your entire coil upside down to free it for running. Synthetic line should not be constantly coiled in the same direction, as doing this tends to tighten the twist. Three-strand synthetic line is often coiled clockwise to reduce a natural tendency to tighten up. It can be coiled in figure eights to avoid kinks when paying out.
- Faking down: This means laying it down in the same manner, except that it is laid out in long, flat bights, one forward of the other, instead of in round

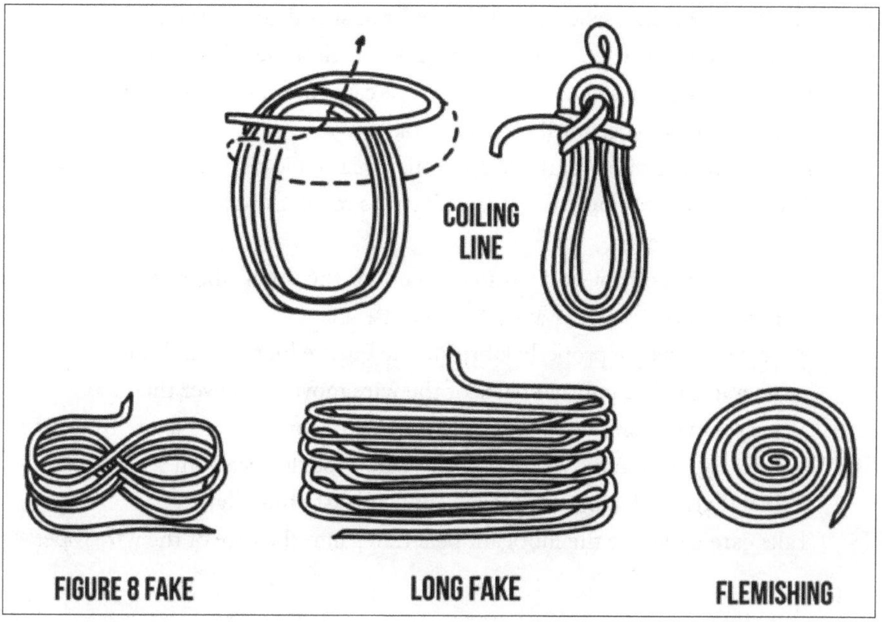

coils. With a long line, this saves the space a large coil might occupy, and faking down a heavy line is much easier than coiling it down.

- Flemishing down: This means to coil it down first, and then wind it tight from the bottom end so that it forms a close mat. Slack ends of boat painters, boat falls, or any other short lines about the decks that are not in continuous use should be flemished down for neatness.

Chafing gear is designed to protect lines at points where they contact the vessel or pier. Leather and pieces of fire hose are common materials used as chafing gear. Chafing gear protects lines from wear due to rubbing or friction and prevents them from weakening or parting.

The practices that should be avoided or observed in the maintenance and care of line are as follows:

DOs

Dry line before stowing it in a cool, dark, well-ventilated space;
Protect line from weather when possible;
Use chafing gear (canvas, short lengths of old fire hose, etc.) where line runs over sharp edges or rough surfaces;

Slack off taut lines when it rains. Wet lines shrink and if the line is taut, the resulting strain may be enough to break some of the fibers;

Lay right-laid lines clockwise on reels or capstans and left-laid lines counterclockwise until they are broken in;

Inspect lines for fiber damage and other wear conditions before each use;

Try to tie knots or hitches in new places as much as possible so as not to wear out the line;

Occasionally end-for-end (swap one end for the other) the line to help reduce excessive wear at certain points;

Wire rope must be properly lubricated to ensure long life and safe operation. The internal parts of the wire move whenever the rope passes over a sheave, winds on a drum, or is put under a load. To limit internal wear and corrosion of the wire rope, lubricate all wire rope at installation, during inspections, and at least annually; and

Take care to ensure the lubricant penetrates into the core of the wire rope.

DON'Ts

Stow wet or damp line in an unventilated compartment or cover it so that it cannot dry. Mildew will form and weaken the fibers;

Subject the line to intense heat or unnecessarily allow it to lie in the hot sun;

Subject a line to loads exceeding its working load limit (WLL). Individual fibers will break, reducing the strength;

Allow line to bear on sharp edges or run over rough surfaces;

Scrub line. The lubricant will be washed away, and caustics in strong soap may harm the fibers;

Put a strain on a line with a kink in it;

Let wear become localized in one spot; and

Unbalance the line by continued use on winch in same direction.

Knots

Knots are used for pulling, holding, lifting, and lowering. When using line for these purposes, it is often necessary to join two or more lines together. Knots and bends are used for temporary joining, and splices provide a permanent joining. In either case, the breaking strength of the joined line is normally less than the breaking strength of the separated lines. The weakest point in a line is the knot or splice. They can reduce the breaking strength of a line by as much as 50 to 60 percent. A splice, however, is stronger than a knot. Because knots decrease the strength of the line, they should always be treated as temporary. If something permanent is needed, a splice or seizing can be used.

Good knots are easy to tie, easy to untie, and hold well. A good knot will not untie itself. In sailing vernacular, a knot is used to tie a line back upon itself, a bend is used to secure two lines together, and a hitch is used to tie a line to a ring, rail, or spar. The knots listed in this section are those most commonly used by Coast Guard crews. Learn to tie them well, for the time may come when the skill to do so could decide the outcome of a mission.

Line Parts

In making knots and splices, the crew member must know the names for the parts of a line and the basic turns employed.

- Bitter End: The free end of a line that is worked with.
- Standing Part: The long unused or belayed end of a line. It is the remaining part of the line, including the end that is not worked.
- Overhand Loop: A loop made in a line by crossing the bitter end over the standing part.

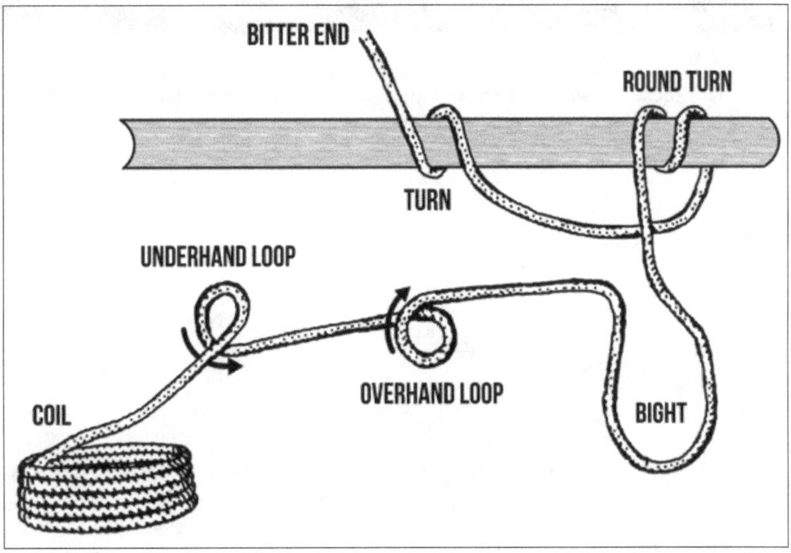

Line turns and loops

- Underhand Loop: A loop made in the line by crossing the bitter end under the standing part.
- Bight: A half loop formed by turning the line back on itself.
- Turn: A single wind or bight of a rope, laid around a belaying pin, post, bollard, or the like.
- Round Turn: A complete turn or encircling of a line about an object, as opposed to a single turn.

Reef Knot / Square Knot

The reef or square knot is one of the most commonly used knots in marlinespike seamanship. Reef knots are primarily used to join two lines of equal size and similar material. Caution should be used if the line is going to be under heavy strain since the reef knot can jam badly and become difficult to untie afterward. Reef knots are best used for small stuff. To tie:

Step	Procedure
1	Tie a single overhand knot.
2	Tie a second overhand knot on top so it mirrors (right and left reversed) the first one. The ends should come out together.
3	Draw tight.

Figure Eight / Stopper

A figure eight knot is an overhand knot with an extra twist. It will prevent the end of a line from feeding through a block or fairlead when loads are involved. It is also easier to untie and does not jam as hard as the overhand knot.

Bowline

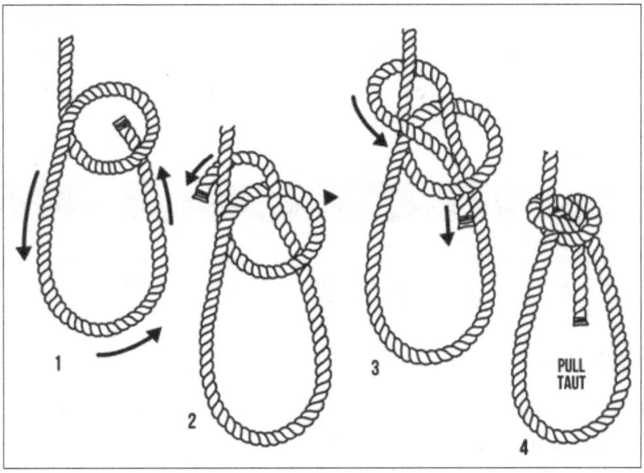

The bowline is a versatile knot and can be used anytime a temporary eye is needed in the end of a line. It also works for tying two lines securely together, although there are better knots for this. An advantage of bowlines is that they do not slip or jam easily. To tie:

Step	Procedure
1	Make an overhand loop in the line the size of the eye desired.
2	Pass the bitter end up through the overhand loop.
3	Bring the bitter end around the standing part and back down through the overhand loop.
4	Pull the knot tight by holding the bitter end and the loop with one hand and pulling on the standing part with the other.

Half Hitch

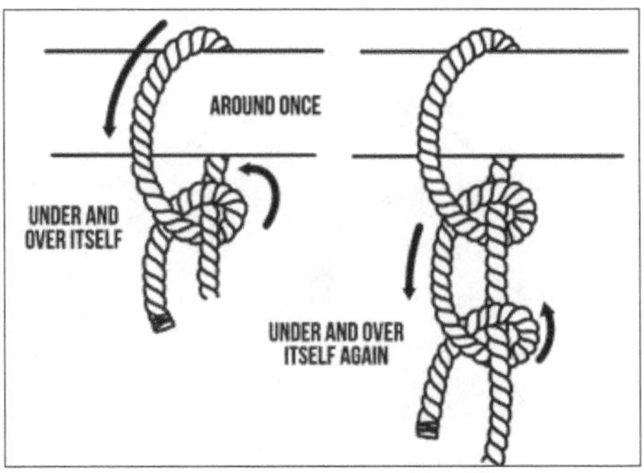

Hitches are used for temporarily securing a line to objects such as a ring or eye. One of their advantages is their ease in untying. The half hitch is the smallest and simplest hitch. It should be tied only to objects having a righthand pull. Since a single half hitch may slip easily, care should be taken in cases where it will encounter stress. To tie:

Step	Procedure
1	Pass the line around the object.
2	Bring the working end around the standing part and back under itself.

Two Half Hitches

To reinforce or strengthen a single half hitch, the line can be tied once more. Two half hitches make a more reliable knot than a single half hitch and can be used to make the ends of a line fast around its own standing part. A round turn or two, secured with a couple of half hitches, is a quick way to secure a line to a pole or spar. Two half hitches are needed to secure a line at an angle where it might slide vertically or horizontally. To tie:

Step	Procedure
1	Take a turn around the object.
2	Bring the running end (bitter end) under and over the standing part and back under itself.
3	Continue by passing bitter end under and over the standing part and back under itself.

Rolling Hitch / Stopper Knot

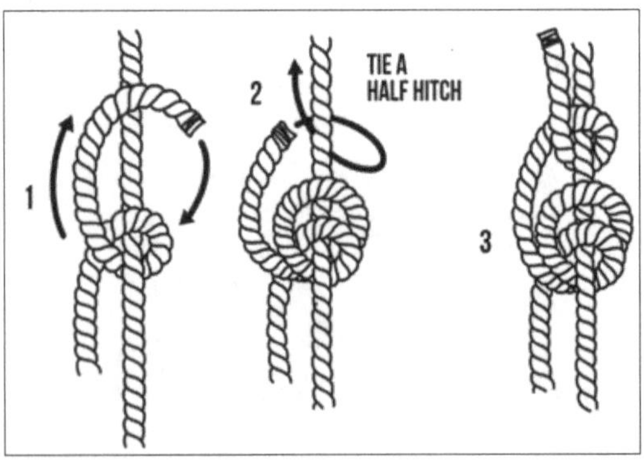

A rolling hitch is used to attach one line to another, where the second line is under a strain and cannot be bent. To tie:

Step	Procedure
1	With the bitter end, "a," make a turn over and under the second line, "b," and pass the link over itself.
2	Pass "a" over and under "b" again, bringing "a" through the space between the two lines on the first turn.
3	Pull taut and make another turn with the bitter end, "a," taking it over, then under, then back over itself.
4	Pull taut and tie a half hitch.

Clove Hitch

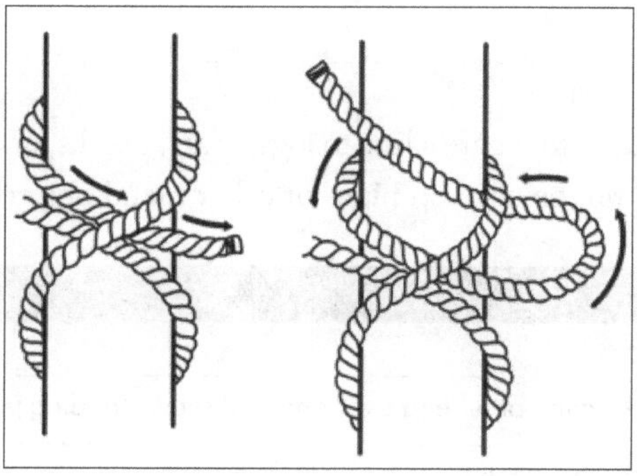

A clove hitch is preferred for securing a heaving line to a towline. It is the best all-around knot for securing a line to a ring or spar. Correctly tied, a clove hitch will not jam or loosen. However, if it is not tied tight enough, it may work itself out. Reinforcing it with a half hitch will prevent this from happening. To tie:

Step	Procedure
1	Pass the bitter end around the object so the first turn crosses the standing part.
2	Bring the bitter end around again and pass it through itself.
3	Pull taut.
4	Reinforce by tying a half hitch.

Slip Clove Hitch

A slip clove hitch should be used in lieu of a clove hitch when a quick release is required. It should be tied in the same manner as the clove hitch, but finish it with a bight to allow for quick release. It is sometimes used for stowing lines and fenders. It should not be used when working with the line.

Timber Hitch

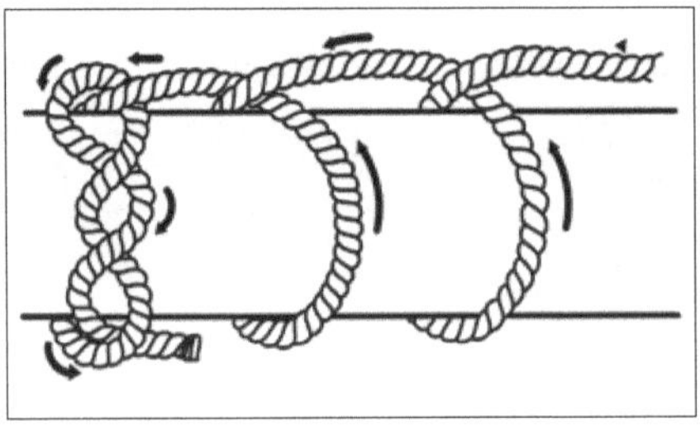

Timber hitches are used to secure a line to logs, spars, planks, or other rough surfaced material but should not be used on pipes or other metal objects. To tie:

Step	Procedure
1	Tie a half hitch.
2	Continue taking the bitter end over and under the standing part.
3	Pull the standing part taut.
4	Add two half hitches for extra holding, if necessary. Unless the half hitch can be slipped over the end of the object, tie it before making the timber hitch.

Double Becket Bend

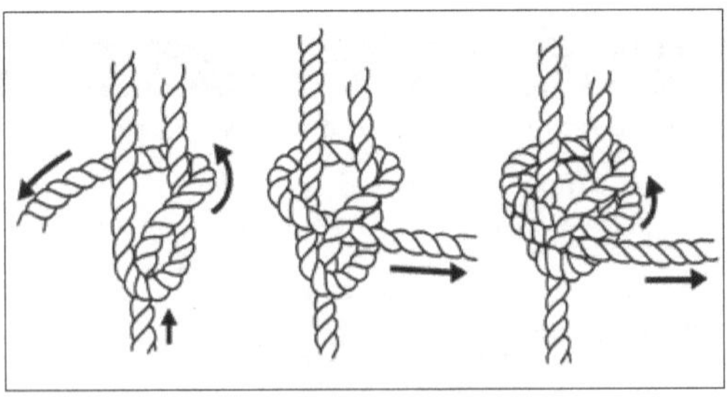

The double becket bend works for joining lines of unequal size. It is tied in the same manner as the single becket bend except for the following variation in step 4 above: Pass line "b" around and under its standing part twice.

Cat's Paw

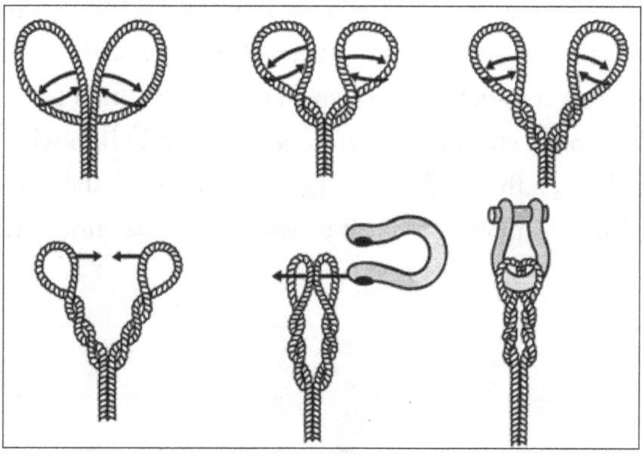

The cat's paw is used to connect a line to a shackle or hook. It can be used to establish a towing rig or as a sling for lifting heavy loads at any angle.

Step	Procedure
1	Make two loops with a rope bite.
2	Twist the loops around themselves.
3	Twist the loops three times.
4	Bring the loops close together.
5	Pass the shackle through the top loops and pull to tighten.

Mousing

A hook is moused to keep slings and straps from slipping out or off the hooks. This is accomplished either by mechanical means or by seizing the hook, using seizing wire or small stuff, from opposite sides. Shackles are moused to prevent the pin from backing out. This is usually done on screw-pin shackles. Mousing is accomplished by taking several turns, using seizing wire or small stuff, through the pin eye and around the shackle itself in such a way so the pin cannot turn.

DECK SEAMANSHIP

Deck seamanship concerns the general work that goes on about the ship's deck and the equipment used. Anchoring, mooring, rigging, and handling heavy weights and cargo, underway replenishment, towing, and a host of other skills are considered deck seamanship.

Line Handling

Line are secured to various deck fittings in order to accomplish mooring, anchoring, towing, and other evolutions and tasks. Proper handling of lines is essential to establishing secure connections and ensuring safety. Snap back is caused by the sudden release of stored energy in a stretched synthetic line when it breaks. This could potentially be deadly. Because of this danger, avoid standing behind or near a line under tension and never stand on or walk over lines under tension.

Deck Fittings

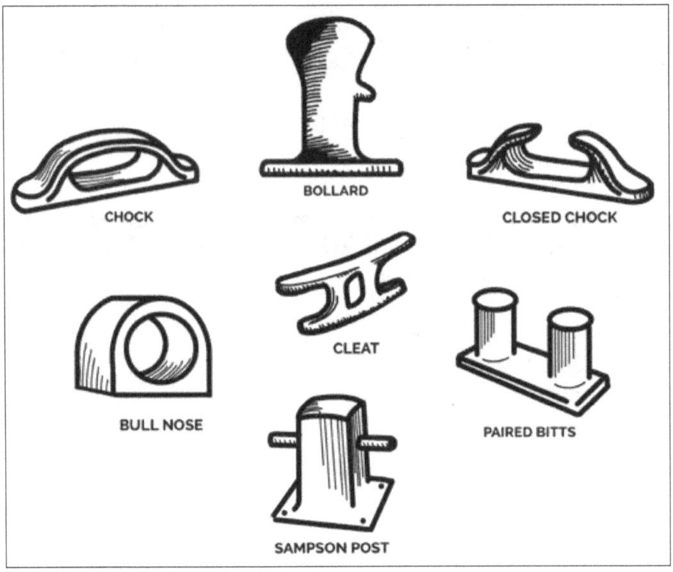

Deck fittings are attachments or securing points for lines that permit easy handling and reduce wear and friction. There are three basic types of deck fittings installed on board cutters and boats—cleats, bitts, and chocks. Cleats are twin horned-shaped devices. Bitts are vertical cylinders, called barrels, arranged in pairs. Both cleats and chocks are installed on both sides of the deck, normally forward and aft each chock. Chocks provide a smooth surface for lines to run over or through and are used with bitts and cleats to help prevent chafing. A bow chock is called a bullnose. Strength of deck fittings will vary widely, but the best attachment points are those fitted with backing plates that distribute the load over a wide area. In addition to cleats and bitts, piers often have bollards, or single posts used to secure mooring lines.

Securing a Line to Paired Bitt. A paired bitt is a set of posts on the deck used as cleat.

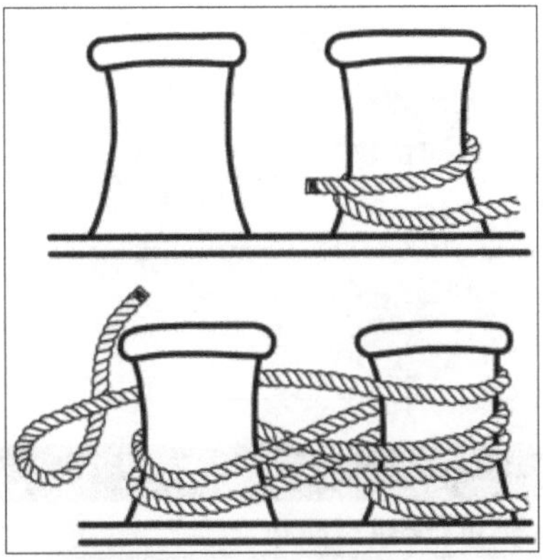

Step	Procedure
1	Make a turn around the horn furthest from the attachment point.
2	Make several figure eights around both horns. Size of line and cleats may restrict the number of turns, but a minimum of three figure eights is the standard.
3	Finish off with a round turn.

Securing a Line to a Sampson Post. A single bitt, or Samson post, is a post on the forward or aft deck of a vessel. It is used as a cleat or tow bitt.

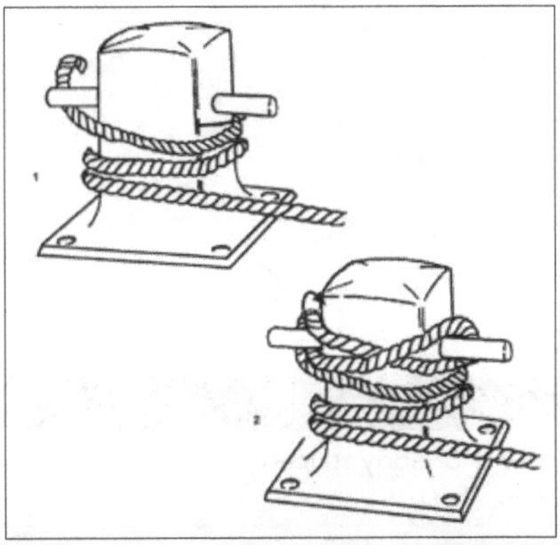

Step	Procedure
1	Make a turn around the base of the bitt.
2	Make several figure eights around both horns. Size of line and cleats may restrict the number of turns, but a minimum of three figure eights is the standard.

Securing a Line to a Cleat.

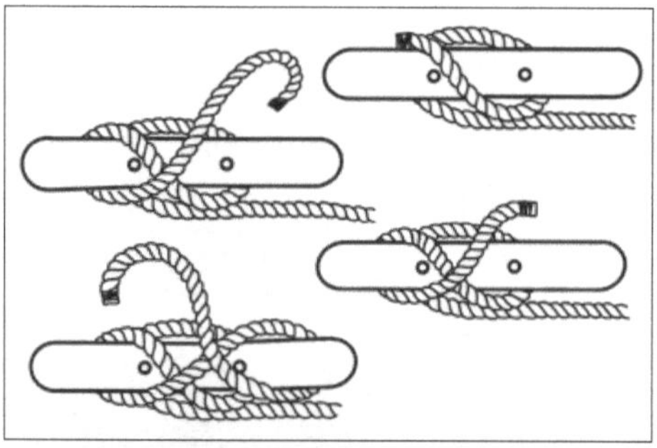

Step	Procedure
1	Make a turn around the base of cleat.
2	Lead line over the top of the cleat and under far horn.
3	Continue to lead line back over the top of cleat and under opposite horn to form a figure eight.
4	Make two or more additional figure eights to secure the line.

Securing a Dock Line to a Cleat.

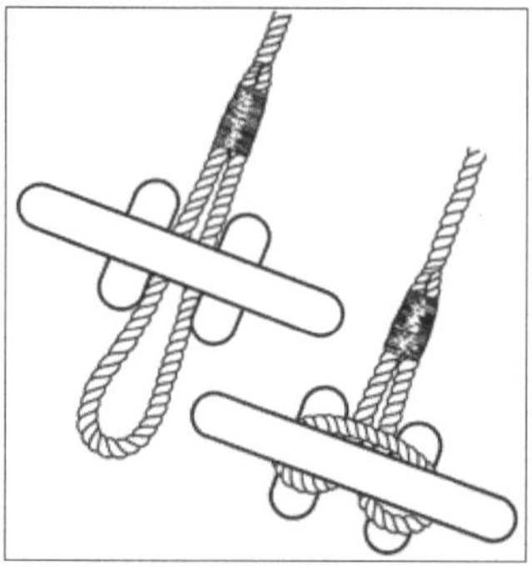

Step	Procedure
1	Use a line with an eye already spliced or tie a bowline large enough to loop back over the horns of the cleat.
2	Feed eye of the line through opening.
3	Loop the line back over both horns and pull line taut.

Mooring

Mooring means to secure a vessel to a pier, wharf, mooring buoy, or another vessel. Mooring systems include the lines, fenders, ship/pier fittings, and related machinery used to bring a ship into a moored configuration. Most moorings are provided in harbors to provide a safe haven, reduce exposure to waves, reduce ship motions, and reduce dynamic mooring loads. Most moorings are made parallel to a pier or dock, but options exist for congested areas and those with minimal infrastructure. In a Mediterranean Moor the stern of the ship is secured perpendicular to a pier or fixed structure with mooring lines or anchor chains and its bow secured to mooring buoys or by its own anchors.

When port loading is congested, vessels may be required to nest, or moor outboard to similar hulls. General safety requirements when mooring include the following:

- Visually inspect all mooring equipment prior to and after each use.
- Avoid using different types of mooring lines from the same pair of bitts.
- Avoid stepping over or straddling lines, wires, or chains. Go around!
- Stay out of the bight of the line.
- Do not stand directly in line of a line under a load or where a line changes direction.
- Listen to all lines under tension. Unusual popping or tearing sounds may mean the line is in danger of failing.
- Never leave lines secured to gypsy heads or capstans.
- Never take a strain on a line unless it is clear of obstructions.
- Use chafing gear.

Mooring Lines

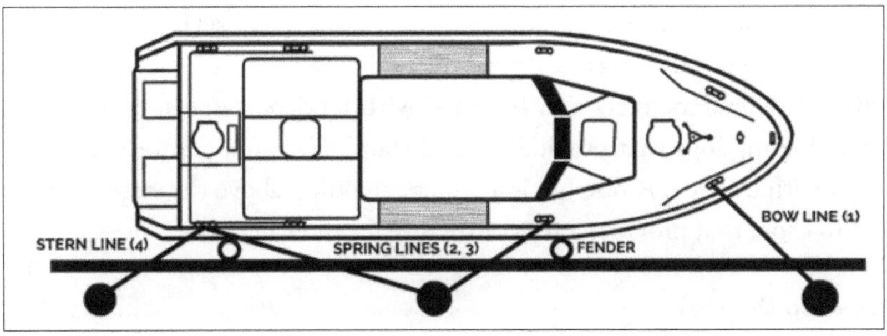

Before using mooring lines to help secure to a dock, crew members need to first know the lines' names and what they do. The bow line and stern line are used to keep the vessel secured to the dock. The forward spring and aft spring are used to keep the vessel from surging forward or aft at the dock. Normally, only these four lines are required when mooring. During times of foul weather, breast lines may be used to provide additional holding strength.

If it becomes necessary to hold position alongside a dock but swing the bow or stern out in order to clear another vessel or obstacle, using a spring line can help to accomplish this. The aft spring is used to "spring out" or move the bow away from the dock, while the forward spring is used to "spring out" or move the stern away from the dock.

Knowing how to rig mooring lines to "slip" can be helpful, particularly when no shore-side line handlers are available. To rig to slip, both ends of the line are secured on board the boat with only a bight around the shore-side attachment point. When casting off, one end is released and the other hauled on board.

Mooring Machinery

Mooring machinery facilitates the handling and securing of mooring lines and includes a windlass, capstan, and winches.

Fenders

Crew members should never attempt to fend a boat off a pier, float, and so on, by hand or by foot, but should always use a fender. Fenders are used to absorb the shock of contact between vessels or between a vessel and pier and should be used whenever there is the possibility of a hull making contact with another object. These devices are slung over the side of a vessel in positions where contact is expected to occur and can be spherical, cylindrical, or rectangular prisms made of resilient materials such as rubber, foam, canvas, wood, line, cork, wicker, or plastic. Fender size varies greatly, and the appropriate size that will keep space between vessel hulls or rub rails and hulls should be used. Fenders are typically tied to rails using a slip clove hitch.

In addition to fenders, piers may be fitted with fenders, camels, or dolphins. Camels are floating logs that provide a vessel standoff from a pier or wharf and usually float with the tide. A dolphin is a piling extending above the water and not connected to shore that provides a fixed structure for mooring.

Line-Throwing Devices

Line-throwing devices are used to pass mooring lines to the pier or another ship and include the following:

- Heaving lines. These are light-weighted floating synthetic lines attached to mooring lines or messengers used to pass the mooring lines to the pier. May be fitted with a latex heaving ball or monkey's fist.
- Line-throwing gun and projectile. Larger ships may be outfitted with line-throwing guns to assist when docking at large distances from the pier or another ship. Prior to using the gun, line-throwing gunners must be properly trained, may need permission from docking authorities, and must provide warning to pier-side individuals of the projectile.

Mooring Procedure

Crews must be briefed and the deck setup with mooring lines and fenders rigged well before any approach to a dock is made. Although common practice is to leave mooring lines attached to the home pier, crews should always have spare mooring lines and movable or roving fenders at the ready while approaching any dock, including the home pier. Crews will always take advantage of wind and current, so to maintain best control, the boat or cutter should approach against the wind and current and moor on the leeward side of a mooring whenever possible. Lines are usually passed from the vessel to the pier using a heaving line or other line-throwing device. Specific line-handling instructions are then given in a loud, clear voice.

Standard Line-Handling Commands

PUT OVER [line number]: Pass the specified line to the pier and provide enough slack to allow line handlers to place the line over the bit, cleat, or bollard.

HOLD [line number]: Do not let any more line out even though the risk of parting may exist.

CHECK [line number]: Hold heavy tension on the specified line but render it as necessary to prevent parting the line.

SURGE [line number]: Hold moderate tension on a line but render it enough to permit movement of the ship.

EASE [line number]: Let a line out until it is under less tension but not slacked.

SLACK [line number]: Take all tension off a line.

TAKE THE SLACK OUT OF [line number]: Take all the slack out of a line but do not take a strain.

SHIFT [line number]: Move a line to the specified location.

HEAVE AROUND ON [line number]: Take a strain on a line.

TAKE [line number] TO POWER: Take the specified line to the capstan or gypsy head and make ready to heave around (DO NOT heave around until told to do so).

SINGLE UP [line number]: Take in all but one bight so there remains a single part to the line. Can also be used to single up all normal mooring lines.

DOUBLE UP [line number]: Pass an additional bight on the specified line so there are three parts to the line. This can also be used to double up all normal mooring lines.

AVAST or AVAST HEAVING [line number]: Stop taking a strain on a line with capstan.

CAST OFF [line number]: When using another ship's lines to secure your ship, it means to cast off the ends of their lines.

TAKE IN [line number]: Allow the pier line handler enough slack to take the line off the fitting and bring the line on board. Used when secured with your own line.

STAND BY YOUR LINES: Man the lines, ready to cast off or moor.

BACK EASY: A command to the capstan operator to ease tension on the line once the stopper is passed. This command is given before up-behind.

UP-BEHIND: Cease hauling on the line and slack it quickly.

Anchoring

An anchor is a type of hook that embeds itself into the sea bottom to hold a boat or cutter in place. The body of the anchor is called the shank, and the flukes are the "teeth" that dig into the bottom as the crown lifts and forces them downward. The anchor's stock is a kind of crossbar that prevents it from flipping over once it is lying on the bottom. Balanced fluke anchors are the most prevalent in the Coast Guard and have a great deal of holding power for their weight. Because these anchors are designed to dig in, bottom characteristics are of prime importance, with sand and clay preferred for anchorages over mud, rock, or coral.

The equipment associated with anchoring is called ground tackle and includes the anchors themselves; the chains used to attach them to the boat or cutter; the windlasses, winches, and capstans used to weigh or recover the anchor; and a variety of shackles, chain stoppers, anchor bars, and detachable links. An anchor windlass is a type of winch used primarily for handling and securing the anchor and anchor chain. The two types of windlasses are the horizontal shaft windlass, which is found on board larger cutters and able to handle port and starboard anchors, and the vertical shaft windlass, which is able to handle only one anchor. A capstans or gypsy head is a rotating drum with a smooth surface keyed to the windlass shaft that is designed to handle lines for mooring. Vertical windlasses have capstans, and horizontal windlasses have gypsy heads. A wildcat is a special type of drum or sprocket rotated by the windlass and constructed to heave in or pay out anchor chain.

The complete anchor system consists of the anchor, the rode, and the various fittings connecting the rode to the anchor. The rode is the line or chain extending from the vessel to its anchor. Boats and some smaller cutters use line or a combination of line and chain. Larger cutters use only chain. Line requires more scope and is usually not heavy enough to create significant catenary in weather conditions that bring the line to moderate or heavy strain, particularly when used to hold a larger vessel. The most commonly used line for rode is nylon. Galvanized chain is added to the line to lower the angle of pull, prevent chafing of the line on the bottom, prevent penetration of bottom sediment into the strands of the fiber line, and allow for easy cleaning of debris and mud. Anchor chains used by cutters are class specific and measured in shots. A fathom is six feet, and a shot is fifteen fathoms, or ninety feet. Chain links are painted to identify the length of chain paid out with a red, white, blue, and yellow pattern. The number of white links correlates with the number of shots.

Setting the Anchor

Having selected a suitable spot, the boat or cutter runs in slowly. When manned and ready, the anchor detail will haul out enough line from the locker and fake it on deck so as to run freely without kinking or fouling. Many an anchor has been lost for failure to attach the rode properly. Seize the pins on all shackles to prevent them from coming out.

On command to "Let go the anchor," the detail will lower it over the side or release the pelican hook holding the chain. The chain will pay out quickly. Expect the vessel to be moving slowly astern so as to avoid paying the chain down on top of the anchor. The brake on a cutter's forecastle may be slowly applied or a working turn made on a boat's forward bitt to assist in controlling the speed. When the anchor and chain are on the bottom, the speed will slow significantly. Scope is a ratio of the length of rode paid out to the depth of the water. Enough rode should be paid out so the lower end of the rode forms an angle of 8 degrees or less with the bottom to help the anchor dig in and give good holding power. The desired scope is reached on the command "Walk out the anchor, veer to [number] shots/feet," which signals the detail to slacken the brake or working turn and veer the line or chain to the desired length as the vessel moves further astern.

When the desired scope is paid out, the command to "set the anchor" will signal the detail to hold the chain or makeup the line in order to create tension and force the anchor to bite into the bottom. After the anchor has gotten a good bite and the proper scope has been paid out, the detail will report "Ship is brought to and anchor appears to be holding," before making the line fast to a deck fitting and setting the anchor watch.

Anchor Watch

Maintain a live watch whenever anchored to monitor the conditions and equipment. Things to watch for include dragging anchor, changes in the weather, and other vessels dragging their anchor or anchoring near your vessel.

Weighing Anchor

When it is time to weigh anchor and get under way, the anchor detail will be manned and ready on the bow. The command to "Heave around on the anchor" is given and the vessel will move forward slowly as the detail takes in the anchor rode or controls the windlass to heave in on the chain, faking the line on the deck as it comes on board. The forecastle detail must keep the bridge informed as to how the anchor chain or line tends and whether it is taut or slack, reporting, "Anchor is tending [clock position] with [heavy/moderate/light] strain." Twelve o'clock is the bow, three o'clock is the starboard beam, six o'clock is directly astern, and nine o'clock is on the port beam. Strain is reported as either heavy, moderate, or light strain.

As the vessel approaches the spot directly over the anchor where the rode tends straight up and down, the detail will report "Anchor is at short stay." When in this position, the anchor cannot dig in and is just short of clearing the bottom. The bridge will order "Weigh the anchor," and the anchor detail responds by heaving

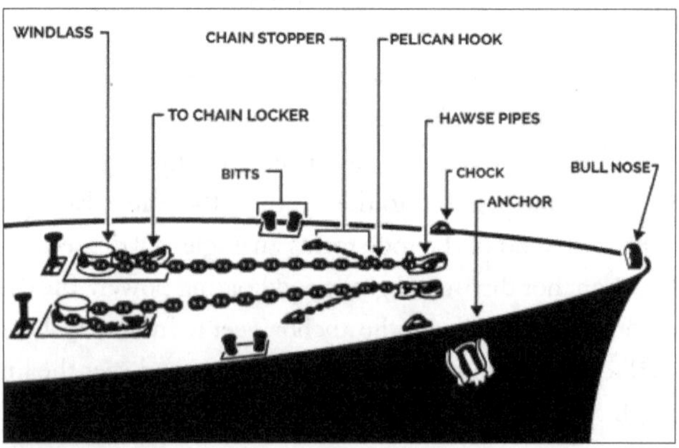

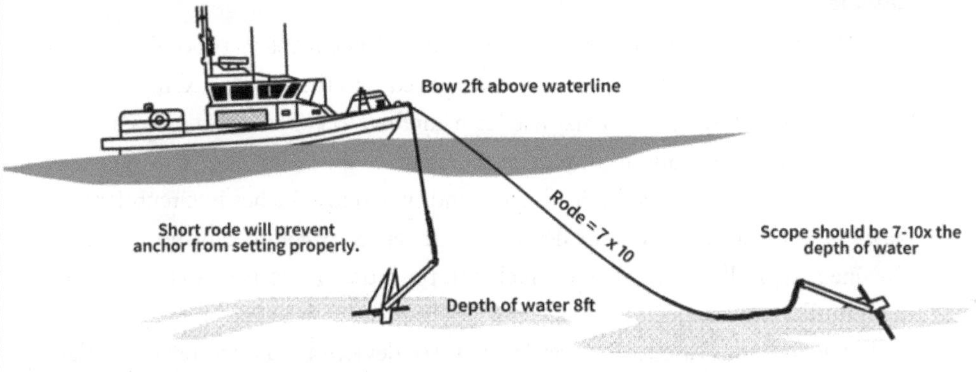

Rode is the length of anchor line and chain. Scope is the ratio of anchor rode in use to the vertical distance from the bow of the vessel to the ocean floor.

around. The anchor will usually free itself from the bottom and the detail will report "Anchor is aweigh." If the anchor refuses to break free, the detail will be ordered to snub the anchor line around the forward bitt or cleat and the vessel will advance a few feet. If this maneuver does not free the anchor, the vessel will need to run in a wide slow circle to change the angle of pull. Once free the anchor is brought to the waterline and checked for cleanliness.

The anchor should be cleaned before bringing it on board, as it may have some "bottom" on it. When in sight, the anchor detail reports "Anchor is clear/foul/shod." Clear means there are no foreign objects entangled in the anchor. Foul means the anchor is fouled on its own chain. Shod means it is balled with mud. If the anchor requires cleaning, it may be dunked up and down in the water or made off so that it is just below the surface of the water before the vessel backs down slowly to drag the anchor through the water until clean. Once clear of any bottom the anchor is hauled on board and housed or secured.

Towing

The Coast Guard is authorized to perform any and all acts necessary to rescue and aid persons and to protect and save property at any time and at any place. Every Coast Guard boat and cutter is equipped to take another vessel in tow or to be taken in tow. Considerations when determining whether to take vessel in tow include the condition of the vessel, the condition of the crew, the location of the vessel, current and forecasted weather, rate of drift toward a navigational hazard or away from assistance, and towing capability of your own boat and cutter.

Towline

Each Coast Guard boat and cutter type has an equipment list that specifies towline length and size. Towlines will have an eye spliced into the tow end. Offshore or in heavy weather, a towing vessel may need five hundred feet or more of towline to keep a towed vessel in step and to minimize the effect of shock-loading. Towlines used on boats are usually double-braided nylon and two to four inches in circumference. Hawsers, or towlines over four inches in circumference, are common on board cutters. Towline is typically stored on a tow reel with the bitter end secured to the reel with small stuff. Many tow reels have mechanical advantage with a hand crank, gear train, or electric motor to ease towline retrieval. These devices are only to retrieve a slack towline, not to take tension. The towline is the hardest worked line on a boat. When in use, it can carry a tremendous strain and is a possible danger to anyone working nearby.

Tow Bitt

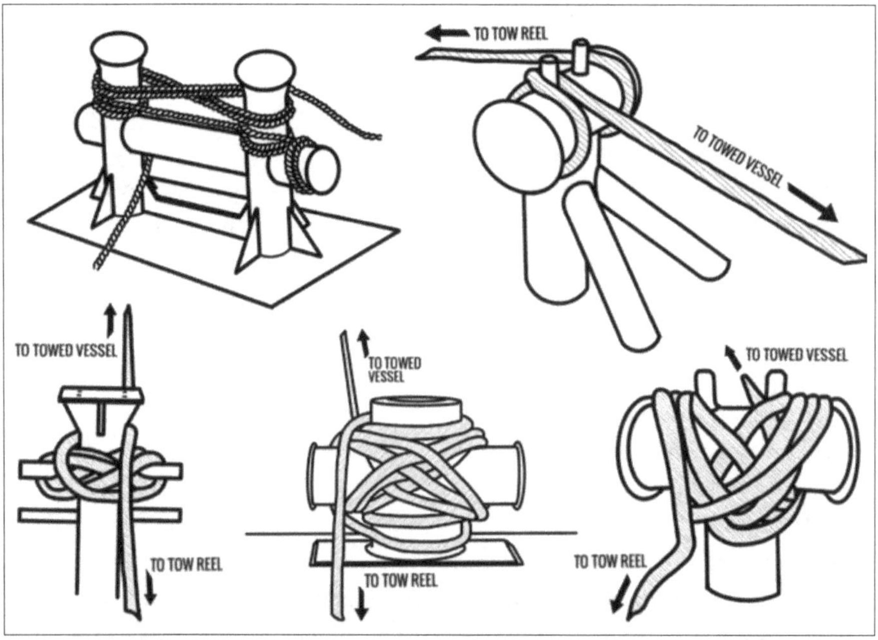

Boats and cutters have different styles of tow bitts installed on their sterns, but regardless the design, towlines should always be secured using one complete round turn and enough figure eights, typically three or more, to create the level of friction needed to prevent additional towline from paying out. Towlines are secured this way so they can be easily and safely adjusted or cast off in an emergency.

Tow Rigs

It is uncommon to attach a towline directly to the bow of a distressed vessel due to the potential for wear, chafing, and damage to the towline. To protect the towline, crews create a tow rig by connecting either a pendant or bridle to the towline's eye or thimble, typically with a Nylite shackle.

A pendant is a single line connected to the towline made long enough so the towline connection is clear of obstructions on the towed vessel. Pendants allow for only a single connection point on the center line of the bow of a distressed vessel. A bridle or "cat's paw" is used when a towed vessel's deck fittings are not right at the vessel's bow or when there are obstructions. It is the preferred rig for heavy-weather towing. When used properly, both legs exert an equal pull on the hull. For smaller, trailerable vessels, a skiff hook is an option, which consists of a line with a snap hook spliced into one end that is connected to the trailer eye bolt on the bow of the disabled vessel using a boat hook fitter with a quick-release safety buckle. Attaching a towline to the trailer eye is a dangerous technique. It requires the towing vessel to maneuver very close to a distressed boat and requires crew members to extend themselves over the side between two vessels, or under the flared bow sections of the distressed boat. Other general hardware for use with tow rigs includes shackles, carabiners, and swivels.

Drogue

A drogue is a device that acts in the water similar to how a parachute works in the air. The drogue is deployed from the stern of the towed vessel to help control the towed vessel's motions. Normally drogues are not deployed when well offshore but rather inshore, where greater control of the towed vessel is required. If necessary to tow a vessel with large swells directly on the stern, it may be more prudent to alter course or lengthen the towline rather than to deploy a drogue.

Tow Approach

All lines, equipment, and connections should be inspected, made ready, and double checked prior to making an approach to a disabled vessel. Minimize loose towline on deck by paying out directly from the reel and, if not equipped, fake the towline carefully so that it will not kink or tangle. In heavy weather, use caution to ensure line is not washed over the side and into the screw or water jet intakes. The type of approach is determined by the design and characteristics of the towing vessel and of the vessel to be towed. How the disabled vessel lies in the water and the direction and rate of its drift are also important. The approach is made to "optimum position," meaning a safe position with the best angle for passing the tow rig.

Passing the Tow Rig

There are two ways to pass a towline, direct and indirect. A towline is too heavy to cast more than a few feet, but in calm conditions or when time is critical it may be passed directly to the distressed vessel when close aboard. In most cases a heaving line is passed to haul over the heavier towline. Using a heaving line allows the towing vessel to remain close aboard the stricken vessel for only the short amount of time needed to pass the heaving line before opening the distance to create a safe working space between the vessels that provides more time to pass the complete towing rig. The bitter end of a heaving line is attached to the towline with a clove hitch, bowline, small carabineers, or snap hook. Slip clove hitches may work best in very cold weather because they are easier to untie.

The throw should be targeted above the center of the vessel so the heaving line has the largest possible target and falls across the vessel's deck. It takes practice to cast a heaving line properly, and throwing one accurately is a perishable skill. Typically two heaving lines are rigged to a towline, one designated primary and the other secondary, to allow rapid follow-on attempts. When making an approach and in a position where you can make the throw, advise the coxswain "I have a shot," and await the order from the coxswain or bridge to "Send it." Prior to making the throw call out "HEADS UP" as a warning to people on board the distressed vessel to take cover and watch out for the toss. When the heaving line has been cast, report whether it has been retrieved, fell short, or missed altogether. If it is in the water, call out "Line in the water" and quickly retrieve the line so it is not fouled in the props or waterjet.

Floating a line is a method used to reach a vessel beyond the range of a heaving line, in an inaccessible position, or in heavy weather. In this method a buoyant synthetic line is floated from upstream or upwind. One end is tied to a ring buoy, float, or fender and the other end to the tow line. This method is only effective if the wind or current can get the float within range of the other vessel.

Regardless the method selected to pass the towline, coordination between both vessels is important and may require sending over a handheld radio first to establish communication. Ensure that the towed vessel understands these procedures before beginning. Once the line is successfully received on the disabled vessel, untie any unused heaving line from the tow rig and report that the rig is ready to be transferred. Stand by to tend the line if opening and closing maneuvers are necessary to maintain optimum position and a safe distance. In opening maneuvers, the towline is paid out. When closing, any slack is hauled in.

On the order to send the rig, order the distressed vessel crew to heave around and tend the line. Report the transfer progress of the tow rig, such as when the towline or bridle is going over the side and when it is hauled on board the distressed vessel. Crew on board the distressed vessel should be directed to lead a single-point tow rig, either a pendant or towline, to a fitting as close to the center line as possible. If led through a secure chock near the center line, the end of the rig can go to a suitable deck fitting. Parts of a bridle should be led to fittings or through chocks equally spaced from the center line. In either case, the eye of the line must be connected to a post, bitts, or cleat so it will not come loose when a strain is placed on the rig.

Stern Tow

To take an initial strain on the towline, a working turn is placed on the tow bitt that allows the line to pay out slowly. The towing boat comes ahead until the line begins to tend aft instead of up and down. This is the initial test of strength and performance of the tow rig and connections. Gradually the towing vessel will increase speed ahead, taking care not to shock load the towline.

When the desired length, or scope, of towline is deployed, the crew is ordered to make up the bitt. Forward motion is slowed enough to slack the towline, and the bitt is made with a round turn and enough turns over the bitt's horns to secure the line from paying out. The towing vessel can gradually increase speed and change course once the bitt is made and both vessels are making way.

The towing watch on board the towing vessel should keep close watch on the action of the tow and immediately report any unusual movements to the coxswain or OOD. Strain on the towline is normally reported by the tow watch as "light," "moderate," or "heavy." If dangerous movements cannot be reduced or controlled, it may be prudent to heave to until sea conditions improve or the source is corrected.

Shorten Tow

It is often necessary to shorten the towline to enter an inlet, cross a bar, tow in a channel, or turn into a basin. The tow is shortened to increase control in confined areas and in current, meaning the tow is stopped to slacken the line so it can be hauled on board as the towing vessel backs down toward it at an angle. As the towline gets slack, the crew removes turns from the tow bitt and pulls in the slack, ready to make up the bitt again when ordered to resume the tow.

Dropping Tow

Disconnecting, or dropping tow, occurs when the towing vessel is relieved by another asset or when power and maneuverability are restored on board the towed vessel. The towed vessel is turned into the prevailing conditions for better control and the tow shortened. Once shortened, the bridge signals for the towed vessel crew to disconnect the rig and let it go into the water. The towing vessel crew then hauls it on board.

Alongside Tow

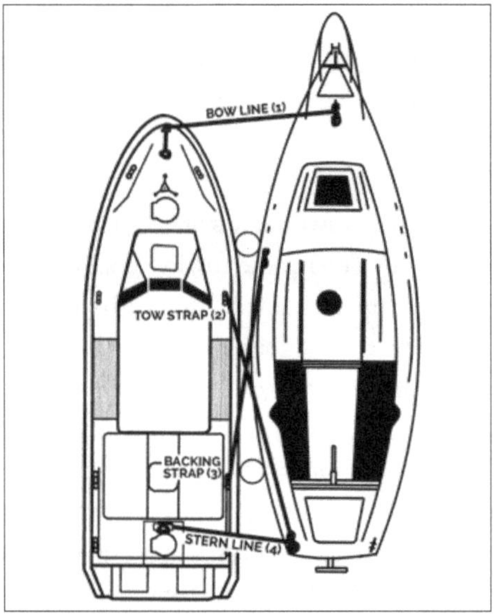

When set up properly, an alongside tow allows two vessels to be maneuvered as one. This is necessary when approaching a dock, mooring, anchoring in sheltered waters, or maneuvering in congested or restricted waters. The tow rig configuration and approach for an alongside tow will be more like that for mooring. If the alongside tow occurs at the completion of a stern tow, two methods of making the transition are available, the drop tow or the transition. In the drop-tow method the towline is completely disconnected and the approach to secure an alongside tow is made as if mooring to a pier, with the coxswain or bridge directing the crew to pass the bow line when alongside. To make a transition, the towline remains connected and is used as the bow line for the alongside tow.

One benefit of using the transition method for an alongside tow is that there is always bow line attached to the disabled vessel, and returning to a stern tow is possible, should the need arise. To complete a transition with a vessel in a stern tow, both vessels are slowed, then stopped and the bitt is broken before the towing vessel slowly backs down and the towline is hauled in. As the distance between the vessels decreases, the crew walks the towline forward to a suitable bow fitting, takes a working turn on the line, and takes in slack to establish a bow line. Once alongside and with the bow line connected, the tow is positioned so that the towing vessel's propellers or waterjets are well aft of the towed vessel's stern. This affords best control for maneuvering in confined areas.

In addition to the bow line, an alongside tow consists of a stern line, tow strap, and backing strap. The stern line holds the sterns together. A tow strap extends from the towing vessel bow to a point aft of the beam on the towed vessel and provides much of the forward pull. The backing strap extends from a quarter location on the towing vessel to a location forward of the beam on the towed vessel and is used when backing to prevent the towed vessel from sliding forward. Fender placement should be checked throughout the evolution and adjustments made so they provide maximum protection at contact points.

Sinking Tow

When it becomes evident that a tow is about to sink, either because it has rolled on one side and is not righting itself or because its decks are submerging, quick decisive action is needed to minimize loss of life. Once abandon ship procedures are initiated, radio communications will likely be lost. The primary action is to break the tow and rescue the people, either from the deck of the towed vessel or from the water. A sinking tow can pull the stern of the towing vessel under unless all crew members pay close attention to the immediate situation. There might not be enough time to disconnect the towline from the towed vessel once it begins to sink. If a tow begins to sink, all towing vessel headway is stopped. Every attempt should be made to have the towed vessel's crew disconnect the towline if possible. If the towed crew is not able to disconnect the tow rig, cut the towline directly aft of the tow bitt. Once free of the tow, make preparations to rescue people who were on board.

WEIGHT HANDLING

Loading, unloading, stowing, and securing cargo and boats on board cutters is complex and requires a thorough knowledge of weight-handling safety, rigging, stability, and griping methodology, much of it specific to the cutter class involved. Weight handling is of particular emphasis on board buoy tenders that are fitted with industrial cranes to set and recover large floating aids to navigation. The safe and efficient movement loads from pier to cutter, from cutter to pier, or when restaging the deck while under way is a critical component for successful operations. Safely and effectively securing loads to prevent weight from shifting on deck while under way is a top priority to ensure stability. The following general safety precautions are observed while handling loads:

- Only authorized personnel shall be allowed in the vicinity of the loading areas during operations.
- Standard crane hand signals shall be used whenever conducting weight-handling operations.
- Ensure that all weight-handling gear is properly sized for the load, paying particular attention to sling angles.
- Ensure that tag/tending lines are properly attached to all loads and positive control is maintained at all times.
- Keep the deck and other work areas clear of stray lines, loose tools, and other gear. Pay particular attention to the load's intended landing area.
- Ensure that each load on deck is secured sufficiently before continuing with loading/unloading operations.
- While pier-side deck loading and unloading evolutions provide great hands-on training opportunities for unqualified personnel, it is imperative that qualified personnel are available and closely supervising break-in personnel under their instruction during these evolutions.

Weight-Handling Rigs

J-Bar Davit

The most basic form of a cargo-handling rig on board Coast Guard cutters is a J-bar davit, a removable metal post in the shape of an inverted J with a maximum outreach of 24 to 60 inches. The J-bar is equipped with a snatch block, meaning a single sheaved block that can be opened on one side to receive the bight of a line. A hoisting line, typically a 3-inch nylon line 120 feet or more in length, is run through the snatch block and used to manually raise and lower loads. The J-bar is moved from side to side manually with 2-inch nylon tag lines, or vang guys, and is anchored with an additional line to prevent unintentional slewing or luffing. It is commonly used in conjunction with a rescue strop to raise and lower cutter swimmers and survivors from the water.

Davit Crane

A davit is any of various crane-like devices used on a ship for supporting, raising, and lowering equipment such as boats and anchors. A davit can be described as a metal post with an angled fixed or an adjustable boom that pivots over a vertical axis. Typically fitted with an electric winch powered by the boat or cutter's electrical system, it may also be operated manually if power fails.

A-Frame Crane

Fitted at the stern of some boats and cutters, the A-frame crane is a type of overhead crane with a single or double girder configuration supported by vertical legs. A-frames employ one winch or may use two winches that work in unison.

Major Cranes

Modern Coast Guard cutters are equipped with a variety of cranes, and one of the biggest and most advanced carried is the 225-foot crane on board the *Juniper*-class oceangoing buoy tenders. Designed to service the largest open-water buoys, a 225's hydraulic telescoping crane is rated to hoist 20 tons with a boom that can telescope out to 60 feet and a winch that can move 60 feet of 1/18-inch 6X37 XIP wire rope per minute.

Rigging

Rigging is a general term for wires, ropes, and chains used to support davits and cranes and to operate weight-handling equipment. Standing rigging describes lines that support but remain in place and do not move. Running rigging describes movable lines. All Coast Guard cutters maintain a registry of lifting appliances, or "rigging log," that identifies all rigging, rigging hardware, and load securing equipment, which includes wire rope, blocks, sheaves, hooks, links, slings, snatch blocks, pelican hooks, gripe chain, and so on. The registry contains the history of all wire rope, rigging hardware, and lifting equipment for its entire service life.

One of the basic principles you must know if you are going to be able to work efficiently in weight handling is that a device you probably would have called a pulley before becoming a sailor is called a block and can be rigged to give you a significant mechanical advantage, thereby saving you a great deal of work and energy. When blocks and lines are combined either to change the direction of an applied force or to gain a mechanical advantage, the combination is called a tackle.

Tag Lines

Tag lines are used to steady loads, to safely position head blocks and saddles, and to vertically hoist/lower tools, minor ATON gear, and so on. A tag line consists of a length of synthetic or natural fiber line with a snap hook, shackle, eye splice, or other appropriate end fitting at one end.

Loading and Unloading

Loading the cutter begins with ensuring all positions are properly manned and staging the appropriate dunnage, gripe down gear, cross-deck, and necessary tag or steadying lines. A minimum of two taglines are used when loading or unloading larger equipment. Personnel on the dock/deck tend these lines to help maintain positive control of the load. Once everything and everyone is positioned, the hoisting purchase, pennant, or sling is connected to load with any hook or shackle used. The load is then slowly hoisted until it clears the gunwale and a cross-deck can be connected. Cross-decks are used to steady loads when being moved horizontally. Once connected, a combination of the cross-deck and crane are used to slowly maneuver the load across the deck—keeping it low to the deck—until it's centered over the appropriate dunnage or saddle. All flat-bottom buoys, concrete and steel stinkers, and all other miscellaneous deck cargo making direct contact with the deck must have dunnage placed between the cargo and the deck to create friction and lessen the chance that the griped cargo will move. The load is then slowly lowered before being griped in place.

Securing Loads to the Deck

The proper securing of cargo, including buoys, is essential to safe and efficient operations. This is accomplished by the use of gripe chains, tensioning devices, web strapping, and wooden saddles/blocks. All equipment stowed on deck must be griped down to prevent them from shifting while under way. The movement of the cutter, especially when sea conditions cause a lot of water to come onto the decks, can cause gripes to become loose; therefore, they require strict attention by watchstanders. While under way, it is recommended that gripes be checked by the relieving watch to ensure there are no problems with the deck load.

Cutter Boat Launch and Recovery

Cutter boat launches and recoveries are among the most dangerous activities conducted in the Coast Guard today. What follows is general guidance on these evolutions. Most cutters use a ship's speed between four and six knots when launching or recovering boats. Small cutters may have to take all way off to safely conduct launch and recovery based on hull design or limited personnel. Engine exhaust and cooling raw water discharge location in the cutter's hull may also require crews to secure the engine on the engaged side of the cutter used in launching or recovering the cutter boat. Regarding headings, there are two practices used by most of the fleet—running down-swell to provide a more favorable ride for the personnel involved and taking the seas on the opposite bow to provide a lee for the cutter boat during launch and recovery. Cutters that operate on rivers must take the current into account. River tenders normally run upriver to stem the current during boat launch and recovery operations.

Sea Painter

The sea painter is a key component for davit-based launch and recovery and is what tows the cutter boat alongside the cutter, not the falls (the wires used to lift or lower boats) or the whip. It consists of two lines—the sea painter and a pendant. The majority of line making up the sea painter is the line attached to the parent cutter forward of the davit/boat deck and is ideally two- to three-inch circumference double-braided nylon. Length of this line varies based on variables in the construction and layout of the cutter.

Davit Launch

Once all positions are manned and ready and tending lines are in place, the boat deck captain (BDC) orders the boat gripes released and the cutter boat is hoisted until it will clear the outboard shoe of the cradle. Once clear, the outboard shoe of the cradle is dropped and the cutter boat slewed out until it is clear of the side. The boat is then lowered to the boat deck rail where the tending lines are secured at the rail. The BDC gives the order to "Lay into the boat," and the boat crew boards the boat at the rail. The coxswain lowers the outboard or outdrive if fitted, energizes the batteries and electronics, conducts a radio check if installed, and attaches the engine kill switch. When ready the coxswain states, “Ready in the boat.” The BDC will then order the slewing out of the cutter boat as necessary to ensure safe lowering then orders the davit operator to winch the boat down to the water. The sea painter tender will adjust the sea painter as necessary to keep the cutter boat positioned under the falls during the launch operation. It is imperative that the boat, once in the water, is towed by the sea painter and not by the davit. The most dangerous time for cutter boat launch operations is at this time, when the cutter boat becomes waterborne. With the sea painter connected the coxswain orders the hook/whip disconnected. For the safety of the boat crew, every attempt should be made to release the whip as expeditiously as possible to prevent shock-loading the winch. The coxswain then starts the cutter boat and energizes navigational lights as required. With the engine running, the coxswain orders the release of the aft and forward tending lines. Once the tending lines have been released and are clear, the coxswain shears the cutter boat away from the side of the cutter by riding on the painter before coming ahead to take the strain off the sea painter ordering “Release the sea painter.”

Davit Recovery

Upon receiving the order to come alongside, the coxswain makes the approach to the cutter, takes station to receive the sea painter, and orders the sea painter past. Once the sea painter has been connected, the coxswain eases back on the throttle and allows the cutter boat to lay alongside the ship. The sea painter tender ensures the cutter boat is towed alongside the cutter in the appropriate location preventing the cutter boat from being towed by the davit. The coxswain then directs the passing or receiving of the tending lines and will normally disembark any passengers, crew, and gear not essential to hoisting. With the words “Ready in the boat,” the BDC orders the davit operator to pay out slack in the whip. Once positive control of the block has been gained by the boat crew, the coxswain orders “Hook it,” then waits for the davit operator to take excess slack out of falls before securing the engine and raising the outdrive or outboards. The BDC then orders the davit to luff out and hoist the cutter to the rail where the davit is slew in to bring the boat against the cutter’s hull and line tenders make off their tending lines. Here the entire boat crew is ordered to “Lay out of the boat” before the BDC orders the davit to slew out to clear the cutter boat from the hull and hoist the boat up to allow for slewing in. Once clear of the outboard shoes or cradle, the davit operator lowers the cutter boat to rest. After gripes are secured to the boat, the davit is secured for sea and the boat is washed and refueled as required.

Stern Launch (Notching)

Stern ramps, or notches, are used by cutters to launch and recover their small boats. National security cutters, 154-foot WPCs, and 87-foot WPBs all have a notch capable of launching and recovering cutter boats. The national security cutters also have an overhead gantry crane able to move various cutter boats from the stern ramp to boat stowage bunks.

Stern Launch Procedures. When all positions on deck are manned and ready, the BDC orders the stern door open and gives permission for the crew to enter the boat. The coxswain starts the engine, ensures the boat is ready for sea, and requests permission to launch the boat. After permission is granted, the coxswain directs a crew member to remove the capture line, if fitted, then signals the BDS to trigger the quick-release connection on the winch retrieval line, releasing the boat. The coxswain then maneuvers the boat clear.

Notching Procedures. After receiving permission from the BDC to enter the notch, the coxswain makes an approach to the cutter and takes station off its stern. The coxswain then maneuvers the cutter boat through the wake of the cutter and into the notch, ensuring to time the seas to maximize the amount of water in the notch and minimize the likelihood of making contact with the notch's sill. When in the notch, the BDC orders the catch line deployed or, in the case of the national security cutter, the catch system automatically performs the task. The cutter boat is recovered once the capture line is engaged. When the boat's engine is secured, the BDC orders the winch retrieval line engaged to hoist the boat to the top of ramp, where a safety strap or chain is installed and the stern door is closed. On board a national security cutter, the boat may be secured in the ramp or moved to one of the cradles on either side.

15

NAVIGATION

Navigation is the science and art of guiding a vessel from one location to another. Mariners throughout history have relied on celestial bodies, currents, predictable winds, and landmarks to travel their intended routes. Modern Coast Guard cutter and boat crews are no exception and, although supported by sophisticated and interconnected electronic systems, must be proficient with traditional navigation principles in order to validate rather than blindly follow what's being shown on their screens.

Regardless of the mode of navigation, the basic process for navigating is the same; only the method and medium upon which position information is plotted, viewed, and evaluated varies. Navigation begins with a known or estimated starting point and a planned direction of travel, or trackline, to a destination that is plotted on an electronic or paper chart and evaluated for safety in relation to factors such as proximity of hazards and draft of the vessel. If turns are needed to avoid land, shoals, or other hazards en route the destination, multiple connected tracklines, termed a route, may be needed.

Once the vessel gets under way and commences its voyage, navigators apply dead reckoning (DR), a method of establishing a vessel's expected position using its course, speed, and time along the planned trackline. At regular intervals, the vessel's position is determined, or fixed, and compared to the expected position established by DR to see if human error, current, wind, or other factors are causing the vessel to deviate from its intended track. On board Coast Guard vessels, integrated electronic navigation systems fix position and provide recommendations for maintaining or regaining track continually and automatically, and fuse in additional inputs from other onboard equipment to provide a greater awareness of the navigational situation. Other manual methods available in the event of GPS or electronic system failures include intersecting three or more visually or radar-acquired bearings or radar ranges to fixed, terrestrial, or celestial objects.

INTEGRATED ELECTRONIC NAVIGATION SYSTEMS

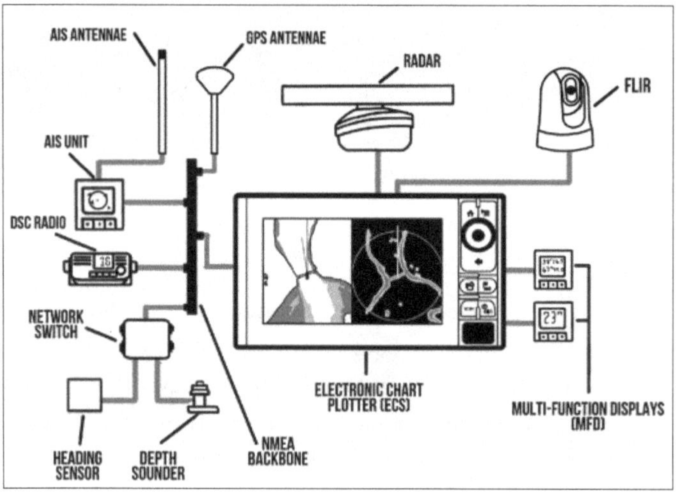

Electronic navigation systems automate the process of integrating real-time positions with a chart display, allowing immediate assessment of vessel position and surroundings. Electronic navigation systems also allow the integration of other operational data, such as ship's course and speed, depth soundings, and radar, into a single or multiple displays. Alarms from sensors and from chart data can warn of potentially dangerous situations well in advance.

Marine instruments are most effective when working together to share and report information. In the early 1980s the National Marine Electronics Association (NMEA) devised a standard interface of language to allow products developed by different manufacturers to communicate. These interfaces were originally designed for one-way communication between two instruments but have advanced to allow a single instrument to send data to several others.

For example, a single GPS may provide information to an electronic chart plotter, a DSC radio, an autopilot, and other systems. These systems, in the event of an instrument failure, are independent and will still function separately.

Multifunction Displays

Electronic navigation systems are characterized by multifunction displays (MFD) that serve to present information from a variety of sources in a graphic format displayed on multiple screens. Capabilities of these systems vary widely, but all have similar characteristics in that they draw information from a variety of sources. Electronic chart plotters integrate radar and electronic navigational charts into their display.

Electronic Chart Plotter

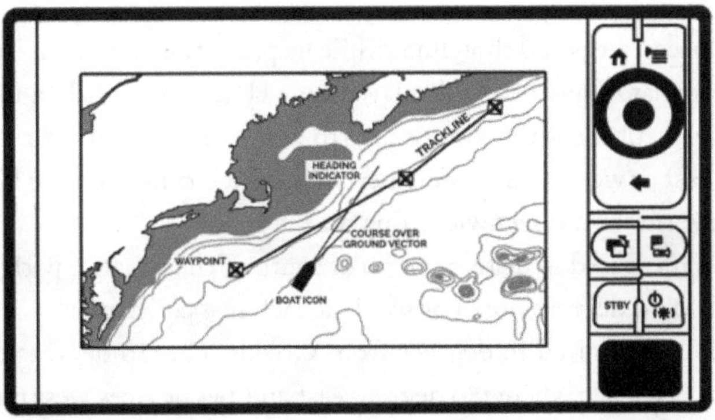

Electronic chart plotters are now a primary maritime navigation tool and have replaced paper charts. They use electronic charts and are interfaced with GPS to allow for a rapid, reliable plot of the boat's position. Chart plotters allow you to see, from an electronic perspective, where your boat is, where you were, and where you are heading. Additionally, they can display a wide variety of data that the operator may find useful. Typical overlays include radar, satellite imagery, depth shading, tidal and current data, and weather. The chart plotter becomes particularly useful when strict navigational tolerance is needed such as with low visibility situations or when transiting through hazardous areas.

Display Matching

Ensuring that displayed information on electronic navigation systems is consistently oriented to the viewer prevents errors in interpretation. When the system displays radar and electronic chart data together in overlay mode, the display must match in scale, orientation, and projection. For example, scaling out the radar to twenty-four nautical miles when the electronic chart scale is at three nautical miles can cause misinterpretations. Unless directed by your unit or otherwise, there is no "correct" display orientation. These are usually considered personal preference. For example, someone may like north at the top because of their affinity to a paper chart, while others may like to see their course as it relates to the electronic chart. Take time to learn what your particular model has available and become comfortable with using them. The course-up mode is preferred for monitoring your progress toward a waypoint, as the destination is at the top of the screen. In north-up, north (000°) is always oriented at the top of the display screen and is useful for voyage planning and when the boat is stationary.

Charts

Charts show ocean areas and shorelines while maps show land masses. Your location, or position, or that of any object displayed on a chart is established using latitude and longitude. Latitude is the distance north or south of the equator. Longitude is the distance east or west of the prime meridian, which runs from the North Pole to the South Pole through Greenwich, England.

Latitude is measured in degrees north or south of the equator, with 0 degrees at the equator and 90 degrees north or south at each respective pole.

Longitude is measured in degrees from Greenwich, 180 degrees east and 180 degrees west. The place where 180 degrees east and 180 degrees west meet, halfway around the world from Greenwich, is called the International Date Line.

Each degree (°) is divided into sixty minutes (′) or nautical miles. A nautical mile measured along the equator or a meridian is roughly 2,000 yards. Any position at sea or place ashore is stated in degrees and minutes north or south and east or west; Cape May, New Jersey, is 38°56.81′ N, 74°53.203′ W.

One of the more vital services a chart performs is to describe bottom characteristics. This is accomplished through combinations of numbers, color codes, underwater contour lines, and a system of symbols and abbreviations. Most of the numbers on

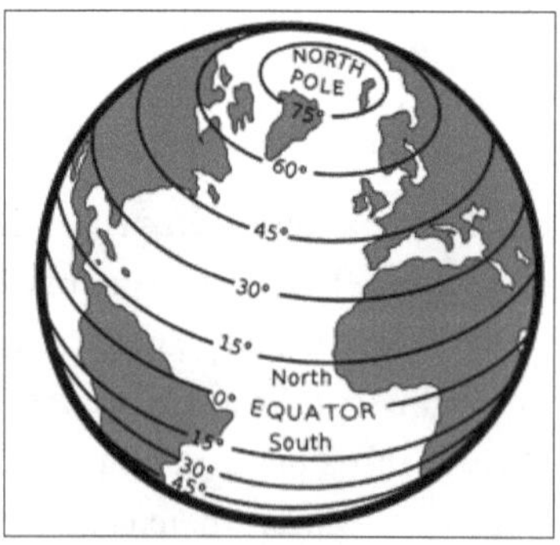

Latitude

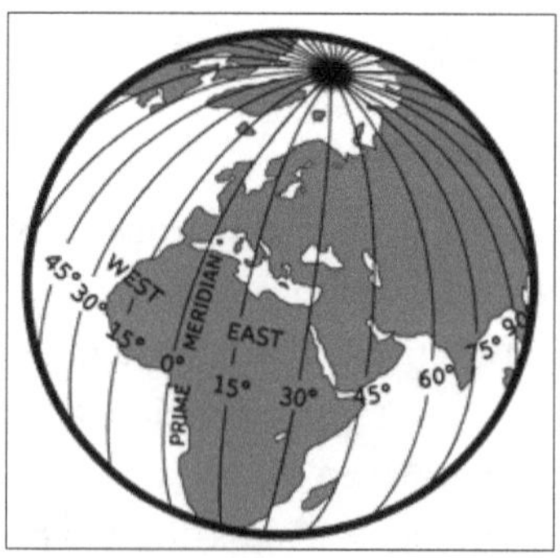

Longitude

the chart represent soundings of the water depth at mean low tide, meaning the average of the lower of the two tides in the tidal cycle. Contour lines, also called fathom curves, connect points of roughly equal depth and provide a profile of the bottom. The depth of water may be charted in feet, meters, or fathoms, with the electronic chart settings indicating which unit is used. A fathom equals six feet.

Symbols and abbreviations used on charts are uniform and standardized but may vary depending on the provider and type of chart. Standardized symbols and abbreviations are shown in "Chart No. 1," published jointly by the National Oceanic and Atmospheric Administration and the National Geospatial-Intelligence Agency.

In 2019 NOAA ended traditional paper chart production to devote more resources to improving electronic navigational charts (ENC). ENCs are layered and interactive, allowing users to query chart elements for more information, and are now the standard chart type used aboard Coast Guard assets. For those requiring a printed chart, NOAA's Custom Chart application allows users to create their own using the latest official NOAA ENC data, customized with location, scale, and safety contours. The application can be accessed at https://devgis.charttools.noaa.gov/pod/.

The U.S. Coast Guard provides notices—"Notice to Mariners" (NTM) and "Local Notice to Mariners" (LNM)—with needed chart correction updates. Electronic charts updates are accomplished by importing updated files into the chart plotter.

Global Positioning System

The Global Position System (GPS) is a navigation system of twenty-four satellites operated by the Department of Defense. In a process called "ranging," a GPS receiver, through its antenna, receives transmissions from any satellites in range and determines the distance between the receiver and each satellite for which a signal is received. Once the receiver has computed the range for at least four satellites, it processes a three-dimensional position that is accurate to about thirty-three meters and reported as a latitude and longitude. Receivers may be installed or handheld and are designed to be integrated or interfaced with other devices such as electronic chart plotters, radar, and radios to automatically provide positional information. Current accuracy exceeds the ten-meter requirement set for harbor navigation and approach.

Differential GPS (DGPS), which used GPS signal corrections broadcast from shore stations to improve position accuracy, has been disestablished as of 2020 in response to improved accuracy of un-augmented receivers.

Latitude (lat) and longitude (long) can be reported by a GPS in three ways and, although written differently, mark the exact same position. This is important to remember as GPS units may be configured differently.

- Degrees, minutes, seconds: 40°26′46″ N, 79°58′56″ W
- Degrees, minutes, tenths of hours: 40°26.767′ N, 79°58.933′ W
- Decimal degrees: 40.446° N, 79.982° W

Convert Seconds and Tenths

One-tenth (0.1) of a degree is 60 minutes. One-tenth (0.1) of a minute is 6 seconds (″). To convert degrees, minutes, seconds to tenths of hours, multiply by 60. To convert tenths to seconds, divide by 60. A conversion table to speed the math is provided in appendix E.

Convert Decimal Degrees

- For the degrees, use the whole number part of the decimal.
- For the minutes, multiply the remaining decimal by 60. Use the whole number part of the answer as minutes.
- For the seconds, multiply the new remaining decimal by 60.
- For example, 110.742 decimal degrees converts to 110°44′31″.

Heading Sensor/Compass

A compass is a device for determining direction on the surface of the Earth. There are three main types: electronic heading sensor, gyroscopic compass, and magnetic compass. Electronic heading sensors have largely replaced the magnetic compass as the primary means of determining a vessel's heading, with a magnetic compass normally serving only as an emergency backup.

An electronic heading sensor is installed on most cutters and boats and is normally mounted low on the centerline. The sensor detects terrestrial magnetism and produces heading data, which is used by electronic navigation system components that need accurate and stable heading input such as an electronic charting system and radar. Heading accuracy of these sensors is typically ±1°. Heading data is shown in three-figure notation (i.e., 000°) and can be adjusted to display as either true or magnetic north, depending on user preference and standard operating procedures.

The gyrocompass is essentially a heavy flywheel driven at high speed by an electric motor and mounted on gimbals so that it is free to move in all directions. It is usually located in a well-protected place below deck on a cutter. Repeaters, or compass cards electrically connected to the gyrocompass and placed on the bridge and in other parts of the ship, show the same readings as the master gyrocompass. The gyrocompass is not affected by either variation or deviation. The motion of the Earth will cause the rotor to move so that its axis lies in a north–south direction. For mechanical reasons and because of the vibrations of the ship, even the best gyrocompass will vary, from time to time, from true north. This gyro error is rarely more than a few degrees, and normally it is constant over a long period of time and on any heading of the ship.

Magnetic compasses use magnetized needles that align themselves with the Earth's magnetic North Pole and are fastened to either a disk or a cylinder marked with the cardinal points of the compass (i.e., north, east, south, and west). The card and needle are supported on a pivot and allowed to freely rotate so no matter how the ship, aircraft, or boat swings, the card is able to realign itself to magnetic north. The moving parts are contained in a bowl or housing provided with a window through which the compass card may be seen. Vessel compasses usually have a flat glass top for all-around visibility and for taking bearings. The lubber's line, a mark in the window of the compass, or on the compass bowl, indicates the fore-and-aft line of the ship or boat. The compass direction under the lubber's line tells the vessel's heading.

Hand-bearing compasses are designed for emergency or survival situations and used in a similar fashion to those installed on board a vessel. To take a bearing, hold the compass approximately eighteen inches in front of you. Line up your target with the sighting notch atop the compass and the lubber line located in the center of the viewing notch. The direction you read on the dial is the direction you are facing.

Direction

Heading is the direction a vessel's bow is pointed. Course is the direction of travel a vessel is to be steered. A bearing is a direction from one's present position to something like a fixed landmark, another vessel, or a waypoint. These directions can be observed as true (T), magnetic (M), compass (C), or relative (R).

Printed nautical charts have one or more compass roses printed on them oriented with true north at the top. There are three main parts to a compass rose: an outer ring representing true north, and inner ring representing magnetic north, and a note defining any variation. Directions on the chart are measured using a straight line from the center point of the compass rose to a number on its inner or outer ring. True north is the direction along the Earth's surface toward the geographical North Pole, while magnetic north is the direction in which a compass

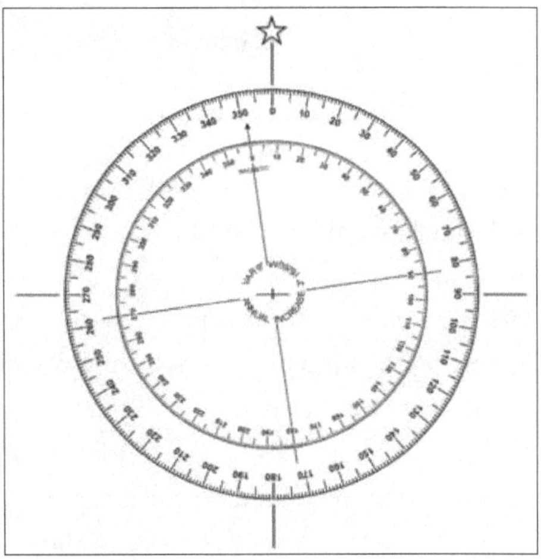

points if it were not subject to any variation or deviation. Variation is the angular difference between true and magnetic north for a particular area. Deviation is the deflection of a compass due to a vessel's construction and its electronics. When the compass bearing is less than the magnetic bearing, deviation is east. When the compass bearing is greater than the magnetic bearing, deviation is west. Errors in variation and deviation are automatically calculated and corrected by a properly calibrated electronic heading sensor. Deviation cards created for and displayed near each compass define the degrees of deviation to be applied.

Bearings are obtained by visual sightings relative to the boat or across an installed magnetic compass, handheld bearing compass, or by radar, and are measured from 000° through 360°. True bearings can be taken on board cutters fitted with a gyrocompass, but for most boats, only magnetic compass and relative bearings are used. When using a compass to take bearings, the object should be sighted across the compass. Relative bearings, in contrast, are based on a circle drawn around the ship itself, with the bow as 000°, the starboard beam as 090°, the stern as 180°, and the port beam as 270°. Relative bearings are used wherever there is no compass.

Radio Detection and Ranging (Radar)

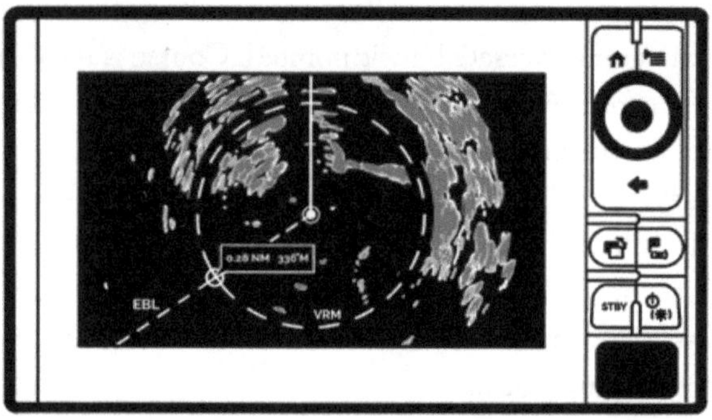

Radar uses radio waves to identify the location, course, and speed of objects. A radar's pedestal transmits pulses of radio waves, then receives those waves that bounce off objects. The returns are processed and contacts are displayed on the radar screen in intensities according to echo strength or gain. The radar does not paint a detailed picture of an object and instead shows them as indistinct masses of color. Knowing the location of land masses and man-made structures such as bridges, docks, and navigational aids helps to differentiate those objects from other vessels on the radar.

The center of the radar screen represents the position of your boat. Depending on the display mode used, the top of the screen may align with your heading, north, or your course. When used properly, the radar enables you to identify your position, land masses and other vessels, determine the distance from your vessel to contacts (range), and determine direction from your vessel to contacts, even in restricted visibility (bearing).

Range Rings. Range rings are used to estimate the distance from the boat to a contact. Range rings display as fixed concentric solid circles emanating out from the center of the screen. The number of rings is automatically determined by the selected range scale. The distance between range rings (interval) is displayed on the screen. Major range scales are indicated in miles and are then subdivided into range rings. Typical range scales are ½, 1, 2, 4, 8, and 16 nautical miles. To measure range to a target using range rings:

- Count the number of rings between the center of the display and the contact.
- Multiply the number of rings by the range ring interval.
- Estimate the distance from the contact to the nearest inner ring and add that number to the range ring distances.

Variable Range Marker. Variable range markers (VRM) measure the distance from the boat to a contact. The VRM displays as a ring around the boat on the radar. Data for the selected VRM is displayed on the screen. Adjusting the size of the ring until it touches the contact will give the distance to the target.

Electronic Bearing Line. Electronic bearing lines (EBL) measure the bearing from the boat to a contact. Radar bearings can be measured in true, magnetic, or relative direction. To obtain a target bearings, the EBL is adjusted until the line crosses the target.

VOYAGE PLANNING

On an electronic or paper chart the navigator plans the route. Where to place the route's waypoints and tracklines is dictated by the position of the intended destination and that of any dangers such as land and shallow water. Also considered are crew, vessel, and environmental factors such as timelines, fuel consumption, and speed. A waypoint is a location. It can be a starting point, destination point, or any point in between. Typically commands maintain a list of regularly visited points of interest, sometimes referred to as a waypoint bank. The trackline defines the direction in which the vessel will travel between waypoints to arrive at its intended location at a predetermined time. Each trackline is labeled with true and magnetic courses for both departure and return and the distance of each track in nautical miles (nm) and yards (yd.). Multiple tracklines form a route.

Publications

Several different publications can be consulted to aid with developing and executing the voyage plan. Instructions for their use are provided within each. These publications include:

- Coast Pilot: A nine-volume text containing sailing directions between points, including recommended courses and distances. Channels with their controlling depths and all dangers and obstructions are fully described. Harbors and anchorages are listed with information on those points at which facilities are available for boat supplies and marine repairs. Information on canals, bridges, docks, and more is included.
- Light List: A seven-volume text identifying and describing all lights maintained or authorized by the Coast Guard.
- Tide Tables: Gives daily predictions of the height of water at almost any place, at any given time. Published annually in four volumes.
- Tidal Current Tables: Provides the times of maximum flood and ebb currents, and the times of the two slack waters when current direction reverses. They also tell the predicted strength of the current in knots. The time of slack water does not correspond to times of high and low tide. The tide tables cannot be used for current predictions.
- Navigation Rules and Regulations Handbook: Sets for the rule of the road required of vessels.
- Notice to Mariners: Published weekly by the National Geospatial-Intelligence Agency. Contains a listing of chart and publication changes and corrections as well as broadcast warnings and advisories.
- Local Notice to Mariners: Printed by the U.S. Coast Guard Navigation Center and organized by Coast Guard districts. Notifies mariners of temporary changes to aids to navigation, special published announcements, and other important information affecting navigational safety.

Speed, Time, and Distance

The most basic navigation and piloting problems involve knowing how far you have gone, how far you have to go, and how long it will take you to get there. Answers to these problems lie in the fixed relationship between distance (D), speed (S), and time (T). There are three basic equations, actually the same equation rewritten, to calculate each specific element. When two elements are known, they are used to find the third, which is unknown.

- DISTANCE (D) = Speed (S) × Time (T)
- SPEED (S) = Distance (D) / Time (T)
- TIME (T) = Distance (D) / Speed (S)

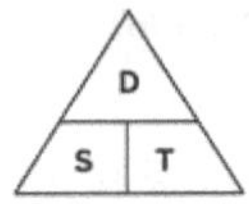

The mathematical relationships and formulas can easily be remembered and applied using the triangle above.

The formulas must have consistent units of measure. Time is expressed in hours and minutes, speed is measured in knots, and distance is measured in nautical miles or yards (i.e., 60′ / 60 = 1.0″).

- Converting minutes into tenths of hours is done by dividing the minutes by 60.
- Converting tenths of hours into minutes is done by multiplying by 60 (5″ × 60 = 30′).
- Converting nautical miles to yards: An actual nautical mile is 6,076.1 feet, or 2,025 yards, which is equal to 1 minute of arc on a nautical chart. For navigation purposes, the current standard for a nautical mile is equal to 2,000 yards. You will use 2,000 yards for navigation purposes, but by doing so be aware you are building in a slight error in distance. Converting yards into nautical miles is done by dividing the yards by 2,000 (i.e., 2,000 yd. / 2,000 = 1 nm; 20,000 yd. / 2,000 = 10 nm).

Three- and Six-Minute Rules

The three- and six-minute rules are tools frequently used during piloting navigation to quickly determine distance or speed. Depending on which rule is used, both are based on two known values, time in minutes (3 or 6) and distance in yards or nautical miles.

Three-Minute Rule. Yards traveled, in 3 minutes, divided by 100, equals Speed; conversely, Speed, in 3 minutes, multiplied by 100, equals yards traveled.

A vessel traveling at a speed of 1 knot (kn) would cover 1 nm in 1 hour. Let's change the time to just 1 minute and see how far the vessel will go (still at 1 kn). Since 1 nm in navigation equals 2,000 yd., and 1 hour is 60 minutes, we can surmise that the vessel would travel 33.333 yd. for every minute at 1 kn (e.g., 2,000 / 60 = 33.333 yd. per minute). Knowing that the vessel will travel 33.333 yd. per minute is not much help until we note that three times that number (or 3 minutes) equals 99.999 yd., which is extremely close to the number 100, and that number is quickly and easily calculated in math formulas. Therefore, for every knot of speed, you will travel 100 yd. in 3 minutes.

Six-Minute Rule. Nautical miles traveled, in 6 minutes, multiplied by 10, equals Speed; conversely, Speed, in 6 minutes, divided by 10 equals nautical miles traveled.

Six is one-tenth of an hour, and each tenth equals 6 minutes. Consider that your vessel would travel 12 nautical miles in one hour. For each tenth (6 minutes) of an hour, you would travel 1.2 nautical miles. Simply stated, if you applied 1.2 nautical miles to each tenth of that hour (× 10), then they would equal 12 nautical miles. Therefore, in 6 minutes, the distance multiplied by 10 equals speed; conversely, in 6 minutes, the speed divided by 10 equals distance traveled in nautical miles.

Alarms

Integrated navigation systems have numerous alarms available, but they provide valuable information only if they are properly configured for the prevailing navigational situation. Too few alarms increase the risk to the safe navigation of the vessel. However, setting too many alarms creates distractions. Alarm management decisions determine the setup and preferred response to alarms and those conditions when an alarm may be silenced or disabled. Common alarms to address during planning include the following:

- Cross-Track Error (XTE): Alerts users upon deviation from the intended trackline. When the vessel's position has exceeded the maximum allowable XTE as established by the coxswain or the unit command, the system triggers an alarm and acts to prompt the operator to immediately address the situation and bring the vessel back onto the intended trackline.
- Depth: A depth sounder will trigger audible and visual alarms at a specified depth of water. A navigational draft, meaning a depth of water that serves as the threshold for operations, is established by commands to ensure timely warning and response prior to a portion of the boat contacting the bottom. Ideally a specified navigational draft will be the total vessel draft plus a safety margin. Factoring negative height of tide may also be needed in some situations.

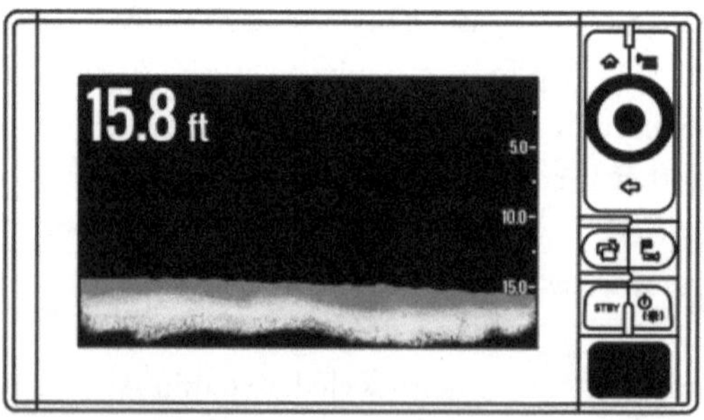

Depth sounder

- Dangerous Radar/AIS Contacts and Targets: Alerts users to any radar or automated identification system (AIS) contacts or tracked targets whose closest point of approach or time to closest point of approach is less than a specified threshold established by the coxswain or command. The alarm indicates a potential risk of collision and provides safety margins of distance and time to assess the situation and take any necessary actions to maintain a safe distance.

Navigation Briefings

Briefs are a critical element of the navigational planning process and ensure safe navigation. They are tailored specifically to operations and serve to ensure all members involved have a common understanding of the risks present and actions to mitigate those risks. Topics addressed in a nav brief should include the following:

- Assignment of crew positions
- Review of charts and intended track / patrol area
- Safe speed for mission and conditions
- Identification of hazards to navigation and how risk will be controlled
- Anticipated traffic and AIS correlation
- Environmental considerations including tides, currents, weather (e.g., winds, precipitation, visibility), and environmentally sensitive sea areas (e.g., marine sanctuaries)
- Maximum allowable deviation from track and confirmation that electronic chart cross-track warnings are aligned with max deviation from track

VOYAGE AND MONITORING

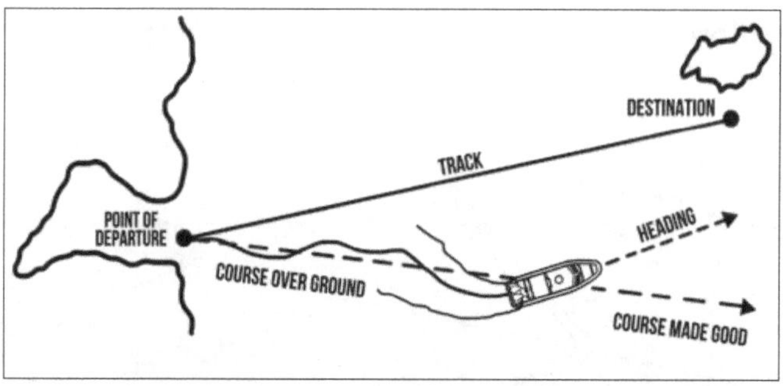

When all equipment is verified for proper operation, desired alarms are set and activated, and all necessary courses have been entered into electronic systems or drawn on a paper chart and evaluated, the planning is finished and the vessel is ready to start navigating. To follow the plan, crews take the following actions:

- Assign helmsman, lookout, and other crew positions.
- Verify position displayed on electronic chart using radar, depth sounder, terrestrial ranges, and ATON.
- Verify alarms set (e.g., XTE, depth).
- Activate planned route.
- Establish estimated time of arrival and begin voyage transit at planned speed.
- Steer or direct helmsman toward the next waypoint using system navigation data and visual and radar information to make good estimated times along the planned track.
- Conduct frequent reports of the navigation situation (i.e., distance left/right of track, time to go to turn, nearest hazard to navigation, depth below keel, recommended course).
- Plot fixes at prescribed fix interval if not shown on the electronic system. Verify position using depth sounder, terrestrial ranges, and ATON. Come to bare steerageway/stop if unable to fix position or maintain plot, or if piloting situation is unclear.
- Plot and label DR course if not shown on the electronic system.
- Recommend course to maintain or regain track. Factor variation and deviation into courses when providing compass course guidance to helmsman.
- Report radar scan "next leg clear," or recommended action based on next leg not clear, before turning.
- Make turns on time to maintain trackline.
- Adjust navigation plan and update remaining estimated times of arrival as needed due to traffic, safe speed, sea conditions, etc.

Fixing Position

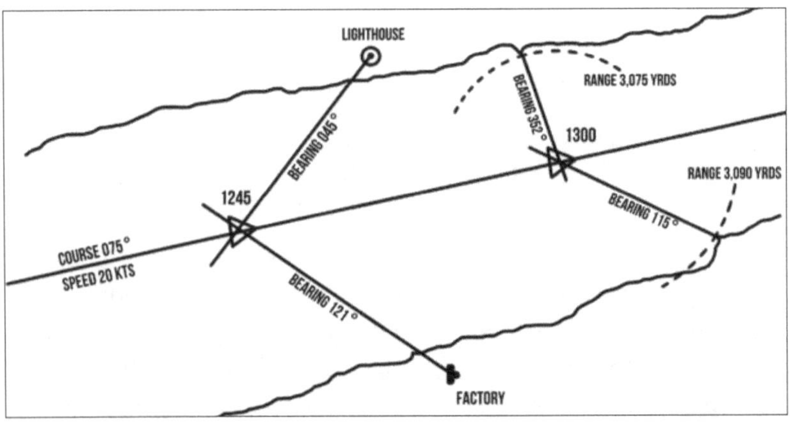

Fixes may be taken by both electronic and visual means at time intervals to coincide with the DR times in order to check for unexpected position differences and update the vessel's course. An ideal fix is either a plotted latitude and longitude reported by a GPS or three or more lines of position (LOP) or ranges separation of at least 60°, but not more than 120° intersecting at a single point.

To fix position with bearings, the bearing to objects is recorded, converted to magnetic or true direction as needed, and plotted. A single observation gives an LOP, not a position, meaning all that is known is the boat is located somewhere along that LOP. Attempt to take bearings to objects as close as possible to the boat as even minor errors in readings are magnified when increasing distance from the object.

When obtaining radar bearings and ranges or visual bearings, the proper procedure is to shoot the object closest to the vessel's beam first because it is changing most rapidly. Objects toward the bow or stern should be shot last because their angle changes more slowly.

Electronic Fix

- Latitude and longitude provided by installed GPS and plotted on an electronic or paper chart.
- Common intersection of three or more radar ranges obtained from simultaneous observations using the VRM or cursor tool.
- Three or more visual or radar LOPs obtained from simultaneous observations using an alidade, handheld bearing compass, or radar EBL tool.

Visual Fix

- Common intersection of three or more LOPs at a common point obtained from simultaneous observations using an alidade or handheld bearing compass.

Terrestrial Ranges

When two charted objects are in a straight line as seen from a vessel, the vessel is located somewhere on a straight line through these objects. This is known as a visual range. Frequently a range will mark the center of a channel. The boat is steered to keep the range markers in line. Ranges may be established navigational aids or natural ranges such as a church steeple and a water tower.

PILOTING

Although used in conjunction with every accepted fix method, piloting provides a viable stand-alone method of navigation even when all other methods fail. Piloting involves directing the movements of a vessel by referring to landmarks, navigational aids, soundings, and local knowledge of the waters and surrounding shoreline. Generally used when entering or leaving port and in navigating along the coast where there are many landmarks, piloting may also prove useful in open water in those regions where the bottom contour makes estimating position possible. Although piloting can be accomplished by the use of local knowledge or seaman's eye alone, it is best supported with accepted fix methods and at a minimum should involve referencing charts that depict shoreline, landmarks, seamarks, and depths, a compass to determine heading and bearings, and a fathometer or lead line to measure depth of water.

Celestial Bodies

Although fixing position using celestial bodies requires skill with specialized tools such as a sextant, many details about your location can be determined with a very basic understanding of certain celestial bodies. Below are details of several of those most commonly used for ocean navigation.

The Sun. The sun rises in the east and sets in the west. If you are north of latitude 23°27′ N, the sun will invariably pass to the south of you on its daily trip across the sky. Latitude 23°27′ N is an imaginary line passing approximately through Marcus Island, Taiwan, the tip of lower California, along the northern shore of Cuba, and through the northern part of the Arabian Sea. The sun follows the upper path at about 21 June each year before starting slowly south. If you are south of latitude 23°27′ S, the sun will always pass to the north of you on its daily trip from east to west. Latitude 23°2′ S passes roughly through Noumea, Rio de Janeiro, and the southern part of Madagascar. These imaginary lines represent the Tropic of Cancer (north) and the Tropic of Capricorn (south).

Stars. Like the sun, the stars move across the sky from east to west. Their position relative to one another remains fixed. Groups of stars are known as constellations. Upon locating one or more stars or constellations, you can use them as markers that tell you where to look for others. You won't find the same stars in the same part of the sky every night. This is because the sun moves westward around the earth at a slightly greater speed than do the stars. Consequently, stars that may be just appearing over the horizon at midnight in one month may be high in the sky at midnight another month. Or they may not appear at all. Some of the most prominent and easily identifiable constellations and single stars are described in the following pages. Train yourself to recognize them.

Orion. The constellation of Orion always rises true east, sets true west, and follows a path directly over the equator. Notice how the three stars forming the

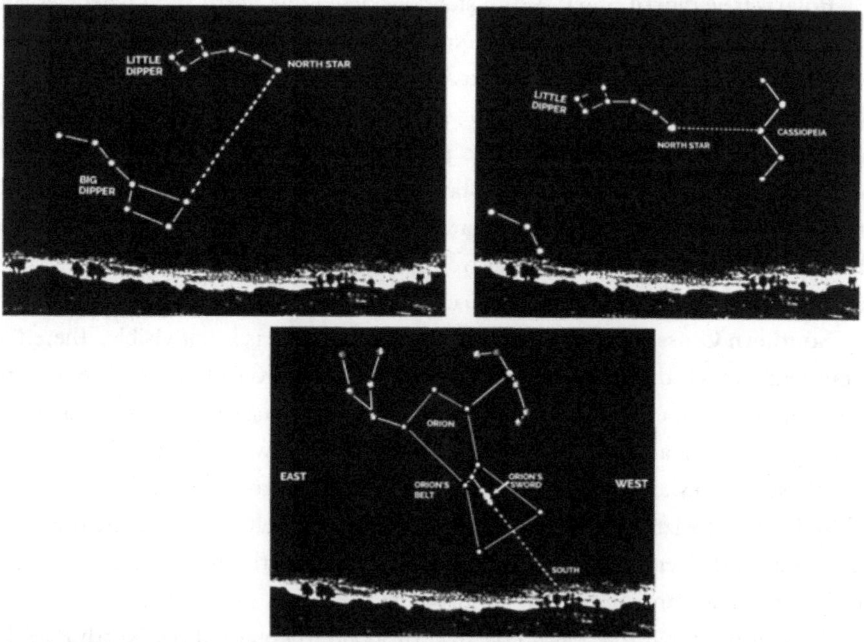

Navigational stars of the Northern Hemisphere

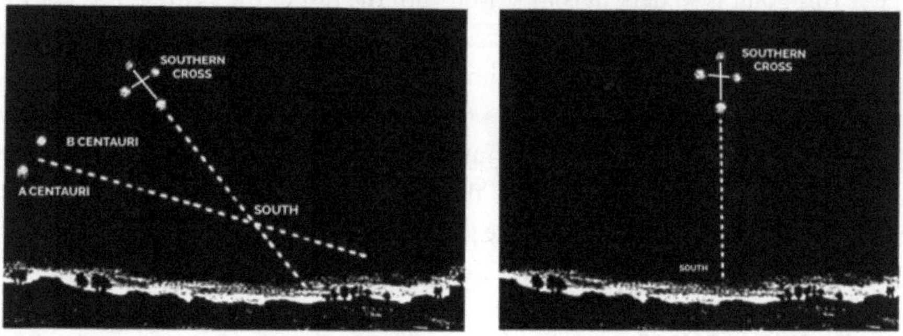

Navigational stars of the Southern Hemisphere

belt are almost parallel to the equator. The two brightest stars of the constellation, Betelgeuse and Rigel, lie approximately the same distance north and south of the line, respectively. Orion can be seen from any position on Earth.

The Big Dipper (Ursa Major). The Dipper is a distinctive constellation containing seven stars. If you are in northern latitudes, it will be the most important and useful constellation for navigational purposes as it "points" to Polaris – the North Star.

Polaris (the North Star). A dim star considered due north from wherever you are since it is almost directly over the North Pole. The two "pointers" of the Dipper, whatever its position, "point" to Polaris. The number of degrees Polaris is above your horizon will always be nearly equal to your latitude. This means that if Polaris is 30° above the horizon, you are in latitude 30° N; if it is 50° above the horizon, you are in 50° N. To estimate the angular distance of Polaris above the horizon, you must first estimate the point in the sky that is exactly overhead (the zenith). From the horizon to that point is 90°. Halfway from the horizon to the zenith is 45°; one-sixth (or one-third of halfway) is 15°, and so on.

Southern Cross. In the Southern Hemisphere, Polaris is not visible. There the Southern Cross is the most distinctive constellation. As you go south, the Southern Cross appears shortly before Polaris drops from sight astern. An imaginary line through the long axis of the Southern Cross points toward the South Pole. The Southern Cross should not be confused with a larger cross nearby, known as the False Cross. The latter, although the stars are more widely spaced, is less bright. It has a star in the center, making five stars in all; the Southern Cross has only four, two of which are among the brightest stars in the sky.

There is no star at the South Pole to correspond to Polaris at the North Pole. In fact, the point where such a star would be, if one existed, lies in a region devoid of stars. This point is so dark in comparison with the rest of the sky that it is known as the Coal Sack.

To find the South Pole, extend an imaginary line southward along the long axis of the Cross. Join the two bright stars to the east of the Cross with an imaginary line. Bisect this line with one at right angles. Where this east line intersects the one through the Cross is (approximately) the point above the South Pole. This point can be used to estimate latitude in the same manner as Polaris is used in northern latitudes—by its height above the horizon.

NAVIGATION RULES

The navigation rules are much like the rules of the road on the highway. They establish a consistent way to navigate safely and avoid collisions when two boats are crossing paths, are on course to meet head on, or when one boat wishes to overtake another. The rules define the roles and responsibilities of vessel operators. If all operators obeyed these rules, most accidents could be avoided. One of the reasons the rules of the road exist is to keep boats from colliding with each other and with other objects. The rules vary slightly depending on whether you are on inland or on international waters. As an example, when operating on inland waters, sound signals are signals of intent, but when operating on international waters, they are signals of action. Some basic tenets of the rules include the following:

- Post a lookout. Designate someone to watch for dangers that may come from any direction.
- Maintain a safe speed. Except where speed is restricted by regulation or the waterway is marked by a "No Wake" or "Slow Speed" aid, you must judge "safe speed" for yourself, taking into account factors including visibility, vessel traffic, your boat's ability to maneuver, and the weather conditions.
- Avoid a collision. The rules define what vessel types must keep out of the way of other vessels and include actions to take when encountering another vessel on the water. The three most common situations you may encounter are overtaking, meeting head on, and crossing the bow of another vessel. In each case, the vessel designated as the "give-way" vessel is required to yield to the other boat, while the boat designated as the "stand-on" vessel maintains its course and speed.

Responsibilities between Vessels (Rule 18)

A power-driven vessel under way shall keep out of the way of

- A vessel not under command;
- A vessel restricted in her ability to maneuver;
- A vessel engaged in fishing; and
- A sailing vessel.

Overtaking Situation (Rule 13)

An overtaking situation involves one vessel passing another when both are moving along similar courses. The give-way vessel can pass on either side of the stand-on vessel. When on inland waters, passing portside requires two short (one-second) blasts on the horn signaling the overtaking boat's intention and two short blasts for the stand-on vessel's agreement. Passing to starboard requires one short blast for intent and agreement. When on international waters, passing portside requires two prolonged (four to six seconds) and two short blasts for the overtaking boat's intention, two short blasts for the stand-on vessel's agreement. Passing to starboard requires two prolonged and one short blast for intention. An agreement signal is only required when operating in a narrow channel or fairway. The following blast pattern signals your agreement: one prolonged / one short / one prolonged / one short.

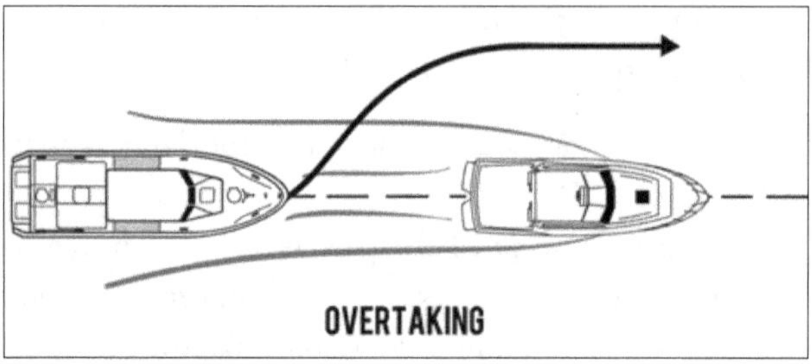

Overtaking situation

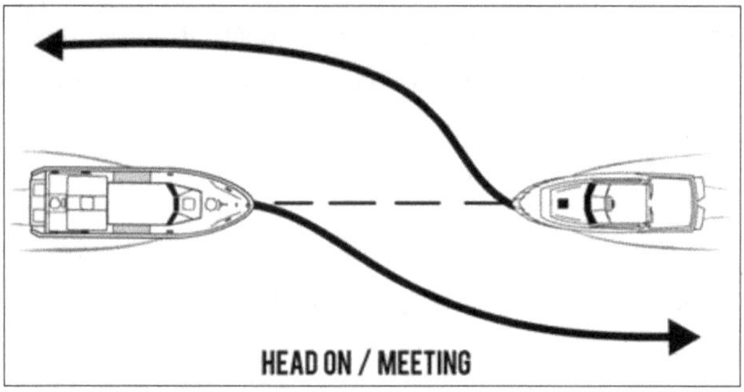

Head-on situation

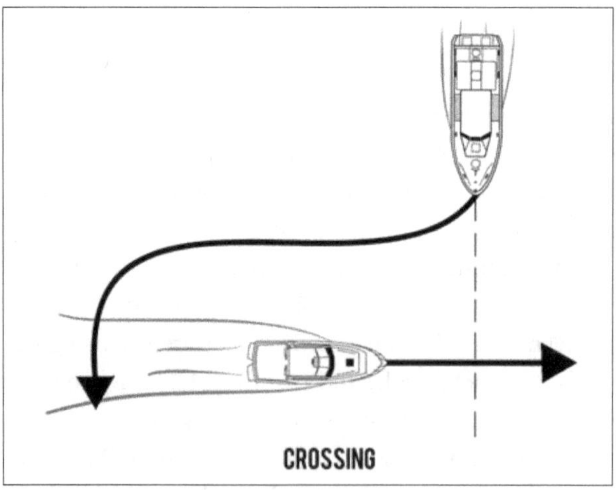

Crossing situation

Head-On Situation (Rule 14)

When meeting another vessel head on, a port-to-port passing is preferred. Passing starboard-to-starboard is acceptable when necessary. Passing port-to-port requires one short horn blast to signal intention and one short blast for agreement. Passing to starboard requires two short blasts from each vessel.

Crossing Situation (Rule 15)

In a crossing situation, the give-way vessel must act to avoid a collision. This may include altering its course to pass astern of the stand-on vessel or slowing down. The stand-on vessel should maintain its course and speed. On inland waters the give-way vessel signals the stand-on vessel with one short horn blast. The stand-on vessel agrees with one short horn blast. On international waters the give-way vessel signals the stand-on vessel with one short horn blast. No agreement signal by the stand-on vessel is required. When meeting another boat, make early and substantial changes to your course. This helps the other operator understand your intentions.

Special Circumstances (Rule 2)

A special circumstance is normally one involving multiple vessels converging on a single location, such as the entrance of a marina. In these situations, a departure from the navigation rules may be necessary to avoid immediate danger.

Navigation Lights

Vessels are required to show navigation lights from sunset to sunrise and during periods of restricted visibility. Understanding these light displays will provide invaluable information to the boater, as these lights can help the boater determine the length of another vessel, its orientation in relation to your boat, and the general direction in which it is headed. All operators must ensure that their vessel's navigation lights are working and that the positioning and distance from which they can be seen meet federal regulations. It is important to know which navigation lights you will see from another boat in certain situations. This can help you determine if you are in an overtaking, meeting, or crossing situation. Whenever you see a red navigation light from another vessel, give way. It is the stand- on vessel. According to the rules of the road, it has the right of way. If you see both the red and green sidelights of another boat, it is coming straight toward you. You should take action to change course in order to avoid a collision.

AIDS TO NAVIGATION

The U.S. Aids to Navigation System, a system maintained by the U.S. Coast Guard, consists of visual, audible, and electronic signals that are designed to assist the prudent mariner in the process of navigation. The aids to navigation system is not intended to identify every shoal or obstruction to navigation that exists in the navigable waters of the United States but rather provides for reasonable marking of marine features as resources permit.

The aids to navigation system used on the majority of American waterways is referred to as the U.S. Aids to Navigation System, or IALA-B. A description of IALA-B follows. The Mississippi River and its tributaries follow a variation of IALA-B called the Western Rivers Marking System. Another system is used exclusively on the Intracoastal Waterway, which runs along the United States coast from New Jersey through Texas. There are also private aids to navigation, which are aids that are neither owned nor maintained by the U.S. Coast Guard. These typically look similar to those in the systems described here and are represented by the same nautical chart symbols, with the addition of "Priv," or "Priv maintd."

Buoys, beacons, and all other aids have a corresponding symbol used on nautical charts. See appendix D for representations of these aids, each alongside its nautical chart symbol, and for day and nighttime graphics of how the buoyage system is used for navigation and piloting.

Lateral Aids

Lateral aids to navigation are generally used for well-defined channels to indicate the route to be followed. These aids show the sides of a navigable channel, mark junctions and forks or splits in the channel, indicate the safe side to pass a hazard to navigation, and mark the safe centerline on wide bodies of water. They are used in conjunction with a conventional direction of buoyage (a local direction of buoyage when approaching a harbor, river estuary, or other waterway from seaward).

"Red, Right, Returning"

You may have heard the phrase "Red, Right, Returning." This expression refers to the fact that when returning (entering a channel from the open sea or proceeding upstream), a boater must keep the red aids on the right (starboard) side of the boat. These aids have even numbers, and the numbers should increase as you proceed inland. When proceeding toward open water (seaward), navigate so that the green aids are to starboard. In this instance you will see odd numbers on the aids, and the numbers should decrease as you proceed toward open water (seaward). If there is no route marked, navigate clockwise around landmasses.

Preferred-Channel Aids

Beacons and buoys that have horizontal bands of both red and green mark the junction of navigable channels. The top-most color indicates the direction of the preferred, or primary, channel. When returning (entering a channel from the open sea or proceeding upstream), keeping the aids with red as the top-most color on the right (starboard) side of the boat will place the boater in the preferred channel, while keeping those aids on the port side will place the boater in the secondary channel. When proceeding toward open water (seaward), keeping the aids with green as the top-most color on the right (starboard) side of the boat will place the boater in the preferred channel, while keeping those aids on the port side will place the boater in the secondary channel.

Non-Lateral Aids to Navigation

Non-lateral aids are the on-the-water versions of the informational, regulatory, and warning signs encountered on the road.

Safe-Water Marks

These aids are used to mark fairways, midchannels, and offshore approach points. They have unobstructed water on all sides. A buoy, lighted or unlighted, may show a red top mark. An appropriate nautical chart must be consulted to determine exact position.

Isolated Danger Marks

These aids indicate an isolated danger that may be passed on all sides. Use caution when approaching.

Ranges

These aids, which are usually shore-mounted, come in pairs to help the vessel operator maintain a straight and safe course within a navigable channel. Each member of the pair is separated from the next in distance and elevation, with the one in front shorter than the one behind it. When the two appear to be vertically stacked, the vessel is on the range line.

Western Rivers Marking System

When on the Mississippi River, on tributaries above Baton Rouge, and on several other rivers that flow toward the Gulf of Mexico, boaters must follow a system called the Western Rivers Marking System, a variation of the U.S. Aids to Navigation System. In the Western Rivers Marking System,

- Buoys are not numbered, and
- Beacons are not numbered, but normally have a mile marker that indicates the distance in statute miles from a fixed point, typically the river's mouth.

Crossing day beacons are used to assist river traffic by indicating where the channel has changed from one side of the river to the other; the aid that marks this channel change will be on the opposite side of the river. When encountered, a vessel operator should head for the aid, using its color as a channel mark.

- Isolated danger marks and safe-water marks are not used.
- Riverbanks are named based on their downstream direction, either "left bank" and "right bank" or "left descending bank" and "right descending bank."
- Because of their somewhat temporary nature, Western River buoys do not have numbers or letters and are not always marked on river charts because they tend to drift due to fluctuating water levels, current speed, and shifting shoals.

Intracoastal Waterway

The Intracoastal Waterway (ICW) runs just inland and parallel to the Atlantic and Gulf coasts from Manasquan, New Jersey, to the Mexican border. The aids to navigation of the ICW closely resemble the green and red buoys and beacons of the lateral system, previously discussed. When operating in the ICW, the color of the aids will typically correspond to the IALA-B scheme (Red, Right, Returning). The yellow squares, triangles, and bands attached to the aids are the primary ICW indicators. There are three types of symbols. When following the ICW from New Jersey through Texas, keep yellow triangles on your starboard and yellow squares on your port, regardless of the color of the aid on which they appear. The yellow horizontal band reflector can be found on any non-lateral aid. It simply identifies the aid as marking the ICW.

Bridges

Bridges are lit to help boaters navigate under them from sunset to sunrise and during other periods of reduced visibility. In general, red lights indicate portions of the bridge where it is unsafe to attempt passage, and green lights indicate areas of safe passage.

16

DAMAGE CONTROL

Damage control refers to all actions taken on board a boat or cutter to limit the damaging effects of an accident, error, attack, or combat engagement. Damage control is concerned not only with battle damage but also with damage from fire, collision, grounding, or explosion. Damage control may be necessary in port as well as at sea. It is not limited to only Coast Guard assets and may be used to save the property of others we are called to assist. Its main objectives are to take all necessary action before damage occurs by making the ship watertight and airtight in those parts that were so designed, removing all fire hazards, and maintaining and distributing necessary emergency equipment; to reduce damage by controlling flooding, fighting fires, and providing first-aid treatment to injured personnel; and to provide emergency repairs or restore services after damage occurs by manning essential equipment, supplying emergency power, and repairing important parts of the ship that were damaged. It is important that you realize your responsibility in damage-control procedures. Learn all you can about your cutter—the whole cutter—not just the part to which you are assigned. In an emergency, you may have to work anywhere on board. The most effective people in any repair party are those who can find their way to any compartment in the dark and close any valves, doors, or hatches by touch alone. Much of this you will learn from drills and practical experience.

Note that regarding search-and-rescue operations, performing damage control solely to save property should be done only if the crew will not be subjected to undue risk.

SOUNDING THE ALARM

Emergencies and casualties must be reported immediately to ensure a timely response from all available hands. Crew members should use any means necessary to alert other personnel to an incident, providing as much information as possible. Sound an alarm by any means, whether word of mouth, portable radio, ship's service phone, or alarm panel. After alerting others, initial actions should be taken to address the emergency. Initial actions vary based on the type of emergency but include providing first aid to injured personnel, using portable extinguishers to extinguish small fires, and isolating the affected space to prevent the spread of damage. If a hazard is known, passing the word to the crew may reduce the number of casualties. If forewarned, personnel may take precautions, such as bracing for shock, verifying proper battle dress, and ensuring watertight integrity.

The general emergency alarm on a cutter is recognized by seven short ringings of the bell followed by a long ring, or the ship horn sounding seven short blasts followed by one long blast.

- General Quarters (GQ) or General Emergency (GE): Sounding of the general alarm plus the words "general quarters" or "general emergency" are passed twice over the general announcing system, for example, "General quarters! General quarters! All hands man your battle stations."
- Fire: Sounding of the general alarm plus the word "Fire" passed twice over the general announcing system, giving class of fire and compartment location, for example, "Fire! Fire! Class Bravo fire in compartment #-##-#, Paint Locker."
- Collision: Sounding of collision alarm plus words "Collision, port/starboard side, forward/aft/ amidships, all hands brace for collision." After a collision, all hands go to general quarters.
- Man Overboard: Word passed twice "Man overboard, man overboard, port/starboard side," followed by five or more short blasts of the whistle.
- Rescue and Assistance: Word passed over the general announcing system, as appropriate: "Plane crash, port (starboard) side," "Away, the rescue and assistance detail, rescue survivors."

When the alarm for one of these sounds, go to your station with a sense of urgency—FORWARD and UP the STARBOARD side, DOWN and AFT the PORT side. Maintain silence.

EMERGENCY ESCAPE BREATHING DEVICES

Emergency escape breathing devices (EEBD) are self-contained, one-time-use emergency breathing devices that provide personnel with approximately ten minutes of breathable air to escape from a smoke-filled or toxic environment. The duration of breathable air will vary depending on individual and environmental conditions. Labored and heavy breathing will shorten the amount of air time, but the EEBD will provide enough time to escape from any space to the weather decks. You can expect to find these stored at the foot of your rack, on the mess decks, and within some workspaces.

An EEBD should be donned immediately at the first indication of a fire or explosion, even if smoke is not detectable. Carbon monoxide is odorless and colorless, and it is possible to breathe a fatal amount before smoke is present. If you suspect an oxygen-deficient atmosphere, don the EEBD and egress immediately. Possible indications of an emergency include, but are not limited to, the sight of smoke, the smell of smoke, fire, alarm, a sudden increase in air temperature, the sounds of an explosion, vibrations from an explosion, or someone nearby having difficulty breathing or losing consciousness. Always don your EEBD first before attempting to escape or helping others. To use:

Step	Procedure
1	Stop work immediately.
2	Locate an EEBD.
3	Remove the EEBD case from the wall mount or storage rack.
4	Remove the EEBD from the orange case.
5	Lift the yellow lever, discarding the cover.
6	Remove the unit by pulling the yellow neck strap upward.
7	Insert the yellow mouthpiece into your mouth.
8	Fit the yellow nose clip to your nose.
9	Don the hood or goggles.
10	Inhale through mouth.
11	Secure the space.
12	Set boundaries.
13	Egress to the nearest weather deck.

WATERTIGHT INTEGRITY

One of the most important aspects of any operation at sea is to keep the vessel watertight. It may sustain any degree of damage, but if the proper degree of watertight integrity is maintained, the vessel will stay afloat. Doors, hatches, and scuttles giving access to all compartments must be securely dogged. They must never be left open overnight or when crew members are not actually engaged in work.

To close a door, first set up a dog opposite the hinges with just enough pressure to keep the door shut. Then set up the other dogs evenly to obtain uniform bearing all around. To open a door, start with the dogs nearest the hinges. This procedure will keep the door from springing and make it easier to operate the remaining dogs. You must obtain permission from the OOD or damage-control central before opening any secured watertight fitting, and you must report back after you have secured the space.

Watertight doors, hatches, manholes, and scuttles should be opened only after making sure that the compartment is dry or so little flooded that no further flooding will be produced by opening the closure. Extreme caution is always necessary in opening compartments below the waterline in the vicinity of any damage. To assess whether the spaces behind watertight closures are flooded, a sounding reading is taken of the compartment suspected of being flooded. To prepare for a sounding, slowly back off the plug or cap on the sounding tube. If a rush of air escapes around the threads while the cap is still under control, tighten the cap immediately. A trickle of water appearing around the sounding plug threads may indicate that the compartment below is flooded. If a compartment is not fitted with a sounding tube, tapping on its bulkhead with a hammer often discloses the presence of water on the other side, and the exact height of the water may be judged by tone variation produced when the bulkhead is struck at different levels. A second method, dangerous but often necessary, is to slowly slack off on the hinged side of some of the dogs that

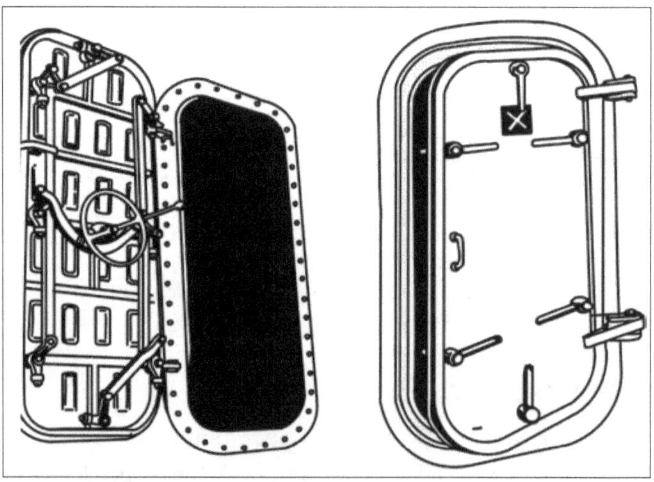

Watertight door

hold a hatch or door closed. There is a slight amount of clearance around the hinge pins, and as the dogs are loosened, any water present will begin to seep between the gasket and the knife edge on the side. Never loosen the dogs on the edge of the door away from the hinges. This action may result in the door buckling or flying open and needlessly flooding another compartment or causing personnel casualties. Secure every fitting after testing.

Material Conditions of Readiness

In order to use compartmentation to its maximum advantage, and to provide for maximum preparedness, all watertight doors, hatches, scuttles, access, valves, and fittings having damage-control value are classified and marked. Ships maintain different material conditions of readiness according to whether contact with an enemy is improbable, probable, or imminent. Each condition represents a different degree of tightness and guarantees the maximum protection against possible damage, with proper regard for the health and comfort of personnel and the essential operation of the ship. All cutters are classified as three-condition ships. These conditions are the number of progressive steps required to achieve maximum protection for the cutter and its personnel against the spread of fire, flooding, smoke, dangerous fumes and gases, and the effects of nuclear, biological, and chemical agents. These material conditions of readiness through which all ships progress are X-Ray, Yoke, and Zebra.

- X (X-Ray): This condition provides the least protection and is set when the ship is in no danger of attack. A black X is placed on all fittings that shall be closed at all times when not in actual operations.
- Y (Yoke): This condition is set and maintained at sea and in port during wartime. For peacetime cruising, Yoke may be modified during regular working hours. A black Y is placed on all fittings that shall be closed for wartime cruising conditions.
- Z (Zebra): This condition provides the maximum protection for ship and personnel for battle and emergency situations and is set when entering or leaving port during wartime. A red Z is placed on all fittings that shall be closed for battle or emergency conditions.

The setting of material conditions X-Ray and Yoke is normally a ship division responsibility and is accomplished by referring to the compartment check-off lists provided. At general quarters, the setting of and maintaining of condition Zebra is the responsibility of personnel stationed within a compartment or area and of the damage-control organization.

Special Classifications

Special classifications are placed on various fittings to meet special requirements. These are described and defined in the following paragraphs.

- Designation W (William): Certain fittings that serve vital systems such as cooling water and fire-main systems are open and operating at all times. They are closed only to prevent the spread of damage. These fittings receive the letter classification W (William) which, in itself, is not a material condition of readiness. When contact with significant amounts of radioactive material in the water is expected, sea suction fittings (classified W), except those essential to main engine operation, are closed.
- Circle X and Circle Y: These classifications are indicated by a black letter within a black circle and indicate that the fittings may be opened without special permission by personnel proceeding to battle stations or as required in routine inspection cheeks. The fittings are secured immediately after use.
- Circle Z: This classification is indicated by a red letter within a red circle. This classification indicates that the fittings may be opened by permission of the commanding officer for health and comfort of the crew during prolonged periods of general quarters. However, they will be guarded for immediate closure if necessary.
- Circle W: This classification is indicated by a black letter within a black circle and shows fittings that are normally open and operating. The fittings are closed only to prevent the spread of chemical, biological, or radiation (CBR) contamination or for defense.
- Dog Zebra: This classification is indicated by a red Z enclosed by a large black D. It is placed on fittings that must be secured to prevent light escaping outside the ship.

BOUNDARIES

A critical component of preventing the spread of damage is determining, ordering, and setting boundaries. Boundaries are set to isolate damage within an affected compartment and to warn personnel of dangers within. Primary boundaries are determined, ordered, and set directly around the affected spaces, while secondary boundaries are set a safe distance from the casualty. There are a variety of boundaries that may be ordered and set to isolate damage:

- Casualty Boundaries: Set to denote an equipment malfunction or deficiency.
- Fire Boundaries: Set to contain fire within the affected compartments, typically at watertight bulkheads, and at decks above and below the fire.
- Smoke Boundaries: Set to contain smoke within the affected compartments. Inner smoke boundaries typically coincide with primary fire boundaries, while outer smoke boundaries coincide with secondary fire boundaries. The space between the inner and outer smoke boundaries is called the smoke control zone.
- Flooding Boundaries: Set to contain flooding within the affected compartments to prevent progressive, or spreading, flooding.
- Toxic Gas / HAZMAT: Set to isolate the compartment(s) affected by securing all supply and exhaust ventilation, and by closing all watertight and fume-tight boundaries.

REPAIR LOCKERS

The damage-control assistant is responsible to the engineer officer for the control of damage. This includes fighting fires, repairing damage, and maintaining CBRN defense. The damage-control assistant is the overall coordinator of damage control matters within the command organization. This responsibility includes the ship's damage-control training program. During emergency situations, the damage-control assistant controls the damage-control problem with the technical advice and assistance of all departments and repair lockers.

When a cutter is placed in commission, damage-control repair lockers are set up in the cutter's organization. The number of such lockers depends upon the size and type of cutter and the number of crew. All available personnel are assigned to repair lockers with the number and ratings of personnel assigned determined by the location of the station and the portion of the ship assigned to that party. Each repair locker contains a number of different ratings in order to provide skilled personnel for any type of work. For example, an engineering repair locker may include machinery technicians, electrician's mates, and damage controlman. Repair parties assigned to areas in which magazines are located will have gunner's mates. Storekeepers are assigned to repair parties that have storerooms located in their areas. Each repair locker has a number of designated positions and is led by a repair party leader who coordinates all damage-control actions in the assigned area.

- Correctly and promptly evaluate the extent of damage in their areas and make accurate "on-scene" reports.

- Control and extinguish all types of fires.
- Repair hull damage and remove flooding and firefighting water.
- Repair electrical and sound-powered telephone circuits.
- Detect, identify, and measure dosage and intensity of radiation, and decontaminate after a chemical, biological, radiological, nuclear, and high-yield explosive (CBRNE) attack.
- Give first aid and transport injured to battle-dressing stations without seriously disrupting other functions.

At each repair locker a number of repair kits are kept ready to be taken to the scene of damage. These kits are commonly called plugging kits, pipe-patching kits, shoring kits, investigator kits, electrical repair kits, and CBRNE monitoring kits. Repairs made to a vessel damaged while under way are termed damage repairs. These are emergency, usually temporary, repairs necessary to allow the vessel to continue its mission and permit its return to port for permanent repairs. Such repairs are made with whatever material is at hand. The important thing is to keep the ship afloat and maintain as much of its operational capability as possible.

GENERAL QUARTERS (GQ)

GQ is sounded whenever battle is imminent or when the highest state of readiness to meet an emergency is desired. For example, the OOD should immediately sound the GQ alarm when a lookout sings out that a periscope has been sighted; when the CIC reports that an unidentified plane has been picked up by radar; or for precaution at dawn and dusk, when there is an increased chance of enemy attack. When personnel reach their stations, they throw off gun covers, break out ammunition, or prepare for action the equipment to which they are assigned. GQ must be set in seconds, not minutes. When each station is manned and ready, the person in charge notifies the control station. When the danger has passed or the drill is over, "Secure from GQ" is announced. No person leaves a station until "Secure from GQ" is sounded or until permission from a control station has been granted.

"Battle dress" is the addition of "flash gear" to the normal day-to-day uniform. Flash gear is designed to help protect the wearer against exposure to short-duration flame, also known as "flash." Battle dress must be worn at all times during GE and GQ, depending on individual watch, quarter, and station bill assignments. Proper battle dress consists of a helmet, antiflash hood, a flash jersey or fully buttoned ODU blouse with rolled-down sleeves, antiflash gloves, bloused pant legs, and standard-issue safety boots. To don:

Battle dress

- Roll down sleeves, tighten cuffs, and button top button of ODU blouse or don a flash jersey.
- Don antiflash gloves, pulling them over the ODU blouse or flash jersey cuffs.
- Tuck pant legs into boots if not already bloused.
- Pull the flash hood over your head, exposing only your eyes.
- Don a helmet.
- Fasten the chinstrap, adjusting the face shield as necessary.

FLOOD CONTROL AND DEWATERING

Flooding occurs when water uncontrollably enters the vessel and may result from battle damage, collision or allision, grounding, ruptured tanks or piping systems, or negligence.

Stability

Flooding changes the vessel's stability, and the added weight of water may cause a loss of righting moment and subsequent capsize or sinking. The effects of flooding include the following.

Free Surface Effect. If a compartment is partly filled with water, the water can flow from side to side as the vessel rolls or pitches. The surface of the liquid tends to remain parallel to the waterline. Liquid that only partly fills a compartment is said to have free surface, and water in such a compartment is called loose water. When loose water shifts from side to side or forward and aft due to turning, speed changes, or wave action, the vessel does not want to right itself.

Free Communication with the Sea. Damage to the hull can create free communication with the sea, which is the unobstructed movement of seawater into and out of the vessel. This is a very dangerous situation that can lead to downflooding.

Downflooding. Progressive flooding of compartments that occurs as water enters the hull is called downflooding. Vessels are designed with sufficient stability and proper righting moments as long as they are not overloaded. These design features cannot compensate for the carelessness of a crew who fails to maintain the watertight integrity of a vessel and allows it to needlessly take on water.

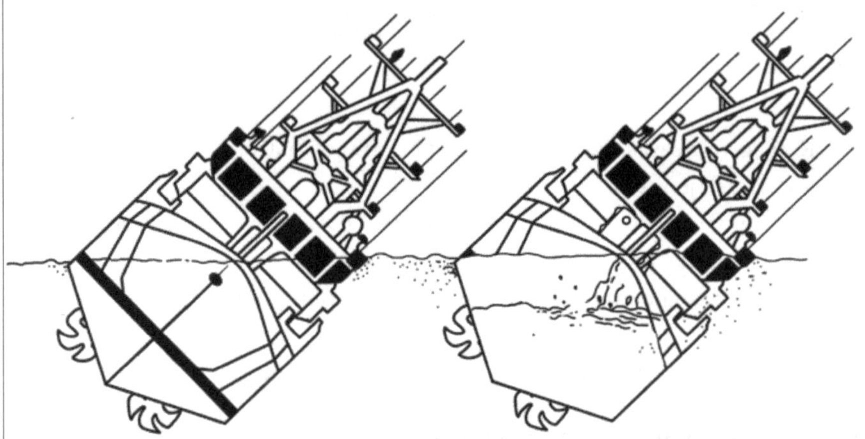

Emergency Hull Repair

When the underwater hull is pierced, there are only two possible courses of action—establish and maintain flooding boundaries or plug and patch the holes.

Flood Boundaries

Collisions, groundings, torpedoes, contact mines, and bombs can create large holes in the underwater hull and cause large sections of hull plating to be destroyed. This type of damage causes complete and extensive flooding and a great amount of wreckage. Such damage, which cannot be repaired by the cutter's crew, must be repaired in drydock. The best action in such incidents may be to allow the space to flood solid to prevent progressive flooding.

Plugging and Patching

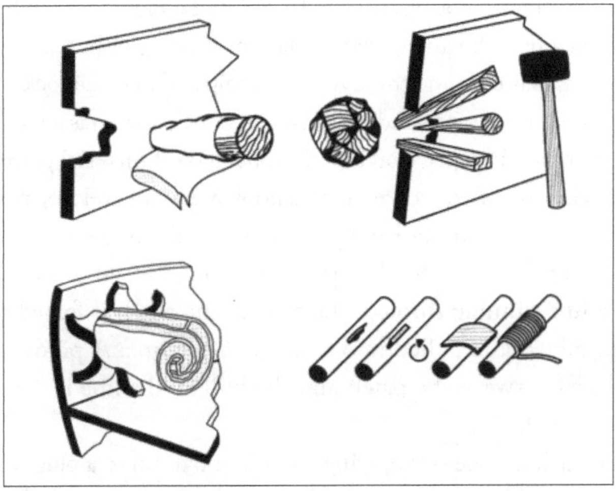

The two general methods of making temporary hull repairs are to put something in it (plug the hole) or to put something over it (patch the hole). In either case, the objective is to reduce the area through which water can enter the vessel or pass from one compartment to another. The shape, size, location, and characteristics of each hole or crack in the hull, deck, or bulkhead of the vessel will vary and must be taken into consideration when determining how to best execute the temporary repair.

For holes or cracks, wrap plugs and wedges with lightweight cloth before inserting to improve grip and fill in gaps. Combinations of plugs and wedges allows for improved conformation to the shape of the hole. Ensure the butt end of the plug or wedge is two inches larger than the hole. Split wedges into smaller sizes if they do not fit tightly. If inserting a plug from outside the cutter, fit the ends of the plug with screw eyes and connect a line to both ends of the plug. Lower the plug from the weather deck using the line connected to the butt of the plug, securing the line to a stanchion. From inside, grasp the line connected to the tapered end of the plug, pulling it through the hole. Draw the line in tight and secure it to a solid stationary structure inside.

Box patches are normally used to cover holes with jagged edges protruding inward. The box is open at one end and has a gasket running along the edges. The gaskets may be made of rubber or of canvas stuffed with oakum. The box is put over a hole from inside the ship and is held in place with shoring. When the compartment is pumped dry, the box may be secured by welding angle clips between the box and the hull plating. Box patches are made locally and, if premade, are stored in repair lockers.

Bucket patches are used to cover holes with jagged edges protruding inward. An ordinary galvanized metal bucket can be used in a variety of ways to stop leaks. It can be pushed into a hole, bottom first, to form a metal plug, or it can be stuffed with rags and put over a hole. It can be held in place by shoring or by using a hook bolt. If applicable, buckets may be found throughout the ship and in repair lockers. Patch plates are normally used for holes protruding outward. A plate patch is a flat piece of wood or metal that is shored into place on top of rubber gasket material placed over a hole. Metal plate patches are often secured with a hook bolt or by welding them into place.

Folding plate patches are normally used on small, eighteen-inch or less holes protruding inward. The folding plate patch is a round wood or metal plate that is hinged with a line running through the center. The patch is folded to fit through a hole then unfolded and pulled into place with the line. A pillow or the rubber material is placed between the patch and the hull, and a line is tied off to a solid structure inside the hull.

Plugging a crack is done by applying a cloth or oakum to a plug or wedge, placing it into crack, and hammering it into place. A hole at each end of the crack is drilled to prevent further spreading. The process is repeated until the crack is mostly clogged, then cloth or oakum are driven into any remaining small gaps between the plugs or wedges.

Damaged Piping

Damaged piping systems are a source of flooding in compartments and may lead to progressive flooding or system failure if left unchecked. When there is a rupture in a piping system, the three immediate courses of action are to isolate the damage, take the affected system off-line, and make temporary repairs. Valves in piping systems are fittings that enable crew members to align source water, oil, and fuel to individual spaces when required. Each protected space is provided with an isolation valve forward and aft of the space to isolate that portion of the distribution main, if required. In manned machinery spaces, these valves are equipped with hand wheels or operating levers located in the adjacent space for backup manual operation.

Structural Damage

Structural damage is especially dangerous, and while it's not always obvious if repairs should be made, the general rule is "If in doubt, shore it." Indications of structural damage include cracking and splitting, groaning or pinging sounds, buckling, bulging, splitting of plating along stiffeners or where bulkheads and decks meet hull plating, bowed frames and stanchions, loose rivets, and cracked seams.

Shoring

Shoring is the process of placing supports against the side of, beneath, or above a structure to provide additional strength and prevent metal fatigue, sagging, and bulging. Shoring is the best way to temporarily support a structure until more permanent measures are possible. Bulkheads and decks, hatches and doors, and equipment that has broken loose are among the most typical items in need of shoring, as well as plugs and patches installed to combat flooding. Generally, there are two objectives of shoring: to reinforce straining bulkheads, doors, and hatches against abnormal conditions, and to support damaged and weakened structures and fittings.

Dewatering

Any pump can be used as a drainage pump if there is power to operate it and if its suction side can be tightly connected through suitable hose or piping to the area to be drained. When used for dewatering, the pump takes suction from flooded compartments and discharges it into the sea. Pump teams typically consist of two crew members.

Submersible pumps are not used to pump gasoline or heavy oils. Operating pumps and equipment with internal combustion engines belowdecks carries the danger of carbon monoxide poisoning. Carbon monoxide is odorless and poisonous and can cause injury or death when inhaled. Do not operate internal combustion engines in an enclosed area without the engine exhaust hose connected and routed safely into the outside skin of the cutter.

Installed Bilge Pump

Bilge pumps are installed in the lowest points of the bilge in key compartments to remove water. They are fitted with a float switch which automatically energizes the pump when the bilge fills to a set level. They can also be manually activated. Although powerful, these pumps are unlikely to keep a vessel from sinking in the event of a significant hull breach but may provide time to effect repairs, slow the flow of water, call for assistance, or prepare to abandon ship.

Eductor

An eductor is used in conjunction with the fire pump. To dewater, it is submerged, either vertically or horizontally, in the flooded compartment. All eductors operate in fundamentally the same manner. Water is forced through a fire main and out through the discharge hose. As pressure of this rapidly moving water passes over the suction opening, it creates a vacuum. The vacuum, or suction, pulls water up through a suction hose, out through the discharge hose, and over the side of the boat. A discharge hose must always lead over the side, and a suction hose must always be placed in flooded areas. If they are inadvertently reversed, a vessel will quickly be filled with water pumped on board through the discharge line instead of dewatering it with a suction line.

Drop Pump

Portable dewatering drop pumps are carried on board most Coast Guard vessels. Depending on the model of drop pump, approximately 100–250 gallons of water per minute (gpm) can be removed from a flooded compartment. Two models are widely fielded—the P-6 and the P-100.

P-6 Pump. Four-stroke gasoline engine rated for 6.5 hp with a 12-foot suction lift and a flow rate of 250 gpm. A removable, portable 2.5-gallon gasoline fuel tank allows operation for 4 hours. A standard design housed in a sealed can, its discharge connection is green in color, and its suction connection yellow. To rig for dewatering:

Step	Procedure
1	Remove the lid of the pump container by opening the seal and pulling the metal bar on the lid.
2	Remove the contents from the P-6 pump container.
3	Connect the fuel hose from the engine to the fuel tank fitting.
4	Roll out suction hose to its full length, ensuring there are no kinks or twists.
5	Attach the suction hose to the suction connection (yellow to yellow) by pushing forward the camlock locking pins. If not seated correctly, the hose will not seal.
6	Secure the suction hose to a stanchion with line.
7	Place the suction hose screen and strainer into the flooded compartment, ensuring it is completely submerged.
8	Roll out the discharge hose to its full length, ensuring there are no kinks or twists.
9	Attach the discharge hose to discharge connection (green to green) by pushing forward the camlock locking pins.
10	Secure the discharge hose to a stanchion with line, ensuring the hose runs straight overboard.

P-100 Pump. Air-cooled, one-cylinder, four-cycle diesel engine rated for 10-hp with a 20-foot suction lift and a flow rate of 100 gpm. To rig for dewatering:

Step	Procedure
1	Move the pump as close as possible to the water source.
2	Remove pump thread protector caps, if installed.
3	Attach a foot valve with a strainer to the male end of a three-inch hard rubber suction hose, using a spanner wrench to tighten.
4	Attach the female end of the hard rubber suction hose to the pump suction connection, using a spanner wrench to tighten.
5	Lower the foot valve / strainer unit into the flooded compartment using a tending line and ensure that the assembly will remain at least one foot under water during pumping operations.
6	Secure the foot valve and strainer unit tending line to the cutter to prevent shifting, ensuring that the suction hose line slopes continuously downward from the suction connection to the water.
7	Attach an eighteen-inch-long jumper hose to the discharge connection
8	Attach a maximum of twenty feet of insulated exhaust hose to the pump exhaust port, ensuring that the hose is properly seated with the clamp connection.
9	Route the exhaust hose downwind, away from the pump, so that the hose is unrestricted and unfouled and away from intake fans.

FIREFIGHTING

The danger of fire is always present. Fire can start in something as benign as an office wastebasket, but if not handled promptly and properly, even the smallest fire has potential to build and spread. If allowed to become a major conflagration, fire may result in the loss of vital systems, severe damage to the ship, and personnel casualties. All-hands participation is required to save the ship—to prevent fire and to bring those that occur under immediate control. The three basic elements in fire fighting are:

- Prevention: Eliminate the possibility of fire before it starts.
- Equipment: Cutters and stations are supplied with the items needed to fight fire. Know where they are.
- Technique: The best equipment is useless if not handled properly. Know how to use it.

Fire Prevention

Fire prevention efforts are essential to protecting a cutter and its crew. Eliminating hazards, proper maintenance and stowage, and effective training of emergency response parties all reduce the potential for fire damage. To minimize fires, personnel must eliminate nonessential combustibles, replace combustible materials and equipment with less flammable items whenever possible, and limit the amount of essential combustibles carried. Clean workspaces and common areas are vital to proper fire prevention. Poor housekeeping practices, improper disposal of smoking materials and dirty rags, and careless work habits can contribute to fires. Personnel should maintain a clean workspace and dispose of all materials properly and promptly.

It is the responsibility of all crew members to recognize, report, and take initial actions when hazardous conditions or fire are recognized in order to prevent shipboard damage and the loss of life. Understanding the science of fire, the different classifications of fire, and the types of fuels that perpetuate fires will enable you to identify the type of precautions, equipment, and extinguishing agents required to successfully fight fires.

Fire is a chemical reaction known as combustion. It is defined as a state, process, or instance of combustion in which fuel or other material is ignited and combined with oxygen, giving off light, heat, and flame. The fire tetrahedron illustrates a fire's complete combustion process. It demonstrates how the three elements of fire—fuel, oxygen, and heat—are related through the chemical chain reaction that enables and sustains the combustion process. If any one of these four elements is removed, the fire will extinguish.

Fires are classified as A, B, C, or D, according to the nature of the fuels involved. Each class has its own characteristics, which determine what extinguishing agents are the most effective. A critical component of the effectiveness of early action is the ability of a crew member to recognize the class of fire based on its characteristics

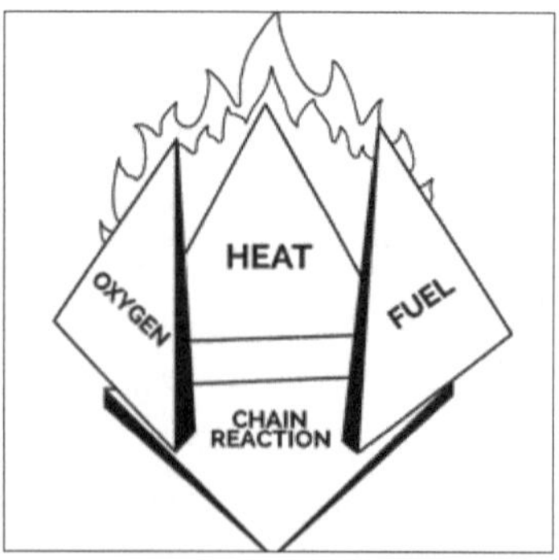

and to use the correct extinguishing agent. Extinguishing agents put out fires by breaking one or more of the four elements of a fire tetrahedron. They work by cooling, smothering, chain breaking, or by a process called oxygen dilution. The different fire classes, the fuel source for each class, the type of extinguishing agent for each class, and the primary effect of each agent are described in the following table.

	Class Fuel Sources	Primary Extinguishing Agent	Primary Effect
A	Common combustible materials such as wood and wood-based materials, cloth, paper, rubber, and certain plastics	(01) Water (02) PKP (dry chemical)	Removes heat
B	Flammable or combustible liquids, flammable gases, greases, petroleum products, and similar products	(03) Aqueous film-forming foam (AFFF) (04) CO2 (carbon dioxide) (05) PKP (dry chemical)	Removes oxygen
C	Energized electrical equipment, conductors, or appliances	(06) CO2 (carbon dioxide) (07) PKP (dry chemical)	Removes oxygen and heat
D	Combustible metals, such as lithium, sodium, potassium, magnesium, and titanium	(08) Water (09) Sand	Removes heat and oxygen

Equipment

Portable extinguishers, fixed fire-suppression systems, the cutter's fire main, drop pumps, and individual firefighter personal protective equipment (PPE) are all available on board cutters.

Portable Extinguishers

Portable firefighting extinguishers are used to extinguish a fire in its early stages while the attack team dons the proper PPE and transits to the scene of the fire. Coast Guard cutters are equipped with a variety of CO_2, PKP, and AFFF extinguishers, which are staged throughout the ship in key locations.

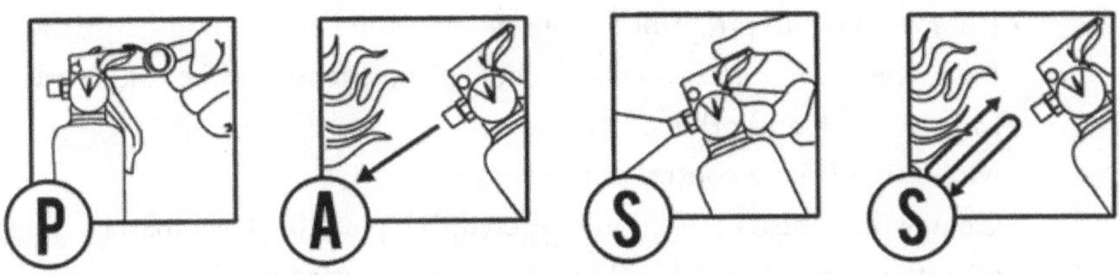

- Carbon dioxide (CO_2) is a chemically inactive gas used to extinguish fires by displacing oxygen. CO_2 is heavier than air, which makes it a suitable extinguishing agent because it tends to settle on or blanket a fire. It is dry, noncorrosive, and does not leave a residue or damage machinery or electrical equipment. CO_2 extinguishers are used for class C and class B fires. They have a maximum effective range of four to six feet and a continuous discharge time of approximately forty to forty-five seconds. To use:
  - › Locate a portable CO2 extinguisher and remove it from its bracket.
  - › Pull the locking pin prior to approaching the fire.
  - › Ground the extinguisher on the deck.
  - › Grasp the horn handle.
  - › Test for agent by briefly depressing the squeeze-grip release lever, releasing a short burst of CO2. If the extinguisher does not work, locate a different CO2 extinguisher.
  - › Carry the extinguisher in an upright position, maintaining positive control and walk toward the fire, maintaining a safe distance from the flames.
  - › Depress the squeeze-grip release lever to discharge the CO2, directing the horn assembly toward the base of the fire.
  - › Move the horn slowly in a side-to-side sweeping motion, advancing toward the fire.
  - › Release the squeeze-grip release lever once the fire is out.

- Potassium bicarbonate / purple potassium powder (PKP), also known as "Purple-K powder," is a chemical powder-based agent that extinguishes fire by separating the chemical reactions involving heat, fuel, and oxygen. Various additives are mixed with the PKP base materials to improve storage, flow, and water repellency. PKP extinguishers are used primarily for class B fires. PKP is difficult to clean up and corrosive if used on class C fires. PKP extinguishers have a maximum effective range of twenty-one feet and a continuous discharge time of eleven seconds. To use:
  - › Locate a portable dry chemical PKP extinguisher and remove it from its bracket.
  - › Remove the pull pin prior to approaching the fire.
  - › Place the extinguisher on the deck, angling the top of the canister away from yourself and others.
  - › Maintain positive control of the nozzle.
  - › Charge the canister by striking the charging puncture lever marked "push" with the palm of your hand in a sharp, downward motion.

  - Test for agent by briefly depressing the squeeze-grip nozzle, releasing a short burst of PKP. If the extinguisher does not work, locate a different PKP extinguisher.
  - Carry the extinguisher in an upright position, maintaining positive control and walk toward the fire, maintaining a safe distance from the flames.
  - Depress the squeeze-grip nozzle to discharge PKP, directing the flow of agent at the base of the flame, using a rapid sweeping motion from side to side across the base of the fire and past both edges.
  - Work from one corner of the fire toward the back until it is completely out, moving closer as the fire is extinguished but not so close as to scatter or splash the burning materials.
  - Release the squeeze-grip nozzle once the fire is out.

- Aqueous film-forming foam (AFFF) extinguishers are typically found near engineering and machinery spaces and galleys with deep-fat fryers. The solution cools the fire and coats the fuel, preventing it from coming in contact with oxygen and suppressing combustion. Portable AFFF extinguishers may be used on class A and class B fires. These extinguishers have a maximum effective range of fifteen feet and a continuous discharge time of fifty-five to sixty-five seconds. To use:

  - Locate a portable AFFF extinguisher; remove it from its bracket.
  - Check the pressure gauge to ensure it is in the "green" range.
  - Hold the extinguisher upright and twist and pull the ring pin, snapping the plastic lockwire seal.
  - Hold the extinguisher by the carrying handle with one hand and the discharge hose behind the nozzle with the other hand.
  - Test for agent by briefly depressing the squeeze-grip release lever, releasing a short burst of AFFF from a safe distance of fifteen feet or more. If the extinguisher does not work, locate a different AFFF extinguisher.
  - Carry the extinguisher in an upright position, maintaining positive control and walk toward the fire, maintaining a safe distance from the flames.
  - Depress the squeeze-grip release lever to discharge the AFFF, starting from fifteen feet away, sweep the AFFF from side to side at the base of the fire, bouncing the agent off surrounding objects to allow it to flow onto the fire, or raining the agent down on top of the fire.
  - Advance toward the fire as the effective range of the extinguisher diminishes.
  - Release the squeeze-grip release lever once the fire is out.

Installed Fire-Suppression Systems

Many cutter platforms, both large and small, are equipped with one or more installed fire-suppression systems that, upon activation, "flood" a space with an extinguishing agent. While there are a wide variety of different installed systems, the general progression upon activation is for the extinguishing agent to flow from cylinders to pressure switches that initiate alarms, secure ventilation, and trigger a time delay for personnel to exit the space. If not bypassed, at the end of the delay the agent will discharge through fixed nozzles into the space.

- CO_2 systems displace oxygen and suffocate the fire and are typically located in spaces generally not occupied by personnel, such as paint lockers and pump rooms.
- Halon and hexafluoropropylene are colorless, odorless, and nonconductive gases that break down the chemical chain reaction to suppress a fire. These systems are typically located in spaces where flammable and combustible liquids are located, such as machinery spaces and paint lockers, and are meant to be activated for fires beyond the scope of manual firefighting equipment and when abandonment of the space is necessary.
- Aqueous potassium carbonate is used for most galley fire-suppression systems, which consist of a range guard and a hood that work in tandem to extinguish burning cooking oil and grease in deep-fat fryers and galley ventilation exhaust ducts. Once activated, aqueous potassium carbonate is released over the range via nozzles and, upon contact with burning surfaces, generates a soap-like front that separates air from the grease or oil to extinguish the fire.

Fire Main

Coast Guard cutters are equipped with dry-riser fire-main systems in which the pipes remain empty until the fire pumps are energized. The system is designed to supply plenty of water for firefighting, sprinkling the magazines, washing the decks, or other purposes for which salt water is normally used. The fire main draws seawater up from the sea chest and distributes it throughout the cutter via a system consisting of pumps, piping with vertical pump risers, longitudinal service mains, cross-connects, branch lines, valves, and fire-hose stations. The fire-main system is designed so that any damaged section of piping can be isolated. That is why there are so many cross connections and shut-off valves at various locations throughout the piping system. Standard fireplugs usually have outlets either 2.5 inches or 1.5 inches in diameter.

Sprinkler systems installed in magazines are supplied by the fire main and spray water in the event of a fire within the compartment or nearby. Fires rarely start within magazines and are more likely to occur in adjacent spaces, threatening magazine contents by heating its boundaries and causing ordnance to "cook-off." If a fire in a nearby compartment breaks out, the sprinkler system helps prevent temperatures within the magazine from rising dangerously high. If a fire breaks out in the magazine itself, the sprinkler system attempts to extinguish the fire, particularly those involving high flash point combustible liquids or ordinary combustibles.

The fire main also supplies seawater to AFFF proportioners, hose reels, and sprinkler systems. AFFF coats fire in a layer of foam, suppressing vapors and inhibiting the chemical chain reaction. Machinery spaces with a high probability of a class B fire are protected by installed AFFF fire-hose stations, and sprinkler heads are installed in helicopter hangars and machinery space bilges.

Drop Pump

Depending on the model of drop pump, approximately 100–250 gpm can be applied to a fire. Two models are widely fielded—the P-6 and the P-100, both discussed in this chapter's section on dewatering. To rig for firefighting, the discharge hose is substituted with a fire hose and nozzle. Water is a cooling agent, and the sea provides an inexhaustible supply for cutters. For many years the principal method of extinguishing a fire was the direct application of a solid water stream at the base of the fire. Although this method is still used, a more efficient method is to apply water in spray form. Water will absorb heat until it changes to steam. The steam carries away the heat, which cools the surface temperature.

A secondary method of water extinguishment is steam smothering. When water changes into steam by absorbing heat, it expands 1,700 times in volume. The large quantity of steam displaces the air from the fuel, which smothers the fire.

Firefighter Ensemble (FFE)

The firefighting ensemble is an assortment of PPE meant to protect a shipboard firefighter from short-duration flame (flash) exposure, heat, cold, steam, water, and falling objects. It can be found in or near repair lockers and is stowed in preassembled stowage bags with boots in the pant legs as a standard to help reduce the amount of time it takes to don the appropriate gear in preparation for fighting a fire. Note the FFE provides protection from external heat but reduces the body's ability to dissipate heat, which results in the possibility of heat stress. Prolonged contact with flames may cause the fire protective coveralls to transmit dangerous heat to the body. The FFE is not a proximity suit like the aluminized firefighting suit used for crash rescue personnel on board cutters with a flight deck. Direct exposure to flames may cause injury. The FFE also does NOT offer protection against chemical, biological, radiological, or nuclear effects. The FFE consists of the following:

- Fire protective garment coveralls that include a Nomex shell, a Gore-Tex vapor barrier, and a quilted removable liner.
- Antiflash hood
- Firefighter's helmet
- Firefighter's boots
- Firefighter's gloves
- Self-contained breathing apparatus (SCBA)
- Multigas detector

Atmosphere Testing Devices

All closed or poorly ventilated compartments, particularly those in which a fire has just occurred, are potentially dangerous. The atmosphere may lack oxygen, contain poisonous gases, or present fire and explosion hazards. Multigas detectors are used to test the atmosphere within a space for toxic gases such as carbon dioxide, carbon monoxide, nitrogen dioxide, and hydrogen sulfide.

Forcible-Entry Tools

To do an effective job when firefighting, you must gain access to the fire. If the normal means of access are blocked, or locked, you will need to use forcible-entry tools to gain access to the area so that you can fight the fire. The tools most frequently used for forcible entry are the sledgehammer, ax, crowbar, wrecking bar, claw tool, hacksaw, bolt cutter, and exothermic cutting torch. The portable pack-type exothermic cutting torch may be used to cut holes in decks or bulkheads so that nozzles can be inserted to fight a fire. The equipment may also be used to cut away debris that would hinder firefighters or rescue workers.

Smoke Curtains

Flame-resistant portable smoke curtains and blankets are stowed in repair stations. There are two kinds. One is a two-panel curtain for use at doorways and in passageways to control the lateral spread of smoke, and the other is a large blanket, fourteen feet square, for use at open hatchways to control the vertical spread of smoke.

Thermal Imager

Thermal imaging cameras enable the user to see through dense smoke and in darkness, detect and display the relative temperatures of objects within the scene, locate the seat and spread of the fire, and move swiftly in search and rescue of casualties.

Firefighting Tactics

Cutters are in constant danger of sudden fire that can spread quickly and place the cutter and its crew in the gravest danger. Fire on board must be handled promptly and effectively wherever it may occur. Every moment of delay increases the danger that the fire will spread. A fire party follows these steps in fighting a fire:

- Locate and report the fire.
- Isolate the area.
- Fight the fire.
- Set a reflash watch.
- Overhaul the area.

Direct Attack

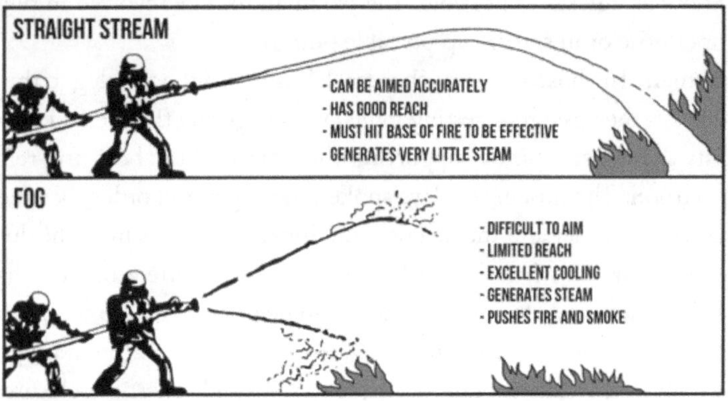

The ideal method of fighting a fire is a direct attack. This technique involves short bursts with a narrow fog or direct stream. Firefighters advance into the immediate fire area and apply water or AFFF directly onto the fire.

Indirect Attack

An indirect attack is used when conditions do not allow firefighters to enter the space. A fog spray is introduced from a cracked doorway or any available penetration. Upon completion, firefighters will then enter the compartment and attack the fire directly. Compartment venting is another means of cooling the space so firefighters may enter. An opening leading directly to an open weather deck area or a large open compartment leading directly outside allows the hot gases overhead to vent.

Firefighters

A hose team is considered "manned and ready" when they have all members dressed out at the entrance to the affected space and with a positive firefighting agent test. Multiple hose teams may be required to maintain firefighting efforts. The capacity of SCBA air tanks and heat stress are major concerns.

- Team Leader: The team leader issues orders for hose advancement, and instructs the nozzleman to attack the fire with the necessary spray pattern.
- Plugman: The plugman connects the hose to the fireplug and, when directed to do so, slowly opens the fireplug valve to activate the hose. The plugman keeps an eye on the hose for a loss of water pressure or a hose rupture. In the event of loss of water pressure, the plugman reports to the on-scene leader and secures the plug. In the event of a ruptured fire hose, the plugman secures the fireplug, replaces the ruptured section of hose, and reactivates the hose as quickly as possible. The plugman may be involved in rigging a jumper hose or in setting up portable pumps.
- Hoseman: The hoseman runs the attack hose from the fireplug to the scene and keeps the hose from getting fouled while fighting the fire. The hoseman assists the nozzleman by supporting the weight of the hose and relieving the tension. The hoseman relays spoken messages and orders between the attack team leader and the nozzleman. The hoseman assumes the duties of the reflash watch when directed to do so. After the fire is out, overhauled, and the space is safe to enter, the hoseman helps clean up the compartment. In the event of a collision or flooding, the hoseman is on the shoring detail and reinforces weakened bulkheads, braces warped watertight closures, and patches holes in bulkheads and piping.
- Nozzleman: The nozzleman mans the attack hose nozzle and extinguishes the fire through the compartment as directed by the team leader. The nozzleman is required to wear a SCBA. Once the fire is extinguished and a reflash watch is established, the nozzleman may be involved in the overhaul of the space, in conducting atmospheric tests, de-smoking the compartment, and investigating for damage.

Locating the Fire

Many fires break out in unattended spaces such as storage areas and gain considerable headway before they are discovered. The first sign of fire may be smoke pouring from a ventilating system outlet or seeping around a door or hatch cover. Smoke may have traveled a considerable distance from the fire, so locating the fire is usually the first job. Repair-locker investigators examine all bulkheads, decks, and vents for abnormal heat. A compartment may be clear, but a hot bulkhead would indicate a fire in the next compartment. Follow up any odor or trace of smoke. Even though the main fire is located, there may be fires in other compartments, and they must all be fought at the same time. When the fire has been definitely located and its extent determined, a complete report of the situation is made to the repair locker.

Isolating the Fire

While the fire party fights the fire, it also sets up fire boundaries to isolate it. If a single compartment is involved, it is isolated from above and below as well as from all four sides. If a number of compartments are involved, the fire party treats them as a unit and protects all adjacent spaces to keep the fire from spreading.

Materials on the deck above are removed or thoroughly cooled to prevent the spread of fire. Usually an inch of water on the deck is enough to absorb the heat from the burning compartment below. Spaces below the fire must not be neglected or paint on the overhead may blister and ignite. Fog should be used to cool the overhead and prevent ignition of materials in such spaces. All ventilation systems in the fire area are secured to prevent heat and unburned vapors being carried to other compartments and to help starve the fire by cutting off its air supply. If there is any doubt that the fire can be kept within the established boundary, a secondary or emergency boundary is set up. Additional hose lines are made ready for use at the secondary boundary.

Fighting the Fire

While the fire is being isolated, it should also be fought. The first step is to de-energize electric circuits in the compartments to protect personnel against shock. All members of the fire party will be dressed in FFEs. Heat and unburned gases build up pressure that should be vented by undogging the door slowly before it is opened completely for entry. Always check fire-hose water pressure at the nozzle before entering any space. A fire can be extinguished by cooling, oxygen dilution, fuel removal, or chemical flame inhibition.

- Removing the fuel: One way to remove the fuel from the fire is to physically drag it away. It is often possible to move nearby fuels away from the immediate vicinity of a fire so that the fire does not extend to these fuels. Sometimes the supply of liquid or gaseous fuel can be cut off from a fire. When a fire is being fed by a fuel line, the fire can be extinguished by closing the proper valve.
- Removing the oxygen: A fire can be extinguished by removing its oxygen or by reducing the oxygen level in the air to below 16 percent. Many extinguishing agents (for example, CO_2) extinguish fire with a smothering action that deprives the fire of oxygen.
- Removing heat: Water is the most effective means of removing heat from ordinary combustible materials such as wood, paper, and cardboard. Cooling with water can ultimately stop the release of combustible vapors and gases associated with the burning of solid fuels. Water is applied to a fire using one of three ways:
  - Straight stream: Used when long reach and penetrating power are critical.
  - Narrow fog (30 deg): Cools a much wider surface than a straight stream and can absorb more heat. Additionally, as fog comes into contact with any surface heated by fire, it becomes steam, providing a secondary smothering effect that further aids in extinguishing the fire. Water fog can also be used to protect firefighters from excessive heat.
  - Wide fog (90–110 deg): Applied with a vari-nozzle and less powerful than narrow fog. Because wide fog covers more area than narrow fog, it may be used most effectively when it is possible to get right up next to the fire where it may be used as a shield by forming a screen of water droplets between a firefighter and the fire.

AFFF is the primary firefighting agent used on board U.S. Coast Guard cutters for flammable liquid fires. Specially equipped firehoses with foam nozzles supplied by the fire main are used by the firefighting team to deliver the agent.

Reflash Watch

Fires that seem to be out may start again from a smoldering fragment of material or through vapor ignition. After extinguishing a fire, a reflash watch is set for each compartment involved and is maintained as long as the possibility of reflash exists.

Overhauling the Area

After all fire is extinguished, the area must be overhauled to prevent a reflash of the fire. To do this, members of the fire party break up all smoldering or charred material and saturate it. Whenever possible, materials are immersed in buckets of water. Compartments are de-smoked and dewatered. All compartments involved are checked for explosive vapors or liquids that may remain. At this time a full report is made to the commanding officer or office in charge on fire and smoke damage and flooding.

CBRNE DEFENSE

CBRNE incidents can occur accidentally, naturally, or be carried out using weapons of mass destruction (WMD) with the intent to cause mass casualties, incapacitation, and potential damage to critical infrastructure on a large scale. CBRNE threats and hazards include weapons; improvised devices; toxic industrial materials; non-traditional agents including riot-control agents; herbicides, pathogens, and toxins. Symptoms of these incidents may take seconds to weeks to appear and result in incapacitation or death. The Department of Homeland Security notes that "one of the gravest maritime risks facing the nation is the potential for a terrorist group to obtain a WMD and detonate it within the confines of a major U.S. port city, military installation, or industrial facility. Closely aligned with this is the potential for the maritime domain to be used as transportation for such a weapon, weapon materials, or its components with an eventual target further inland." Early recognition of threats allows you to take primary and secondary defensive measures, ensuring exposure to CBRNE threats is kept as low as reasonably achievable.

Each cutter has a CBRNE bill that sets forth the duties and responsibilities of cutter personnel to defend against and respond to a CBRNE incident based on available shipboard CBRNE equipment and the platform's physical characteristics. The three general principles of these bills are

- Contamination Avoidance: Avoid CBRNE hazards by acting on intelligence and using detection and monitoring equipment, and so on.
- Control and Mitigate the Threat: Protect personnel, equipment, and the cutter from unavoidable CBRNE hazards by ensuring that PPE is properly donned, setting GQ and material condition Zebra, and initiating countermeasure washdown system, and so on.
- Contain: Prevent the spread of CBRNE contamination by establishing control zones, setting up decontamination (DECON) stations, and so on.

CBRNE Contamination

Contamination is the deposit, absorption, adsorption, or adherence of radioactive material or of biological or chemical agents by structures, areas, personnel, or objects. Additionally, contamination is when food and drink is made unfit for consumption by humans due to the presence of environmental chemicals, radioactive elements, bacteria and organisms, and decomposing material or waste.

Duration

Persistence is an expression of the duration of the effectiveness of a CBRNE agent. This depends on the physical and chemical properties of the agent, the weather, methods of dissemination, and terrain conditions. Persistent agents, when released, remain able to cause casualties for twenty-four hours to several days or weeks or even indefinitely. Persistent agents generally include radioactive materials, biological pathogens that can live on surfaces, blister agents, viruses, and thickened nerve agents. Nonpersistent agents, when released, dissipate or lose the ability to cause casualties after ten to fifteen minutes. Nonpersistent agents are commonly liquids or vapors and are usually lethal and generally include choking agents, blood agents, and nerve agents. Radiation released after the initial burst of a nuclear device can be considered a nonpersistent agent.

Route of Entry

CBRNE agents enter the body through absorption, injection, inhalation, ingestion, and penetration. Absorption describes skin contact with hazards and through cuts, wounds, and punctures of the skin. Injection is the introduction of an element into the system through forcible entry into skin or organs. Inhalation occurs through the mouth, eyes, nose, or other mucosal membranes, and via the respiratory tract to the lungs. Ingestion occurs through the gastrointestinal tract by way of contaminated food, drink, and swallowing contaminated mucous from the nasal area. Penetration is through the body as with gamma, X-ray, and neutron radiation.

Chemical Agents

The use of weaponized chemicals can produce powerful casualty or harassing effects, but with proper protection, such as clothing and gas masks, you have an excellent chance of surviving any chemical attack. There are three types of antipersonnel chemical agents:

- Casualty Agents: Can kill or seriously injure large numbers of people and include nerve agents, blister agents, choking agents, and blood agents.

- Incapacitating Agents: Not lethal but produce effects that harass and temporarily incapacitate persons. Principal types are tear gases and vomiting agents.
- Riot-Control Agents: Cause temporary irritation as experienced with pepper spray.

Chemical agents can be dispersed in a variety of ways, including bursting chemical munitions, bombs or missiles that disseminate chemical agents by venting instead of bursting, spray devices such as those mounted on aircraft, aerosol generators, and accidental releases.

Biological Agents

Living organisms or their toxic products can be weaponized to destroy or contaminate food sources, such as crops and domestic animals, and to inflict disability or death on the populace by spreading an epidemic. These agents come in three basic forms: bacteria, viruses, and naturally occurring toxins. A biological agent attack is most often conducted by spreading the agent with aerosolizing devices or sprayers; introducing it into ventilation systems, food, or water sources or product tampering; contaminating items commonly touched by the general public; or infiltrating the population with infected contagious people. It is nearly impossible to see, smell, or taste any biological agent, and because of their various incubation periods, there may be no immediate effects on people, animals, or crops to indicate that an attack has been made. It may take hours to days for symptoms to appear. The only way to detect and identify a biological agent is for trained experts to examine specimens of air, food, water, or human and animal blood and excretions. Most are not fatal. They will cause illness if inhaled, absorbed, or ingested, but the victims will recover. The best defense against them is to keep yourself and your living areas clean and report any sickness immediately. Most biological agents will die or lose their effectiveness after a few days of exposure to ordinary weather conditions, but there are some that may remain dormant for a long period.

Radiological Incidents

Radiation attacks use radioactive particles to cause the breakdown of molecular bonds in the human body. The radiation is composed of invisible alpha and beta particles and gamma rays. Alpha particles can be stopped by a thin sheet of paper; the other particles travel at the speed of light and therefore are more difficult to stop. Particles are released in the form of ionizing radiation, which is invisible and may leave no immediate physical indication of its presence. Any living tissue in the human body can be damaged by ionizing radiation. Health effects from acute exposure appear quickly, thus your exposure must be kept as low as reasonably achievable. In radiological incidents, radioactive materials are released into the environment and contaminate air, water, surfaces, soil, plants, buildings, people, or animals. These typically result in alpha and beta radiation causing damage to living tissue once ingested but may involve absorption of penetrating gamma radiation. Contamination normally occurs when radioactive material in powder, dust, or liquid form encounters a person's skin, hair, or clothing. The primary concern with external contamination is that if not effectively removed or cleaned up, it could become internal contamination and a potential medical event. The three types of radiological devices of most concern are

- Radiation Dispersal Device: Incorporate radioactive material into a conventional explosive device or some type of dispersal equipment that spreads radioactive powder, liquid, liquid mist, gas, or solid.
- Radiation Exposure Device: A device placed in a public area with the intent of exposing people in the vicinity of the device to radiation, usually gamma, emitted by a radioactive material contained inside. These devices have high dose rates and can be detected from hundreds of yards.
- Toxic Industrial Radiation: Radioactive material manufactured, used, transported, or stored by industrial, medical, or commercial processes, such as spent fuel rods and medical testing. Since these have legitimate commercial and medical uses, these are not as tightly controlled as other radioactive materials.

Nuclear Attacks

Nuclear attacks are easily recognizable. A large initial blast or explosion is accompanied by a fireball that consumes nearby areas and releases gamma and neutron radiation during the initial three to five minutes post-explosion. Heavy shielding is required as a protection, depending on distance from the burst and the "yield" or size of the bomb. The blast and heat effects generate most of the damage and casualties. After the initial radiation is released, alpha and beta radiation become the primary concern. This residual "fallout" can contaminate large areas. The two types of nuclear devices of greatest concern are

- Nuclear Weapon: A military explosive device incorporating weapons-grade nuclear material that produces a massive explosion, followed by an intense fireball.
- Improvised Nuclear Device: A "homemade" nuclear weapon that incorporates nuclear and explosive materials. Produces an explosion and fireball but at a much smaller scale than a conventional nuclear weapon.

Nuclear incidents may also occur accidentally or through acts of sabotage at locations such as nuclear power plants.

In a nuclear attack, cutters may be exposed to air, surface, or subsurface bursts. An air burst produces blast, heat, intense light, and initial radiation but no fallout requiring decontamination of the ship or personnel. In an air burst, 85 percent of the damage is due to shock, 10 percent to thermal radiation, and 5 percent to initial or residual radiation. A surface burst will produce shock, heat, intense light, and initial and residual radiation. The fallout, especially over water, will spread radioactive contamination over a wider area. A subsurface blast normally produces no heat or light and very little initial radiation. The danger in this type of explosion is in the intense underwater shock and heavy residual radiation from the highly contaminated base surge of earth or water created when the column formed by the explosion falls back to the surface.

Attack without alarm is always possible, and the only warning will be a very bright flash or the sky lighting up. When you see the flash, close your eyes, cover your face with your hands, crouch and bend down, grasp something solid and hang on. Otherwise, the shockwave caused by the rapid movement of air or water away from the explosion could blow you over the side. The force of the shockwave can easily wreck ships, damage buildings, and kill or severely injure people, even at considerable distance. Thermal radiation from an air or surface burst will produce skin burns and eye damage only among those directly exposed. You can reduce the chance of burns by dropping out of direct line of sight of the fireball and covering exposed skin surface. You may suffer from flash blindness, but your eyesight should recover after a short period.

Radiation Sickness Symptoms

A radiological or nuclear attack may cause radiation sickness, also known as radiation poisoning. If the dose is fatal, death usually occurs within two months. Signs and symptoms of radiation sickness include nausea, headache, fatigue, weakness, vomiting, hair loss, minor bleeding and lesions, convulsions, coma, skin burns, diminished organ function, or death. Radiation is measured in units that account for exposure (air dose), total absorbed dose, or dose equivalent.

Medical Pretreatment and Medical Countermeasures

The following are the general medical countermeasures to take for each type of agent:

- Nerve agents: Inject Nerve Agent Antidote Kit (NAAK), Antidote Treatment Nerve Agent Autoinjector (ATNAA), or Convulsive Antidote for Nerve Agent (CANA). NAAK is a single-use, autoinjector device that provides for automatic intramuscular injection for nerve agent exposure.

The kit consists of one atropine injector that blocks the enzymes that make muscles contract and one pralidoxime chloride injector that allows muscles to relax. You may receive a maximum of three self-injections at ten- to fifteen-minute intervals, when symptoms persist or recur.

- Blood agents: Don a protective face mask to prevent further damage and seek medical attention—no self-aid/buddy-aid treatments are available.
- Blister agents: Flush eyes with water.
- Choking agents: Don a mask to prevent further damage and treat for shock, as necessary.
- Incapacitating agents: Victims may need to be temporarily confined due to their lack of judgment or inability to make responsible decisions.
- Tear agents: Flush eyes with water—DO NOT rub eyes. Don a protective face mask.
- Vomiting agents: Don a protective face mask, and keep it on in spite of coughing, sneezing, salivating, or nausea.
- Biological agents: Pre-exposure vaccines (e.g., anthrax and smallpox), antivirals, and antibiotics.
- Radiation: Potassium iodide (KI) blocks the thyroid's uptake of radioiodine. To be effective, administer KI within three to four hours of exposure. Coast Guard assets maintain stocks of KI based on the threat assessment, such as proximity to nuclear plants or fallout zones.

Personal Protective Equipment

Chemical, biological, radiation, or nuclear (CBRN) PPE is designed to shield or isolate you from biological, chemical, radiological, and some nuclear incidents. PPE protects your respiratory system, skin, eyes, face, hands, feet, head, body, and hearing and consists of both clothing and respiratory protection to ensure safety from vapor, liquid, and solid forms of CBRN agents.

You must take personal defense measures to limit exposure and cross-contamination. The priority of operation is self, others, cutter. In the event of an attack, don PPE or, if unavailable, use anything that covers your skin (e.g., foul-weather gear, flash gear, goggles) and nose and mouth (e.g., cloth, rags). You must don your own PPE before helping others, including casualties. Assess if you may have been contaminated or exposed to an agent. If exposure is expected, administer proper medical countermeasures and report to the DECON station. If contamination is not expected, follow preplanned evacuation routes to your watch, quarter, and station bill assignment. Listen to announcements over the 1MC regarding the status of missions, operations, response activities, and contamination.

CBRN PPE is contained in standardized kits issued to units based on their assigned missions and expected operations in contaminated environments. You must be aware of the type of WMD PPE on board your unit, where the kits are located, and how to properly don and doff the equipment. The typical contents of a USCG WMD kit include

- M50 mask
- M61 replacement filters
- Canteen
- Tychem F coveralls or the Joint Lightweight Integrated Suit Technology chemical protective coat and trousers
- Chemical protective gloves
- Chemical protective footwear
- Duct tape
- Chemical detection armband

The most important CBRN PPE component is the protective mask, which should be donned prior to securing any other form of PPE, such as suits, coveralls, overboots, gloves, helmets, and so on. To don the M50 mask:

Step	Procedure
1	Remove glasses, if wearing.
2	Remove the mask from the carrier.
3	Close your mouth and eyes and hold your breath.
4	Position the mask in front of your face.
5	Insert chin into chin pocket while holding mask.
6	Grasp the tab and pull the harness over your head, ensuring the center of the harness patch is centered on the back of your head and that straps are flat against your head.
7	Tighten chin straps.
8	Clear the protective mask by covering the air inlet port on the filters and the outlet valve with your hand.
9	Exhale forcefully then breathe in so the mask collapses against the face.
10	Exhale forcefully then breathe in so the mask collapses against the face.
11	Check for leaks and tighten mask if it does not have an airtight seal.
12	Resume normal breathing.

Mask filters should not be replaced while inside warm or hot zones unless absolutely necessary. This minimizes the chances of trace amounts of contamination entering the mask. To change the filter:

Step	Procedure
1	Hold the mask to your face with one hand while holding the chin cup with the other hand.
2	Close your eyes and take a deep breath.
3	Buddy removes the old filter by squeezing the locking tabs and rotating the top of the filter toward the front of the mask.
4	Buddy aligns the locking lugs of the new filter, rotating the top of the filter toward the back of the mask.
5	Place your hand over the outlet valve cover and blow hard to clear the mask.

Passive Detection and Monitoring

Several technologies are employed by Coast Guard crews and boarding teams to detect CBRNE environments as they perform their normal duties. These require minimal training to operate and respond to. In general, upon the detection of any CBRNE contaminate, those present will stop, move outside the contaminated environment, alert specialized CBRNE teams for further investigation of the threat, and close off the area to prevent others from entering.

Personal radiation detectors (PRD) are pager-size devices worn by operators to detect radiological and nuclear threats. Maritime variant PRDs signal a threat is present by alarming the user if the sea environment's radiation level exceeds the natural amount and by localizing the threat for follow-on identification of the isotope by specialized teams.

A chemical detection armband is an easy-to-use, inexpensive, reusable device that can be attached to the forearm or other easily observed location and holds up to ten one-time-use disposable cassettes. Each cassette detects a particular toxic chemical or class of chemicals and changes color upon detection of the chemical agent.

Decontamination

The process of making any person, object, or area safe by absorbing, destroying, neutralizing, making harmless, and removing chemical, biological, or radioactive agents and material is termed decontamination (DECON). The type and extent of DECON is determined by the physical state, concentration, amount, and length of time a contaminant is present as well as by environmental considerations such as ambient temperature, humidity, and so on. The goals of DECON are to prevent further harm to victims, protect personnel from exposure, and prevent the spread of contamination.

DECON Methods

DECON methods include physical removal, chemical neutralization, and natural decay. Physical removal uses mechanical action to carry the agent away, including using water, detergent, scrub brushes, firehoses, and so on. This method does not change the toxic properties of chemical, biological, or radioactive agents; it only moves it from one place to another. Care must be taken to ensure that contamination is not simply moved from one part of the cutter to another. Chemical neutralization involves decomposing a chemical or biological agent through a chemical reaction. Biological and chemical agents can be neutralized by strong oxidizers. The chemical or biological agent should be physically removed upon neutralization. Radiation contamination cannot be neutralized, it can only be physically removed. Natural decay is a passive DECON method where the toxicity of chemical agents and biological toxins diminish over time due to weathering or the effects of the environment. Natural decay is not applicable to radioactive agents, which have a long period of decay, sometimes hundreds of years.

DECON Principles

The four principles of DECON are speed, need, priority, and limited transfer.

- Speed: DECON victims as soon as possible or practical. Direct exposure to some CBRNE hazards may be fatal within minutes. The sooner equipment is decontaminated, the less likely it is to absorb or spread the agent to other surfaces. Flushing surfaces with water during initial DECON is generally the easiest method.
- Need: DECON only what is necessary. Cutters have a limited amount of resources and should only expend them as specifically required.
- Priority: DECON the most essential items first, including exposed skin if contact has occurred. Once PPE is donned, begin to DECON clothing and equipment. In many cases, removing the clothes off personnel exposed to liquid or vapor agents will remove a majority of the contamination.
- Limited Transfer: DECON near the areas where contamination occurred (i.e., the warm zone). This limits the spread of contamination to other areas.

Water used to wash down contaminated surfaces should be isolated and collected to prevent further cross-contamination through movement to other surfaces. To reduce the likelihood of further contamination, and regardless of the CBRN agent, DECON should always proceed top to bottom, fore to aft, and windward to leeward.

Shipboard DECON

Decontamination of a cutter begins at sea and consists of a saltwater washdown of the entire topside. This will remove 98 percent of contamination if the surface was wet prior to contamination and washdown commences while it is still wet. Washdown should be completed immediately after contamination occurs. Hose squads wear protective clothing and gas masks. Saltwater washdown commences even if the sea is contaminated, in which case it will continue until the ship reaches a clear area.

Preliminary decontamination of personnel is done immediately as exposed personnel are washed down under fire hoses or topside showers. Decontamination is organized into stations. All contaminated clothing is removed in the undressing area. Each person scrubs thoroughly with soap and water in the washing area. The clean dressing area provides uncontaminated clothing. Radiation detection devices at each section ensure that contamination is not spread. Decontamination priority is generally the repair party and damage-control teams, followed by uninjured personnel and injured personnel. Injured personnel are usually handled last because of the extra time required for medical personnel to assist them.

17

SEARCH AND RESCUE

The mission and purpose of the Coast Guard's search-and-rescue (SAR) program is to prevent death or injury to persons and loss or damage to property in the marine environment. The success of our SAR program is measured by the success of each and every SAR mission that we perform.

The statutory authority for the U.S. Coast Guard to conduct SAR missions is contained in Title 14, sections 2, 88, and 141 of the U.S. Code. The code states that the Coast Guard shall develop, establish, maintain, and operate SAR facilities and may render aid to distressed persons and protect and save property on and under the high seas and waters subject to the jurisdiction of the United States. It also states that the Coast Guard may use its resources to assist other federal and state entities. SAR may be considered a mandated function, but no specific level of performance has been cited under legislative authority. Nevertheless, judicial rulings have made it clear that once the Coast Guard undertakes a particular mission, we must conduct that mission with due diligence, we must not worsen a situation by our actions, and we must meet a reasonable standard of performance. Moreover, it is within our service's own ethos to carry out each mission to the best of our ability. The ultimate goal of the Coast Guard's SAR program is to prevent loss of life in every situation where our actions and performance could possibly be brought to bear.

SEARCH-AND-RESCUE FUNDAMENTALS

A search is an operation to locate persons or property in distress. A rescue is an operation with the primary purpose of retrieving persons in distress and delivering them to a place of safety. Rescue may include providing medical care to survivors and actions to save property from damage or loss.

Incident to rescues are often actions to prevent environmental damage and remove hazards to navigation. The rescue of persons in distress is the highest priority SAR mission. Missions solely for saving property or for other purposes such as preventing environmental damage will always give way to saving a person's life.

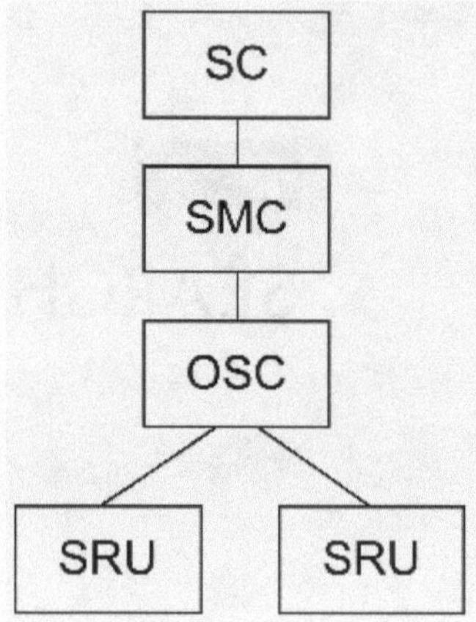

SAR Organization

The National SAR Plan designates the Coast Guard as the aeronautical and maritime SAR coordinator for the waters over which the United States has jurisdiction, such as navigable waters of the United States and land areas other than the continental United States and Alaska.

District commanders serve as regional SAR coordinators and have overall responsibility for establishing, staffing, equipping, and managing the SAR system, identifying SAR resources, developing SAR plans, and suspending search efforts.

For each mission a SAR mission coordinator (SMC) is assigned, normally at the Sector Command Center level. The SMC alerts and dispatches resources; establishes a search action plan, if needed; and communicates with those on scene through the Sector Command Center to maintain situational awareness and coordinate efforts.

The on-scene coordinator (OSC) oversees the scene. The OSC communicates with the Sector Command Center and directs individual search-and-rescue units (SRU) in carrying out search action plans and rescue operations.

SRUs are the assets on scene who physically perform the search-and-rescue tasking received from the OSC.

SAR Program Standards and Requirements

Certain standards and requirements have been developed for Coast Guard units to ensure the response to SAR is immediate and effective.

SAR organization

SAR Readiness. Each Coast Guard unit with a SAR readiness responsibility shall have a suitable SAR resource ready to proceed within thirty minutes of notification of a distress.

SAR Mission Response. There will be no greater than a two-hour total response time for any one response unit within a sector's or unit's area of operation to arrive at any location within the area of operation. This time is calculated from time of notification of the Coast Guard until the time of arrival on scene, allowing thirty minutes of preparation time and ninety minutes from under way to on scene, for a total of two hours.

Notification

The SAR system is activated upon first notification of an actual or potential SAR incident. The report may come directly from the distress party, someone who was witness to the distress, relatives reporting a family member overdue, or from an electronic signal, such as an emergency position-indicating radio beacon. If the reporting party is excited or panicked, speak calmly to ease their concerns, be confident and professional, and collect essential information. Realize this may be the only contact made with the distressed craft or reporting source (e.g., the radio broke, power was lost, or the boat sank). It is vital to immediately record the following information:

- Location of the emergency
- Nature of distress
- Number of persons onboard
- Description of the craft

The location of the emergency is the single most important piece of information to obtain. Without a location, it is impossible to send the search-and-rescue resource. The location can be provided in many different ways, and often "local terminology" concerning position can be confusing. The exact location of the disabled vessel should be understood before continuing to gather information from the reporting source. Position can include latitude and longitude of the incident, bearing and distance from the incident to any points of land or landmarks, or last known position of the incident or distressed vessel. If the vessel is under way, find out its course, speed, and intended destination.

Upon receiving a request for assistance, the incident is classified as one of three emergency phases:

- Uncertainty: Exists when there is knowledge of a situation that may need to be monitored, or to have more information gathered but that does not require moving resources.
- Alert: Exists when a craft or person is experiencing some difficulty and may need assistance but is not in immediate danger or in need of immediate response. Apprehension is usually associated with the alert phase.
- Distress: Exists when grave or imminent danger requiring immediate response to the distress scene threatens a craft or person. Immediate response shall be initiated, if feasible, to any known situation in which the mariner is in imminent danger. This response may be provided by regular Coast Guard, Coast Guard Auxiliary, or other federal, private, state, local, or commercial entity resources. The SMC may use all sources of assistance in a distress situation without concern for conflict with private enterprise.

Briefing

A mission briefing shall be conducted among the crew of all SRUs prior to launching on a SAR case. Gather as much information as possible before launching so you'll be sure to select the right resources and bring any additional equipment needed, such as a dewatering pump, stokes litter, handheld radio, extra life jackets, medical gear, and so on.

Preparations En Route

SRU crews should make every effort to prepare for the search prior to arrival in the search area. Preparations include briefing and assigning lookouts and establishing a relief/rotation schedule if appropriate. Charts of the area should be prepared for use, with the search area boundaries, search pattern, and turning points plotted. Homing, monitoring, and on-scene radio frequencies should be tuned. Depending upon the resource used and the weather conditions, equipment necessary to perform the rescue should also be prepared for immediate use.

There is always a possibility that the search object is outside the assigned search area; therefore, searching while en route to the assigned search area is one way the SRU can increase the search area size. By posting lookouts while en route to and from the search area, effective trackline searching can be accomplished and the results passed to the SMC.

Check in with the SMC or OSC fifteen minutes before arriving on scene to report your status and limitations and to obtain a search update or any additional instructions/tasking.

Initial Actions on Scene

Report on-scene weather to the SMC when you arrive at the last known position as well as any weather changes as your search progresses. On-scene weather affects both sweep width and track spacing of planned searches. If no search action plan or specific tasking has been provided by the SMC, commence a drifting search pattern using the last known position as your commence search point. If the search object is not sighted at the last known position and no debris field is visible, commence a square search (SS), sometimes called an expanding square. If a debris field is sighted, commence a sector search (VS) to concentrate your search efforts in its vicinity. Deploy a datum marker buoy (DMB) at the commence search point to measure currents and assist with developing follow-on searches. Make sure you record and report the time and position where it was inserted and the serial number for a self-locating DMB. You will typically use a fender or life ring with a strobe light as your DMB.

Visual Distress Signals (VDS)

A distressed vessel has a limited number of VDSs, if any at all, and experienced mariners usually do not activate these signals until they actually see or hear an SRU. You should consider the use of audible signals such as an air horn or whistle when searching to encourage the distressed vessel to use its VDSs while the SRU is in the area. Pyrotechnic devices may be used to make the SRU's presence known to the distressed vessel. However, any signal used, whether visual or audible, must not be mistaken by other SRUs as a distress signal.

Aural Searches

Surface SRUs should make every effort to reduce background noises and occasionally stop the engines to permit faint calls for help to be heard. If possible, lookouts should be posted away from the engines and radios.

SEARCH ACTION PLAN (SAP)

Once the SMC determines the search area, a systematic search for the object must be planned. The SMC will consider weather, search area size, search object size, numbers of SRUs available, search area location, and time limitations in deciding which pattern to use. Once a SAP has been developed and communicated by the SMC, make every effort to follow the SAP and to remain on track during any assigned search. Subsequent searches assume you searched as planned. When the SAP needs to be modified, report the problem to the SMC along with recommendations to correct the problem. Situations requiring an SAP to be modified include hazards to navigation in the search area, insufficient water depth, inappropriate track spacing based upon prevailing weather conditions, and the position of the sun in relation to the lookouts. Anytime the SAP doesn't make sense on scene, report it to the SMC along with your recommendations.

Search Patterns

Each search pattern has a specific use. Some are better suited for small, confined areas, while others are better suited to larger areas. Some follow the drift or datum and others specified waypoints. Regardless of the type of search pattern selected, SRUs are expected to understand the SMC's SAP and be able to complete the assigned search. Search patterns are designated by letters. The first letter indicates the general pattern group:

V = Sector
S = Square
T = Trackline
P = Parallel
C = Creeping Line

The second letter indicates the number of SRU's assigned to that pattern:

S = Single Unit
M = Multiunit

The third letter indicates specialized patterns or instructions; the two most commonly used are

R = Return
N = Nonreturn

Sector, Single Unit (VS)

A VS is a drifting pattern used when the search area is small, the location of the search object is well known, and a concentrated search is desired. The SRU goes to the best-known location of the search object. This becomes the center (datum) of the search area. Datum should be marked with a buoy, life ring, strobe light, and so on. The first leg will begin from datum in the direction of drift for one track space (S). If there is no drift, the first leg will be true north, 000°T, the second leg is 120°T, the next is 240°T, and so on. On the 3rd, 6th, and 9th legs, steer on your marker, and once passed, return to your base course. All course changes are 120° to the right and the length of all legs are equal to one S. A second search (with the same SRU–VS) is performed by rotating the search pattern 30° to the right beginning at datum. A second search (with a second SRU–VM) is performed by rotating the search pattern 90° to the left beginning at datum.

Expanding Square, Single Unit (SS)

An SS is a drifting pattern used when the search area is small, the location of the search object is known within relatively close limits, but some doubt exists about the distress position, and a concentrated search is desired. The SRU goes to the best-known location of the search object. This becomes the center (datum) of the search area. Datum should be marked with a buoy, life ring, strobe light, and so on. The first leg will begin from datum in the direction of drift for one track space (S). If there is no drift, the first leg will be true north, 000°T; the second leg is 090°T, the next is 180°T and so on. All course changes are 90° to the right. Search leg length is increased by one track space on every other leg. Multiply your track space (S) by the numbers shown in the search pattern below to determine leg length. A second search is performed by rotating the search pattern 45° to the right.

Drifting Pattern Track Spacing and Timing

When conducting drifting square (SS) or sector (VS) search patterns, refer to the following tables to establish initial track spacing and timing.

INITIAL TRACK SPACING (NM)

Search Object	Good Conditions Winds < 15 kt Seas < 3 Feet	Poor Conditions Winds ≥ 15 kt Seas ≥ 3 Feet
Person in Water (PIW)	0.1	0.1
Vessels < 15 ft	0.5	0.2
Vessels ≥ 15 ft	1.0	0.5

Timing Chart

This chart shows the time required to run each of the first two legs of an SS pattern and each leg of a VS at different speeds according to the initial track spacing. To calculate the time required to complete one sector search (VS), multiply the leg time by nine.

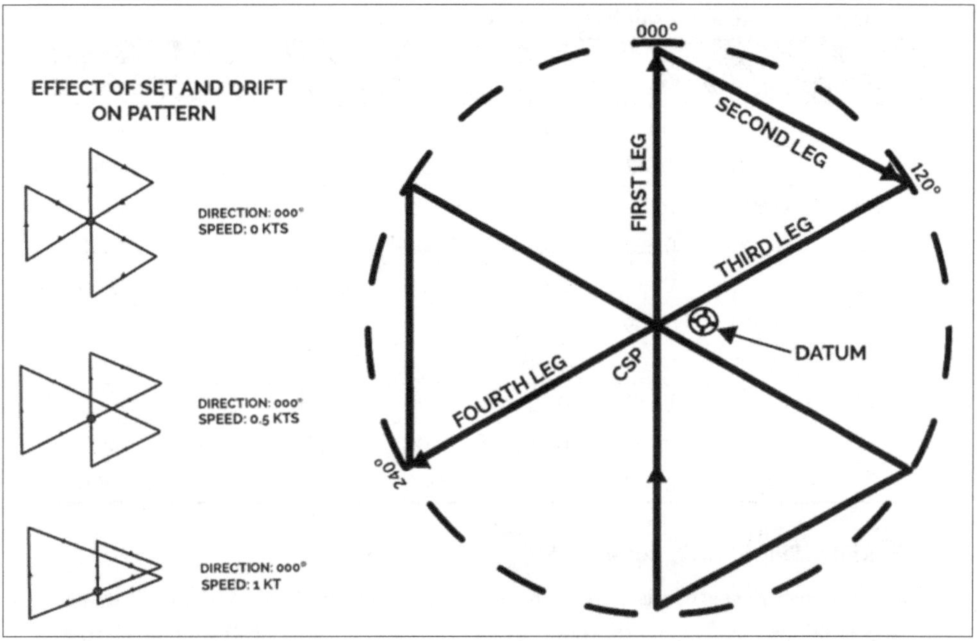

Sector, single unit (VS)

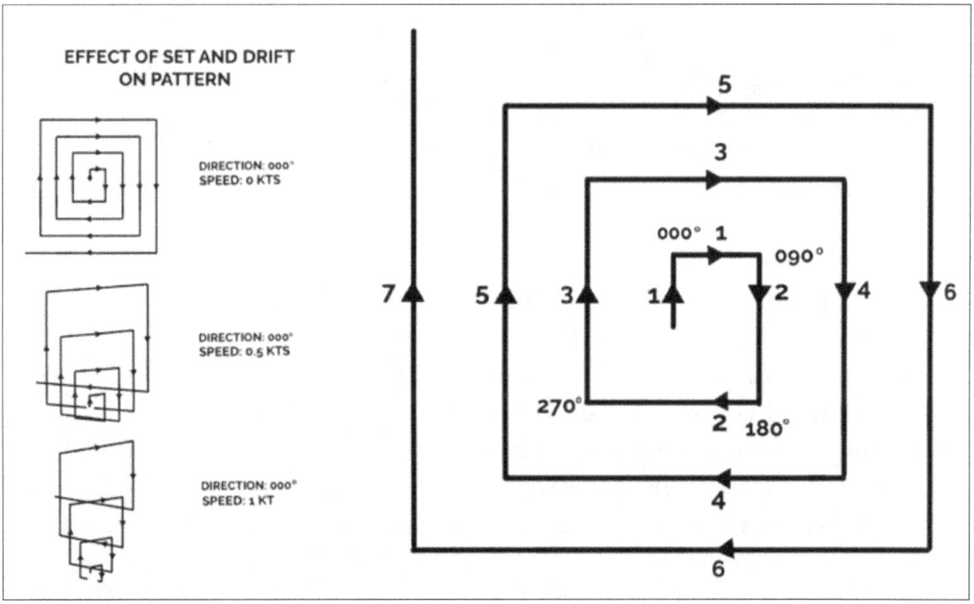

Expanding square, single unit (SS)

SPEED / TIMETABLE IN MINUTES : SECONDS

Track Spacing	6 kn	7 kn	8 kn	9 kn	10 kn	11 kn	12 kn
0.1	01:00	00:51	00:45	00:40	00:36	00:33	00:30
0.2	02:00	01:43	01:30	01:20	01:12	01:05	01:00
0.3	03:00	02:34	02:15	02:00	01:48	01:38	01:30
0.4	04:00	03:26	03:00	02:40	02:24	02:11	02:00
0.5	05:00	04:17	03:45	03:20	03:00	02:44	02:30
0.6	06:00	05:09	04:30	04:00	03:36	03:16	03:00
0.7	07:00	06:00	05:15	04:40	04:12	03:49	03:30
0.8	08:00	06:51	06:00	05:20	04:48	04:22	04:00
0.9	09:00	07:43	06:45	06:00	05:24	04:55	04:30
1.0	10:00	08:34	07:30	06:40	06:00	05:27	05:00

Trackline, Single Unit, Return (TSR)

The TSR precision pattern is used when the intended route of the search object is known and a rapid and reasonably thorough coverage of the missing craft's intended track and the area immediately adjacent is desired. This is usually the first search effort in an overdue case since it is assumed that the search object is near track and either it will be easily seen or the survivors will signal. The SRU searches out one-half track space in the direction of the intended track from the origin to the destination and ends one-half track space on the other side of the origin.

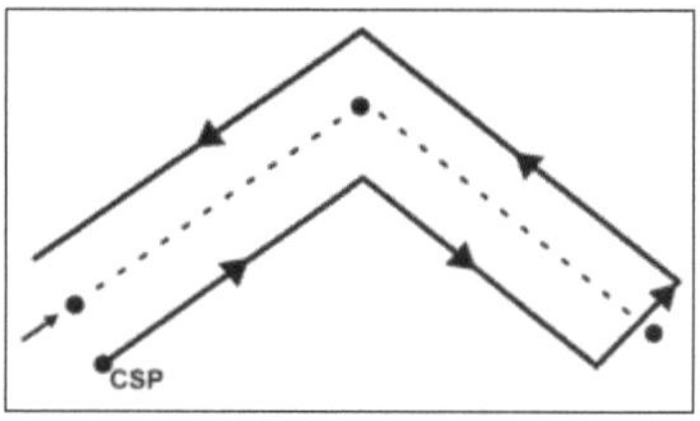

Parallel, Single Unit (PS)

The PS precision pattern is used when the search area is large, the location of the search object is approximate, and uniform coverage is desired. The SRU searches parallel to the long side (major axis) of the search area, moving over one track on the return. This process continues until the entire search area has been searched or the search object is located. The commence search point is located one-half track space inside the search area in the specified corner.

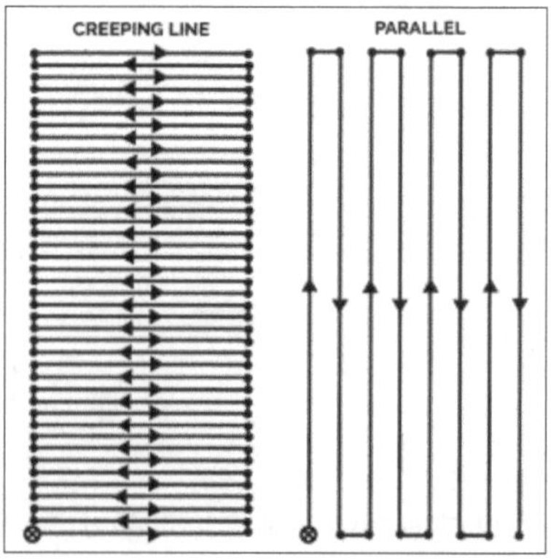

Creeping Line, Single Unit (CS)

The CS is a precision pattern used when the search area is large and the location of the search object is approximate, but there is a greater chance that the search object is at one end of the search area versus the other (i.e., debris was found on one end during a previous search), and uniform coverage is desired. The SRU searches parallel to the short side (minor axis) of the search area, moving over one track on the return. This process continues until the entire search area has been searched or the search object is located. The commence search point is located one-half track space inside the search area in the specified corner.

Sightings

While searching, you may encounter objects that need to be investigated. If you need to divert and leave the search pattern, do so only after recording or marking the position and progress in the search pattern to which you are assigned. Once sighting identification is complete and the results are negative, return to the position recorded and continue the search. Notify the SMC of how much time is spent evaluating sightings.

RESCUE

Once you have found the search object and reported it to the OSC, stop for a moment and think the case through. A very important question at this time is, can you, with the personnel and equipment that you have with you, complete this mission successfully? On a number of cases crews have thought they could but actually couldn't. The results of their actions complicated the case. If, for instance, you feel that a larger SRU is required for towing, or that more pumps are required, or that you will need a medical evacuation, you must request support. Don't try to do it all yourself if you feel that you could use some help. If you feel the case is within your capabilities, then the following information should be considered.

Rescue Planning

During most cases, the actual rescue of survivors is the easiest part of the mission. SRU's are normally able to recover the survivors without any problems. Sometimes, however, the SRU is unable to perform the rescue themselves, and rescue planning becomes essential. Examples of when an SRU might not be able to perform the rescue include the following:

- Too many survivors for one SRU to carry
- Weather conditions exceed the capabilities of the SRU

- Survivors located by a fixed-wing aircraft
- The distance from shore exceeds the capabilities of the SRU

When the SRU cannot perform the rescue, the SMC must develop an effective rescue plan. Rescue planning includes the following elements:

- Evaluating the survivability of the incident
- Evaluating the environment
- Selecting the rescue method
- Selecting the rescue facilities
- Developing the optimum and attainable rescue plan

Medical Evacuation (Medevac)

Always think high risk when considering a medevac. A flight surgeon must be consulted prior to performing the mission. The flight surgeon may recommend a medevac; however, it is not an order. The final decision on performing a medevac always rests with the leadership on scene. If you take survivors on your boat, never leave them alone and ensure they are properly wearing a lifejacket. People who are saved are sometimes panic stricken, thinking irrationally, or in a state of shock. If left alone, survivors might do something that would threaten the safety of your boat, your crew, and their lives. Make them comfortable and keep a member of your crew with them. Remember to handle all survivors with care. Rough handling while trying to get them on board your boat could make a serious injury fatal. Treat all survivors for shock, whether they show the symptoms or not. This will make them more comfortable. If any of the survivors have been seriously injured, you should consider a medevac by helicopter or call to have an ambulance awaiting your arrival.

Firefighting

Although the Coast Guard clearly has an interest in fighting fires involving vessels or waterfront facilities, primary responsibility for maintaining necessary firefighting capabilities in U.S. ports and harbors lies with local authorities. The Coast Guard renders assistance as available, based on the level of personnel training and the adequacy of equipment. Coast Guard personnel shall be prepared for and respond to fires on board Coast Guard vessels as discussed in this manual's chapter on damage control. For all other marine firefighting situations, commanding officers of Coast Guard units shall adopt a conservative response posture. Coast Guard personnel shall not engage in independent firefighting operations, except to save a life or in the early stages of a fire to avert a significant threat without undue risk. For fire on board commercial vessels and at waterfront facilities, Coast Guard personnel shall not actively engage in firefighting except in support of a regular firefighting agency under the supervision of a qualified fire officer.

Maritime SAR Assistance Policy

The Maritime SAR Assistance Policy represents more than a decade of development of relationships among the Coast Guard, Congress, the commercial towing industry, and the Coast Guard Auxiliary. The Coast Guard supports and encourages efforts of private enterprise and volunteerism to assist mariners. Coast Guard resources normally do not provide immediate assistance in non-distress cases if alternative assistance is available and will not unnecessarily interfere with private enterprise. A Coast Guard resource may assist in a non-distress situation when no higher priority missions exist and no other capable resource is reasonably available. In non-distress cases when specifically requested assistance such as a commercial firm, marina, or friend is not available, a request for assistance will be broadcast. If a commercial provider is available and can be on scene within a reasonable time, usually one hour or less, or an offer to assist is made by an alternate responder, no further action by the Coast Guard beyond monitoring the incident will be taken. Otherwise, a Coast Guard Auxiliary facility, if available, or a Coast Guard resource may be used. Three principles that guide assistance to vessels not in distress are

- The first responder on scene with the vessel requesting assistance normally will provide assistance;
- If a Coast Guard resource or Auxiliary facility takes a disabled vessel in tow, the tow will normally terminate at the nearest safe haven; and
- Once undertaken, there is no requirement to break the tow although the assisted vessel may be released to another provider who appears capable and the transition can be done safely.

General Salvage (Other than Towing)

When commercial salvors are on scene performing salvage, Coast Guard units may assist them within the unit's capabilities if the salvor requests. When no commercial salvage facilities are on scene, Coast Guard units should only engage in salvage other than towing when limited salvage operations such as ungrounding, pumping, and damage-control measures can prevent a worsening situation or complete loss of the vessel. Any salvage operations shall not unduly hazard Coast Guard personnel and will only be performed at the discretion of the unit command.

Debrief the Survivors

Make sure you have accounted for everyone before leaving the scene. If there is someone missing, question the survivors to see if they can give you any leads to the search. Have a member of your crew take notes so you will be able to recount the incident later. If the survivors were in the water for any length of time, have them describe in detail their emergency and survival procedures. This information might be helpful in later SAR cases. Once the survivors are cared for, then and only then can you shift your attention to any property in danger. Again there are a few things that you must consider before taking action:

- Are you in danger of losing the property? Can the damage be repaired by the crew?
- Can the property be towed? What is the best method of towing for this property—alongside or astern?

If you decide to tow the vessel, maintain communications with the people on board the boat that you are towing. Stay informed of their needs. They may have needed a tow in the beginning, but after you take them in tow, they may need a pump. Keep the operator informed of your intentions. Always keep in mind that people are far more important than property. If a survivor needs medical attention, you must forget about towing and get the victim ashore as quickly and safely as possible. Keep your station informed of your intentions. Let them know where you are and where you will moor. Keep them informed of any changes. After you have brought the survivors and the property back safely, make sure you retrieve all of the gear that you may have loaned the operator.

Completing the Mission

The SAR system is deactivated or shut down under one of the following situations:

- Case Closed: The distressed person or craft is located and necessary assistance is rendered;
- False Alarm: The person or craft reported to be in distress has been determined not to be in distress and does not need further assistance; and
- Active Search Suspended: Suspending the search after exhausting all possible hope for locating survivors with the information known. In many cases, only the SAR coordinator is authorized to suspend a search. However, depending upon the type of case, the SMC might be delegated the authority. Consult your District SAR Plan for further guidance.

Disposition of Lifesaving Devices

Private property that becomes the possession of the SAR system should be safeguarded, inventoried, then returned to its rightful owner and a receipt obtained. Emergency position-indicating radio beacons should be recovered and the signal secured whenever possible at the time of a rescue. Beacons left adrift at the conclusion of a SAR incident continue to transmit. The signal produced may prevent another distress beacon from being properly tracked or heard. Life rafts, lifeboats, and lifesaving float devices are involved in a number of SAR cases. The preferred action is to recover and deliver lifesaving vessels ashore. If conditions and circumstances do not permit a safe recovery by on-scene rescue units, rescue personnel should make every effort to mark the lifesaving vessel. The marking shall clearly indicate that the Coast Guard has investigated the lifesaving vessel.

18

WEAPONS

The Coast Guard employs various naval guns, weapons systems, and small arms to fulfill its many ports, waterways, and coastal security, law enforcement, and defense missions. This chapter introduces the weapons in general use on board cutters, boats, and aircraft.

WEAPON SAFETY RULES

Safety is the most important aspect of weapons handling. The following are the four weapons safety rules that apply at all times to any weapon, whether on the range, a boarding, a ship, at a station, or even at home.

- Treat all firearms as if they are loaded. You must always know the condition of your firearm. This rule intends to prevent unintentional injury or damage due to careless handling of weapons or perceived low risk. Treat every weapon with the respect due a loaded weapon.
- Always maintain proper muzzle control. You must be fully aware of your muzzle position and where you are pointing it. When at the range, the safe direction is down range. In the line of duty, the safe direction is any direction that does not cross the path of another person or vessel. This rule intends to minimize accidental discharge due to an individual being startled, bracing in a fall, or experiencing a rush of adrenaline.

- Keep your finger off the trigger until on target and ready to shoot. Keep your trigger finger indexed along the frame or receiver until the command to fire has been given or the decision to shoot has been made. This rule intends to minimize accidental discharge due to an individual being startled, bracing in a fall, or experiencing a rush of adrenaline.
- Know your target and what is beyond it. Positively identify your target and target environment.

NAVAL GUNS

Large guns installed on the bows of cutters are called naval guns. Almost all naval guns are now classified into a single mark series, regardless of their caliber. When an ordnance item is approved for testing, it is assigned a mark number. It retains this mark number whether it is finally approved for service use or if it is rejected. Modifications to the mark design are treated in a similar manner. Gun Mod numbers generally indicate changes to the original design for new guns or a change made to a finished gun. For example, a change in the rifling pattern from the original design might be designated as Mod 1, and new guns built with this rifling would be so designated. Guns more heavily modified would be given new mark numbers.

Mk 110 57-mm gun on bow of the Coast Guard's first national security cutter, USCGC *Bertholf*

Authority to Fire

The authority to fire a cutter's gun comes from the commanding officer, officer in charge, or patrol commander.

Gun-Mount Crew

The gun-mount crew consists of the mount captain and loaders. The mount captain is stationed in the ammunition-handling room at the gun control panel. It is the mount captain's responsibility to set the gun up for the desired mode of operation, then monitor it in case of a malfunction. In case of a malfunction or misfire, the mount captain supervises and directs the corrective action. The two loaders are stationed in the ammunition-handling room during loading and unloading operations. Their primary duties are to load and unload the gun, clear misfires, and assist in corrective maintenance.

Mk 110 Mod 0 57-mm Gun

This weapon is a multipurpose, medium-caliber, fully automatic gun housed in a stealth cupola on board national security cutters. The MK 110 gun mount can fire salvos at up to 220 rounds per minute and has a range of nine miles and is designed to engage air, surface, shore, and missile threats. The ammunition comes

Coast Guard cutter *Bertholf*'s 57-mm gun fires at an unmanned aerial vehicle drone

equipped with a smart programmable fuse with six modes: contact, delay, time, and three proximity fuse modes. The gun mount includes a 120-round automatic loader drum. Additional rounds are fed to the gun by the mount captain and designated ammunition loaders, who are stationed belowdecks. The maximum firing rate is 220 rounds per minute. The gun is remotely controlled and fired using the Mk 160 gun computer system (GCS), but as a redundancy measure the crew can also operate the gun using instrument panels in direct contact with the gun.

General Characteristics

Date Deployed: 2008
Range: 9.0 nm with conventional ammunition.
Type Fire: Maximum 220 rounds per minute, automatic
Magazine Capacity: 120 rounds conventional
Caliber: 57 mm 70-caliber Mk 38 Mod 0 Barrel, barrel length of 3,990 mm (57 mm × 70)

Mk 160 GCS

The Mk 160 receives target data from shipboard sensors to compute a ballistic solution. From this, it provides gun orders and selects projectile types. The GCS also generates the fire order to the gun mount. The Mk 160 GCS consists of a gun console computer, computer display console, recorder-reproducer, signal data converter and processor, gun-mount control panel, and velocimeter.

Mk 75 76-mm Gun

Mk 75 76-mm deck gun fires at a surface target

This is a lightweight, remote-controlled, fully automatic 76 mm gun housed in a turret carried on board medium endurance cutters. The Mk 75 can fire at a rate of 80 rounds per minute and has a range of ten miles and is designed to engage air, surface, shore, and missile targets or threats. The gun has an 80-round magazine. Additional rounds are fed to the gun by the mount captain and designated ammunition loaders, who are stationed belowdecks. The maximum firing rate is 120 rounds per minute. The gun is remotely controlled and fired using the Mk 92 fire control system.

General Characteristics

Date Deployed: 1978
Range: 10 nautical miles (11.5 statute miles, 18.4 km)
Type Fire: Maximum 80 rounds per minute, automatic
Caliber: 76 mm (3 in.)

Mk 92 Fire-Control System

The Mk 92 is capable of simultaneously detecting, tracking, and engaging multiple air and surface targets. The system consists of a combination antenna system and radar, track and illuminating radar, weapons control processor, data exchange auxiliary console, and a planned-position indicator display console.

Mk 15, Phalanx Close-In Weapon System

This is a fast-reaction, detect-through-engage, radar-guided, 20 mm gun weapon system. The close-in weapon system (CIWS) provides cutters with an inner layer

Coast Guard Cutter *Bertholf*'s close-in weapon system fires on and destroys an aerial target MK 25-mm machine-gun system

point defense capability against antiship missiles, aircraft, surface, and shore-based threats. Phalanx is the only deployed close-in weapon system capable of autonomously performing its own search, detect, evaluation, tracking, and engagement.

General Characteristics

Date Deployed: 1980
Type Fire: Maximum 4,500 rounds per minute, automatic
Magazine Capacity: 1,550 rounds
Caliber: 20 mm

Mk 38 Mod 2/3 25 mm Machine-Gun System

This weapon is a fully automatic, remotely and manually trained and elevated, stabilized machine-gun system carried on board fast-response cutters. The Mk 38 can fire 180 25 mm rounds per minute or 500 7.62 mm rounds per minute in single-shot or burst modes and has a range of 2,500 yards. Designed for to counter slow, low-flying aircraft and surface targets, the Mk 38 Mod 3 is unique from other naval weapons because its surveillance system moves separately from the gun system. The gun is controlled by a C2 subsystem consisting of the commander safety panel and the remote-operation console.

General Characteristics

Date Deployed: 2005
Range: 2,500 yards
Type Fire: 25 mm: Maximum 180 rounds per minute, automatic
Caliber: 25 mm (1 in.).

Mk 25-mm machine-gun system

SMALL ARMS

The Coast Guard classifies small arms as any firearms .50-caliber or smaller. Weapons larger than .50-caliber are considered to be a weapon system.

Crew-Served Weapons

Crew-served weapons are issued to a crew of two or more in order to efficiently load and fire, as opposed to an individual service weapon, which only requires one person. The weight and bulk of these weapons and their ammunition also necessitates multiple personnel to move. They are fired from mounts installed on board boats and cutters. Maximum rates of fire for these weapons are generally not achieved in use, as sustained fire at that rate will wear out the bore within a few thousand rounds, necessitating replacement.

Mk 19 40-mm Grenade Launcher

A belt-fed, blowback-operated, air-cooled, fully automatic crew-served 40 mm grenade-launched weapon is used by units in Patrol Forces Southwest Asia and those working under combatant commands. The Mk 19 can fire 60 grenades per minute and has a maximum range of 2,419 yards. The nearest safe distance to launch the grenade is 75 meters. The weapon operates on the blowback principle, which uses the chamber pressure from each fired round to load and recock the weapon. The primary ammunition for it is the high-explosive dual-purpose M430 grenade. On impact, the grenade can kill anyone within a radius of 5 meters, wound within a radius of 15 meters, and punch through two inches of armor plating. The ammunition comes in cans that hold a 32- or 48-grenade belt weighing 42 and 60 pounds.

M2HB .50-caliber Heavy Machine Gun

This gun is a belt-fed, air-cooled, recoil-operated, fully automatic crew-served machine gun chambered for .50-caliber ammunition. The M2 fires from a closed bolt and can fire up to 600 rounds per minute with a maximum range of 8,100 yards. In addition to full automatic, the M2HB can be selected to fire single shots or at less than 40 rounds per minute, or rapid fire for more than 40 rounds per minute. Slow and rapid firing modes use 5–7-round bursts with different lengths of pause between bursts. It is commonly known as "Ma Deuce" in reference to its M2 nomenclature and has been in use longer than any other firearm in the U.S. inventory except the .45 ACP M1911 pistol, also designed by John Browning.

M240 7.62-mm Heavy Machine Gun

This gun is a belt-fed, air-cooled, gas-operated, fully automatic crew-served machine gun chambered for 7.62 mm ammunition. The M240B can fire up to 950 rounds per minute and has a maximum range of 4,074 yards. The weapon fires from the open-bolt position. Ammunition is fed into the weapon from 100-round bandoliers containing a disintegrating metallic link belt. The weapon features a fixed headspace, which permits rapid changing of barrels. Slow and rapid firing modes use 5–7-round bursts with different lengths of pause between bursts.

Precision Weapons

Precision weapons are employed by precision marksmen to minimize the risk of collateral damage during boarding operations and in stopping noncompliant vessels with disabling fire.

M107 Long-Range Sniper Rifle

The long-range sniper rifle is a semiautomatic, air-cooled, box magazine–fed rifle chambered for .50-caliber ammunition. It operates by means of the short-recoil principle, rather than gas. Designed for long-range precision fire out to two thousand yards, its ammunition is fed via a ten-round magazine.

Firing a mounted .50-caliber machine gun

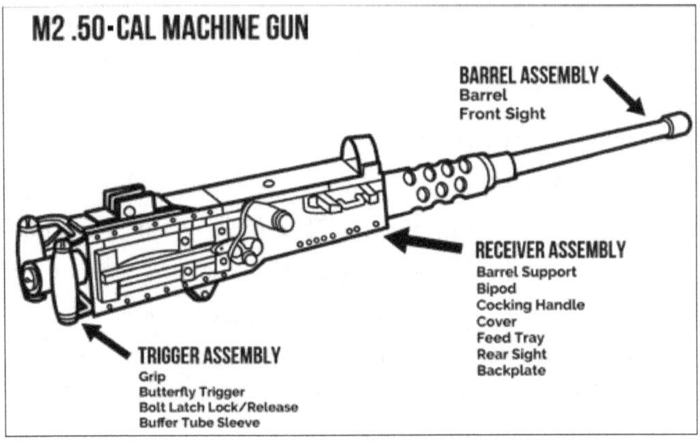
M2 .50-CAL MACHINE GUN
BARREL ASSEMBLY
Barrel
Front Sight
RECEIVER ASSEMBLY
Barrel Support
Bipod
Cocking Handle
Cover
Feed Tray
Rear Sight
Backplate
TRIGGER ASSEMBLY
Grip
Butterfly Trigger
Bolt Latch Lock/Release
Buffer Tube Sleeve

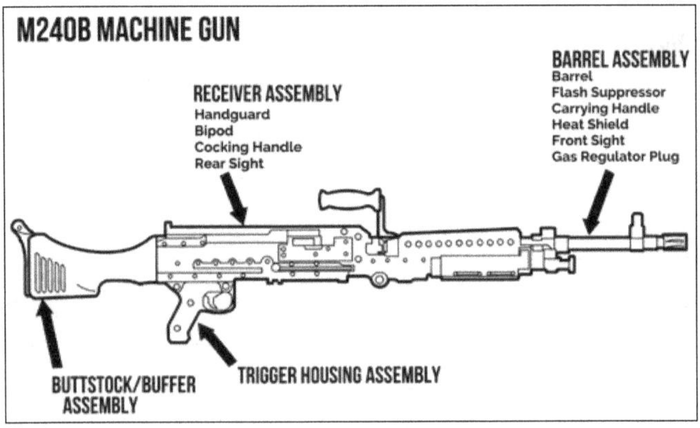
M240B MACHINE GUN
RECEIVER ASSEMBLY
Handguard
Bipod
Cocking Handle
Rear Sight
BARREL ASSEMBLY
Barrel
Flash Suppressor
Carrying Handle
Heat Shield
Front Sight
Gas Regulator Plug
BUTTSTOCK/BUFFER ASSEMBLY
TRIGGER HOUSING ASSEMBLY

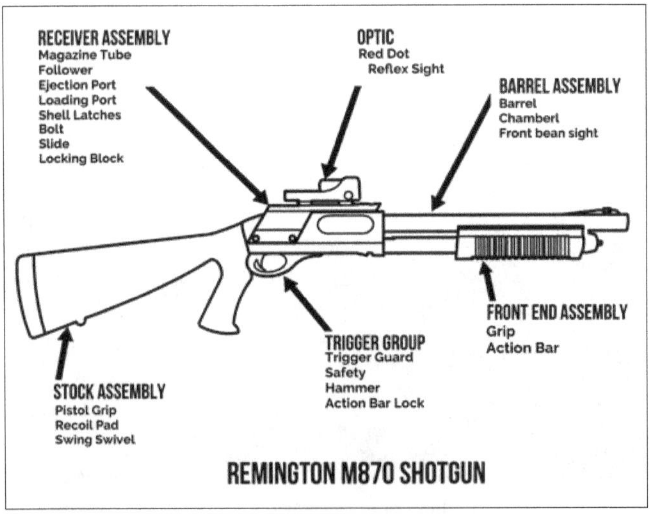
RECEIVER ASSEMBLY
Magazine Tube
Follower
Ejection Port
Loading Port
Shell Latches
Bolt
Slide
Locking Block
OPTIC
Red Dot
Reflex Sight
BARREL ASSEMBLY
Barrel
Chamberl
Front bean sight
FRONT END ASSEMBLY
Grip
Action Bar
TRIGGER GROUP
Trigger Guard
Safety
Hammer
Action Bar Lock
STOCK ASSEMBLY
Pistol Grip
Recoil Pad
Swing Swivel
REMINGTON M870 SHOTGUN

M110 Semiautomatic Sniper System

The M110 semiautomatic sniper system is a lightweight, direct gas-operated, air-cooled, magazine-fed, shoulder-fired weapon chambered for 7.62 mm ammunition. It is designed for precision semiautomatic fire out to one thousand yards.

M14T Enhanced Battle Rifle

The M14T enhanced battle rifle is a lightweight, air-cooled, gas-operated, magazine-fed, shoulder-fired weapon chambered for 7.62 mm ammunition and is designed for precision semiautomatic fire out to six hundred yards.

Individual Service Weapons

Individual service weapons are issued to individuals for duty use while serving as sentries, on boarding teams, and as crew on board boats, cutters, and aircraft when performing law enforcement, ports, waterways, and coastal security, and defense missions.

M4 Carbine

The M4 carbine is a lightweight, gas-operated, magazine-fed, shoulder-fired firearm capable of firing all NATO 5.56 mm ammunition in semiautomatic, burst fire, or fully automatic modes through the use of a selector lever. It is used to engage targets at ranges out to five hundred meters with a rate of fire between seven hundred and nine hundred rounds per minute. Ammunition is fed into the weapon by thirty-round magazines. It is equipped with multiple MIL-STD-1913 rails to facilitate the mounting of various accessories for tailoring to operational requirements, to include optics.

Mk 18 Mod 0 Close-Quarter Battle Rifle

This rifle is a lightweight, gas-operated, magazine-fed, shoulder-fired carbine chambered for 5.56 mm ammunition and capable of semiautomatic, burst fire, or fully automatic modes through the use of a selector lever. The Mk 18 is a variant of M4 with a short, 10.3-inch barrel upper receiver intended to provide individuals operating in close quarters the capability to engage targets with accurate lethal fire.

M870 Shotgun

The M870 is a manually operated, pump-action, single-shot, shoulder-fired 12-gauge weapon. It features a bottom-loading, side-ejecting receiver and a tubular magazine under the barrel. Ammunition types include buckshot for self-defense and copper sabot rounds for disabling fire in noncompliant vessel situations.

Sig P229R DAK

This weapon was the standard-issued personal-defense weapon for Coast Guard sentries and boarding teams. It is a short-recoil, semiautomatic, magazine-fed, double-action pistol with an intermediate reset point and is chambered for .40-caliber cartridges.

Glock 19 Gen5 MOS

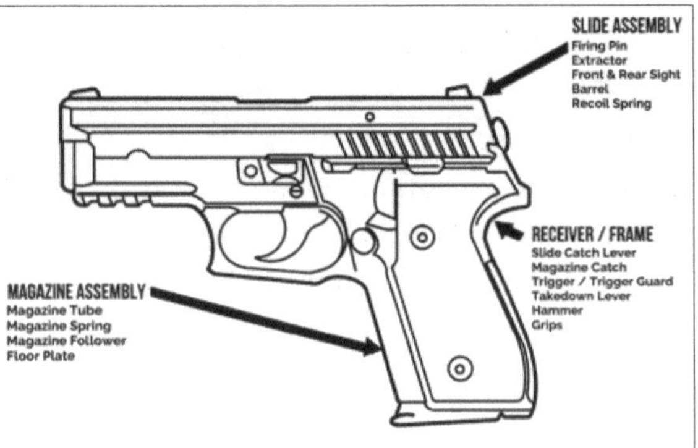

SIG P229R DAK

This Glock was selected in 2020 to replace the Sig P229R DAK as the standard-issue personal-defense weapon for Coast Guardsmen. Chambered for 9 mm ammunition, the Glock 19 is a safe-action, self-loading, recoil-operated pistol with an automatic firing pin and trigger safety.

LESS THAN LETHAL

Less-than-lethal weapons are intended to be less likely to kill a living target than conventional weapons such as firearms with live ammunition. Unintended or incidental casualties are risked wherever force is applied, but nonlethal weapons minimize the risk of casualties.

PepperBall Launcher System

This system uses compressed air to launch and deliver pepper irritant projectiles that break upon impact. When inhaled, the powder within the projectile leads to coughing, shortness of breath, and, in some instances, vomiting. The PepperBall launcher holds 180 projectiles in its hopper and can fire 8 per second out to a range of 60 feet.

19

LAW ENFORCEMENT

Law enforcement is the oldest of the Coast Guard's missions, dating back to our founding as a system of cutters in 1790. Today the service is the lead federal agency for maritime law enforcement, and its commissioned, warrant, and petty officers possess authorities to prevent, detect, and suppress violations of federal law on, under, and over the seas and at adjacent shoreside facilities.

BOARDINGS

Boardings are the primary means used by the service to accomplish its at sea law enforcement mission. A boarding is an armed intervention on board a vessel to prevent, detect, or suppress violations of federal laws, international conventions, or treaty. No Coast Guard operations or activities conducted unarmed are referred to in any way as boardings. Vessel boardings may be conducted by any Coast Guard unit, but they are generally performed by boat and cutter crews and deployable specialized forces. Stations perform many of the boardings that occur inshore and out to the twenty-four nautical mile line of the contiguous zone, while patrol boats conduct boardings from the twelve nautical mile border to the boundary of the exclusive economic zone, and major cutters on international waters and the high seas. Virtually every type of vessel afloat is boarded, from small rowboats to supertankers. Most vessels boarded in the course of general maritime law enforcement, though, are either pleasure craft such as sailing vessels or sports fishing boats, fishing vessels, or small freighters.

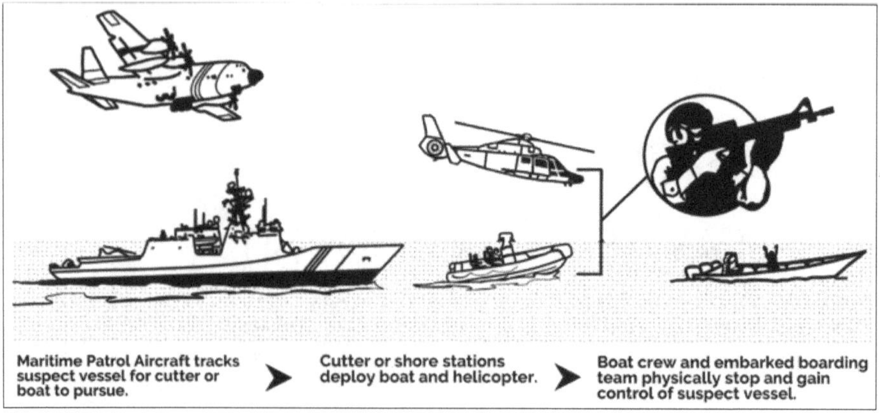

Boarding Teams

A boarding team consists of at least two Coast Guardsmen but may be six or more, depending on factors such as the tonnage and condition of the involved vessel, the number of persons thought to be on board, and whether there is suspicion of criminal activity. The boarding officer, under the supervision of their commanding officer or officer in charge, is the federal officer in charge of the boarding team responsible for the proper conduct of the boarding.

Although all Coast Guard commissioned, warrant, and petty officers have law enforcement authority granted by Congress under Title 14 of the U.S. Code, the boarding officer is the individual whose authority is used to perform the boarding. All others assist the boarding officer and are referred to as boarding team members. Only those personnel who possess the requisite knowledge, judgment, maturity, attitude, and aptitude are assigned law enforcement duties.

Equipment

Every person in the boarding team must wear the proper Coast Guard uniform and an approved personal flotation device. In addition, boarding team members wear specialized equipment for their law enforcement activities:

- Personal defense weapon (PDW): During law enforcement boardings, all Coast Guard boarding team members are armed with the standard PDW, a Sig P229R DAK pistol, and extra magazines. In September 2020 the Glock 19 was selected to replace the P229R.
- Body armor: Body armor consists of a vest designed to minimize ballistic injuries from shots fired at the body of the wearer. Soft armor composed of

layers of high-strength woven fibers is worn concealed under the uniform blouse or in an external carrier and is intended to stop pistol ammunition from penetrating. Plate armor is worn in external carriers during missions with heightened risks to stop rifle rounds.

- Duty belt: Also known as a "gun belt," the duty belt holds the officer's PDW, magazine pouches, and other pouches to carry required and mission-specific equipment.
- OC spray: OC spray is a chemical agent contained in a thumb-activated spray can that, when discharged onto the skin of a suspect, creates a burning sensation and leads to severe eye, nose, and throat irritation.
- Expandable baton: This telescoping metal baton serves as an impact weapon.
- Handcuffs and key: These are the preferred restraint device when dealing with a suspect who needs to be restrained.
- Flashlight: Intrinsically safe flashlights are standard to limit the risk of igniting explosive or combustible substances.
- Radio: The boarding team will carry an encrypted radio to communicate securely with the parent boat, cutter, or unit.
- Riot shotgun: When the boarding officer has an articulable suspicion of an increased threat, one or more boarding team members may carry a riot shotgun.
- Rifle: At the discretion of the commanding officer, boarding team personnel may carry the M4 or Mk 18 on a boarding where there is reason to believe the persons on board pose a substantial threat.

Boarding Kit

In addition to the personal equipment, the boarding team carries a kit containing the supplies and equipment anticipated necessary for a boarding. The contents of the boarding kit will be determined by the boarding officer and will depend on the type of boarding conducted. A typical kit will contain a Report of Boarding (CG-4100) pad, the boarding officer job aid kit, pens, pencils, notebooks, flashlight with extra batteries, tape measure, multitool, digital camera, rubber or latex gloves, hearing protection, disposable masks, evidence bags, property bags, an inspection mirror, a narcotics identification kit, and field sobriety test sheets.

Access Kit

In cases in which the boarding officer suspects that the boarding vessel has hidden compartments concealing drugs or other contraband, the boarding team may bring a kit containing equipment to locate, open, and inspect compartments. The kit includes such things as a power drill, a borescope for looking through a drill hole into a dark compartment, saws, crowbars, and sledges.

BOARDING PROCEDURE

The boarding process begins with the Coast Guard unit establishing the identification and nationality of the vessel to be boarded and verifying its position to establish jurisdiction. The unit will send any identifying information obtained for the vessel and crew to a communications center ashore, which runs the information through federal databases to determine whether there is any derogatory information, such as violations issued during prior boardings.

Freeboard, or the distance from the water to the top of the vessel's gunwale, is critical to establish in order to plan delivery and recovery methods for the boarding team. For low freeboard vessels, the team is able to step across gunwales and directly onto the targeted vessel's deck. High-freeboard ships require climbing accommodation ladders or Jacob's ladders lowered by the vessel's crew. Noncompliant high-freeboard vessels whose crews will not lower a ladder are especially difficult and may require deployable specialized forces teams to either fast rope onto the vessel or embark with specialty ladders.

High freeboard

Once the person in charge of a patrolling Coast Guard unit decides to conduct a boarding, the subject vessel is told to prepare for a boarding, and procedures for embarking the boarding team are explained. The boarding team forms on board the cutter or boat, checks the available intelligence, double-checks their equipment, and then proceeds to embark the vessel.

When the boarding team is over the gunwales of the other vessel, the team members take prearranged stations about the deck to establish and maintain positive control for the boarding. The boarding officer will meet with the vessel's master, explain the purpose and the process of the boarding, and answer any questions. The boarding team will then conduct an initial safety sweep, a brief survey of the vessel to ensure that it is seaworthy enough to continue with the boarding and to identify any hazards. The boarding officer then conducts an administrative inspection, checking for compliance with a wide range of federal laws and regulations.

If there are no violations detected during the boarding, the master is thanked for cooperating, and the boarding team departs.

CRIMINAL ACTS AND REGULATORY VIOLATIONS

If, during the course of the boarding, the boarding officer develops probable cause to believe that the vessel is engaged in criminal activity, the scope of the boarding may be expanded to include a complete search of the vessel for evidence. If evidence of criminal activity is found, such as narcotics or human trafficking, the crew and vessel are detained and brought into port for further disposition. The Coast Guard Investigative Service or another federal agency will normally assume the lead for such criminal cases and brief them to the U.S. Attorney's Office for federal prosecution. Depending on the nature of the crime and where it occurred, state or local agencies may instead be offered the lead and assume custody of the suspects and evidence. All violations involving federal fisheries are documented and referred to the National Oceanic and Atmospheric Administration for disposition.

If only minor regulatory violations are found, the boarding officer may issue the master a written warning or a notice of violation with a recommended summary settlement amount to be paid. For serious regulatory violations or when the master declines a notice of violation, the boarding officer will refer the matter to the Violation Case Coordination Center for assessment of a class I administrative civil penalty by a Coast Guard hearing officer.

Violations involving persons required to hold a Coast Guard–issued Merchant Mariner Credential and license may also be referred to a Coast Guard marine investigator who may initiate suspension and removal proceedings where the continued status of the license is decided by an administrative law judge.

Use of Force

As uniformed federal law enforcement officers, Coast Guard personnel are authorized to use appropriate force to defend themselves and others and to carry out their duties. The use of force by Coast Guard personnel is governed by Coast Guard policy, federal law, and case law.

Only that force reasonably necessary under the circumstances may be used. Force shall not be used where assigned duties can be discharged without it. Excessive force may never be used; however, nothing in the application of the Coast Guard use-of-force policy shall be construed as to require personnel to meet force with equal or lesser force. In most instances, boarding teams will have to make immediate use-of-force decisions based on their training and experience. In general, Coast Guard personnel may use force in the following circumstances:

- For self-defense
- To compel compliance with an order
- To prevent a federal crime
- To effect a lawful arrest or prevent an escape
- To protect property

Noncompliant Vessels

Force may be used to stop or disable vessels that are noncompliant with a lawful order to stop and heave to for boarding. In such situations the vessel's engines or other propelling machinery is targeted rather than the people on board. Specialized pursuit boat crews and precision marksmen on board armed helicopters respond to these incidents. Once disabled, the suspect vessel is boarded and those on board are restrained to allow the boarding to proceed as described above.

Limited Access Areas

Specialized security boarding teams, tactical boat crews, and cutter crews may use force to enforce security zones and naval vessel protection zones, both types of limited access areas intended to safeguard vessels, harbors, ports, and waterfront facilities from destruction, loss, injury, sabotage, or other subversive acts. Rather than inspect for compliance with laws and regulations, a security boarding team embarks a protected vessel to secure its vital areas and systems, notably the bridge and engine rooms. Tactical boat crews armed with crew-served weapons patrol the waters surrounding the protected vessel or facility. A patrol commander embarked on a cutter with its weapons mounted will normally oversee the operation. These crews are trained to use force to disable noncompliant vessels that violate limited access area boundaries and to recognize and disable or destroy suspect vessels that pose an imminent threat of attack, such as those fitted with improvised explosives.

20

SERVICING AIDS TO NAVIGATION

One of the Coast Guard's statutory missions is the care and maintenance of aids to navigation (ATON). Through its ATON mission, the Coast Guard promotes safe waterways and an efficient marine transportation system. The Coast Guard has statutory responsibility to operate and maintain a system of maritime aids to facilitate navigation and to prevent disasters, collisions, and wrecks. To fulfill this mission, the Coast Guard operates and maintains fixed and floating ATON that are placed along coasts and navigable waterways as guides to mark safe water and to assist mariners in determining their position in relation to land and hidden dangers.

Aids to navigation (ATON)

Fixed ATON include lighthouses, towers, and other structures that are directly affixed to the ground or seabed. Floating ATON includes buoys and markers anchored to the seabed by a concrete or metal sinker connected by a metal chain or mooring. Today the Coast Guard manages 45,664 federal fixed and floating ATON.

Mariners expect each aid to navigation encountered will be operating in accordance with its *Light List* description. It's a matter of navigation safety, of legal necessity, and of professional and personal pride to properly inspect, service, and maintain all assigned ATON. Mission success is measured by the aid availability rate, or the percentage of time that ATON are functioning correctly.

SERVICING AND DISCREPANCY RESPONSE

Coast Guard ATON units are responsible for the servicing and maintenance of ATON by conducting both routine servicing based on the last service dates of the ATON and nonroutine servicing of ATON within their area of responsibility. The nonroutine servicing process includes responding to and addressing discrepant ATON, which are aids that are not functioning properly due to, for example, a weather-related event such as a hurricane or an equipment failure. Timely response to and correction of discrepant ATON is a high-priority task for the Coast Guard. The Coast Guard has a tiered approach to address ATON discrepancies that accounts for the importance of the ATON relative to the waterway and the nature of the discrepancy. The servicing unit response ranges from immediately after notification up to seventy-two hours or as soon thereafter as weather and resources permit.

During the routine servicing process for floating ATON, the primary purpose is to check the buoys' positions, their condition, and ensure the correct operation of the buoys' signal hardware. As part of this process, the Coast Guard may extract the buoys from the water and bring them on board an ATON vessel to check the condition of their mooring chain, hull, and lighting equipment. If necessary, the mooring chains are cleaned and repaired and nonfunctioning lanterns are replaced. After the planned repairs are made, the buoys are placed back in their assigned position in the water.

When ATON units conduct routine or nonroutine servicing of fixed and floating ATON, they also collect data on the condition of the ATON. These data provide a "snapshot" of the ATON's condition at the time of servicing and include the aid's geographic position; the last date that the ATON was serviced; the next-scheduled service date; and other detailed information about the aid, such as an assessment of the physical integrity of the ATON. If warranted, ATON units can initiate action for repair or replacement of ATON if necessary. The information gathered by ATON units during their servicing activities is entered into a Coast Guard database—the

Source: GAO analysis of U.S. Coast Guard information; GAO (photos). I GAO-20-107

Fixed aids

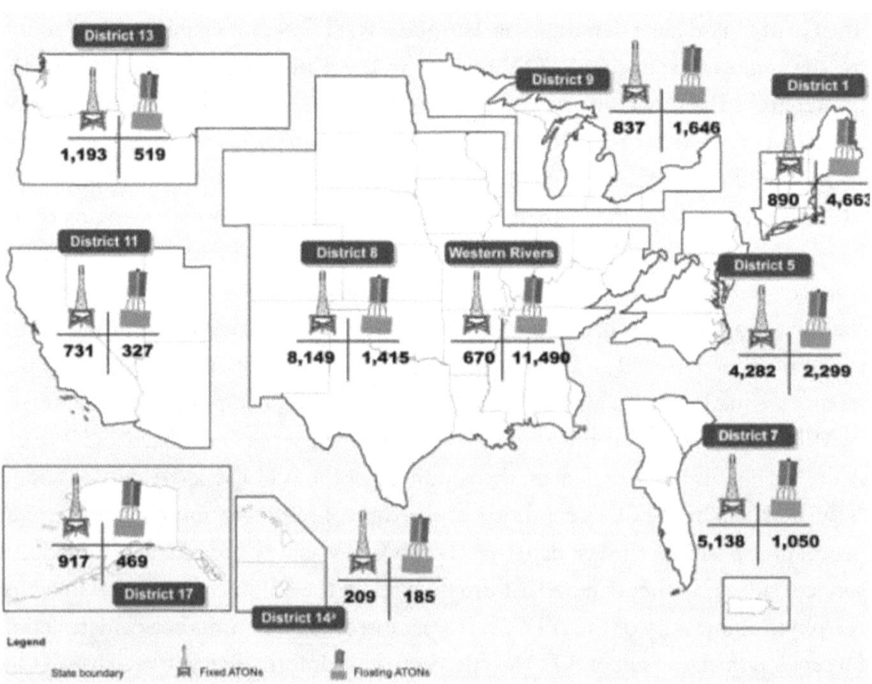

ATON number and locations

Floating ATON

This **unlighted steel buoy** weighs 2,600 pounds and is used as an aid for mariners in smaller waterways, including rivers, tributaries, and bays.

This **6' X 16' lighted foam buoy** weighs 3,640 lbs. and is used as an aid for mariners in protected and semi-exposed waterways with moderate protection from strong environmental exposure, including rivers, tributaries, and bays.

This **8' X 26' lighted steel buoy** (depicted in the foreground) weighs 11,800 lbs. and is used in protected and fully exposed waterways with greater environmental forces such as large rivers, inlets, and the ocean.

Source: GAO analysis of U.S. Coast Guard information; GAO (photos). | GAO-20-107

Floating aids

Integrated ATON Information System (I-ATONIS)—that is used to track and monitor fixed and floating ATON.

The designated team leader for an ATON servicing mission must identify what equipment is used at the aid and what service is required. Most minor aids consist of a single solar panel, one or more secondary batteries, a lantern, and associated hardware. These aids will have the following maintenance performed either annually, biennially, or triennially, depending on its scheduled visit:

- Solar Panel: Visual inspection, clean
- Secondary Batteries: Visual inspection, load test
- Power Cable: Visual inspection
- Lantern: Visual inspection, clean
- Minor Aid Hardware: Relamp, minor aid quick check
- Self-Contained LED Lanterns: Visual inspection, clean

Protective Equipment

The following personal protective equipment is worn at the aid site:

- Hard hat
- Safety-toe boots

- Gloves
- Eye protection
- Personal flotation device
- Personal fall protection equipment

Equipment must be deenergized whenever possible while installing or replacing components.

Minor Aid Components

With the addition of the appropriate lantern, the components described below are used on buoys, structures, and nighttime ranges with single lights. Rotating beacons and high-intensity nighttime ranges will have the same lighting equipment but will use larger batteries and more solar panels. In most cases, these components can be replaced by a self-contained LED lantern. Other sites may have an LED lantern in place of the lamp changer and flasher and may retain the legacy solar power system.

Solar Panels

Solar panels are used on most minor ATON to charge batteries that operate the light. Depending on the power requirements of the aid, a 10-, 20- or 40-watt solar panel will be used. Some minor aids may use multiple solar panels mounted on an array or around the superstructure or lantern stand of a buoy. During every visit the team will check to be sure the tilt angle is properly set for aid and that it is properly oriented to the south. The team will also ensure that there is no evidence of water intrusion around the edges of glass such as discoloration of the solar cells, that wires are free of cuts and abrasions, and that the panel is clean. Repairs and replacement of damaged equipment occurs on site.

Batteries

Solar-powered ATON are powered by 12-volt secondary batteries. The solar panels recharge the battery during the daytime and the battery operates the load at night. Batteries must be installed in a battery pocket or battery box on buoys, or in a battery box or indoors on fixed structures. All ATON batteries are tracked from cradle to grave and must be disposed of properly.

Wear chemical gloves and goggles whenever handling batteries as they contain electrolyte, a dilute sulfuric acid. If electrolyte comes into contact with the skin, wash immediately with water. If electrolyte comes into contact with the eyes, flush with water for fifteen minutes and consult a physician immediately. Secondary batteries may also vent hydrogen when charging. Allow the battery pocket, battery box, or battery enclosure to air out for five minutes before entering or servicing the battery. The battery should be the last connection made at an aid. Do not allow "+" and "–" wires to touch as the current is high enough to melt wiring, possibly damage the battery, or cause personal injury.

Signals

Signals include LED lanterns, traditional non-LED lanterns, and sound signals.

LED Lanterns

LED stands for light-emitting diode. LED lighting products produce light up to 90 percent more efficiently than incandescent light bulbs.

Carmanah LED

Carmanah 700 Series Self-Contained LED Lantern. The Carmanah 700 series self-contained LED lanterns are available in four versions: the 701/701-5, 702/702-5, 704-5, and the 708. The first version has multiple power system sizes, but only one intensity setting. The 704-5 and 708 have two and three intensity settings, respectively, but only one power system size. Proper selection of the intensity and solar system sizing must be made prior to installation. These are programmed and fully charged at the servicing unit prior to installation.

Servicing should be performed in accordance with the standard cycle established for the aid. Ensure that the lens and solar panels are clean, and wipe with a cloth dampened with mild soap and water, if necessary. Cover the lantern with a shroud, jacket, box, blanket, and so on, to simulate darkness. The lantern should flash on rhythm after a few seconds. It should stop flashing after the cover is removed in the daytime. While covered, observe the LEDs through the lens. Dark sectors indicate that an LED cluster is not operating, necessitating replacement of the optic head.

Vega VLB-36 Self-Contained LED Lanterns. The Vega VLB-36 is a self-contained LED lantern containing three solar panels and a lead-acid battery. There are three power system sizes—small, medium, and large—with multiple intensity settings. Proper selection of the intensity and power system sizing must be made prior to installation.

Vega VLB-67 Stand-Alone and Self-Contained LED Lanterns. These lanterns are available in four system sizes: stand-alone (used with legacy power system), small, large, and large with two batteries as self-contained LED lanterns. Proper selection of the intensity and power system size must be made prior to installation.

Non-LED Lanterns

Lights using incandescent light bulbs are not self-contained and instead consist of a flasher, daylight control, wiring kit, lantern bracket, lamp changer, and lamps. These are assembled at the servicing unit and bench tested for twenty-four hours prior to installation at the aid site. Flashers are used to regulate battery voltage to the lamp, sense lamp failure, provide a flash rhythm, and turn off the light during the daytime. Flashers are available in four versions (CG-181, CG-493, CG-481, and CG-504), with each used to operate lamps of different wattage at varied flash rhythms. Lamp changers are used to extend the service interval by stocking the lantern with six operable lamps. The clockwork mechanism is spring wound and triggered by a solenoid activated by the flasher. Lamp changers are available in two versions, the CG-6P and CG-6PHW. Daylight controls are photoresistors that change resistance with respect to light levels. As the light level decreases, the

155-mm lantern

resistance of the daylight control increases. The photoresistor is used on the flasher to turn the light off in the daytime. A flasher without a daylight control will operate continuously, twenty-four hours a day.

155-mm Lanterns. These omnidirectional lanterns are used on buoys and structures. The lens is replaceable and available in clear, yellow, red, and green.

At every scheduled annual or biennial service visit, replace all extinguished lamps and the operating lamp. Rotate remaining good lamps to the forward positions, and use new lamps to fill the remainder of the lamp changer. Wipe lamps with a medicinal swab or a clean rag dampened with denatured alcohol. Replace the daylight control.

250-mm Lanterns. These omnidirectional lanterns are used on stable platforms. The lens and lens cover are replaceable. The beam color of white, yellow, red, or green is determined by the lens cover. Sectored lens covers can be used, and condensing panels can be added for range applications.

300-mm Lanterns. An omnidirectional lantern is used only on stable platforms. The lens is replaceable and available in clear, yellow, red, and green.

14-inch Range Light (RL-14). Two versions of the RL-14 are currently installed in the field: the Tideland Signal RL355 and the Carlisle & Finch RL10668; the latter is considered the Coast Guard standard 14-inch range light. Parts between these beacons are not interchangeable. The beam color can be white, yellow, red, or green when using the appropriate colored spread or nonspread lens.

VRB-25 Rotating Beacon. The VRB-25 is the Coast Guard's standard 12-volt rotating beacon. It is composed of a direct-drive turntable assembly with six rotating lens panels, it has a user selectable speed range of 0.67 to 15.9 rpm, and it is capable of operating up to 110-watt lamps. The beacon is used only on structures not subjected to noticeable vibration such as lighthouses, dolphin-type structures and steel structures.

BUOY DECK EVOLUTIONS

Servicing a floating aid to navigation from an unstable buoy deck is always a hazardous operation. Each buoy brought on board will present a slightly different situation depending upon any combination of sea conditions, weather conditions, length of chain, size and type of buoy, experience of the conning officer, experience of the deck crew, fatigue level, and so on. Considering these factors, if the crew is trained to work every aid using a standard procedure, the evolution is safer and more efficient. The buoy deck evolution described below has been adopted as the required Coast Guard standard method for servicing buoys, although the equipment available on board a cutter or boat will cause slight variations.

Hoisting the Buoy

Following the brief, the buoy tender or boat maneuvers close aboard the buoy and maintains a steady position. "HOOK IT WHEN YOU CAN" is ordered and the buoy is hooked and moused. Lift the buoy until the crane supports its weight. Slew the davit inboard, snugging the buoy against the hull. Attach a cross-deck or tagline secured to the deck opposite the buoy port to the buoy bail. Lift the buoy and slew the crane inboard while keeping the cross-deck or tagline taut, ensuring the load is kept close to the deck.

Pass the pelican hook on the buoy mooring chain and ease the cross-deck or tagline until the chain fetches up on the pelican hook. Lower the buoy to the deck and onto the dunnage, then gripe the buoy. Prepare to break the mooring.

Pulling Chain

Whether relieving the aid, conducting a mooring inspection, or just doing routine servicing, at some point you will need to pull chain. Attach the hoist to the mooring chain. DO NOT clear the pelican. Take a pick with the crane, and haul the chain to short stay, faking the chain on deck. Reset the pelican and take a pick to haul the sinker to the water's edge. Reposition the hoist to bring the sinker aboard. Use a cross-deck or tagline to control the sinker. Lay out dunnage, bring the sinker on board, set it on the dunnage, and gripe down.

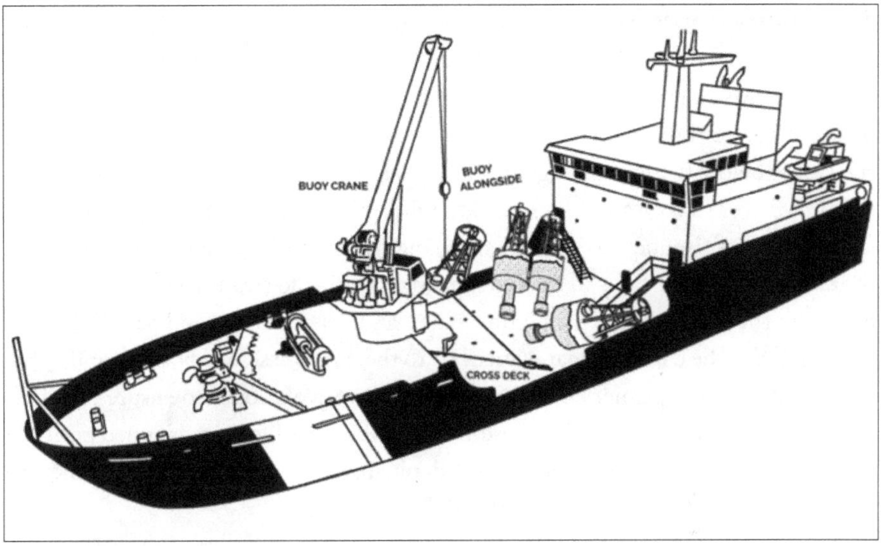

Servicing buoy

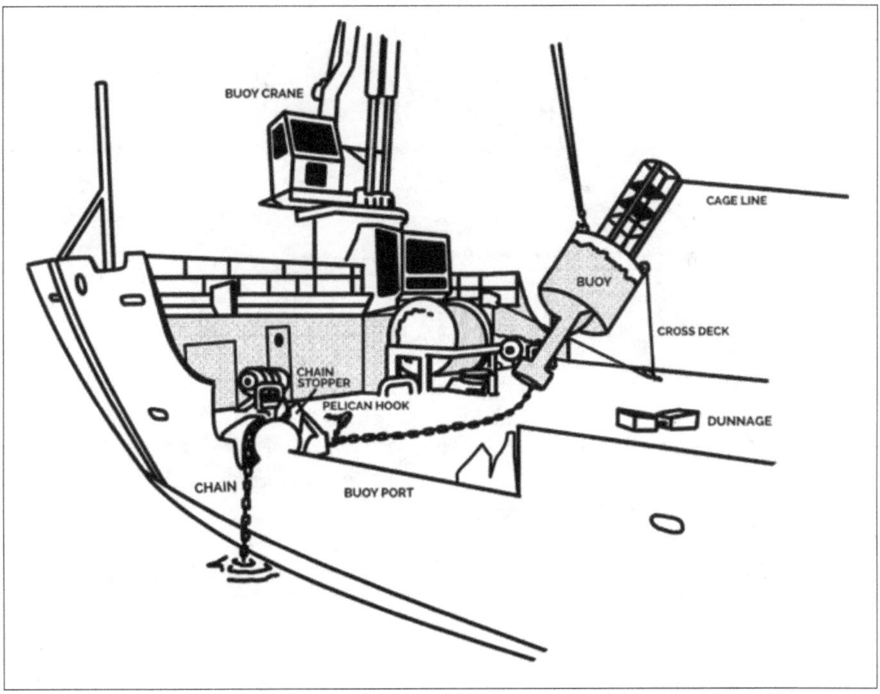

Cross-deck

Positioning the Buoy

After planned work has been accomplished, it is time to reset the buoy and mooring. Stage the sinker at the edge of the buoy port with the chain faked out. Pass the pelican hook to the mooring chain above the sinker, leaving enough chain to hang the sinker below the water's edge. Request to hang the sinker. Use the crane to lift the sinker over the rail and lower until the sinker is suspended on the pelican hook. Clear the hoist. Request to shackle and hang the buoy. Shackle the buoy. Position the crane to place the buoy in the water, using a tagline to steady the buoy. Once the buoy is in the water, secure the tagline to a cleat. Clear the hoist and pass "READY ON DECK." The cutter or boat maneuvers to the buoy's assigned position and "SET IT" is ordered. The primary objective of this entire evolution is to ensure the aid is watching property and placed as close as reasonably possible to its assigned position. At the command, trip the pelican hook deploying the sinker. Slip the tagline and release the buoy, passing "BUOY AWAY."

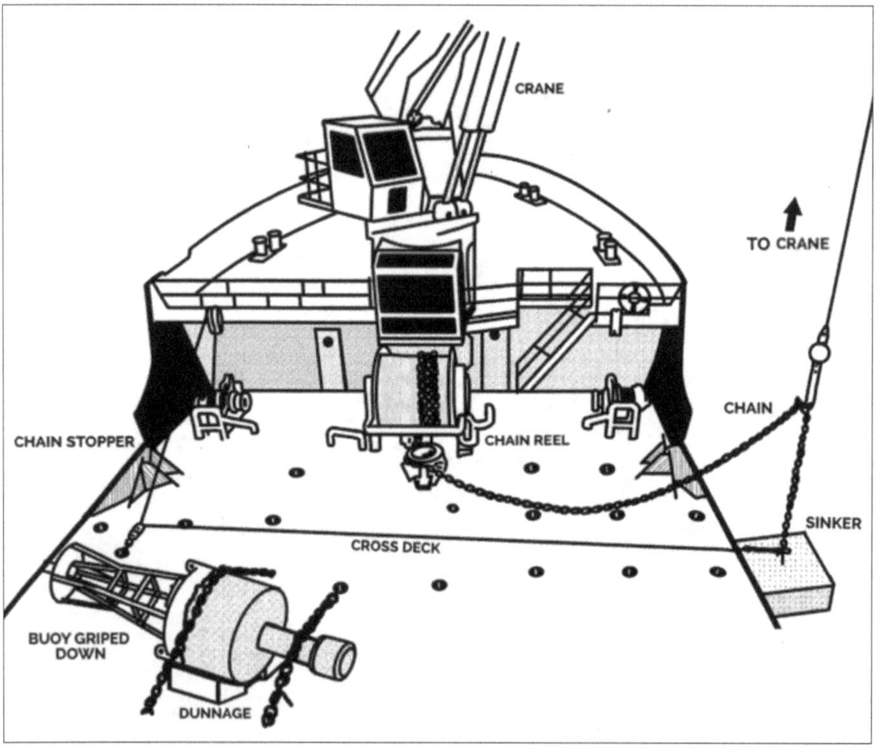

Pulling chain

JUMPING BUOYS

Transferring personnel from a boat to a lighted buoy and back again requires steady boat handling and good communication between coxswain, crew members, and passengers. Since environmental conditions affect the boat differently than the buoy, approaching the buoy and timing the transfer requires practice and coordination. If transferring equipment between boat and buoy, one person is placed on the buoy and then the equipment transferred rather than trying to jump to the buoy carrying the equipment. Servicing personnel need to have both hands free to grab onto the buoy during the transfer. Once servicing personnel are safely on the buoy, the coxswain makes additional approaches as needed to transfer equipment. The use of tag or tending lines is highly recommended to ensure equipment is not lost. If conditions permit, the coxswain may tie the boat to the buoy to assist with personnel and equipment transfers. If the coxswain ties off, ensure the boat and aid are not damaged from repeated contact with one another, the buoy is not dragged off station by the extra strain on the mooring, and the added tugging on the buoy by the boat's weight is not hampering the work being performed on the aid.

Once service to the buoy is complete, the boat approaches the aid to recover equipment and servicing personnel. If conditions change while servicing the aid, the coxswain reevaluates conditions and passes the new plan for approach and pickup to the servicing crew. The boat does not leave the area while a crew is servicing a buoy. The boat remains on scene to provide support as needed. Boat crews cannot tie off to an aid and abandon the boat to service an ATON. At a minimum, one certified coxswain and crew member remain in the boat at all times.

SERVICING FIXED STRUCTURES

While fixed ATON structures do not move like buoys, they still offer a challenge to approach and service. Remember, many aids mark shoals and other coastal hazards. Often there is only one access point, which is not always positioned in the best location for environmental conditions, and surrounding hazards might not allow access to the aid from shore or from some directions. It might also be necessary to drop personnel off at one location but transfer equipment at another. Limited or no access from shore, ease of equipment transportation, and speed of response compared to shoreside access make boats the preferred method for accessing most shoreside aids. Detailed information regarding service by boat, preferred access points, charts, local knowledge, and so on, is maintained in the aid's record. If no information exists in the aid remarks or if the information is no longer valid, ensure that new or updated information is included in the aid remarks so the next crew to visit the aid benefits from your knowledge and experience.

Prior to climbing a structure, conduct a thorough visual inspection. Do not attempt to climb a structure or steps that appear rotten or weak. If using an extension ladder, be sure that it has enough support angle when leaned against the structure to safely support the climber. Once the servicing technician is on the tower, all hands should stand clear. Wear a hard hat and eye protection when working on or around a structure. Tools should be equipped with a lanyard to prevent them from falling. It is common to erect an A-frame or sheer legs, sink a deadman in the ground, or secure a tackle or snatch block directly to a tree or to the structure to allow equipment and materials to be hoisted from the beach or boat.

When a shore aid is serviced, first check the initial voltage of the batteries and while under load. Replace batteries as needed. Next check the wiring throughout the structure. Check the flasher and lamp changer, if installed, and change lamps if necessary. Check the voltage output and diode on the solar panel. Check the panels for damage and cleanliness and ensure that they are facing south. Clean, inspect, level, and focus the lantern assembly. Check the structure for leaks and make those repairs that are within the capabilities of the work party. Document any remaining discrepancies for additional resources to correct at a later date. Clear away brush and trees obscuring the light and daymark.

Wildlife

Wasp's nests are often found on shore structures and in battery boxes with the screens missing. A good spraying of the battery box with insecticide, which can be applied from a distance, will often remove the wasp or hornet problem. In any event, be careful when working on structures. A sudden wasp sting could cause you to lose your balance and fall. Snakes can also pose a problem at certain times of year. Be careful that you do not place your hands or step where you cannot see. Sunny rocks and platforms are a favorite early-morning and evening haunt of snakes when the temperatures begin to drop. A little caution and respect will prevent accidents. Poison ivy, poison oak and poison sumac will move rapidly into the fringe of clearings around structures. Be familiar with local plants' appearance and wear clothing that will prevent direct exposure.

Clearing Brush

When clearing brush, you must be constantly aware of the location of all persons in the work party. Do not work too close to other people. When felling trees, take care in choosing a landing site and escape route before commencing. Make sure a large enough clearing exists before a tree is permitted to fall. Always be sure of your footing and make sure that brush, vines, and so on are clear from overhead before swinging an ax or brush hook. All persons who use chainsaws, brush cutters, and similar tools must have received adequate training and wear proper protective equipment to include safety glasses, hard hats, and chaps. All tower structural parts and materials should be precut or preassembled as much as possible, painted, and be ready to be erected on the spot. Units should carry extra towers, sand anchors, guy wires, and lumber to repair or replace towers or structures as needed.

Historic Places

Many Coast Guard lighthouses are listed as eligible for the National Register of Historical Places. Prior to any significant work at a historical site it is necessary to obtain approval from the district and the state historical preservation officer. Significant work involves the removal of buildings and the replacement of items such as windows, railings, and so on.

21

AIR OPERATIONS

Operational response is the primary mission of Coast Guard aircraft. For this reason, aviation capabilities are founded primarily on search and rescue, enforcement of laws and treaties, marine environmental protection, and defense readiness. Various aircraft types and models also perform a logistic role, including the transport of personnel and cargo. Air stations maintain at least one Bravo Zero (B-0) aircraft at all times to allow a rapid response to any tasking or emergency.

AIRCRAFT

Information concerning the operational limits and flight characteristics of the Coast Guard's fixed-wing and rotary aircraft is in chapter 4 of this volume, "Cutters, Boats, and Aircraft." Coast Guard aircraft have the distinctive Coast Guard painting design and carry a VHF-FM radio for contacting maritime vessels. Coast Guard Auxiliary fixed-wing aircraft may be available to assist.

Auxiliary Aircraft

Auxiliary aircraft are commonly known as general aviation aircraft. They are mostly single-engine land planes, either high wing or low wing. There may be some twin-engine aircraft, seaplanes, or helicopters. Auxiliary aircraft have no special painting design, but all are required to have their Federal Aviation Administration registration numbers on the fuselage or tail. The Coast Guard logo and lettering are not permitted;

however, the facility decal is required. The aircraft may also carry the Auxiliary logo decal aft of the wings, or the word "RESCUE" on the bottom of the wing or fuselage in twelve-inch letters for visibility at low altitudes, or both the logo and the word. From the surface, an Auxiliary aircraft looks like any other civilian airplane.

Aircraft Capabilities

Aircraft are equipped to succeed in their missions and have different configurations for specific assignments. To the maximum extent possible they carry detection aids, infrared sensor equipment, surface-search radar, and rescue equipment to maintain a search-and-rescue and law enforcement response capability at all times. The requirements for an aircraft to be considered fully mission capable for each identified mission type as well as the expectations for aircrew are outlined below. All are day and night capabilities.

Search and Rescue

- Electronically detect and locate a 406 MHz distress beacon
- Communicate on international distress frequencies
- Mark datum and determine drift
- Visually mark the position of a search object
- Deliver life rafts to survivors and dewatering pumps to vessels
- Communicate with survivors
- With rotary-wing aircraft, provide safe transport for survivors and search using the electro-optical/infrared sensor system

Enforcement of Laws and Treaties

- Record visual images of targets of interest
- Detect and classify a surface target while maintaining a covert distance and altitude
- Communicate on civil maritime frequencies
- Rotary-wing aircraft may be equipped to deliver a boarding team by fast rope

International Ice Patrol

Mission is specific to fixed-wing aircraft:

- Detect and accurately determine the position of a small iceberg in zero visibility

Airborne Use of Force

Mission is specific to rotary-wing aircraft:

- Maintain encrypted communications with law enforcement entities
- Clearly show the aircraft represents law enforcement
- Communicate on civil maritime frequencies
- Surveil targets of interest covertly and from beyond the range of small-arms fire
- Temporarily prevent a noncompliant vessel from operating under its own power or from maneuvering freely via the application of warning shots and disabling fire if necessary
- Apply deadly force

Air Intercept

A mission unique to the national capital region and specific to rotary-wing aircraft:

- Maintain communication with command-and-control and operational assets
- Communicate on civil aviation frequencies
- Visually provide compliance directions to a target of interest

AIRCREW

The exact configuration of the crew varies based on the aircraft size and mission; however, the following are assigned to all flights:

- Pilot: The pilot in command conducts all activities related to flying the helicopter. The final authority regarding any aircraft is always the pilot in command.
- Copilot: The copilot is second in command of the aircraft during flight operations and is responsible for performing en route navigation and communication tasks.
- Aircrew: The aircrew is responsible for all operations in the aft cabin of the aircraft (e.g., hoist operations, surveillance).

Additional crew and specialists are assigned based on mission needs. For example, a navigator is assigned to any flight that will be flown at night and a rescue swimmer to one tasked with search and rescue.

- Rescue swimmers are trained to deploy from a helicopter to recover an incapacitated victim from the maritime environment, day or night.

COMMUNICATION

Communication between a boat and an aircraft can be done by voice radio or a variety of visual signals. Aircraft are equipped with VHF-AM, VHF-FM, and UHF radios with encryption. Air-to-surface and surface-to-air visual signals may be used when a radio is not available.

Air-to-Surface Signals

An aircraft may use the following signals to direct a boat to a place:

- Circle the vessel at least once.
- Cross the vessel's projected course close ahead at a low altitude while rocking the wings (opening and closing the throttle or changing the propeller pitch may be used instead of rocking the wings).
- Head in the direction in which the vessel is to be directed. An aircraft may show that assistance of the vessel is no longer required by crossing the vessel's wake close astern at low altitude while rocking the wings.

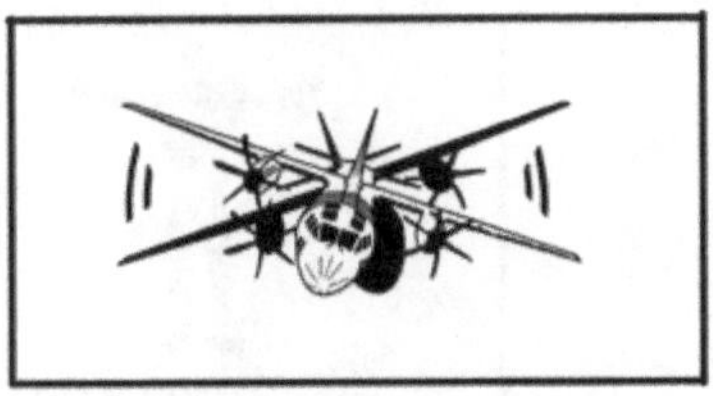

MESSAGE RECEIVED AND UNDERSTOOD
(ROCKING THE WINGS)

MESSAGE RECEIVED AND NOT UNDERSTOOD
(CIRCLING)

AFFIRMATIVE
(PITCHING NOSE UP AND DOWN)

NEGATIVE
(YAWNING LEFT AND RIGHT)

Ground-to-Air Signals

Make these symbols a minimum of one meter wide and six meters long. If you make them larger, keep the same 1:6 ratio. Ensure the signal contrasts greatly with the ground it is on. Place it in an open area easily spotted from the air.

When an aircraft is close enough for the pilot to see you clearly, use body movements or positions to convey a message.

GROUND OPERATIONS

The airport ramp can be a complex environment with air station personnel, passengers, trucks and other vehicles, airplanes, helicopters, and wildlife. Whether departing to or from the ramp, the pilot must maintain a high level of awareness that requires preparation to maximize safety. This includes being familiar and competent with safety protocols for engine starts, ramp and taxi operations, loading and unloading cargo and passengers, and securing aircraft postflight.

NUMBER	MESSAGE	CODE SYMBOL
1	REQUIRE ASSISTANCE	V
2	REQUIRE MEDICAL ASSISTANCE	X
3	NO OR NEGATIVE	N
4	YES OR AFFIRMATIVE	Y
5	PROCEED IN THIS DIRECTION	↑

Ground-to-air signals

Ground-to-air hand signals

Flight line

Ramp and Taxiing

Ramp traffic may vary from a deserted open space to a complex environment with heavy private, commercial, or military aircraft. Powerful aircraft may produce an environment, from exhaust blast or rotor downwash, that could easily cause a light airplane to become uncontrollable. Some ramps may be staffed by personnel to assist the pilot in managing a safe departure from the ramp to the taxiway. These personnel use standard hand signals.

Taxiing is the controlled movement of the airplane under its own power while on the surface between a parking area and the runway. An essential requirement in conducting safe taxi operations is situational awareness of the ramp, parking areas, taxiways, runway environment, and the persons, equipment, and aircraft at all times. Depending on the airport, parking, ramp, and taxiways may or may not be controlled. At small, rural airports these areas may have few aircraft, which limits the potential hazards; however, as the complexity of the airport increases, so does the potential for hazards. When an aircraft is being taxied within twenty-five feet of obstructions, a two-person taxi crew is required. One serves as a taxi/signalman and wing walker and the other as an additional wing walker. Aircraft are at no time taxied within five feet of obstructions.

Engine Starts

An aircraft engine is only started by a pilot or crew member qualified by their command to perform engine starts. Before starting, a fire watch is posted outside the aircraft to monitor each engine as it is started.

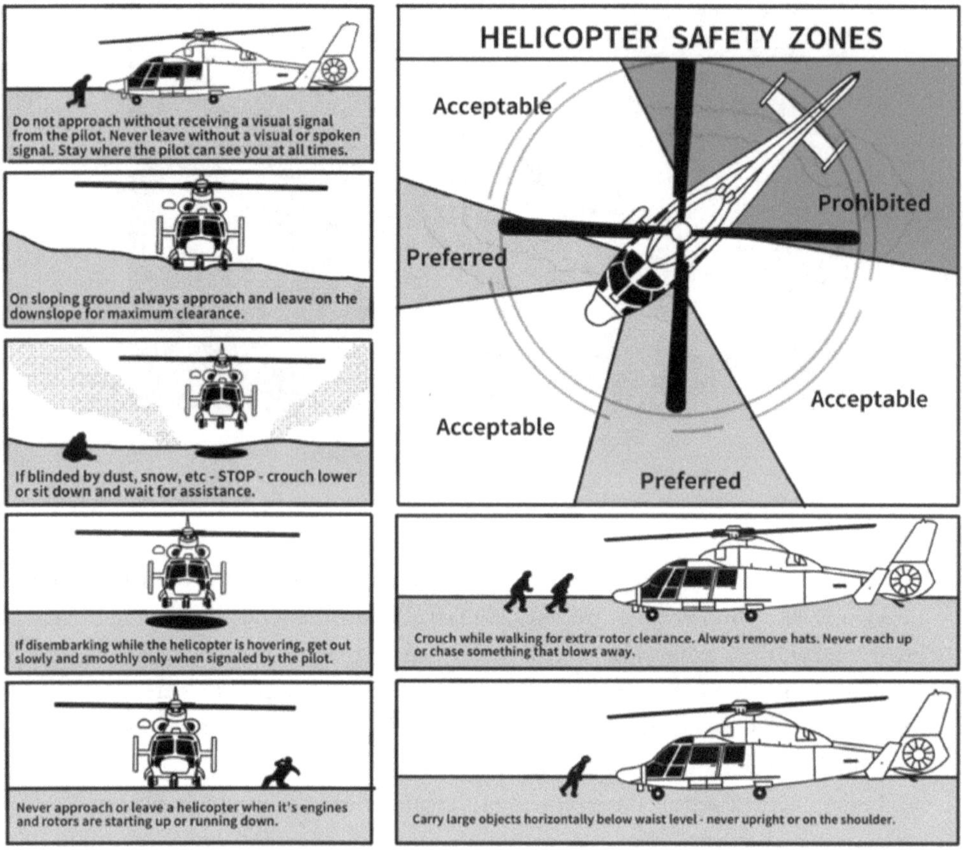

Safely approaching helicopters

Loading and Unloading

Whenever cargo or passengers may affect the weight and balance of the aircraft, and whenever hazardous cargo is involved, loading and unloading operations are supervised by a qualified loadmaster. Normally the aircraft's engines should not be running and the propellers or rotors should not be turning while loading or unloading is in progress or when personnel are entering or leaving the aircraft.

Care must be taken to ensure an adequate safety zone is maintained around any turning propellers or rotors and exhaust blast areas during any engines running evolution. Care must also be taken to prevent any foreign object from becoming dislodged and damaging the aircraft or cargo, or injuring personnel during the loading and unloading process. If deemed safe by the pilot in command, personnel may enter or depart the aircraft with engines running and propellers or rotors turning under the supervision of an aircrew member.

Securing Aircraft

The aircraft should be hangared or tied down, flight controls secured, and security locks in place. The type of tie-downs varies significantly from chains to well-worn ropes. Chains are not flexible and as such should not be made taut as to allow the airplane some movement and prevent airframe structural damage. Tie-down ropes are flexible and may be reasonably cinched to the airplane's tie-down rings.

Hangaring is not without hazards to the airplane. The crew must ensure that enough space is allocated to the aircraft so it is free from any impact to the hangar, another aircraft, or vehicle.

HELICOPTER LANDING ZONES

The urgency of a helicopter rescue can result in flight or ground crews using a landing zone that is less than adequate. This action has directly resulted in rotor strikes with terrain, vessels, cables, wires, foreign object debris, and, tragically, rescue personnel on the ground. Select an adequately sized landing zone, ideally a level one hundred feet by one hundred feet or larger with a grass or hard surface. Management of dust by wetting down the landing surface or snow by compaction can minimize the amount of blowing particles that reduce visibility during landings and takeoffs. The rotor wash from an unmanaged dust or snow-covered landing zone could result in a brownout or whiteout as the aircraft attempts touchdown, resulting in spatial

Landing zone

disorientation for the pilot. A landing zone covered in tall grass may contain hidden hazards, such as large rocks, and the grassy surface can develop wavelike motions resulting in disorientation for the pilot.

A visual indication of wind speed and direction can be very useful for the pilot. A windsock, flagging, smoke signal, or dirt thrown into the air as the helicopter initiates a high orbit over the scene can serve as a wind indicator. A helicopter achieves optimum performance when landings and takeoffs are made into the direction of the oncoming wind. Ground personnel should anticipate this in their selection of a landing zone and in the staging of rescue apparatus or personal equipment.

Hot loading, or loading and unloading with the engine running and blades engaged, may sometimes be required for efficiency, such as for survivor shuttles. This practice poses greater risk for personnel, particularly from possible rotor droop if engine rpm is reduced. Anticipate the situation and secure loose debris and gear in advance. Shutting down the aircraft to load personnel or a rescue subject dramatically reduces the risk and increases the safety of everyone working near the aircraft.

EXTERNAL LOAD TRANSPORT

The helicopter sling-load method of carrying cargo and equipment overcomes many of the obstacles that hinder other modes of movement. Sling-load operations are used extensively in the ship-to-shore movement of cargo. Cargo hooks are attached to the bottom of the aircraft and are operated manually or electronically by the aircrew or, in an emergency, manually by the ground crew. During hookup the cargo sling or cargo net apex fitting is attached directly to the cargo hook. Only single point

Transporting cargo

loads can be accommodated by Coast Guard aircraft, which use hooks rated at two thousand pounds mounted on the underside of the helicopter.

The helicopter approaches the load with the pendant or hook prepared and maneuvers in position over the load as directed by the signalman and aircrew. Once the helicopter is in a stable hover and correctly positioned, a static wand is used to ground the cargo hook and the ground team connects the hook to the load. When the load is ten to twenty feet higher than surrounding obstacles, the signalman gives the takeoff signal in the direction of departure.

HELICOPTER RESCUE SWIMMER DEPLOYMENT

The decision to deploy the rescue swimmer is initiated by the pilot in charge, but the swimmer has the authority to decline deployment if they assess the situation as beyond their capability. Deployment of the swimmer must be thoroughly coordinated and briefed by all crew members, who must consider on-scene environmental factors, marine life, and the ability of the swimmer to safely complete the mission. Swimmers may be deployed by freefall, sling, harness, direct, or, when touchdown is possible, disembark.

HELICOPTER HOIST RESCUE

Helicopter hoist is a widely used and efficient procedure for both insertion and extraction from a rescue site. The first civilian helicopter hoist rescue was performed by Dimitry "Jimmy" Viner, chief test pilot for Sikorsky, near Fairfield, Connecticut, on November 29, 1945. The rescue was initiated following a violent storm to save two men stranded on an oil barge who were in danger of being washed overboard. An early Sikorsky production model R-5 helicopter pulled off the rescue in Long Island Sound, a short flight distance from the Sikorsky factory.

Helicopter hoist rescue uses a hoist cable attached to a rescue hook assembly. The terms "hoist" and "winch" are often incorrectly used in referring to rescue hoists. For clarification, "hoists" are used for lifting and "winches" are used for pulling. Helicopter rescue hoists are either electrically powered or hydraulically powered. Externally mounted hoists can be a fixed installation above the aft doorway as on the MH-60 or on a movable boom controlled by the hoist operator, which is stowed against the helicopter fuselage during forward flight on the MH-65. The hoist operator works in the open aft doorway, secured by a safety tether strap that runs from the operator's harness to a fixed hard point in the aircraft. The hoist operator uses a handheld pendant hoist control with a visual readout display of the amount of cable in use. The pilot is also equipped with hoist controls on the collective, which can override the hoist operator's position.

Hoist Cable

The hoist cable is truly a lifeline and must be treated with respect. Helicopter hoist cables are 3/16-inch diameter manufactured with 133 individual stainless steel wires. Hoist cable lengths vary, but 245 feet of usable cable is a common working length with a mean breaking strength of 3,330 pounds and rated working load capacity on the hoist of 600 pounds.

Hook

The hook assembly is based on an old 1963 design that consists of two gated hooks and an eyelet. The large hook is the only portion of the rescue hook to be used for hoisting personnel. The smaller hook is used for handling equipment or light cargo. The smaller equipment hook on the opposite side of the rescue hook is not intended to be used to hoist personnel. The equipment ring is an additional attachment point intended for cargo hoisting and may be used to suspend chemical light sticks during night operations.

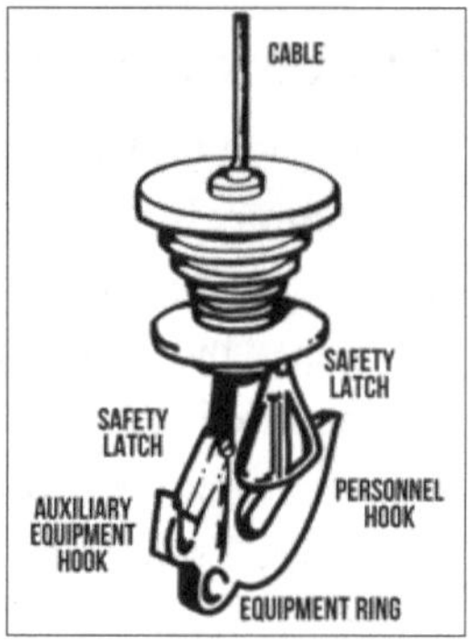

Static Discharge

Helicopters generate static electricity as air particles move over the surface of the fuselage and rotor blades. Blowing snow or dust increases the interaction with the aircraft and the risk of static discharge. Forward flight causes some of the static to bleed off the tail of the aircraft. However, in a hover, the static does not bleed off.

If static is present, the hoist cable becomes grounded when touched by ground personnel, who may receive a harmful shock as the static electricity is discharged. Rescuers can prevent this static shock by always allowing the hoist cable to contact the deck, ground, or structure before they touch it. A dead man's stick may also be used to discharge static electricity prior to the hook actually making contact with the surface.

Cable Entanglement

During helicopter hoist rescue operations, cable entanglement is a very real concern as the cable reaches the deck, ground, or structure. Conscious efforts should be made to prevent entanglement of the hoist cable, or the load on the line, with fixed objects. The aircrew is responsible for managing the cable. The amount of slack in the cable should be enough to avoid sudden and unintended tension while minimizing the opportunity for snags or injuries to the boat crew. In the event of excessive slack, the boat crew must be prepared to clear the cable away from personnel and snag hazards and notify the aircrew. Minimize handling the hoist cable to the maximum extent practical. Unless required to handle the rescue device or to sweep the cable away from personnel, rigging, or other snag hazards, the boat crew should not handle the hoist cable. Do not coil the cable in the rescue device. Always unhook the rescue device before moving it inside the boat.

Trail Line

A trail line may be employed by personnel on the ground or a vessel to assist with hoist hook delivery; however, use of such a line to stabilize a litter during a hoisting evolution is very important. Using a tag-line is a standard procedure for a "static hoist" evolution, where the aircraft remains in a hover above the rescue scene. The rescue litter on the end of a hoist cable can quickly develop an uncontrollable spin in certain configurations. Using a handheld tended tag line between an attendant on deck and the litter can eliminate rotational spinning. The connection point at the litter includes a quick release to permit the hoist operator to detach the line once the litter reaches the aircraft doorway. In the event the helicopter makes an emergency maneuver during a hoisting, the tag line connection is designed to fail from the additional applied force.

Trail line

The standard tag line is constructed with 120 feet of 3/8-inch polyethylene or polypropylene line. It can be deployed from the aircraft hand-over-hand with a 5-pound weight bag attached. Ground rescuers should wear gloves when handling a tag line during hoisting, and the line is not to be anchored. Tag lines are nonconductive and will not ground the litter, so the litter should be permitted to make contact with the ground in order to safety dissipate static electricity.

Preparations for a Boat Hoist

Before the helicopter arrives, the boat crew will complete action on the following general preparations:

- Navigation: Check charts for hazards that would prevent the boat from maintaining course and speed until the hoist is complete.
- Communication: Establish communication with the helicopter as early as possible to exchange information and instructions.

- Protective Gear: Helmet, goggles, and hearing protection shall be worn. Gloves are optional and, if worn, should be as form-fitting to the hand as possible to reduce the possibility of becoming fouled with any of the hoisting equipment.
- Loose Gear: All loose gear is stowed or secured on deck (e.g., hats, cushions, life rings, etc.).

Hand Signals

One boat crew member is designated to give hand signals to the hoist operator. Communication between boat crew members on deck should occur prior to giving a hand signal to the helicopter. For example, once a rescue device is on deck, prior to giving the thumbs-up signal for device removal, check with other crew members to make sure they are ready to proceed. The hand signals authorized for communications between the boat and helicopter are presented at right.

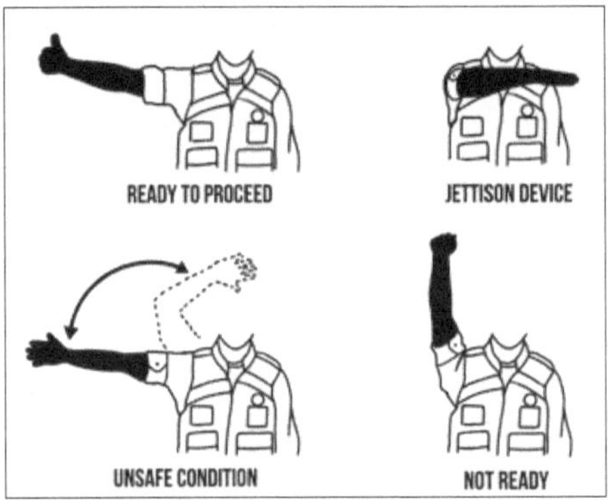

Boat to helicopter hand signals

Helicopter–Boat Positioning

Boat and aircraft positioning for hoists

Hoists from standard boats are normally made from the stern. The pilot normally will direct the coxswain to assume a certain course and speed with a relative wind speed of fifteen to thirty knots and 35–45 degrees off the port bow. The rescue device will be lowered from the right side of the aircraft. The helicopter will approach the boat from astern (downwind) and hover off the stern/port side. This method of approach allows the pilot and hoist operator located on the right side of the aircraft a full view of the boat during the evolution.

Delivery of the Rescue Device

Delivery of a rescue device from the helicopter to a vessel in distress or for training will be accomplished by one of the following three methods. After the rescue device is delivered and if previously agreed to in the aircraft brief, a boat crew member will disconnect the hook before moving away from the delivery/hoisting location.

Direct Delivery

During direct delivery, the rescue device is lowered directly to the deck of the vessel.

Trail-Line Delivery

During trail-line delivery, a five-pound or heavier weight bag is attached to the trail line and lowered from the helicopter to the vessel. The helicopter will then back off to a safe hoisting distance while paying out the trail line. The non-weighted end of the trail line is attached to the rescue device, weak link first, and lowered to the vessel. Boat crew members will tend the trail line by hand-over-hand method, exerting enough strain to guide the rescue device to the delivery point on deck. Too little strain causes the device to be lowered nearly vertically, which makes device management more difficult for both the hoist operator and crew on deck. The crew on deck then must work harder to pull the device toward the boat. A second crew member should assist the crew member hauling in the trail line by coiling the trail line in a safe place on deck, ensuring the line does not get fouled around an object or get blown overboard.

Indirect Delivery

Indirect delivery is designed for delivery of a portable dewatering pump. The trail line, weighted bag attached to the weak link first, is delivered from the helicopter to the vessel. The helicopter will pay out the trail line as the helicopter backs off and establishes a low hover with the rotor blades and downwash clear of all rigging. The hoist operator will then attach the end of the trail line to the pump container and deploy it to the water. The boat crew member will then pull the pump aboard.

Rescue Devices

Below are rescue devices used by helicopter crews during hoisting operations.

Collapsible Rescue Basket

The standard, collapsible rescue basket is designed for lifting one survivor at a time during a hoist rescue. The basket has integrated flotation attached to the frame. Once the rescue basket is lowered to a subject, it can be easily used by a conscious and mobile individual without direct rescuer assistance. When a rescuer is available to provide loading assistance, they instruct the subject to keep all body parts inside the rescue basket and not attempt to get out of the rescue basket until directed by the helicopter crew. When the rescue basket has been hoisted up to the aircraft, the helicopter crew simultaneously pulls the basket into aircraft cabin while slowly reeling out the hoist cable until the rescue basket is at a safe position within the cabin to safely unload a subject.

Rescue Strop

The rescue strop or "horse collar" is designed as a retrieval device for uninjured personnel. The rescue strop fits around the chest of a single subject and is connected back to the rescue hook assembly. The strop is not meant to be used on victims who have sustained significant spinal injury unless the circumstances do not permit appropriately securing the injured subject in a more appropriate device. The strop is a buoyant device made with closed-cell foam constructed with a high-visibility external cover. Deploying a rescue strop to a subject without the aid of a rescuer is not recommended. Rescue-strop recovery without proper application has resulted in subjects falling out of the device during extraction with fatal consequences.

Stokes Litter

The Stokes litter is a stretcher with a flotation collar and chest pad for buoyancy. The helicopter will provide the Stokes litter used for hoisting. A five-pound ballast weight located at the foot end provides stability. A permanently mounted hoisting sling attaches the litter to the helicopter hoist cable. For restraining patients, five securing straps, including a chest pad, are supplied. The Stokes litter is used to transfer an injured or unconscious person in any weather condition. It is generally used when the patient's condition prevents use of the basket. When the patient is placed in a litter, a crew member must tighten all straps to keep the person securely bound to it.

Casting Off

When a person or equipment is secured in the rescue device, the designated boat crew member will give the hoist operator a "thumbs-up" hand signal. The hoist operator will then commence lifting the rescue device. During this procedure, the boat crew must ensure the rescue device is not caught on anything attached to the boat. When a trail line is employed, a boat crew member will tend it until the weighted end is reached. Then the crew member should gently release the weighted bag over the side of the boat on which the hoist was conducted. The bag should never be thrown upward toward the rotors. Once the trail line is cast off, the boat will maneuver to starboard and away from the helicopter unless there is another hoist to be conducted.

Shearing the Cable

Immediately report any fouled cable or equipment to both the boat and to the aircrew via the "unsafe condition" hand signal. If injury or damage is not imminent, attempt to clear the fouled cable or device. If unable to safely clear the foul, communicate the "intension to jettison device" hand signal to the flight crew and communicate that a shear is necessary. The boat crew on deck must take cover and be aware of falling equipment when the cable is sheared.

SHIPBOARD OPERATIONS

Helicopter operations from cutter flight decks place both the cutter and flight personnel in a unique and demanding environment. Close coordination and proper actions are required by all personnel at all times. Predeployment training and workups are required prior to aircraft deployments on board cutters. Space is very limited, with aircraft required to share takeoff, landing, and maintenance spots. Therefore, helicopters may need to be repositioned frequently and expeditiously, often on a windy, wet, and moving deck in proximity to ship's structure. When not being repositioned, aircraft must be chained to the flight deck to prevent unintended movement due to wind and ship motion.

Maintenance

Conducting aircraft maintenance on board a cutter requires detailed planning and coordination. Most shipboard helicopter maintenance must be conducted on the flight deck. Flight decks are exposed to the elements and are in constant motion due to changing sea states. As a result, helicopter maintenance can be extremely hazardous at all times but especially at night, in inclement weather, or when flight operations are in progress.

Fueling

Helicopter fueling operations on ships are classified as either cold refueling (engines off) or hot refueling (rotors/proprotors turning or engines operating). Cold refueling may be accomplished by pressure or gravity. Hot refueling is limited to pressure fueling only.

Corrosion Prevention and Control

The shipboard environment is inherently corrosive. Embarked aircraft will require cleaning and treatment for corrosion more frequently than shore-based aircraft. When at sea, embarked units must place special emphasis on the importance of a corrosion prevention and control program and ensure it receives priority for timely accomplishment along with other required maintenance.

Landing

Takeoff and Landing

Each cutter and embarked aircrew will establish the pre-takeoff sequence that best supports their operations and the assigned mission. Helicopter engines units shall not be started without the cutter's permission. Cutter personnel coordinate with the embarked aircrew to ensure they have the winds required for a safe engine start and rotor engagement.

Flight Deck Status Lights

The flight deck status light provides information to the aircrew and flight deck personnel. When red, the cutter has attained the proper launch envelopes, which indicates to the flight deck permission to start engines. When amber, the flight deck has permission to engage rotors. When green, the flight deck has permission to launch the helicopter.

Shutdown

Engine shutdown and rotor disengagement on board a cutter can be more difficult and more hazardous than ashore. High winds and a rolling/pitching flight deck, coupled with the geometric design of ships and their superstructures, can cause unusual wind patterns and helicopters may experience rotor coast-down times in excess of five minutes. The cutter will strive to obtain the minimum winds across the deck to reduce the possibility of helicopter component damage and personnel injury.

22

SWIMMING

All Coast Guard personnel must know how to swim. The day when you need to use your swimming ability may come sooner than you think. All the life rafts and life vests in the world will not replace the need to swim under certain circumstances. In many cases just being able to swim will not be enough—you have to be a *good* swimmer.

Immediately upon entering the water in a survival situation, become oriented to the surrounding area. Try to locate your boat, aircraft, floating objects, and other survivors. Try to board a lifeboat, raft, overturned boat, or other floating platform as soon as possible to shorten the immersion time. Body heat is lost many times faster in the water than in the air. Do not attempt to swim unless it is necessary to reach a fellow survivor or a floating object that can be grasped or climbed onto.

WATER ENTRY

The cross-and-cover technique is the optimal body position to enter the water and minimize possible injury. To enter the water, stand erect, look at the horizon, and use the cross-and-over technique:

Cross and cover

Step	Procedure
1	Place your nondominant hand over your helmet to reduce the possibility of the helmet coming off upon impact with the water.
2	Make a tight posture, tuck the elbows close to the body, and reach across with your dominant hand. Grab the top of the ballistic plate to prevent injury to your chin at water impact. Place and keep the elbow close to the body.
3	Step off; do not jump. Immediately after stepping off, cross the legs at the ankles.
4	Keep your body vertical by continuing to look at the horizon. Maintain this position until just after impact, then immediately slow the descent by making one large, powerful downward sweeping motion with your hands while simultaneously kicking.
5	Immediately begin in-water emergency procedures.

Conduct the following steps upon entering the water:

Step	Procedure
1	Orient yourself in the water or in the downed helicopter, capsized boat, or flooded space.
2	Swim away from hazards (prop, hull, etc.), as necessary.
3	Manually activate the personal flotation device (PFD) if it did not automatically inflate by pulling the beads or lanyard.
4	If positive buoyancy is not achieved, remove the helmet, vest, and firearm and open your inflatable PFD.
5	When you have positive buoyancy after automatic or oral inflation, activate the strobe and Velcro to your helmet, if applicable.
6	Activate the personal locator beacon and attach it to the Velcro on the helmet, if applicable.
7	Use available means to establish communication with parent unit such as radio, mirror, whistle, or pyrotechnics.

Mustering

If multiple personnel are in the water, muster together and maintain physical contact until rescued. Use only one strobe at a time to preserve battery life. Coordinate use of pyrotechnics, if applicable.

Raft Technique. Huddle as closely together as possible to conserve heat and protect high heat loss areas (head, groin, armpits) with two or more persons. To execute the raft technique, put your arms over each other's shoulders so that the sides of the chests are together and, if possible, intertwine legs.

Chain Technique. Form a chain by wrapping your legs around an individual's waist from behind and paddle in unison. Rotate the cadence among survivors to raise spirits as part of long-term survivability.

HEAT ESCAPE LESSENING POSTURE

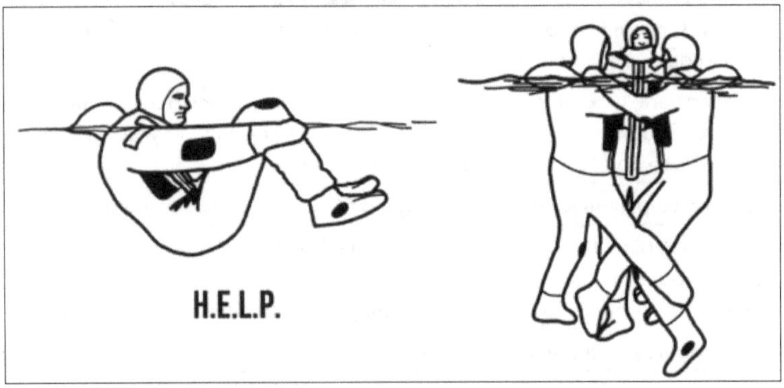

The body position assumed in the water is very important in conserving heat. Float as still as possible, using the heat escape lessening posture (HELP) to minimize exposure of the body to cold water. To execute the HELP technique, cross the legs at the ankles and draw the knees up to the chest, keeping the face forward and out of the water. Cross arms, keeping the upper arms tucked close to the sides of the body and the lower arms crossed over the chest.

STAYING AFLOAT WITHOUT FLOTATION

Remaining on the surface of the water without a flotation device and in a position that allows comfortable breathing without tiring is an important skill. The facedown method is effective when wearing restrictive or negatively buoyant clothing. You can rest on the surface of the water using minimum energy to continue breathing and catch your breath following vigorous swimming or to conserve energy. In cold water, quickly assess the situation and use whatever is at hand to provide buoyancy to keep your head out of the water. In cold water, use the survival floating techniques described below only as last resort to enable you to catch your breath or to implement some form of flotation.

In Extremis Flotation

Shirts may be inflated by tying the cuffs and collar, blowing air into the opening, and then holding the garment under water. Trousers make better floats than shirts. After the trousers are removed, they should be floated on the surface with the fly turned up. A single knot should be tied in each leg, then one side of the waist of the trousers should be grasped with each hand, and the garment should be worked around on the surface until the legs are at the back of the head and neck. When this position is reached, the trousers should be flipped over the head, and the waist brought down smartly on the surface, trapping a good pocket of air in each leg. The waist should then be gathered under the water and held in one hand.

Treading Water

Treading water allows you to check the surface for floating objects, other survivors, rescue craft, and so on. Tread water to catch your breath following sudden entry into cold water, to activate flotation equipment, to shed unwanted bulky equipment, or to signal rescue craft. Keeping your head out of the water requires considerable effort, especially when fully clothed. Rapidly remove negatively buoyant equipment. In cold water, clothes offer thermal protection, and you must decide if there are benefits to leaving them on.

Being able to tread water is a vital skill and should be practiced until you can do it for minutes at a time, not just a few seconds. It is required for every situation in which you need to raise your head to look around, or call for help, or use your hands. There are several methods of treading. Practice all of them, and by changing from one to the other, you will be able to stay afloat longer. The most commonly used methods are mentioned here.

Frog Kick

Stand upright in the water. Draw legs up together with the knees spread out to the sides. Separate and extend legs. Finally press the legs together until the feet are ten to twenty inches apart. The movement of drawing, spreading, and pressing them is then repeated. Pushing the legs out and pressing them together gives an upward thrust to the body.

This kick can be practiced while hanging onto the edge of a pool. Its advantage is that the slow movement is not as tiring as some of the other treading kicks and it is adjustable; you can tread low in the water or thrust up high by this same kick.

Scissors Kick

This is similar to walking and is easy to learn. It is best practiced at first on land. Lie on your side with feet together and legs extended, and then draw them slowly up in a semi-tucked-up position. Separate the legs, one moving forward, the other backward, then squeeze them together. The movement should be slow, steady, and continuous. Practice for a while, holding on to the edge of the pool or a firm object. The scissors kick is easy to learn and is a natural movement, effective but not tiring. It also leads to the sidestroke, which is important in distance swimming and lifesaving.

Sculling

Sculling with the hands enables you to control your body in the water in a number of positions, with or without the help of your feet. Sculling and treading combined will save more energy than using the feet or hands alone. Sculling has the advantage of teaching a sense of balance in the water. It develops the muscles of the hands and arms needed for pushing against the water, and it teaches people how to get a "grip" on the water. One method of sculling while holding oneself upright in the water is to weave the hands vigorously back and forth in front of and to the sides of the body in a figure-eight movement. This forces the water downward and keeps the body up. Another method is to drop the hands down near the hips in the water, then turn them to about a 45-degree angle away from the body, thumbs downward. With the palms of the hands, push the water away from the body until the hands are about two feet out from the sides of the body. Immediately rotate the hands so that the thumbs are up, and then return to the starting position, pulling water toward the sides of the body. When the hands reach the sides, they are again rotated until the thumbs are down, and the movements are repeated. The emphasis is on the push-away part of the stroke. A third method is to place the hands palms down on the water, press down on the water several feet in depth, and recover again. All these methods, or combinations of them, will be useful at different times.

Drownproofing

Place the face in the water, chin at chest, with the back of the head just breaking the surface. The upper back and shoulders are underwater, horizontal to the surface, and the arms are at the surface with the elbows bent and hands separated slightly. Bend the waist with the hips underwater, lower than the upper body, and the legs dangling beneath.

To breathe, pivot at the neck and lift your chin off your chest until your mouth clears the surface. Keep your waist bent, shoulders in the same near-horizontal position to the surface. As your mouth clears the surface, exhale quickly and forcefully through your mouth and nose. Inhale a deep, full breath through your mouth. After inhaling, lower your head to the resting position with your chin on the chest. Make sure every breath is a good, complete exchange of air deep into the lungs. Support your head while breathing with a broad sculling motion of the arms. Coordinate the sculling motion of the arms with the breathing to provide maximum lift when needed.

SURVIVAL SWIMMING

Carefully think out and plan a decision to swim in open water. Swimming even short distances in open water may be dangerous if the water is cold, if there is a strong current, or if there is rough wave action. In the case of accidental water entry, it is often most logical to stay near the site of water entry to aid in recovery and to conserve energy by floating. If at any time you feel disoriented underwater, follow the bubbles to the surface.

If the decision is made to swim in a survival situation, mastery of swimming strokes increases the chances of survival by offering the most efficient propulsion with the least expenditure of energy. The energy saved by efficient swimming may be needed later to produce body heat, climb into a raft, or activate signal and rescue devices.

Breathing

In general, try to keep your breathing regular, don't hold your breath too long, and don't pant. Breathe through your mouth when your face is out of water. Exhale by blowing out through your mouth and nose to keep them clear of water. Try to take air in fast, blow it out slowly.

Breaststroke

The breaststroke is generally considered the best survival stroke when swimming in open water and is frequently used as an approach stroke for water rescues. The advantages of this stroke include good forward visibility and controlled breathing when swimming in choppy seas, a powerful kick while wearing boots or shoes, and an efficient energy-conserving glide.

Sidestroke

The sidestroke is useful when towing equipment, or a victim, or to swim if one arm is injured. It provides good sideward visibility but very little forward visibility.

Combat Sidestroke

The combat sidestroke is a variation of the sidestroke. It is faster than the normal sidestroke, offers good forward and sideward visibility, and has excellent controlled breathing when swimming in rough seas. It is identical to the normal sidestroke except the head position and breathing. During this stroke, rotate your head to the side and inhale when recovering the top arm, then place your face in the water as the top arm and kick propel you forward. Repeat this breathing and head action with each stroke.

Backstroke

The backstroke has an effective kick while wearing boots and an energy-conserving glide. Disadvantages include difficulty seeing where you are going and the inability to control your breathing in rough seas.

Crawl Stroke

The crawl stroke is the fastest of all strokes and is effective in survival situations when speed is required. It may also be used if your legs are injured. The crawl stroke offers poor forward visibility and is fatiguing to swim with operational clothing.

SWIMMING IN DANGEROUS WATERS

There are several emergency techniques that should be studied and practiced.

Underwater

The ability to swim underwater is very important in avoiding surface hazards such as floating debris, oil, or flaming oil or gasoline. It has lifesaving value, for it enables a swimmer to rescue someone who has gone under. There are occasions, too, when it is necessary to recover articles lost overboard in shallow water. Almost any swimming stroke can be used underwater. The breaststroke and the sidestroke lend themselves particularly well to underwater swimming. The main thing is to learn breath control. Practice holding your breath and keeping your eyes open as much as possible.

Shark-Infested Waters

A lot of research has been done on shark attacks in recent years. During World War II, people in the water were told to splash wildly to drive away sharks, but more recently it has been discovered that sharks seem attracted to frantic movements. Sharks are extremely dangerous and unpredictable predators and have extraordinary sensory receptors. Species of sharks vary in their feeding habits, with the result that some are considered more dangerous than others. Sharks seem to sense fright exhibited by creatures in the water. They also seem to be attracted to bright colors, but most of all, sharks are attracted by blood. Scientific experiments indicate that sharks exhibit a characteristic "S" twist before striking, and immediately before biting they roll their eyes upward, losing sight of their prey. These experiments indicate that during the last phase of the attack, the sharks are guided by sensors located in their snouts. Although there is no good advice on how a swimmer might avoid shark attacks, except by avoiding sharks, the information that is currently available indicates that the following actions may reduce the risk of shark attack.

- Avoid bleeding into the water.
- Wear shoes and other protective clothing, particularly clothing of dark, drab shades; save the bright colors to show above the surface to attract rescuers.
- Remain calm and remember all species of sharks are not the same, nor do all sharks have the same level of aggressiveness.
- Swim slowly and quietly to minimize attracting the attention of sharks.
- In the presence of nonattacking sharks, do not strike at them because that may provoke an attack that might not otherwise occur.
- If the water is clear enough to see sharks swimming beneath the surface, watch for a telltale "S"-like twisting by a shark, which may indicate that an attack is imminent.
- If a shark is about to attack, its jaws may possibly be diverted at the last moment by thrusting an object such as a piece of driftwood toward the shark's mouth.
- Do not confuse sharks with dolphins and porpoises, both of which are mammals, known in fact and legend to be friendly and helpful to sailors in the sea.

Heavy Weather and Surf

Along coastlines and on reefs, waves steepen and break, turning into what are called "breakers." Surf consists of more than one of these breakers. Coast Guard boats often cross bars in leaving and entering inlets and harbors and often encounter broken waves at sea, on bars, and in inlet channels; seldom, however, do Coast Guard boats go deliberately inshore into the surf, except during special emergency operations. You are most likely to encounter surf after the foundering of a Coast Guard boat or if you are swept overboard from a boat near the beach. Surf conditions in these cases are probably going to be very rough.

You should be wearing a PFD. The PFD-equipped swimmer's best chance of avoiding injury is to make sure that the PFD is properly secured and then to assume a rolled-forward position to protect the neck, back, and limbs. Take deep breaths whenever possible, and keep the eyes open underwater to help you reorient to the surface.

A non-PFD-equipped swimmer must avoid being caught on the surface by a large, plunging breaker. Such a swimmer must dive under approaching crested waves and then surface soon enough after the wave has passed to get another breath before diving under the next wave. Dives should be as deep as possible to avoid getting caught in the surface water as it curls over from the back face of the wave. Waves can be flattened by onshore winds yet retain all the force of much higher seas. The swimmer in such surf must watch for approaching seas and rise with them or dive as necessary to conserve air and strength. A swimmer must move quickly onto the beach to avoid the back rush of water after each wave.

Strong inshore feeder currents will set parallel to the shoreline, usually in the direction of the wind and surf. These currents are formed by waves that have been pushed onto the beach and that must find an outlet to return to the level of the sea. These outlets are marked by lower wave heights, slower wave speeds, confused current swirls, and foam. When these feeder currents turn to sea, they are called "rip currents." They flow seaward at angles from 45 to 80 degrees from the inshore feeder currents. Swimmers must avoid these rip currents. If the swimmer seaward of the breaker line can determine the direction of the wind or the angle at which the surf is approaching the beach, it is possible to pick the area where a rip current is least likely. That area will be just downwind of the place in the surf line where the waves seem lower and break farther to sea. Should a swimmer get caught in a rip current, the best strategy is to swim gently at right angles to the direction of the current, never against it. Since most rip currents angle away from the beach, swimming at right angles to the current, down the beach and in the general direction that the wind and surf is moving, will enable the swimmer to make slow progress toward the shore without becoming exhausted. Once clear of the current, the swimmer should head directly toward the beach. Because of their unique profession, Coast Guard personnel should learn to swim comfortably in surf, studying its movements and rhythms. Under emergency conditions, experience and knowledge of the surf may make survival possible.

Rocky Shore Landing

Avoid places where the waves explode with a high, white spray. Swim slowly in making your approach; you will need your strength to hold onto the rocks. If you must land on a rocky shore, look for a place where the waves rush up onto the rocks. After selecting your landing point, advance behind a large wave into the breakers. Face shoreward and take a sitting position with your feet in front, two or three feet lower than your head. This will allow your feet to absorb shocks when you land or strike submerged boulders or reefs. As the next wave approaches, take a sitting position with feet forward. Repeat the procedure until you land.

Reef Crossing

Cross a reef the same as if you were landing on a rocky shore. Keep your feet close together and your knees slightly bent in a relaxed sitting posture to cushion blows against coral.

LIFESAVING

The Coast Guard employs helicopter rescue swimmers, cutter surface swimmers, and boat swimmers to assist and recover fatigued, unconscious, entangled, or injured survivors from the water. Their use may significantly reduce some of the dangers inherent in maneuvering rescue vessels close to survivors. Cutter and boat swimmers wear a swimming harness with a tending line, when available, to allow another crew member to tend the harness with a retrieval line whenever the swimmer is in the water. An untethered cutter or boat swimmer should be deployed only as a last resort as there is no means to retrieve the swimmer should they need assistance.

Cutter rescue swimmer

Tending Line Commands

Communication between the crew members on the cutter or boat and the swimmer can be impacted as the distance between the two increases. To compensate, tending line commands are used to avoid the need for shouting. The signals primarily assist with communication but when used can also aid the boat swimmer in the recovery of a person in the water. The signals are based on the OATH signaling system.

OATH SIGNALING SYSTEM

Meaning	Tugs/Pulls	Swimmer Visual Signal
I am OK / Are you OK?	One	Tap top of head
Advance	Two	One arm raised over head
Take in slack / I am ready to return to the boat	Three	One arm waving over head
Help / Do you need help?	Four	Two arms waving over head

Swimmer Rescues

Effective lifesaving in the water depends upon the rescuer's presence of mind, knowledge of the methods that may have to be used, and strength and skill in carrying them out. Drowning people usually panic. Their one idea is to keep their heads above water. The whole technique of lifesaving is centered upon that fact. If a drowning person grabs or tries to grab some part of the rescuer's body or clothing, the rescuer should sink, taking the victim under water. Under water, the victim usually will let go. Under no circumstances should a rescuer strike a drowning person. A drowning victim's system has already had sufficient shock to cause severe physical reactions, and a severe blow may cause heart failure.

Lifesaving Approaches

There are several ways to approach a drowning person.

- Frontal approach: The rescuer should swim slowly toward the victim and attempt to calm and reassure the victim by talking. The victim should be told exactly what the rescuer is going to do and must be instructed to follow orders. Then the rescuer should grasp with the right hand the victim's left wrist, turn the victim's body slowly, and use one of the carries described below.
- Rear approach: If the victim is too excited to pay attention to directions, the rescuer should swim behind the victim, grasp the chin with the right hand, apply pressure to the back with the left hand, and use one of the carries.

- Underwater approach: The underwater approach is by far the safest because the drowning person does not have a chance to get a grip on the rescuer. The rescuer should swim within ten feet of the victim and surface-dive to a depth at which the victim's legs can be easily reached. If the victim is facing the rescuer, the rescuer should turn the victim in the opposite direction by pressure on their upper legs. The rescuer should slide the right hand up the drowning person's back and grasp the victim by the chin, applying pressure to the back with the left hand. The victim should be brought to the surface as quickly as possible and one of the carries used.

Lifesaving Carries

- Hair carry: This is the easiest of the carries because it allows the most freedom of movement on the part of the rescuer. The rescuer turns on the left side, slides the right hand up the back of the victim's head to the top and grasps the hair or collar tightly, using the left arm and legs for swimming sidestroke. The rescuer may swim on either side, changing hands in order to rest.
- Head carry: The rescuer swims on the back, holding the victim's head above the water with both hands meeting under the victim's chin.
- Cross-chest carry: The rescuer turns on the left side, places an arm over the victim's right shoulder, across the chest and under the left arm. The victim's body should be supported on the rescuer's right hip, and the rescuer should swim sidestroke, using the free left arm and both legs. This carry may be done from either side but should not be used unless the victim is in bad condition, nor should it be used for long distances because it is the most tiring.
- Tired swimmer's carry: If the victim has enough control to obey orders, the victim should be ordered to turn on the back, face the rescuer, spread the legs, and place both hands on the rescuer's shoulders with the arms stiff. The rescuer assumes the position for the breaststroke and swims, pushing the victim ahead. The breaststroke is useful because it leaves the arms and legs unhampered for swimming and is only slightly more tiring than ordinary swimming.
- Carries from the water: The most common carry is the fireman's carry. On reaching shallow water, the rescuer stands the victim up temporarily, places the right arm between the victim's legs, and throws the helpless person's body over the right shoulder. In this position, the victim's right wrist is grasped in the right hand of the rescuer, whose left arm is free.

23

FIRST AID

Those at sea are often far removed from advanced medical care and reliant on themselves, the Coast Guard, and other professional and volunteer responders to provide first aid for sudden illness or injury. This aid includes both initial interventions and transport to higher levels of care for serious conditions, such as performing cardiopulmonary resuscitation (CPR) while making way toward a waiting ambulance on shore and complete treatment of minor conditions such as applying a bandage to an abrasion. The Coast Guard equips its crews, boats, cutters, aircraft, and facilities with the equipment needed to deliver meaningful first aid but relies on its personnel to know how and when to use it. Rather than establish a single service-wide standard for formal and recurrent medical training for its crews, the responsibility is largely left to individuals, with the service authorizing personnel to deliver care up to the level of their competence. There are many sources of such training, to include courses in CPR, first aid, basic life support, and wilderness medicine from groups including the American Heart Association and the Red Cross. Increasingly, the first aid kits carried on board Coast Guard assets adhere closely to the U.S. military's Tactical Combat Casualty Care guidelines, making these courses valuable to understanding and applying the unique features of associated equipment. Although attending training and gaining practice and operational experience are individual responsibilities, these are all vital to saving lives at sea, whether your own or of a person you are called to assist. This section provides information on the standard first aid kits approved by the Coast Guard and those fielded on board its assets along with the information needed to employ their contents to render lifesaving aid. In every case, operate within your knowledge and skills and do no harm. If you were the one injured, would you want a responder to do what you're thinking of doing? If it risks doing harm, don't do it.

SCENE SAFETY

When coming upon a victim, take care to protect yourself by surveying the scene to be sure there are no hazards such as live electrical wires or loose gear that may have caused the injury and may be dangerous to you. Never enter an enclosed space, whether on a cutter or a fishing vessel, where a person is unresponsive. There may be deadly gases present that contributed to his or her collapse. Secure any hazards before starting treatment. An injury to you while administering first aid will only complicate an already difficult situation. After ensuring that the scene is safe, control it. Move bystanders away. Establish a working order with your team or crew, designating a lead to maintain the overall picture and those who will physically render aid to any victims.

UNIVERSAL PRECAUTIONS

Bodily fluids are potentially infectious and may contain bloodborne pathogens such as the hepatitis B virus, hepatitis C virus, and HIV, which causes AIDS. Universal precautions prevent direct contact with bodily fluids and include non-latex gloves and eye protection. Face masks like the N-95 should be worn in any instance of known or suspected respiratory infection like tuberculosis or COVID-19. Gloves should be changed and hands washed with soap and water or alcohol-based hand cleaners whenever moving to treat a new person.

MEDICO

When working to aid the seriously injured while under way, seek medical advice by contacting the sector or district command center with a radio or cell phone and requesting MEDICO. Medical advice transmitted by Coast Guard facilities comes from qualified medical officers or flight surgeons so make the call early to allow watchstanders to notify and conference in these personnel. MEDICO is an international term meaning the passing of medical information by radio or other electronic means. Medical advice is available through many sources that include Coast Guard and Department of Defense medical providers, medical firms, and hospitals contracted by shipping companies and international service organizations such as the International Radio Medical Centre. Despite your distance from shore, resources are readily available to help answer questions and guide you through treating even the most severe trauma. If the patient's condition demands it, a medical evacuation, or medevac, may be requested to get resources on the scene to transport the patient to higher levels of medical care.

STANDARD FIRST AID KITS

The most widely fielded standardized first aid kit carried on board operational Coast Guard assets is the Boat Response Aid Kit, which can be found on board nearly every shore-based and cutter boat. A small, compact, and rugged yellow hard case that is airtight, waterproof, and sealable, the Boat Response Aid Kit contains four pouches labeled "BLEED," "ABD/BURN," "SPLINT," and "EYE" that organize supplies to treat each injury type per the pouch label. The pouches are attached by hook and loop to allow removal from the case and better facilitate on-scene care. Other standard kits used by operational units include the USCG Aircraft First Aid Kit, carried on board fixed- and rotary-wing aircraft, and the Individual First Aid Kit, carried by personnel assigned to deployable specialized forces units. In addition, most merchant vessels and all life rafts are required by law to carry Coast Guard–approved first aid kits with standardized contents and instructions. If you are on board a vessel, there is a very good chance a first aid kit is nearby.

Contents and Uses

Below are the contents required of Coast Guard–approved first aid kits and their intended use, although all are designed to be adaptable to many different situations.

- Ammonia inhalants: Break one and inhale for faintness, fainting, or collapse.
- Aspirin, phenacetin, caffeine tablets: Chew up and swallow two tablets every three hours for headache, colds, minor aches, pains, and fever. Maximum of eight in twenty-four hours.
- Bandage compress, 4", 2", and 1": Apply as a dressing over wound. DON'T touch the part that comes in contact with wound.
- Bandage, gauze, 2": For securing splints, dressings, and so on.
- Burn dressing: Petrolatum gauze bandage designed for covering a bodily injury resulting from exposure to heat, caustics, electricity, or certain radiations.
- Eye patch: Apply as dressing over inflamed or injured eye.
- Forceps: Used to remove splinters or foreign bodies. Don't dig.
- Ophthalmic ointment: Apply in space formed by pulling lower eyelid down, once daily for inflamed or injured eyes. Don't touch eyeball with tube.

- Splint: Pad with gauze and mold to limb to immobilize broken bones. Held in place with bandage. Do not attempt to set the bone.
- Tincture of iodine: Remove protective sleeve, crush tube, and apply swab end. DON'T use in or around eyes.
- Tourniquet: For control of hemorrhage on the arms and legs.
- Triangular bandage: Use as arm sling, tourniquet, bandage, splint wrap or padding, environmental mask, or bandanna.

Additional items carried in first aid kits supplied to operational Coast Guard crews, such as the Boat Response Aid Kit, include

- Abdominal emergency trauma dressing: For dressing abdominal eviscerations, traumatic amputations, burns, and large-pattern wounds. Consists of an elastic wrap equipped with a 12" × 12" sterile nonadherent pad and securing device. Can also be used to sling and swath an arm, to secure splints to an extremity, or to perform in any other function requiring an elastic wrap.
- Bacitracin ointment: Used to treat bacterial skin infections or to prevent infection of minor burns, cuts, or scrapes.
- Burn relief gel: Made with aloe vera, tea tree oil, and lidocaine, the water-based gel quickly helps to halt the burning sensation and prevent further contamination of minor burns, including sunburns.
- Chlorhexidine wipes: Used to clean the skin after an injury and to clean the hands before a procedure.
- CPR mask: Breathing barriers help prevent disease transmission when performing CPR.
- Emergency survival blanket: Aluminized and reflective blanket to allow retention of 90 percent of body heat
- Emergency trauma dressing / Israeli bandage, 70" × 6": Used to apply immediate and focused direct pressure to control hemorrhage, particularly on areas like the chest, head, and neck, where a tourniquet cannot be used.
- Hyfun chest seal, 6" × 6": Used to seal penetrating chest injuries.
- Isopropyl alcohol prep pad: Used to decrease germs in minor cuts and scrapes.
- Lip balm: Concentrated ointment to keep lips chap-free.
- Nasopharyngeal airway device: A hollow plastic or soft rubber tube used to establish an airway via insertion through the nose when the airway is obstructed. Carried in Individual First Aid Kits.
- Nitrile gloves: Used to protect against exposure to bloodborne pathogens. Puncture-resistant and eliminate the risk of latex allergy reactions.

- Polycarbonate eye shield: For penetrating eye trauma.
- QuikClot / combat gauze: A brand of hemostatic wound dressing containing an agent that promotes blood clotting. QuikClot devices are impregnated with kaolin, an inorganic mineral that accelerates the body's natural clotting ability and produces no heat, as was experienced with early versions of the product.
- S-rolled gauze, 4.5" × 4.1 yd.: Packaging configuration allows controlled application as a complete unit or in a continuous linear feed. Used as backing gauze for hemostatic agents and in conjunction with a compression bandage to control hemorrhage.
- Saline: Cleans and rinses areas of the body because of a wound such as a cut, scrape, or burn, or contact with something irritating such as a chemical, medicine, or pollen. Also used to moisten and lubricate absorbent wound dressings for traumatic wounds, cuts, bruises, and minor burns prior to the removal from the wound area.
- Sting and bite relief towelettes: Provide temporary relief of pain and itching associated with minor burns, scrapes, insect bites, and stings. Active ingredients are benzocaine and isopropyl alcohol.
- Trauma shears: Stainless steel with serrated lower edge to cut through materials. Clothing lift lip and rounded tip to allow cutting clothes without threat to patient.

PATIENT ASSESSMENT

Assess the victim to determine what injuries were suffered, the degree of urgency for the situation, and what help you need. Be systematic so nothing is missed. First determine the patient's level of consciousness. Most people should be able to promptly and spontaneously state their name, location, and the date and time. Any inability to do so shows an altered level of consciousness indicative of an illness or injury. For example, cardiac arrest leads to a sudden loss of consciousness because blood stops flowing to the brain. In cardiac arrest, the brain isn't the problem—the heart is. However, the loss of consciousness is the indicator of the problem. The patient may be awake but confused and may reply to your questions with only a grunt, moan, or movement, or may respond only to pain. When a person is unaware of their surroundings and unable to respond, they are described as unconscious. Unconsciousness is a medical emergency, and it is often a sign of a serious health issue that could result in death or permanent disability. Unconsciousness can occur suddenly, or it may develop gradually, depending on the cause. A person might feel sick, dizzy, short of breath, have chest pain, or feel lethargic or faint before becoming unconscious. Causes of unconsciousness include drowning, alcohol, epilepsy, exposure, diabetic emergency, overdose, oxygen deficiency, trauma, infection, poisoning, or stroke. If you find someone unconscious, or they become unconscious while you are treating them, check if they are breathing and feel for their pulse. "Look, listen, and feel" is a simple memory aid to use when checking for breathing or respiration. Lower the side of your face to the patient's and look to see if the chest rises and falls. Listen for breathing sounds from their mouth and nose, and feel for breath against your cheek.

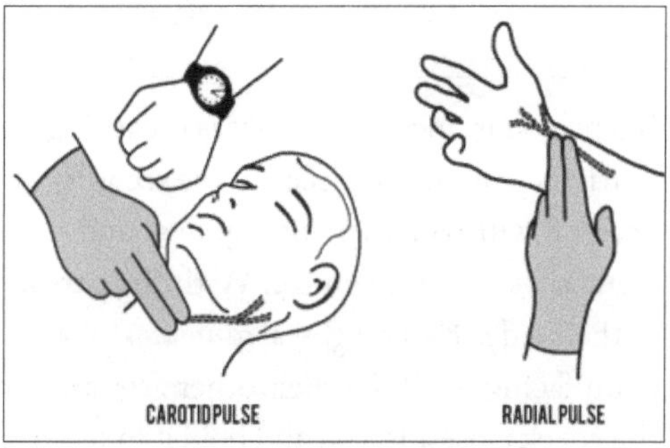

Feel for a pulse by placing two fingers on the inside of the wrist below the thumb (radial pulse) or on the neck just under the jaw (carotid pulse). Apply just enough pressure so you can feel each beat. Do not push too hard or you will obstruct the blood flow. If you feel a pulse, the victim's heart is beating.

Breathing must be restored before any other first aid action. Clear the airway if it is blocked and commence CPR if the heart has stopped beating.

If the patient is breathing, loosen or remove clothing and look for bleeding, wounds, fractures, or burns. Normal adult respiration is between sixteen and twenty-four breaths per minute. A normal pulse for an adult is regular, strong, and between sixty and one hundred beats per minute. Notice the color of the skin. Bleeding from the nose and ears is often a sign of a fractured skull. Bloody froth coming from the mouth or nose indicates damaged lungs. See if the victim carries any bracelet, tag, or card concerning conditions or allergies to drugs and medication.

If exposure is suspected, cover the patient with a warm blanket if the temperature is too cold or place an ice pack on them if the temperature is too hot. Do not in any circumstance place anything in their mouth, administer medication, or move their neck or back. Patient movements should be careful, deliberate, and the minimum required. If possible, avoid moving the patient until that person is examined and all injuries are protected by properly applied splints, dressings, and so on. If a head or neck injury is suspected, immobilize the patient's spinal cord prior to movement by securing the patient to a backboard. For conscious patients, always explain the move procedure in advance.

Recovery Position

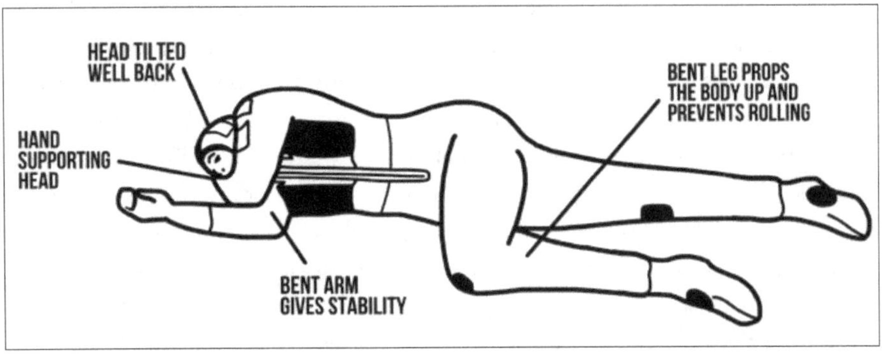

If a person is unconscious but is breathing and has no other life-threatening conditions, they should be placed in the recovery position. Putting someone in the recovery position will keep their airway clear and open. It also ensures that any vomit or fluid won't cause them to choke. With the person lying on their back, kneel on the floor at their side. Extend the arm nearest you at a right angle to their body with their palm facing up. Take their other arm and fold it so the back of their hand rests on the cheek closest to you and hold it in place. Use your free hand to bend the person's knee farthest from you to a right angle. Carefully roll the person onto their side by pulling on the bent knee. Their bent arm should be supporting the head, and their extended arm will stop you rolling them too far. Make sure their bent leg is at a right angle. Open their airway by gently tilting their head back and lifting their chin, and check that nothing is blocking their airway. Stay with the person and monitor their condition until help arrives.

Shock

Shock is a depressed physiological or mental state. Shock syndrome, a set of symptoms which occur together, can change throughout treating an injury and are unique for every casualty. Signs and symptoms may develop rapidly or be delayed for up to several hours after the apparent cause. Several types of shock exist; therefore, recognizing and treating shock immediately is important. Some syndromes do not appear in every casualty, nor are they equally noticeable. Symptoms may include restlessness, fainting, thirst, nausea, weakness, dizziness, and changes in mental status or behavior such as anxiety, fright, or agitation. Signs include weak and rapid pulse; shallow, rapid, and irregular breathing; cold, clammy, or sweating skin; dilated pupils; and unconsciousness.

Initial treatment for shock includes limiting a patient's activity, ideally having the person lie down and remain alert for the signs and symptoms of shock. If unconscious, appropriate treatment is to seek higher-level medical care and, if needed, institute resuscitation procedures.

If CPR is not necessary, the victim should remain lying down, should be kept warm, if not already overheated, and should be checked for other injuries. If there is not a head injury or breathing trouble, place the victim flat on their back and elevate the lower extremities eight to ten inches.

Hypothermia

Boat Response Aid Kit Items: Emergency Survival Blanket, CPR Mask

The leading cause of death in maritime accidents is hypothermia, a condition in which heat loss following total immersion in cold water or exposure to cold temperatures leads to an abnormally low body temperature. If not recognized and treated promptly, hypothermia can rapidly turn survivors into fatalities. Survivors removed from the water and left untreated may suffer further critical loss in body temperature, bringing on death even after being rescued. Those who suffer significant trauma and blood loss must, after the bleeding has been stopped, also be treated for hypothermia. Victims may lose consciousness. Breathing is slow and shallow and the pulse faint or unobtainable. Body tissues feel semirigid, and the arms and legs may be stiff. Successful treatment hinges on bringing the body temperature to normal. If the victim has been immersed in cold water, remove the wet clothes and wrap the patient in warm blankets and move them into a warm environment. Take care to warm the body and not the arms and legs as cold blood from the extremities can cause cardiac arrest if circulated to the body too fast. Be alert for further complications such as shock or cardiac arrest. It is very important to handle the victim gently and to get the victim to a higher level of medical care as soon as possible.

Resuscitation

Boat Response Aid Kit Items: CPR Mask

When a person stops breathing, known as respiratory arrest, seconds count. Death can occur within four to six minutes after respiratory failure, and it is imperative to start resuscitation immediately. Resuscitation is a general term that covers all measures taken to restore life or consciousness to an individual, to include artificial respiration and CPR. Artificial respiration involves rescue breathing maneuvers such as mouth-to-mouth and mouth-to-nose. Cases will be encountered in which the person has stopped breathing but their heart is still beating and they require only rescue breathing. Cyanosis, meaning blueness of the skin, lips, fingernails, tongue, or mucous membranes of the mouth, are signs that indicate the patient is not breathing. Cardiac compression is a method used to restore normal blood flow to the brain.

CPR uses both chest compressions and artificial respirations to revive a victim in cardiac arrest. If the patient has no pulse, they are in cardiac arrest and CPR must be performed. The recommended sequence for a victim with no pulse is to initiate chest compressions before giving rescue breaths. This sequence is easily remembered as C-A-B, which stands for Compressions, Airway, Breathing. This approach reduces delay to first compression and pumps oxygenated blood to the body.

Rescue Breathing

This is always your first step of treatment of an unconscious patient who is not breathing:

- Place the patient on their back and kneel beside the head.
- Place one hand under the chin and the other hand on the forehead, so that your thumb and forefinger can pinch the nose closed.
- Tilt the head back gently by lifting the chin and pushing the forehead down. This will open the airway in most cases. If neck or back injuries are suspected, use a jaw thrust to open the victim's airway by placing the heels of both hands on each side of the head, grasping the lower jaw, and moving it forward away from the body in a slight thrusting movement.

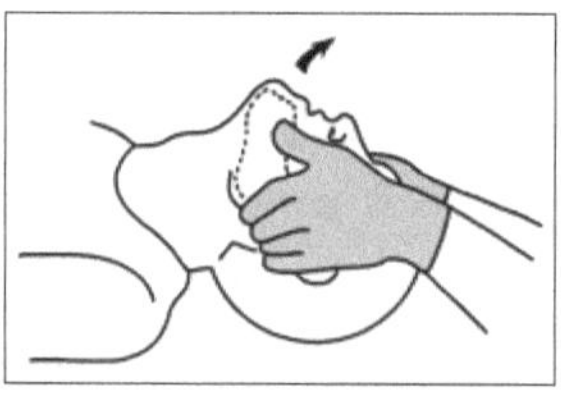

- Look, listen, and feel. Bend over the victim so that your ear is a few inches from the mouth and you're looking at the chest. Look for the chest to rise while listening and feeling with your ear for the movement of air.
- Fit a CPR mask or barrier to the patient's mouth.
- Take a deep breath, place your mouth over the mask or barrier and blow, giving two full breaths.
- Watch for the patient's chest to rise. As soon as this happens, remove your mouth from the mask or barrier and allow the air to escape naturally after each breath. Give two breaths initially then repeat.

If the chest does not rise, reposition the patient's head to further open the airway, ensure you have an airtight seal between the mask and the patient's mouth, and try again. If the chest still fails to rise, look into the victim's mouth and remove any foreign bodies, vomit, or blood clots, then try another breath. If the chest still will not rise, the choking sequence should be followed. Airways and tubes should be used only by personnel trained in their use or by medical officers.

Cardiopulmonary Resuscitation

After it has been determined that the heart has stopped, external cardiac compressions should be started and combined with rescue breathing to give CPR. Cardiopulmonary resuscitation must be started without delay. You have only four to six minutes in which to initiate this rescue technique. After that time, irreparable damage begins to occur in the victim's brain. Perform CPR as follows:

- Find proper hand placement. Place the heel of one hand on the center of the breastbone and the heel of the other hand on top of the first.

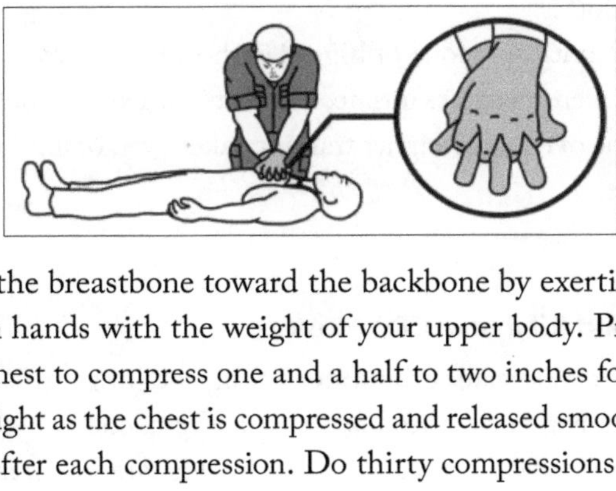

- Compress the breastbone toward the backbone by exerting downward pressure on hands with the weight of your upper body. Pressure should cause the chest to compress one and a half to two inches for adults. Keep elbows straight as the chest is compressed and released smoothly, allowing full recoil after each compression. Do thirty compressions at a time, at a rate of 100 to 120 per minute.
- Children and infant breastbones are not as strong; therefore, external cardiac compressions should be done with one hand for children and two fingers for infants. Compress the breastbone one to one and a half inches for children and half to one inch for infants at a rate of one hundred times per minute.

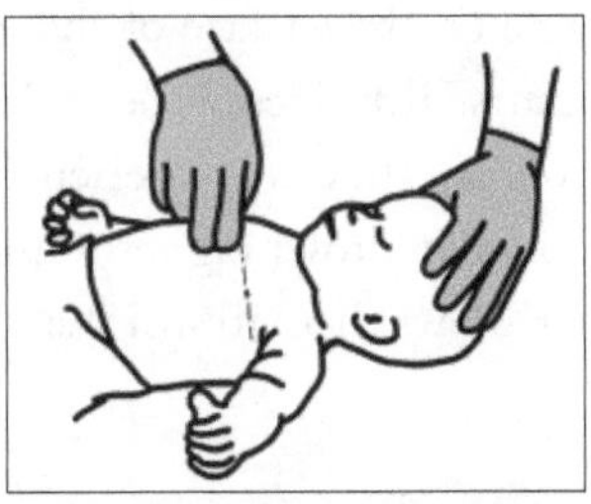

- If not trained in CPR, continue chest compressions at a rate of 100 to 120 per minute.
- For those trained in CPR, open the airway and deliver two rescue breaths.
- A cycle of thirty chest compressions to two rescue breaths is repeated at a rate of one hundred times per minute in adults and children. Timing is critical and the tempo must be maintained. A helpful way to maintain this tempo is to count aloud as you compress the chest (one, two, three, four, etc.). During two-rescuer CPR, the compressor must pause after the thirtieth compression to allow the rescuer performing the rescue breathing to administer a breath. If your timing is right, you should be completing four cycles per minute.

While performing CPR, the victim's stomach may become distended with air. This is especially true if the airway is not entirely clear. It is not dangerous and should be left alone. Often during CPR the patient vomits, particularly in drowning incidents, so you must be ready to turn the head to one side and clean out the mouth with fingers or a cloth.

Continue CPR until someone of higher authority takes over, until the patient recovers, until the scene becomes unsafe, an automatic external defibrillator is ready to be used, someone of equal or higher training takes over, or until you are physically unable to continue.

Drowning

Boat Response Aid Kit Items: CPR Mask

Drowning is suffocation caused by water filling the lungs. It is one of the most common medical emergencies faced by maritime rescuers. The waving, splashing, and yelling that Hollywood prepares us to look for is rarely seen in real life. Except in rare circumstances, someone who is drowning is physiologically unable to call out for help, wave, or assist in their rescue. An instinctive drowning response forces them to extend their arms laterally and press down on the water's surface so they can lift their mouths out of the water to breathe. Unable to hold their mouths above the surface for long, they only quickly inhale and exhale. Unless rescued, those who are drowning can only struggle on the surface of the water from twenty to sixty seconds before submersion occurs. This doesn't mean that a person yelling for help and thrashing isn't in real trouble. They are experiencing distress, a reaction not always present before the instinctive drowning response, and one that doesn't last long, but unlike true drowning allows them to still assist in their own rescue as they can grab lifelines, throw rings, and so on before instinct forces their movements. Sometimes the most common indication that someone is drowning is that they don't look like they're drowning. One way to be sure is to ask them, "Are you all right?" If they can answer at all, they probably are. If they return a blank stare, you may have less than thirty seconds to get to them.

Properly administered, CPR may successfully revive a drowning victim. To treat a person in a drowning event, evaluate compressions, airway, and breathing and initiate CPR if indicated. If revived, treat for shock and hypothermia, as appropriate.

Mammalian Diving Reflex

Medical researchers have also identified a "mammalian diving reflex" that occurs on contact of the trigeminal nerve of the face with cold water. After an involuntary breath, the cold water inhaled causes the body to rapidly cool and the heart rate and metabolism to slow dramatically, decreasing the body's oxygen requirement and leading to a protective hypothermia. This condition allows individuals to be revived without brain damage beyond the four- to six-minute maximums associated with a typical cardiac arrest. Children who fall through the ice, because of their small size, drop core temperature quickly and have a chance of survival up to fifty minutes after the loss of a detectable heartbeat. Although unlikely to be fully revived on the scene, a cold-water drowning victim given CPR even after a significant amount of time in the water may survive if rapidly transported to higher levels of medical care. Actions to rewarm the body in such cases should be limited to prevent the release of excessive amounts of lactic acid that will occur when metabolism increases in response to an increase in body temperature. If rewarming is too rapid, the amount of lactic acid released into the bloodstream could become toxic and be fatal. Higher levels of medical care can administer the appropriate drugs to balance the release with a controlled and deliberate rewarming process.

Heart Attack

Boat Response Aid Kit Items: CPR Mask

Heart disease is the leading cause of death in the United States. Heart attacks are caused by a lack of blood supply to the heart muscle. A heart attack is always considered a medical emergency since the victim is in significant danger of going into cardiopulmonary arrest and dying. Whenever a heart attack is suspected, be prepared for the possibility of sudden cardiac arrest and the need for CPR. Symptoms of a heart attack in men include severe, crushing type of pain under the breastbone, arms, neck, and jaw, profuse sweating, shortness of breath, nausea and vomiting, and bluish discoloration of lips, fingernails, and skin. Women often do not experience these signs and will describe indigestion, weakness or fatigue, and back pain. If a heart attack is suspected, keep the patient calm, loosen any tight-fitting clothing, and place the patient in the position of most comfort. Determine if the patient is on any type of medication for a heart condition such as nitroglycerine or aspirin. If so, determine if the medication was taken as prescribed. Seek immediate medical assistance.

Stroke

Boat Response Aid Kit Items: CPR Mask

A stroke is any interruption of the blood supply to the brain. This is most commonly a result of bleeding or clotting affecting the blood vessels of the brain. All strokes are extremely serious regardless of how severe they appear to be. Early recognition and treatment are critical to minimize or prevent more serious complications. Symptoms of a major stroke are unconsciousness, confusion, dizziness, numbness or weakness to one side of the body, seizures, and impaired vision. The acronym FAST is a good way to remember the symptoms of a stroke.

F—Facial drooping: Ask the person to smile, check for an even smile.
A—Arm weakness: Have person raise arms and see if one drops or can't be held in place.
S—Speech difficulty: Slurred speech, hard to understand.
T—Time is critical: If the symptoms above are observed, get medical assistance quickly.

To treat, seek medical assistance immediately. Treat for shock and, if the victim has difficulty breathing, help the person maintain an open airway. Be prepared for the possibility of sudden cardiac arrest and the need for CPR.

Choking

Choking occurs when an object obstructs the airway. The obstruction may even be the tongue in an unconscious victim. This is relieved by the airway-opening methods explained in the CPR section. Most commonly it is caused by a piece of food in adults or small toys or food in children. The universal choking sign is one or both hands clutching the throat. Ask the victim, "Are you choking?" If the victim can speak and is getting enough air, encourage them to relax and try to cough the object up. Do not attempt to remove the object yourself. In complete obstruction, or in partial obstruction where the victim is not getting enough air, back blows and abdominal thrusts should be performed. Stand behind the victim and tilt their body slightly forward.

Give five back blows in an upward motion between the shoulder blades, then wrap your arms around his or her waist. Make a fist and place the thumb side of your fist just above the navel but below the lower margin of the ribs. Grasp your fist with your other hand and press into the victim's abdomen in an inward, upward motion five times. Repeat until the object is coughed up or until the victim goes unconscious.

Infants who are choking will be unable to cry, cough, make noise, or breathe. Hold the baby face down along your thigh with their head lower than their bottom and hit them firmly on their back between the shoulder blades up to five times. If back blows do not dislodge the blockage, turn the baby over so they are facing upward. Place two fingers in the middle of their chest just below the nipples and push sharply downward five times. Continue with back blows and chest thrusts until the object is coughed up or until the infant goes unconscious.

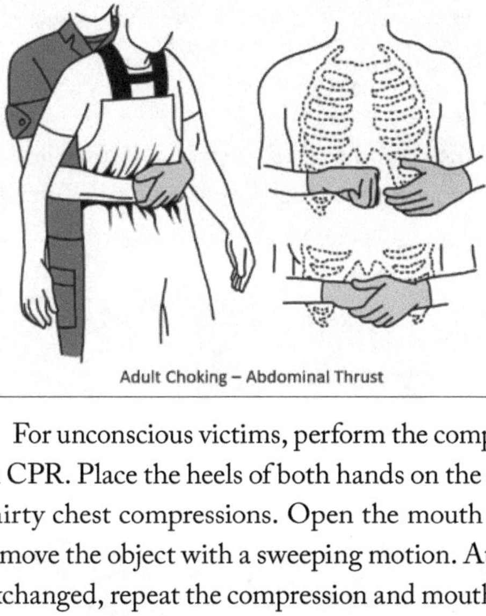

Adult Choking – Abdominal Thrust

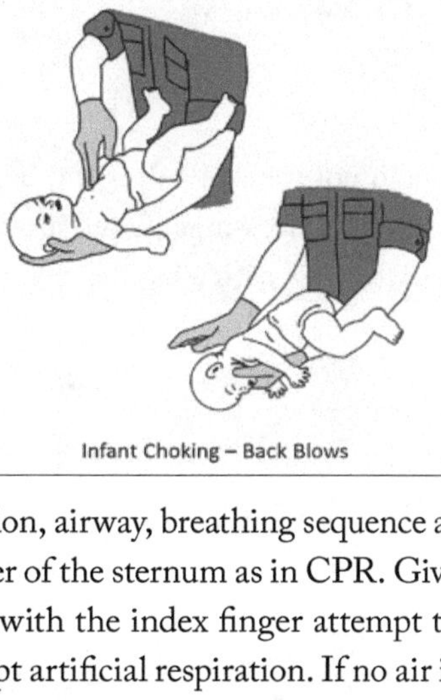

Infant Choking – Back Blows

For unconscious victims, perform the compression, airway, breathing sequence as in CPR. Place the heels of both hands on the center of the sternum as in CPR. Give thirty chest compressions. Open the mouth and with the index finger attempt to remove the object with a sweeping motion. Attempt artificial respiration. If no air is exchanged, repeat the compression and mouth sweep sequence until the obstruction is cleared or until you are relieved by another rescuer.

Bleeding

Boat Response Aid Kit Items: CAT Tourniquet, S-Rolled Gauze, 6" Emergency Trauma Dressing, Combat Gauze Z Fold

Hemorrhage, or bleeding, is the escape of blood from arteries, veins, or capillaries because of a break in their walls. Arterial bleeding is characterized by spurts, venous bleeding is steady, and capillary bleeding trickles from the body. Some wounds may need a tourniquet initially while others can be controlled with only a dressing.

Control of a severe hemorrhage is always urgent. With only ten pints of blood in the human body, arterial bleeding can cause death in a short amount of time. The quicker the bleeding is controlled, the greater the chances of survival for the victim.

Dressings and Direct Pressure

Treatment for most wounds begins by covering the wound with a sterile dressing and applying continuous, firm, and direct pressure.

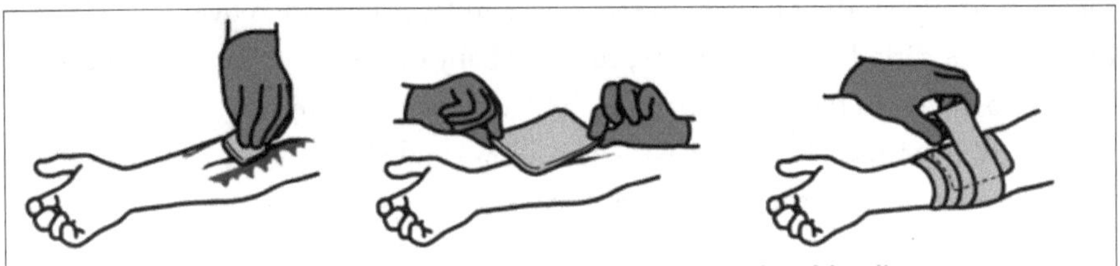

Clean skin around wound.

Place dressing over wound and apply pressure. Place additional dressings over the original as needed.

When bleeding stops, secure dressings in place with a bandage.

Do not remove a dressing if it becomes blood soaked; instead reinforce it with additional dressings placed atop the original. Once bleeding is controlled, secure the dressing with a bandage wrap.

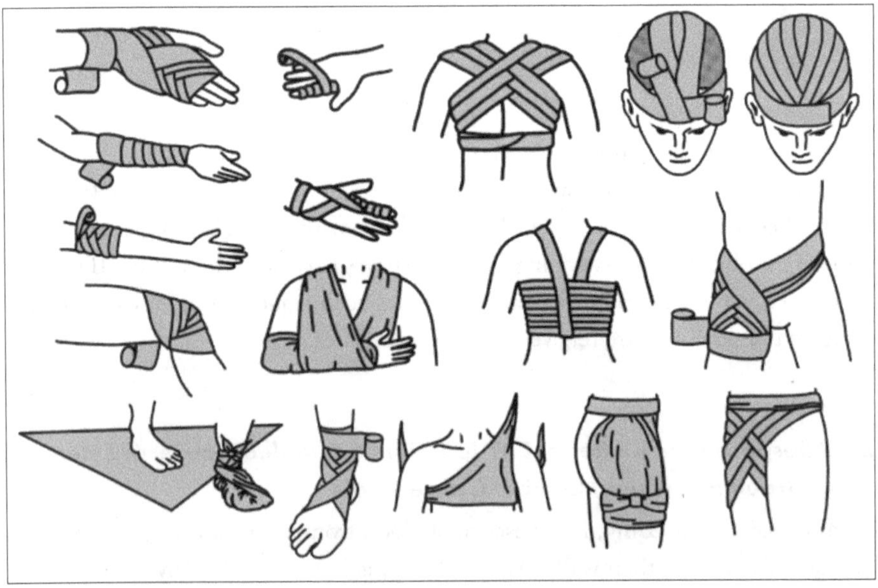

Hemostatic Agent

Gauze containing a hemostatic agent to help with the body's clotting ability is packed into a wound, particularly to control severe bleeds in areas where a tourniquet cannot be applied. These areas, or "junctions," include where the neck meets the shoulders, the arms meet the armpits, and the legs meet the groin. Don't pack anything into a penetrating injury to the chest.

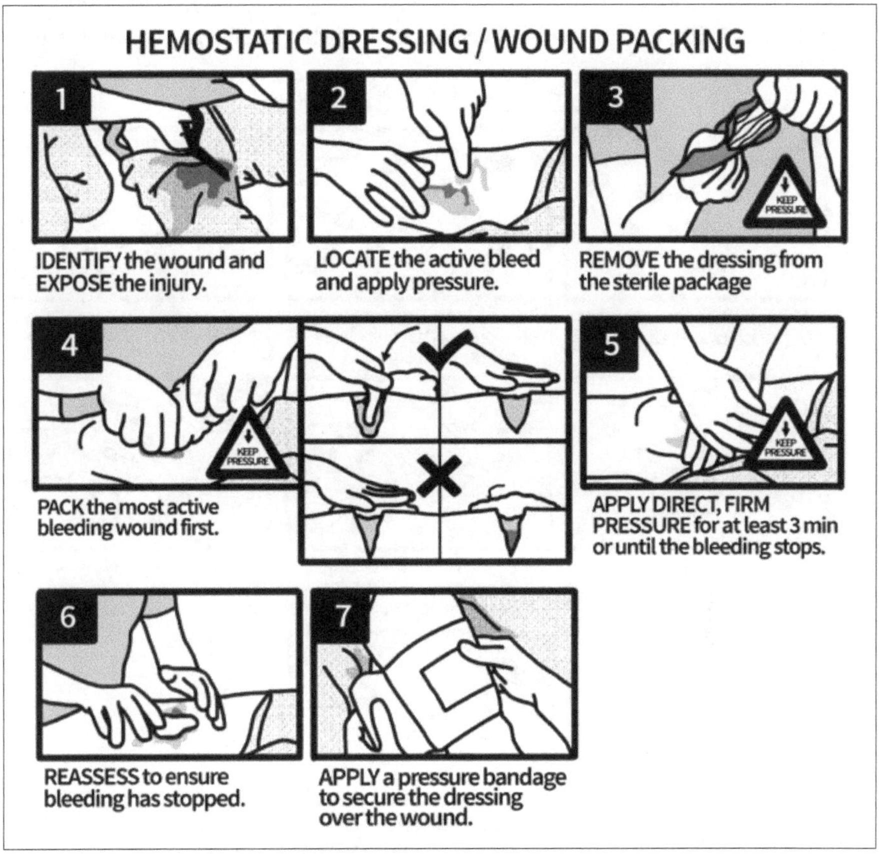

Pack the junctions by getting the agent down to the source of the bleed and filling the wound while applying constant, consistent pressure. Keep packing the wound until no more will fit, then hold pressure, observing for any active bleed. You're trying to put as much of the agent in the wound as possible while the gauze is conforming to the shape of the wound. If bleeding continues after packing the wound, you'll need to pull the gauze and try again. If you lack enough hemostatic gauze to pack the entire wound, just pack more standard gauze in behind and continue to hold pressure. When the bleeding has stopped, reinforce the packed gauze with a pressure bandage

Pressure Bandage

A pressure bandage can replace direct hand pressure on most parts of the body. Apply the pressure bandage by placing the center of the bandage or strip of cloth directly over the pad. Hold the pad in place by circling the bandage ends around the body part, twisting over the wound and tie it off with a knot directly over the pad.

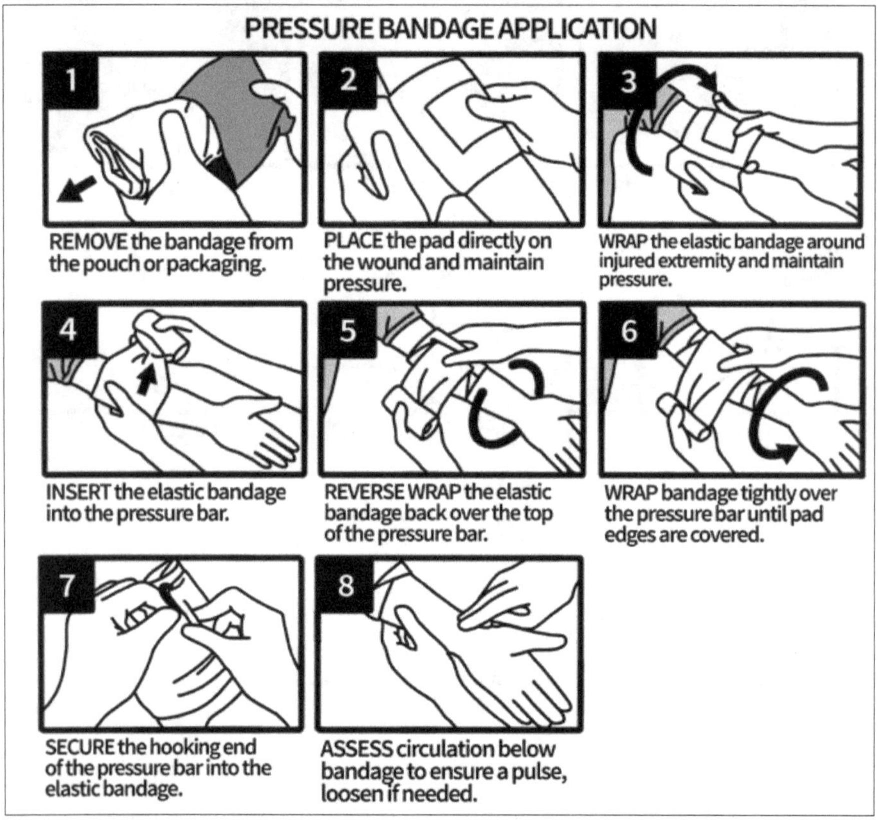

Pressure Points

If bleeding persists after applying direct pressure or if there is severe arterial bleeding, pressure can be applied to points in the body where a major artery flows over a bony prominence. By applying pressure to these areas, blood flow to the wound is further reduced.

Below are pressure points that can be used:

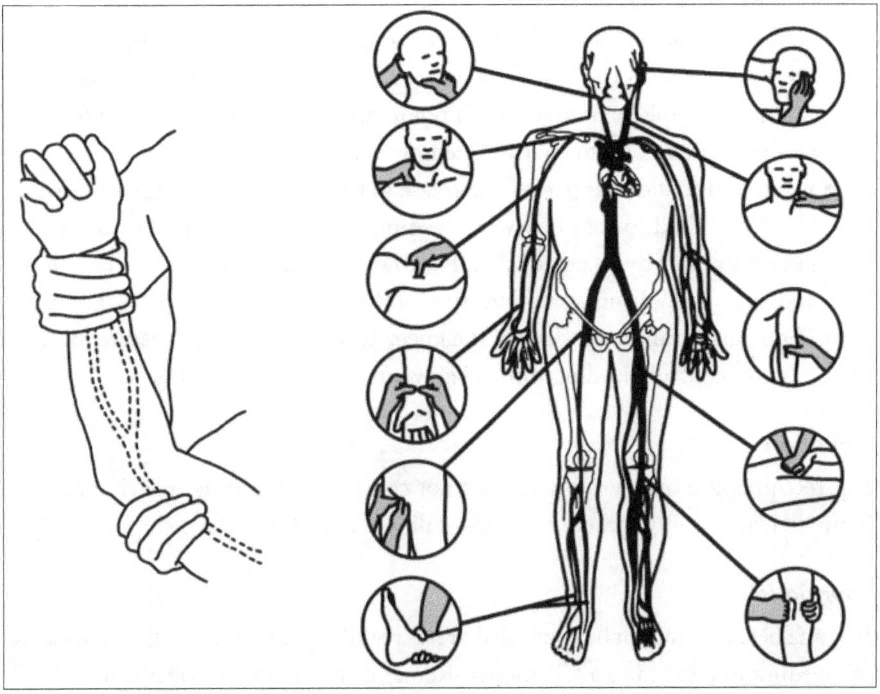

- Temporal: Scalp or head. Use this pressure point for no longer than thirty seconds as it may cut off blood to the brain.
- Facial: In the "ridge" along the lower edge of the bony structure of the jaw. Use only for a minute or two to help slow blood flow from a cut on the face.
- Carotid: Begin at the trachea at the midline of the neck. Slide your fingers to the sight of the bleeding and feel for the pulsations of the carotid artery. Place fingers over the artery and thumb behind the neck. Apply pressure by squeezing fingers toward the thumb. Never apply pressure to both sides of the neck at the same time. Apply pressure for only a few seconds as this procedure cuts off blood circulation to the brain.
- Subclavian: Deep behind the collar bone in the "sink" of the shoulder. Push thumb through the thick layer of muscle at the top of the shoulder and press the artery against the collarbone.
- Axillary: Under the upper arm. Press the artery just under the upper arm against the bone from underneath.
- Brachial: Groove on the inside of the arm and elbow; two locations, near the elbow joint. Apply pressure to the point, grasp the victim's arm with

the thumb on the outside of the arm and fingers on the inside. Press fingers toward the thumb.

- Radial and ulnar: Radial point located on the forearm close to the wrist on the thumb side of the hand; ulnar point located on the little finger side of the hand. Apply pressure to both points to control bleeding of the hand. Use the radial point to control bleeding of the wrist.
- Femoral: Front center part of the crease in the groin area, pelvic basin; two locations. Used to control severe bleeding on the lower extremity and any bleeding caused by leg amputation. Place heel of the hand directly on the point and apply a small amount of pressure to the artery across the pelvic basin.
- Popliteal dorsalis pedi: Back of the knee. Top of foot. Apply pressure to the point to control bleeding from a leg wound or the foot and toes.

Elevation

If direct pressure over a dressing does not control the bleeding, try elevating the injured area, but only if there is no possibility of a spine injury.

Tourniquet

If severe bleeding cannot be controlled or it is readily apparent the victim is in danger of bleeding to death, use a tourniquet. Remember that a tourniquet is useful only on the extremities—the arms and legs. A tourniquet is a constricting band placed around an extremity, then tightened until bleeding has stopped. Although your first aid kit carries a purpose-built tourniquet, any wide-gauge material such as a triangular bandage and a rigid object, such as a stick or screwdriver for tightening, can be used to the same effect.

Place the tourniquet high and tight on the extremity. Due to their elasticity, arteries can retract up and into a limb. Putting a tourniquet below where the artery is will not stop the bleed. When you secure the tourniquet, tighten it until you stop ALL of the bleeding. Regularly reassess for any bleeding and ensure you can't feel a pulse below the tourniquet. Attach a note to the victim giving the location of the tourniquet and the time that it was applied. Always leave the tourniquet exposed to view and never hide it with a splint or bandage. If it is not possible to attach a note, write the letter "T" on the patient's forehead with a grease pen or other suitable marker, and show the time it was applied. After making the decision, and applying a tourniquet, DO NOT LOOSEN OR REMOVE IT. Continue to treat for shock and obtain higher-level medical care.

"Seal the Box"

The trunk of the body from the collarbone to the pelvis and all the way around is referred to as "the Box." We can't tourniquet and we don't pack bandages or gauze into wounds in this area. Instead we "Seal the Box."

A chest wound is especially risky because the normal function of the victim's vital organs could be disrupted if the wound is open or penetrating—sometimes referred

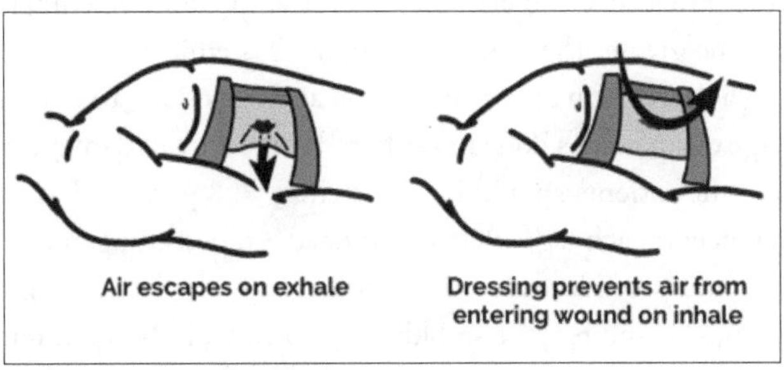

to as a sucking chest wound—and allows pressure to build inside the chest cavity as the victim breathes. A chest wound is often characterized by heavy bleeding, frothy blood, shortness of breath, or difficulty in breathing, and the chest may not rise normally when the person inhales.

Standard chest seals are "vented" to allow any pressure built up inside the chest to be released passively so that trapped air doesn't cause tension pneumothorax, a condition in which an excess of pressure on the lungs, heart, trachea, and other internal structures compromises their normal function. Before applying a chest seal, ask the victim to exhale and hold it, or wait for an unconscious victim's chest to fall, so that their chest is flat with as much air as possible out of the wound. Forcing air out before sealing the wound will allow the person to breathe more easily after the wound is sealed.

After dealing with any sucking chest wounds, look for other wounds inside the box. If an object is protruding from a wound, don't try to remove it, instead stabilize it by placing heavy dressings around it. Cover protruding intestines with a dressing. For all others, apply pressure over a dressing to control and stop bleeding. If blood soaks through the dressing, apply additional dressings on top of the original one. Regularly reassess victims for any increased difficulty breathing, chest pain, or change in heart rate that might indicate a dangerous build-up of pressure.

Burns

Boat Response Aid Kit Items: 4" × 4" Burn Dressing, Burn Relief Gel, Abdominal Emergency Trauma Dressing

Burns can range from minor irritations to life-threatening and disabling injuries. Burns can be caused by thermal heat, chemicals, sunburn, electric shock, or radiation. In general, the size of the burn is more important than the degree of the burn as the size raises the possibility of infection. Burns are classified by depth or degree of skin damage. The three general classifications of burns are first degree or superficial, second degree or partial thickness, and third degree or full thickness. Burns of the respiratory tract are a separate classification. The greater percentage of the body that is burned, the greater the risk of developing systemic complications, the most common being a decrease in blood circulation and infection. To determine roughly what percentage of a patient's body has suffered some type of damage, the following estimates for adult patients should be used: chest = 18 percent; back = 18 percent; each arm = 9 percent; each leg = 18 percent; head = 9 percent; genitals = 1 percent.

General first aid procedures for all burns begin with eliminating the source of the burn. Extinguish and remove smoldering clothing, but do not remove charred clothing that may be sticking to the burn. Remove any jewelry. For burns resulting from electrical shock, ensure the patient is no longer receiving an electrical shock. With the source of the burn eliminated, cool the burn, provide pain relief, and cover the burn to prevent infection. Cool water has proved both comforting and effective. If it is not possible to put the victim under running water, immerse the affected areas in water, or pour great quantities of water over the burns. Continue until no pain is felt when the burned part is withdrawn from the water. Be careful not to overcool. If the victim begins to shiver, stop cooling and dress the burns. Burn dressings are petrolatum gauze bandages that are applied over the burned surface and an area extending two inches beyond it. The burn dressing should be allowed to remain in place and be covered by a dry sterile dressing that can be changed as needed. If no higher level of medical care is available, the burn dressing should remain in place unless signs of infection develop, in which case the dressing should be removed and the burn treated as an infected wound. In addition to these general steps, the following procedures apply to each classification of burn.

First Degree / Superficial

This is the mildest form of burn that involves only the outer layer of skin. These produce redness, increased warmth, tenderness, and mild pain. To treat, immerse in cool water or apply burn gel until pain is relieved, then cover loosely with a burn dressing.

Second Degree / Partial Thickness

This burn extends through the outer layers of the skin and involves the inner layers, but not enough to prevent rapid regeneration. It may produce blisters and is characterized by severe pain, redness, and warmth. To treat, immerse in cool water or apply burn gel until pain is relieved, then cover loosely with a burn dressing. The skin covering a blister is better protection than any bandage and should therefore be left alone. Do not break open any blisters. If the blister breaks, press out fluid with a bit of sterile gauze and apply a bandage.

Third Degree / Full Thickness

Third-degree burns penetrate the full thickness of the skin, destroying both the outer and inner layers. Severe pain may be absent because nerve endings have been destroyed. Color may range from white and lifeless to black and charred. Any full-thickness burn requires hospitalization, and anyone who suffers a full-thickness burn or a burn that covers more than 10 percent of the body needs to be admitted to a hospital with a special burn unit for treatment. Healing requires many months, and usually results in scarring of the skin tissue. Skin grafts are generally required to achieve full healing.

To treat, cool the burn with water. Do not remove clothing unless it is smoldering. Do not place ice or apply ointments to the burn. Loosely cover the burn with a burn dressing. If affected, separate fingers or toes with dry, sterile, non-adhesive dressings. Treat for shock even if not apparent. Monitor the airway. Transport to higher-level medical care.

Chemical Burns

Chemicals in contact with the skin or other body membranes may cause chemical burns by direct chemical destruction of the body tissue. This kind of injury can be caused by acids, including nitric acid, sulfuric acid, hydrochloric acid, and by caustic alkalis. Wash the chemical away completely as quickly as possible, using large quantities of water. Continue flushing the burn for at least twenty minutes. If the chemical is a dry powder, do not apply water but remove the agent by "dusting" and removing the clothes. Do not apply ointments to the burn. Loosely cover the burn with a burn dressing. Treat for shock even if not apparent. Monitor the airway.

Respiratory Tract

Burns of the respiratory tract, or inhalation burns, are very serious and demand advanced medical care. These burns may be diagnosed by singed eyebrows, eyelashes, or nostril hairs, hoarseness, sore throat, coughing of blood, drooling, or a harsh vibrating noise when breathing called stridor. Many who escape burns to their skin during a fire suffer inhalation burns caused by smoke, which may be hot, mixed with steam, and very toxic. First aid treatment options are limited in these cases and consist of maintaining an open airway and transporting to higher-level medical care.

Eye

Boat Response Aid Kit Items: Polycarbonate Eye Shield, 2" Surgical Tape, Eye Wash

Eye injury is normally the cause of great anxiety for a victim, many times causing more concern than more serious injuries to other parts of the body. Patients who have experienced a blinding injury become totally dependent upon their rescuer and should never be left alone. Constant contact and continuous conversation with them should be maintained to reduce anxiety. Do not allow patients to rub or apply pressure to the injured eye. If the patient wears contact lenses, don't take them out as the removal could worsen the injury. The only exception is in the event of a chemical burn. The following describe the symptoms and appropriate treatments for various eye injuries.

Small Object on the Eye or Eyelid

The eye will often clear itself of tiny objects, like eyelashes and sand, through blinking and tearing. If not, examine the eye in a well-lighted area. To find the object, have the person look up and down, then from side to side. If you cannot find the object, grasp the lower eyelid and gently pull it down to look under the lower eyelid. To look under the upper lid, place a clean cotton swab on the outside of the upper lid. Grasp the eyelashes and gently fold the lid over the cotton swab. If the object is on an eyelid, try to gently flush it out with clean water. If that does not work, try touching a second cotton swab to the object to remove it. If the object is on the surface of the eye, try gently rinsing the eye with clean water. If available, use an eye dropper or a bottle of eye drops, such as artificial tears, positioned above the outer corner of the eye. Do not touch the eye itself with the dropper or bottle tip. A scratchy feeling or other minor discomfort may continue after removing eyelashes and other tiny objects. This should go away within a day or two. If discomfort or blurred vision continues, get medical help.

Object Stuck or Embedded in the Eye

Leave the object in place. Do not try to remove it. Calm and reassure the person. Cover the affected eye with the polycarbonate eye shield or a dry sterile bandage and secure with surgical tape. Covering both eyes will help prevent eye movement. If the object is larger than the shield will accommodate, place a clean paper cup or something similar over the injured eye and tape it in place. This prevents the object from being pressed on, which can injure the eye further. Get the patient to higher-level medical care as soon as possible.

Chemical in the Eye

Turn the patient's head so the injured eye is down and to the side. Holding the eyelid open, allow running water to flush the eye for fifteen minutes. If both eyes are affected or if the chemicals are also on other parts of the body, have the patient take a shower. Another way to wash chemical substances from the eye is to have the victim open and close the eyes several times while the face is immersed in a pan of fresh water. If the patient is wearing contact lenses and the lenses did not flush out from the running water, have them try to remove the contacts after the flushing.

Keep flushing the eye with clean water for at least twenty minutes. Seek higher-level medical care right away.

Eye Cut, Scratch, or Blow

Gently apply a clean, cold compress to the eye to reduce swelling and help stop bleeding. Do not apply pressure to control bleeding. If blood is pooling in the eye, cover both eyes with a clean cloth or sterile dressing. Seek higher-level medical care right away.

Eyelid Cuts

Carefully wash the eyelid. If the cut is bleeding, apply gentle pressure with a clean, dry cloth until the bleeding stops. Do not press on the eyeball. This is because the cut may go all the way through the eyelid, so there could also be a cut in the eyeball. It is usually safe to press on the bone around the eye. Cover with a clean dressing or polycarbonate eye shield. Place a cold compress on the dressing to reduce pain and swelling.

Splint

Boat Response Aid Kit Items: 3" ACE Wrap, SAM Splint, Triangular Bandage, Duct Tape

A fracture is a broken or cracked bone. There are two types of fractures, compound or open and simple or closed. In a compound fracture the bone has broken the skin and an open wound is present. The bone may protrude from the wound, leaving little

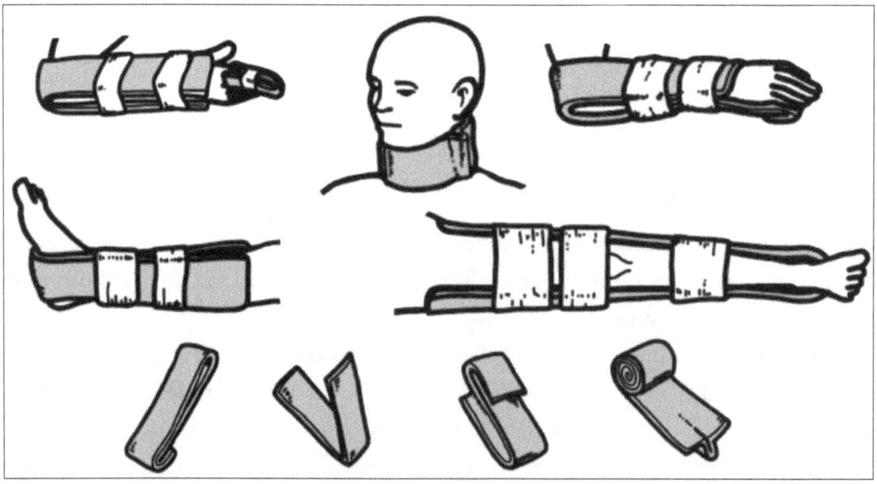

doubt that there is a fracture. In a simple fracture, no open wound is present, but the bone is broken or cracked. Indications that a fracture has occurred include pain, swelling, and discoloration at the injury site, misalignment, deformity, or disability of the injured part, and a patient's statement that a "crack" or "snap" was heard. Every suspected fracture should be treated as if it were a fracture until it is proven otherwise. Do not attempt to straighten broken limbs and limit all unnecessary handling of the injured part.

There are 206 bones in the human body. Several of these bones, if broken or injured, require very specific treatment based on the sensitive nature of their functions or their proximity to organs or arteries.

Extremities

Use a splint to immobilize the fracture. Selecting exactly the proper splint is less important than achieving immobilization. Apply splints over clothing if the victim is to be moved a short distance or if higher-level medical care will be reached soon; otherwise, apply well-padded splints after the clothing has been cut away. Splints vary according to the area and nature of the fracture, but there are general rules to follow. For a bone fracture, place the splint so that the joint above and the joint below the fracture cannot move. For a joint fracture, place the splint so that the bone above and the bone below the fracture are immobilized. Leave tips of fingers and toes exposed and check them often for circulation adequacy. Be alert for the development of shock during treatment.

- Forearm: Place two well-padded splints, top and bottom, from elbow to wrist and bandage in place. Hold the forearm across the chest with a sling.
- Upper arm: For a fracture near the shoulder, put a towel or pad in the armpit, bandage the arm to the body, and support the forearm in a sling.
- Middle upper arm: Use one splint on the outside of the arm, shoulder to elbow. Fasten the arm to the body and support the forearm in a sling.
- Elbow: Do not move the arm at all. Splint it as it is found.
- Thigh: Due to the large artery and muscle mass, a thigh injury is a major injury, and medical assistance must be sought immediately. If higher-level medical care is unavailable, use two splints, an outside one from armpit to foot and an inside one from crotch to foot. Fasten the splints around the ankle, over the knee, below the hip, around the pelvis, and below the armpit. Tie both legs together. Do not move a patient until this has been done This injury is often associated with major trauma, and bleeding may occur

if the thigh bone severs the adjacent femoral artery. The patient should be monitored closely for signs of shock, and the leg should not be manipulated. Check pulse in feet.

- Lower leg: Use three splints, one on each side and one underneath. Always pad the splints well, especially under the knee and at the ankle bones. Use a pillow under the leg with the edges brought around in front and pinned, and then add two side splints.

Nose

If an injury to the nose occurs, stop the bleeding. If conscious, have the patient sit and lean forward applying gentle pressure to the sides of the nose. Apply a cold compress or an ice pack over the nose to ease pain, reduce swelling, and assist in stopping the bleeding. Place an unconscious victim on their side to keep airway open.

Jaw

If an injury to the jaw area interferes with a victim's breathing, have the patient lean forward to assist with draining of blood/saliva rather than lying on their back. If patient is unable to sit, then they should be placed in the recovery position on their side.

Collarbone

On the injured side, place the forearm across the chest, palm turned in, thumb up, with hand four inches above the elbow. Support the arm in this position with a sling. Fasten the arm to the body with several turns of bandages around the body and over the hand to keep the arm close against the body.

Rib

A broken rib can be very painful and very dangerous because of the possibility of puncturing a lung, a symptom of which is coughing up frothy, bright red blood. Immediate medical assistance should be sought.

Spine

Any actual or suspected damage to the spine requires definitive care and careful management. Permanent disability, paralysis, or death can result from a spine injury. Severe pain in the back or neck after injury should be treated as a fractured spine. If the victim cannot move their legs, feet, or toes, the fracture is probably in the back;

if the victim cannot move his fingers, it is probably in the neck. Treat for shock. Keep the victim flat and do not move the head. If transportation is necessary, place hands on both sides of the head to maintain alignment of the spine and immobilize face up on a backboard, rigid stretcher, door, or wide frame by sliding the board underneath the body. Never try to lift such a victim with fewer than four people. Move gently, purposefully, and once.

Pelvis

Treat a victim with a pelvis injury for shock, but do not move unless absolutely necessary. If a victim must be moved, handle the same as for a victim of a fractured spine. Bandage legs together at ankles and knees; place a pillow at each hip and fasten them in place. Fasten the victim securely to the stretcher.

Electrocution

A person knocked unconscious by electrical current may need artificial respiration or CPR at once. If the victim is still in contact with any object that might carry current, secure power. Pull or push the victim away from contact with a dry rope, piece of clothing, pole, oar, or board. Don't touch the victim until they are clear of contact. Commence artificial respiration or CPR. Seek medical assistance.

Exposure

Boat Response Aid Kit Items: Emergency Survival Blanket, Bandages

Continuous exposure to environmental conditions, both cold and hot, can cause a variety of injuries. To treat any exposure injury, it is essential to get the victim into a climate-controlled space as soon as possible. It is almost impossible to give effective first aid treatment while the victim is still exposed. Remember, however, that there may be other life threats that must be treated before the victim can be moved.

Seasickness

Seasickness is the nausea and vomiting caused by the motion of a raft or boat. It can result in extreme fluid loss and exhaustion, loss of the will to survive, others becoming seasick, attraction of sharks to the raft or boat, and unclean conditions. To treat seasickness, wash both the patient and the raft or boat to remove the sight and odor of vomit. Keep the patient from eating food and have them lie down and rest until the nausea is gone. Using the horizon as a focal point has helped many to overcome seasickness. Give the patient seasickness pills, if available.

Saltwater Sores

These sores result from a break in skin that is exposed to saltwater for an extended period. The sores may form scabs and pus. Do not open or drain. Flush the sores with fresh water, if available, and allow to dry. Apply an antiseptic, if available.

Dehydration

Extremely warm weather significantly increases the loss of fluids. An average, healthy adult requires two or three liters of fluid a day to replace these losses. Fluids are lost from the body in several ways. The most obvious is through the kidneys. The less obvious loss occurs through perspiration from the skin and respiration through the lungs. One vital element of body fluids that must be maintained is electrolytes. "Electrolyte" is a scientific term for salts, specifically ions, which are important because the body uses them to maintain voltage across cell membranes and to carry electrical impulses for moving the muscles. When water and electrolytes are not replaced, the body experiences dehydration. Drinking alcohol and caffeine increase dehydration. At first there is thirst and general discomfort, followed by an inclination to slow physical movement and a loss of appetite. As more water is lost, an individual becomes sleepy and experiences a rise in body temperature. By the time the body loses 5 percent of body weight in fluids, the individual begins to feel nauseated. When 6 to 10 percent of body fluids are lost, symptoms progress to include dry mouth, dizziness or lightheadedness that does not allow a person to stand or walk normally, fainting, headache, rapid heart rate, and confusion. To treat a dehydrated patient, immediately remove the patient from further exposure to heat and seek medical attention. Mild dehydration cases will become serious if the level of activity and environmental conditions do not change.

Heat Rash

Heat rash is prevalent among those living and working in warm, humid climates or in hot enclosed spaces. Heat rash develops when sweat ducts in the skin become blocked and result in the breakdown of the body's ability to perspire. Heat rashes may be very itchy and may interfere with sleep, resulting in decreased efficiency and increased cumulative fatigue, making the individual susceptible to more serious heat disorders. Heat rash also accelerates the onset of heat stroke. To treat, the patient should be immediately removed from further exposure to excessive heat and positive action taken to prevent the onset of more serious disorders. The heat rash should be washed with a gentle soap, rinsed, and patted dry with a towel. Clothing should be worn that allows the skin to breathe. A cool, wet towel may be applied.

Heat Cramps

Heat cramps are painful contractions caused by excessive salt and by water depletion. Heat cramps may occur as an isolated occurrence with normal body temperature or during heat exhaustion. Recently stressed muscles are prone to heat cramps, particularly those muscles in the extremities and abdomen. To treat, remove the patient from the hot environment and supply cool drinking fluids. An electrolyte drink may speed up recovery. Gentle stretching and gentle massage of the affected muscle groups may help. Re-exposure to heat should be avoided for at least twelve hours. Prompt medical assistance is recommended for severe or persistent conditions.

Heat Exhaustion

Heat exhaustion typically occurs in warm, humid environments where body fluids are lost through heavy sweating. Fluid loss can result in a decrease of blood flow to vital organs. Symptoms include dizziness, headache, nausea, weakness, clumsy/unsteady gait, and muscle cramps. An individual who collapses in the heat and continues to perspire freely almost surely has heat exhaustion. The presence of sweating usually rules out heat stroke. Move the patient to a cool place and loosen clothing to make as comfortable as possible. Keep the patient quiet and lying down, with the feet and legs somewhat elevated. If the patient is conscious and able to swallow, give them water to drink. If available, an electrolyte beverage is helpful. Replacement of the salt and water that has been lost by sweating is probably the most important part of the treatment and often brings rapid recovery. If recovery is not prompt, get medical attention as soon as possible.

Heat Stroke

Heat stroke is a major medical emergency that results from the complete breakdown of the body's sweating and heat regulatory mechanisms as the body's temperature rises above 105°F (40.5°C). Heat stroke or "sun stroke" is caused by operating in bright sun or working in a hot environment, such as an engine compartment. The victim may have preliminary symptoms such as headache, nausea, dizziness, or weakness, but very often the first signs are sudden collapse and loss of consciousness. Breathing is likely to be deep and rapid, and the pulse is strong and fast. Convulsions may occur. Heatstroke may cause death or permanent disability, and, at best, recovery is likely to be slow and complicated by relapses. The longer the victim remains overheated, the greater the danger of death. To treat, move the patient to a cool place, remove the clothing, and place the patient on their back with the head and shoulders slightly raised. Sponge or spray the body with cold water, and fan the victim so the water will evaporate rapidly. Get the victim to a medical facility as soon as possible.

Frostbite

Frostbite is the development of ice crystals within body tissues following prolonged exposure to cold temperatures. Frostbite is most likely to develop in air temperatures less than 20°F (–6.6°C). Factors that contribute to frostbite include cold stressors such as wind, air temperature, and exposure to water; any restriction of blood flow; any lack of appropriate protection; and skin exposure. A frostbite victim will complain of painful cold and numbness in the affected area. Waxy white or yellow-white, hard, cold, and insensitive areas will develop. The first aid treatment for frostbite is rewarming the frozen tissue, but only if there is no danger of the part refreezing. Get the victim into the warmest available place as soon as possible. Undress the victim. If feet or legs are frostbitten, do not allow the victim to walk. Do not handle the frostbitten area unnecessarily, and do not exert any pressure against it. Keep sterile gauze or cloth pads between the toes and fingers, and keep frostbitten parts slightly elevated. When the frostbitten area is warmed up, it immediately becomes red and swollen, and large blisters develop. Severe frostbite causes gangrene, in which body tissues, and sometimes even bone, are permanently destroyed. If deep tissue has been destroyed, the injured part may have to be amputated. Obtain medical assistance as soon as possible.

Frostbite may also be caused by contact with certain chemicals that have rapid freezing action. Injuries caused by these substances are often referred to as chemical burns, but the body tissue is actually frozen rather than burned. Such injuries are treated as described above.

Immersion Foot (Trench Foot)

Trench foot, or immersion foot syndrome, is a serious condition that results from feet being wet for too long. Symptoms include swelling, discoloration, skin tissue falling off, and sensations of numbness, persistent itching, and tingling. These symptoms may only affect a portion of the feet but in severe cases can extend over the entire foot, including the toes. Over time trench foot can lead to complications, including amputation, inability to walk, gangrene, and permanent nerve damage. To treat, expose the injured part to warm, dry air and elevate the affected foot to encourage circulation that prevents new blisters and wounds. Be careful not to rupture blisters. Ibuprofen (Advil) can help alleviate pain and swelling. If symptoms fail to improve, seek medical attention.

Toxic Gases

The first senses affected by toxic gases are those that control a person's judgment and decision-making ability. Once dangerous fumes affect a person, they may not be able to help themselves. If toxic gas exposure is suspected, see if the victim is responsive by calling out to them. If able to move, direct the conscious victim to fresh air and get medical help immediately. If the victim is unconscious, do not try to assist them alone. Needless casualties occur from people trying to help someone overcome by lethal fumes, only to become victims themselves. If there is no response, immediately call for assistance and wait in a clean atmosphere until help arrives.

When responding to maritime casualties, there may be hazardous chemicals, gases, vapors, or fumes involved that might be immediately dangerous to life and health. The safest place is upwind and up-current. Publications like the *North American Emergency Response Guidebook* identify potential hazards and recommendations for safe distances and appropriate protections.

Poisoning

When poisoning occurs, medical assistance should be sought immediately. Attempt to identify the involved substance and estimate the quantity taken, and immediately contact the poison control center at (800) 222-1222 or, when outside of cellular range, seek MEDICO from a Coast Guard command center.

Bites and Stings

Victims may suffer many different types of bites and stings depending on the area of operation and the life that exists there. Although some bites and stings will cause only minor skin irritation, others may be fatal and require immediate evacuation to higher levels of medical care. Most bites and stings will result in swelling of the affected body part. For this reason, all jewelry should be removed from the affected area or limb to prevent greater damage.

Marine Life

The following describes the effects and proper treatment for various fish bites/stings encountered:

- Shark and Barracuda: Generally loss of large amounts of tissue. Prompt and vigorous action to control hemorrhage and shock are required to save a victim's life. Control bleeding with pressure dressings, if possible. If not, use tourniquets. Seek medical help immediately.

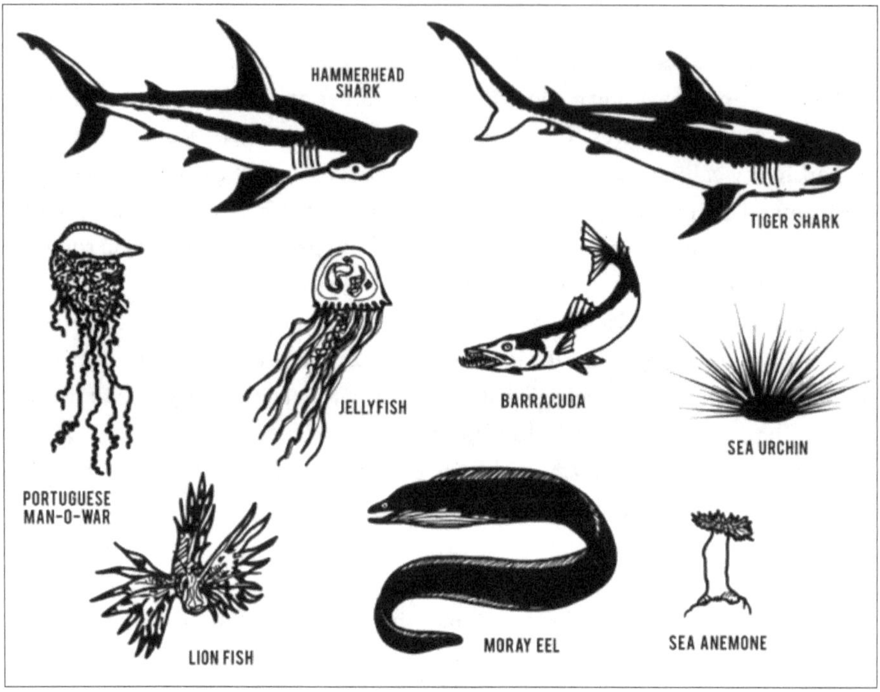

- Fish Stings: Stings can be delivered by lionfish, puffer fish, scorpion fish, and Caribbean fire coral. Symptoms include burning sensations, redness, swelling, rash, blisters, abdominal cramps, numbness, dizziness, shock, and cardiac arrest. Individuals extremely sensitive to fish stings may rapidly go into shock and require immediate evacuation to save their life.
- Portuguese Man-of-War and Jellyfish: Symptoms include a burning sensation, redness, and jelly-like matter from tentacles stuck on the body. Remove any remaining tentacles to prevent further damage. Rinse the area with clean fresh water or salt water, wash the skin with rubbing alcohol, and, if available, apply ammonia, vinegar, or a soothing lotion. Apply an ice pack to reduce pain. If the sting is serious, treat for shock and seek medical attention.
- Stingray Injuries: Typically a small open wound with swelling. Immediately irrigate the wound with cold salt water. Most of the toxins will wash out and the cold water will reduce the pain. Immerse the wounded area in hot water for thirty to sixty minutes. Keep the water as hot as a patient can tolerate without injury. Apply hot compresses to wounds in areas not lending themselves to complete immersion. Apply a sterile dressing after the soaking.

- Sea Snake / Coral Reef Snake Bite: Symptoms include headache, blurred vision, sensation of thick tongue, difficulty speaking or swallowing, thirst, sweating, vomiting, aching, stiffness, tenderness, and paralysis. Seek medical help and antivenom immediately. Wrap the limb with elastic bandage and immobilize with a splint.

Insect Bites and Stings

Wash insect bites with soap and water. Itching can be relieved by using cold-water compresses. Ice, applied to bee stings, will prevent pain and swelling. Some people experience severe allergic reactions to bee and other insect stings. The person stung should be carefully observed and given immediate medical attention if allergic reactions are either observed or are part of that person's medical history. Ticks can be dislodged by applying mineral oil to the tick; be careful not to crush the insect or leave its jaws embedded in the flesh. Certain types of ticks carry Rocky Mountain spotted fever or Lyme disease, which can be fatal. Inflamed bites require medical assistance.

Snakebites

Poisonous snakes in the United States include "pit vipers" such as rattlesnakes, copperheads, and water moccasins. Identifying the snake, in some cases, is not easy. Pit viper bites may produce tissue swelling at the bite that gradually spreads to surrounding areas. Swelling occurs between three to five minutes following the bite but may continue for as long as an hour and be so severe as to burst the skin. Symptoms include severe pain, severe headache, thirst, and bleeding from internal organs into the intestines that causes blood in urine or stools. Bites of cobras or kraits may produce an irregular heartbeat, generalized weakness and exhaustion, shock, severe headache, dizziness, incoherent speech, stupor, mental confusion, difficult or labored breathing, numbness, nausea, and possible unconsciousness. Extreme pain is not characteristic. Seek medical attention and an antivenom immediately. A snakebite victim should lie down and keep calm. Remind the patient that very few people are ever disabled by snakebite. Keep the victim at rest and immobilize the affected part below the level of the heart. For a bite on the arm or leg, place a restricting band two to four inches above the bite. The restricting band should be tight enough to stop the flow of blood in the veins but not the arteries. Treat for shock. Administer CPR as required.

Spiders and Scorpions

Brown recluse and black widow spiders are probably the most common of the poisonous spiders. The black widow spider bite causes pain almost at once, and the pain spreads quickly from the region of the bite to the muscles of the back, shoulders, chest, abdomen, and limbs. The pain is usually accompanied by severe spasms of the abdominal muscles. A bite from any other poisonous spider may be felt as a sharp sting, but it is not usually accompanied by severe pain. About half an hour after the bite, the victim begins to feel painful muscle cramps near the bite, which soon spread to other muscles of the body. Shock usually develops at about the same time, although it may be delayed for as much as an hour or two. The victim becomes restless and anxious and may be very thirsty. Nausea and vomiting may occur. Scorpions have a stinger and venom sac in the tail. Their sting causes immediate, intense pain. The area near the sting will become numb and paralyzed. First aid measures for spider bites and scorpion stings are limited. Clean the wound and surrounding area. Treat the victim for shock and keep them quiet and warm. Severe muscle cramps and pain may sometimes be relieved by warm-water baths. Do not apply suction and do not incise or cut the wound in an attempt to remove the poison as it does no good. Seek medical care.

Anaphylactic Shock

Anaphylactic shock is a rapid, extreme allergic reaction. Sensitivity reactions can occur within seconds of contact and can result in death within minutes. Anaphylactic shock can be caused by eating fish or shellfish, ingesting particular types of berries, or taking oral drugs such as penicillin. Insect stings from yellow jackets, hornets, wasps, and so on; injected drugs; exercise; cold; and inhaled substances such as pollen or dust may also cause sensitivity reactions. Anaphylactic shock can be identified by visual changes to the subject's normal appearance and by changes in vital signs. Symptoms include itching; hives or a raised rash; flushed skin; swelling of lips, tongue, feet, throat, or hands; wheezing, shortness of breath, and coughing; nausea and vomiting; abdominal cramps; diarrhea; headache; altered mental status; or loss of consciousness.

Anaphylactic shock requires medication to counteract the allergic reaction to the substance. If the victim carries an epinephrine kit, assist them in administration. The victim should be treated for shock and, if necessary, administered CPR.

Scuba Incidents

Boat Response Aid Kit Items: CPR Mask

There are two special problems usually seen in scuba diving accidents, arterial gas embolism and decompression sickness, sometimes called "the bends." Arterial gas embolism, or air bubbles in a diver's blood, is most often found in divers who hold their breath during ascent. This typically happens following an equipment failure or some other underwater emergency. Divers can develop arterial gas embolism in very shallow waters. The onset of symptoms is often rapid, and a victim's senses may become distorted. Victims may have convulsions and can quickly lose consciousness. Decompression sickness may occur as the result of coming up too quickly from a deep, prolonged dive. Rapid ascent defeats the body's ability to filter escaping gases through the lungs, resulting in nitrogen gas bubbles in the bloodstream. The onset of decompression sickness is usually slow for scuba divers, taking from one to forty-eight hours to appear. The symptoms and signs of decompression sickness include deep pain to the muscles and joints, choking, coughing, labored breathing, chest pains, and blotches on the skin. If a diver experiences either mild or severe symptoms on surfacing, immediately begin transport to the nearest recompression facility, place the diver on his or her left side with head down, and treat for shock.

To facilitate treatment at the recompression chamber, attempt to obtain information on the depth of the victim's diving activities, the number of dives performed, the victim's overall medical condition and history, the first occurrence of the victim's symptoms, and any problems that may have occurred during the dive.

24

SURVIVAL

Survival is the "preservation of one's own life under conditions of immediate peril." To preserve one's own life at sea requires the ability to live through extreme conditions of emotional and physical shock and hardship for an indefinite period of time. When faced with an open-water survival situation, it is important to remember that environmental obstacles are as much mental as physical. Before accumulating information on the use and operation of survival systems, it is important to first understand the psychological barriers to the will to survive that must be overcome. The most predominant psychological barrier to survival is fear: fear of the unknown, fear of discomfort, and fear of one's own weakness. Fear of the environment in an open-water situation leads us to fear our own chances of survival, and even though we overcome these fears to some extent, a lack of confidence in our ability may weaken our will to survive. Studies of survivors and their experiences show that the successful survival of any situation depends on several factors. The survivor must be mentally prepared for the possibility, be in good physical condition, have the proper equipment available and know how to use it, be properly dressed, and be thoroughly familiar with vessel egress procedures. The key to these experiences is developing a survivor's "attitude"—in other words, to develop those traits and characteristics that will enhance one's chance of survival.

In the next few pages you will learn techniques that will help you to extend your chances of survival whether you find yourself in peril at sea or on land. The danger of falling overboard, capsizing, or sinking is always present while under way, as is the possibility of your aircraft ditching into the sea. Few people can stay alive for a long period of time in the water without some type of survival equipment. Fear, fatigue, and exposure are the enemies of water survival. The desire to live, the ability to think clearly, and the proficient use of available equipment can make the difference between life and death. All Coast Guardsmen have a responsibility to be knowledgeable of survival tactics and in the use of survival equipment not only for their own survival but for that of their crews.

SURVIVAL FACTORS

Although statistically rescues occur within a short time following a mishap, you must be prepared for a prolonged survival situation. The most important factor of any emergency is to recognize that you have a problem and that you are in danger. Once you recognize that danger is present, you can prevent the situation from getting out of hand. Don't let your pride or ego get in the way of clear judgment. When you have determined that danger is present, communicate it immediately, announcing the following elements:

- MAYDAY, MAYDAY, MAYDAY.
- Name and description of aircraft or vessel (three times).
- Location or position (avoid nicknames of places).
- Number of persons on board.
- Nature of distress and intentions.

Keep in mind that recognition is an ongoing process. You are not out of danger until you are home and telling your friends ". . . there I was . . . ."

The most inhospitable location on our planet is the open water. You must be prepared for hot, temperate, cool, or cold conditions and be aware of the dangers of too much exposure to the sun. When you become aware that a situation is developing into one where you may have to enter the water, there are several preparations you can make to increase your chances of survival:

- Don a personal flotation device (PFD).
- Don a boat or aircrew survival vest.
- Put on as much warm clothing as possible, making sure to cover head, neck, hands, and feet.
- Before entering the water, button up clothing, turn on signal light if dark.
- Avoid entering the water if possible. If it is necessary to jump into the water, use the cross and cover method discussed in chapter 22, on swimming, in this manual.

Activating your personal locator beacon, one of which is carried in all Coast Guard survival vests, should be one of your first tasks upon entering the water. Although these devices are capable of broadcasting on 121.5 MHz for immediate signaling of nearby boats and aircraft with installed directing finders, not all carry such equipment. Activating the 406 MHz signal, which takes time to reach satellites and be relayed from monitoring command centers to rescue assets, should be a top priority to ensure continuous transmission of your real-time position to all with access to the SARSAT system.

As a survivor on the open sea, you will face waves and wind. You may also face extreme heat or cold. To keep these environmental hazards from becoming serious problems, take precautionary measures as soon as possible. Use the available resources to protect yourself from the elements and from heat or extreme cold and humidity. In open water, rescue efforts may be hampered or even postponed because of inclement weather. As time progresses, drift associated with wind and sea currents makes pinpointing the location of survivors more difficult. If land is in sight, you must make a decision whether to swim or row for land or to wait. Swimming or rowing away from a "last known position" has inherent risk, and survivors are often safer staying with the boat or other visible debris. Shorelines are dotted with various hazards including sharp rocks, strong current, and breaking waves. Unless land is close and conditions appear favorable, you should not try to swim to shore.

In hours of darkness, it becomes critical that the appropriate signaling devices are employed. Flares and lights are crucial to get the attention of rescuers. Be selective about what signaling devices you use based on the conditions and about when you use them based on the proximity and orientation of rescue assets to your position. Pyrotechnic signals especially are finite resources; if they are not seen when expended, they are simply wasted.

SURVIVAL EQUIPMENT

The purpose of survival equipment is to sustain life and to promote rescue. Recent technological advances have shortened the duration survivors spend in open water and attributed to increased successes in survivor recovery. However, the possibility of being in an extended open-water survival situation still exists. Being mentally prepared and using Coast Guard approved and correctly maintained equipment will increase your chances for survival. It is important to learn how to operate the equipment discussed in this chapter before it is actually needed in a survival situation. The equipment that has been provided for you is the finest in the world and is continually tested and improved so that the Coast Guard's own will lack nothing they could reasonably expect to have if they find themselves in a survival scenario. It is your responsibility to care for your issued gear. This care should involve cleaning and frequently inspecting the equipment for wear and tear or damage.

Personal Flotation Device

The term "personal flotation device" is a general name for the various types of devices designed to keep a person afloat in water. PFDs include life preservers, vests, cushions, rings, and other throwable items. They are available in five different styles, each with its own purpose and flotation characteristics.

A usable PFD is approved, in good serviceable condition, and of appropriate size for the intended user. Each crew member must wear a usable PFD appropriate for the weather conditions and operations in which they will be performing.

Hypothermia Protective Clothing

The Coast Guard outfits its boat crews with hypothermia protective clothing that is designed for prolonged exposure to the elements, making them more effective. The clothing must be worn properly and maintained in accordance with its maintenance schedule in order for it to maintain its effectiveness.

Hypothermia protection: wicking, insulation, and moisture barrier

First Layer—Wicking

Staying dry is an essential factor to maintaining body temperature. Clothing worn next to the skin must carry or "wick" moisture away from the body. Cotton clothing can pose particular problems because it absorbs and retains moisture, which robs the body of heat through evaporation. Wool has good insulating properties even when wet, but it is less than ideal because it stays wet. Modern synthetic wicking fibers such as polypropylene do not retain moisture. They will actually draw moisture from the skin and transport it to an absorbent outer layer. This clothing works well by itself, or it can be combined with a second layer for extreme cold.

Second Layer—Insulation

The insulating effect of a fabric is related to how much air it can trap. This is why a loose-knit or fuzzy material is better than material that is tightly knit. It is also why two thin layers of a given material are better than one thick one. The second layer traps air, which retains body heat, while absorbing excess moisture from the first layer. Wool or cotton thermals are an acceptable second layer if worn over a wicking layer, but a number of synthetic fleece or pile garments do a much better job. An example of this is the fleece coverall.

Third Layer—Moisture Barrier

The outer layer should stop wind and water, so the inner layers can work as designed. Choices for the third layer include the following:

- Rain Gear: Worn in inclement weather and when members are exposed to sea spray but don't need additional thermal protection beyond the uniform.
- Anti-Exposure Coverall: Worn when members are exposed to intermittent sea spray or rain and thermal protection is required. If properly worn with the straps appropriately adjusted, the water will not flush out of the suit and will delay the onset of hypothermia.
- Dry Suit: When worn with a PFD and proper undergarments, a dry suit offers mobility and superior protection against the effects of wind, spray, and cold-water immersion.
- Immersion Suit: An immersion suit, or survival suit, is a type of waterproof dry suit intended to protect the wearer from hypothermia if immersed in cold water or otherwise exposed after abandoning a vessel, especially in the open ocean. Immersion suits have integral footwear, a hood, and built-in gloves. Suits may also include an inflatable pillow that is attached high on

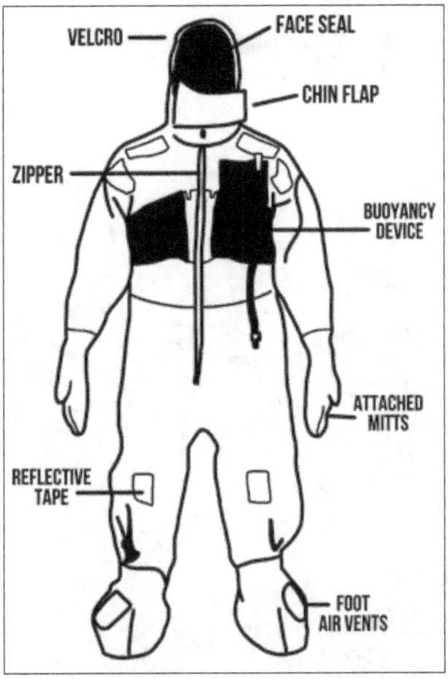

Immersion suit

the back and that, when inflated, allows the head to remain above water and keeps wind and seas from striking the face. Suits are also provided with straps and clips to allow multiple survivors to connect to one another in the water, to keep survivors together.

Survival Vest

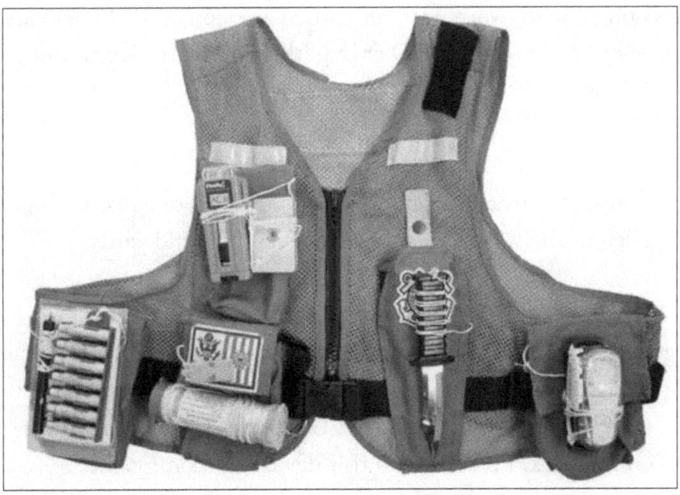

Boat crew survival vest

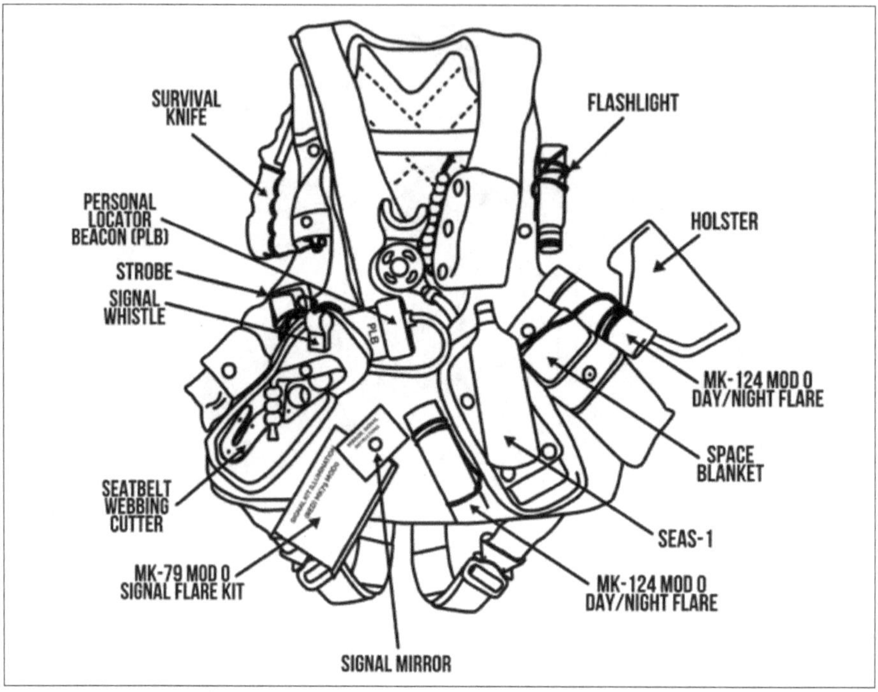

Aircrew survival vest

The equipment carried in boat crew and aircrew survival vests or pouches provides a means to signal your position on the surface of the water, day or night. These vests are combined with, worn over, or worn as a pouch on PFDs and are designed not to interfere with wearing a PFD or hypothermia protective clothing. Contents include the following:

Whistle

The whistle is used to attract the attention of personnel of a rescue ship. In still air and perfect silence, the whistle is audible for one thousand yards.

Signal Mirror

The emergency signal mirror is approximately two by three inches and is made of hardened plastic and an aluminized reflective coating. The mirror reflects light and can be seen up to forty nautical miles away. Directions on operating the signal mirror are printed on the back. Practice with this device is recommended.

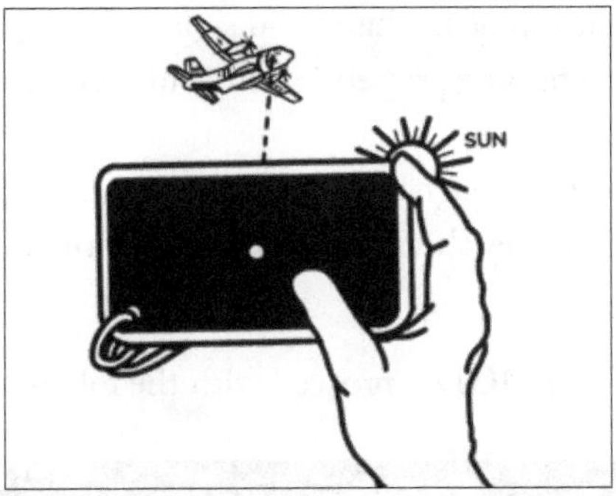

Emergency signal mirror

Strobe Light

The strobe light is a high-intensity visual distress signal with a visibility of three nautical miles and a battery life that allows eight hours of continuous or sixteen hours of intermittent use. During a survival situation, keep the strobe light in a position high above your head or attached to your helmet. Turn the strobe light on only when you are certain there is a rescue party or aircraft in the vicinity.

MK-79 MOD 0 Signal Kit

The MK-79 Signal Kit consists of a Mk-31 launcher and seven Mk-80 flare cartridges that, when fired, produce a 4.5-second, 12,000-candlepower red pyrotechnic star at heights of 250 to 650 feet. Safety precautions specific to this kit include the following:

- Never aim loaded projector at personnel or aircraft.
- Do not load the launcher without first placing the firing trigger into the angulated safety slot.

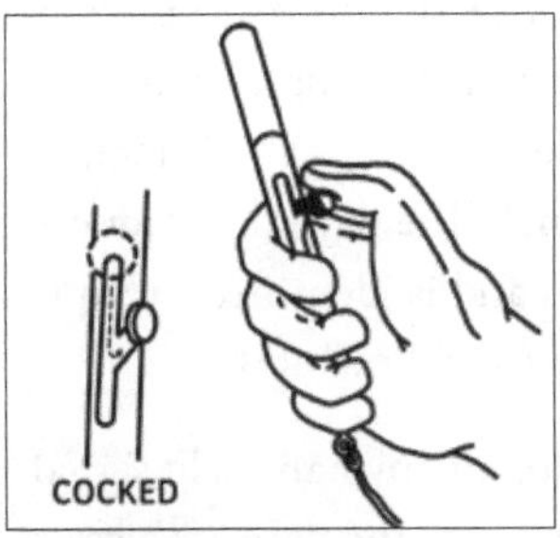

Cocking the MK-79

- When a cartridge misfires, make two more attempts, waiting thirty seconds between attempts to eliminate the possibility of a hang fire. Discard misfired signals in the water or bury them if on land.

- Do not look in the top of any handheld signal.
- Handle signals containing primers carefully to prevent a blow to the primer and possible ignition.
- The plastic tabs over the signals protect the percussion primers from being struck accidentally. They should be kept intact until just before loading the signal in the projector.

To operate the MK-79 MOD 0 proceed with the following:

Step	Action
1	Remove the bandolier and projector from the plastic envelope.
2	Turn the thumb trigger screw clockwise to ensure that it is secure.
3	Cock the firing pin of the projector by moving the thumb trigger screw to the bottom of the vertical slot and slipping it to the right so it catches at the top of the angular slot.
4	Bend the protective plastic tab away from the signal in the bandolier to allow attachment of the projector.
5	Place projector over threaded portion of signal and rotate clockwise until the signal is tightly seated.
6	Hold the projector overhead with arm fully extended. The projector should be pointed away from the body, other personnel, and approaching aircraft.
7	While firmly gripping the projector, fire the signal by slipping the thumb trigger screw to the left of the (angular) safety slot and into the firing slot and then release thumb trigger screw. This will be one continuous motion.

MK-124 MOD 0

The MK-124 Mod 0 signal flare is a dual-purpose, day-or-night distress signal also used to indicate wind direction to helicopters performing pickup of downed aircrew members. The night end of the flare produces a 20-second red flame of 3,600 candlepower visible for 8 nautical miles and is identified by an inscription and two ridges that encircle the signal. The day end produces 20 seconds of orange smoke visible for 0.5 nautical miles and is identified by an inscription. Safety precautions specific to this flare include the following:

- Do not look into the top of any handheld signal.
- NEVER ignite both ends of the flare at the same time.
- Before activating, hold at a 45-degree angle, pointing device downwind.
- Whenever possible, wear gloves to protect your hands.
- Both the day and night end will drip burning phosphorus, which will damage rafts and life preservers and injure personnel.

After deciding which end to use, operate the signal as follows:

Step	Action
1	Remove black rubber cap.
2	Slide plastic lever in the direction of the arrow until fully extended.
3	Hold the signal downwind, at a 45-degree angle from the horizon and over the side of the raft or away from debris to prevent burning.
4	Using your thumb, pull down on the extended tab to ignite signal.*
5	After using one end of the signal, douse in water and save in case the other end is needed.

* If the flare end does not immediately activate, shake the canister vigorously. If you are using the smoke end and it only emits flames, briefly immerse in water. If this occurs on land, briefly press the flare into the ground; this should snuff out the flame and enable it to emit smoke.

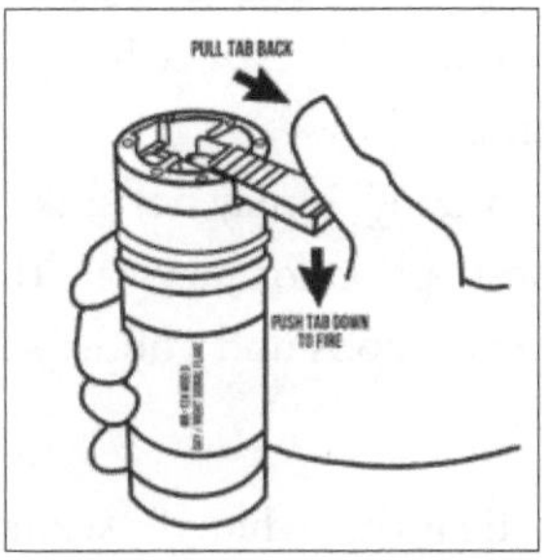

Firing MK-124

Survival Knife

The survival knife is a basic tool used to free a crew member from entangling lines. It is also used to cut material blocking a path in escaping a capsized or sinking boat. It should be a fixed-blade design with a blunt tip made of corrosion-resistant material. The blade should be checked periodically for sharpness.

Personal Locator Beacon

The personal locator beacon (PLB) is a smaller version of the common ship-mounted 406 MHz emergency position-indicating radio beacon. This personal transmitter is capable of broadcasting a distress signal that can be received and tracked worldwide, guiding emergency response resources to the transmitting position for rescue. This device is the primary distress signal and should be activated immediately to signal for help. Once activated, do not turn it off. Once search vessels or aircraft have reached the transmitting position area, use other signaling equipment (radio, signal mirror, flares, etc.) to vector them to the position.

The PLB is a personal transmitter capable of broadcasting on both 406 MHz and 121.5 MHz. The international satellite-based search-and-rescue system (COSPAS SARSAT) monitors 406 MHz and is able to provide a position accurate to within three nautical miles within ninety minutes. Once the rescue platform is in the vicinity, the 121.5 MHz transmitter provides a signal, allowing the resource to home in on the vessel or individual in distress. Since PLBs may vary in style and operation, as new models are produced, each crew member should read and understand the PLB owner's handbook.

EGRESS FROM A CAPSIZED BOAT

The key to surviving capsizing is practicing what you will do when you are suddenly under water, upside down in the dark, and only have one breath in you. Proper preparation and your initial response will largely determine the outcome. The crew must recognize the conditions that make capsize likely and control the preconditions that will make a capsizing lethal.

If the hull is intact after capsizing, it may not sink for some time, even in rough seas, but quick egress is the best option for survival. The crew will be able to escape if panic is avoided. Some preparations and good-to-know information ahead of time include the following:

- Perform frequent visualization "what-if" scenarios (nighttime, seating locations, etc.).
- Learn the boat's interior. Initially the crew will be disoriented due to being upside down, water in their noses, and without lighting, so knowing the interior will help orient the crew under duress.
- Stow all loose gear and have all equipment and doors operating properly for ease in escaping.
- Know the location and use of all survival equipment. Check it regularly to be sure that it is adequate, in good repair, and that all signaling devices work.
- Be ready to grab a sturdy support to prevent being thrown about.
- Cold water decreases the length of time most people can hold their breath under water. Immersion in cold water may give a sensation of tightness in the chest and also causes a gasp reflex, which may impact their ability to hold their breath.
- Remember, if you are wearing an inherently buoyant PFD or an inflatable PFD that has been activated, the PFD's buoyancy will try to bring you to the surface of the water but the cabin will be in the way. In theory, if you

remain upright, you will be able to walk out of the boat on the overhead, which is now acting as the deck. In most cases it will be difficult to remain upright; therefore, your PFD may "pin" you to the overhead (deck). Consider removing the PFD, if necessary.

Escape Procedures

If the boat you are on capsizes and you are not entrapped, make every effort to egress as quickly as possible. Your escape route is a commonsense decision. Do not think of egress routes in terms of primary, secondary, or tertiary egress routes where there are preferred exits. Instead consider where you are and your initial response given your position in the boat.

Procedures for egress from a cabin boat are below:

Step	Action
1	Remain calm.
2	Brace for impact/roll.
3	Wait until all violent motion has stopped.
4	Maintain reference point with one hand on armrest, handrail, etc.
5	Open window/door.
6	Release restraint (if applicable).
7	Continue to maintain reference points while exiting through door/window.
8	Exit through door/window by pulling your body using a hand-over-hand technique—DO NOT kick, as this can disorient or injure a crew member following you out the same exit.
9	If necessary, remove excess gear (PFD, LE belt, etc.).

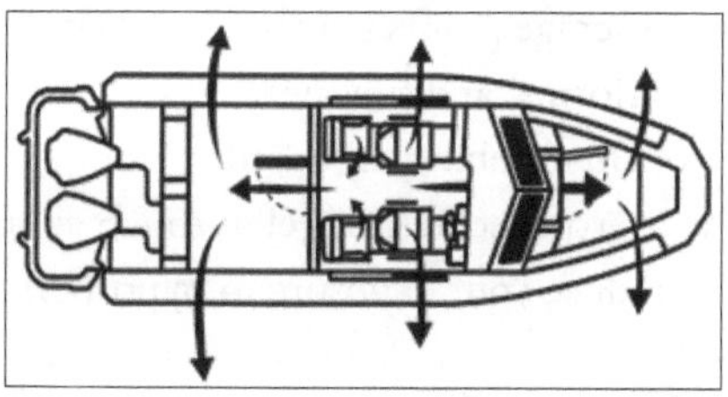

Postcapsize Procedures

Survivors from a capsized boat should attempt to stay with the boat or other visible floating debris. This not only offers a larger target for search assets but also affords more protection against hypothermia as water robs your body of heat faster than air.

Step	Action
1	Climb onto hull of vessel (if possible).
2	Take personnel muster.
3	Activate PLB.
4	Visually scan for possible rescue vessels. If in range, personnel act as a team to signal using pyro, signal mirror, and so on.
5	Inventory survival gear.
6	Stay with the boat until rescued or boat sinks.

If unable to board the overturned hull, survivors should huddle and tether themselves together and stay with the overturned vessel. Survivors should make every effort to stay with the overturned vessels without physically tying themselves to the vessel.

Climbing onto an Overturned Hull

Climbing onboard an overturned vessel will provide greater protection from the elements and hypothermia. Once on board, your chances of being spotted by rescuers greatly increases because it is a larger object that can be seen from greater distances. The majority of capsized vessels will inherently float and not sink for some time. The hull will float bow-up with the stern lying just beneath the surface of the water due to the vessel's hull design and weight of the engines. Although there is no standard procedure for climbing onto an overturned vessel, general procedures are listed below:

- Look for the easiest access point to climb on board; this is often found toward the stern of the vessel since it will be lower in the water than the bow.
- Use the natural design of the boat's hull for hand-holds, that is, the keel, chine, and through hull drains.
- Be aware that some leverage points may be slightly below the surface, such as rub rails, swim platforms, and lifelines.
- Assist other survivors in climbing on board.
- Once on board the overturned vessel, get as much as your body out of the water as possible to reduce your exposure to hypothermia.

Remaining Inside a Capsized Boat

Make every effort to egress from a capsized boat quickly; it is your best option to survive, but if for some reason someone cannot exit the capsized boat:

- Remain calm and stay within an air pocket.
- Unless certain of rescue efforts on the surface, activate PLB. There is no guarantee that other crew were able to escape and initiate rescue.

- When hearing rescuers, attempt to communicate to them by shouting or tapping on the hull.
- Conserve oxygen by remaining calm and minimizing physical activity. If possible, get out of the water to reduce risk of hypothermia.
- Remember that rescuers should arrive soon.

AIRCRAFT DITCHING

Ditching an aircraft into the sea is a maneuver that cannot be practiced and is extremely hazardous. It is only contemplated when no other realistic option remains in an emergency. Even if the sea state is slight, there is a strong possibility that an aircraft will fail to remain upright and will sink rapidly. A large, fixed-wing aircraft will likely suffer a loss of structural integrity on impact with the water surface. For helicopters, sudden-onset emergencies will often be accompanied by considerable structural conditions outside of the norm, such as tail rotor failures or very high vibrations. All of these contribute to the already considerable pressure of getting the aircraft down and evacuating it. The following is a list of generally accepted considerations and techniques for ditching that should be applicable to virtually any aircraft type.

Facilitating Rescue

Position. Transmit location coordinates repeatedly on the final descent. If fitted and possible, consider activating the aircraft's emergency locator transmitter (ELT) or a PLB during the descent to ensure they are transmitting prior to impact.

Shipping. Survival post-ditching will depend on how quickly you are picked up. Locate shipping, visually or using the radar, and try to ditch close to a ship. If there is a ship in the vicinity, touching down beside and slightly ahead is your best option as large vessels can take a considerable distance to slow down and are not maneuverable enough to take immediate avoiding action should you touch down right in front.

Aircraft. Other aircraft may be able to follow the ditching aircraft in the descent and maintain visual contact with the ditched aircraft and radio contact with controlling authorities and other aircraft until the limit of endurance is reached.

Reduce Weight. If there is a choice, a lighter aircraft allows a lower approach speed and will probably remain afloat higher in the water and for longer, thus facilitating occupant evacuation. Burning off or dumping fuel also has the advantage of increasing buoyancy in some aircraft types by creating a larger air mass held within the fuel tanks.

Propellers. Where possible, consider moving passengers away from seat rows in line with the propellers. The propellers may break up on touchdown and debris could penetrate the fuselage.

Cabin Preparation. In the time available, all loose items should be secured, particularly in the flight deck. At impact, loose items may fly around causing injuries, and once water enters the fuselage, loose items may float and impede occupant evacuation. Any available clothing items should be worn to improve chances of survival in the water. A briefing should be given to passengers, especially with respect to exits, to the extent that available time permits. Life jackets should be donned by all but not inflated until outside the aircraft.

Fixed-Wing Egress. Once the aircraft has come to a complete stop, initiate the evacuation. The expected attitude of the aircraft in the water is almost always "nose up," so the rear doors will not be usable for evacuation. Depending on the aircraft type, egress routes include the crew entrance door, paratroop doors, emergency exit door, and overhead emergency escape hatches. Ensure all available survival gear is taken from the aircraft.

Helicopter Egress. Once on the surface, the engines should be shut down if running and the main rotor stopped. In most cases, the rotor brake should not be used as the transfer of energy through the aircraft could be enough to capsize it. Once the rotor has stopped, the occupants should be instructed to begin evacuation. The primary objective will be to deploy and board the life raft without undue delay.

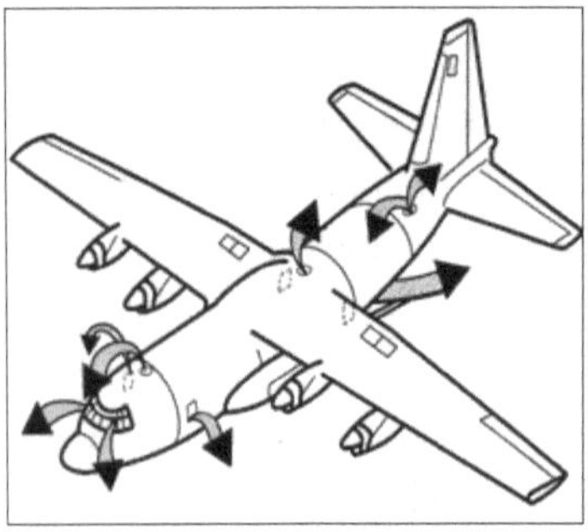

Fixed-wing aircraft egress

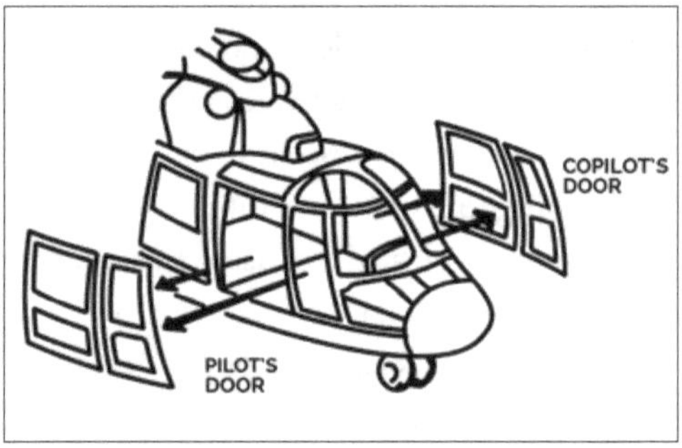

HH-65 egress

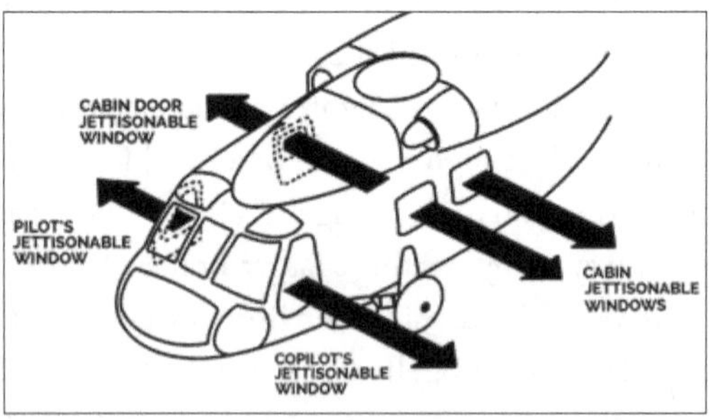

HH-60 egress

LIFE RAFTS

The discussion here applies to almost any commercially available raft. When properly stowed, a life raft is designed to automatically float free from its storage rack and inflate in the event of capsizing or sinking. As the raft container is released and drifts away, the inflation cable, attached to the raft-end of the fifty-foot painter line, is pulled tight. When this occurs, a CO_2 cylinder will automatically discharge and inflate the life raft. The painter line will remain attached to the rack by a weak link, which requires five hundred pounds of force to separate. Separation will also occur by heaving around on the painter line or by the stress exerted on it from the raft's buoyancy if the boat/cutter sinks to a depth greater than fifty feet.

Manual Inflation and Deployment of Cutter Life Raft

Depending on the situation, if time permits before sinking, manually deploy the life raft. Use the following procedure to deploy the raft for use by the crew:

Step	Action
1	Cut or untie the painter line from the weak link.
2	Trigger the hydrostatic release and remove the raft container from the stowage rack.
3	Secure the painter line to a cleat or bitt.
4	Drop the raft into the water on the leeward side of the cutter.
5	Pull the remaining painter line (approximately fifty feet) from the raft container to actuate the inflation assembly. As the raft inflates, tend the painter line to keep the raft close. Fend off the inflating raft to prevent damage to the raft from the cutter.
6	Time permitting; place extra equipment and supplies aboard the raft such as signals, portable radios, immersion suits, water, and food.
7	If practical, pull the raft alongside the embarkation net / Jacobs ladder and board the raft directly.
8	Set a watch on the cutter and painter line. If the cutter begins to sink, cut the painter line to free the raft to drift.

Life Raft Boarding

Once the life raft is in the water you will need to get on board. This is done by climbing on the boarding ladder or ramp, depending on the type of raft, and pulling yourself up using the straps attached to the raft for this purpose. Due to the high risk of accidentally puncturing inflatable rafts, extreme care must be taken when boarding or assisting personnel out of the water and into the raft. This is especially important if such persons are wearing survival vests. Once in the life raft, all persons should seat themselves on the floor and remain in that position if possible. All sharp objects should be collected and stored to prevent puncturing the raft fabric. After you have boarded your raft and adjusted the tube pressures, a new way of life begins. Regardless of how close to land or the shipping lanes your aircraft may have ditched, you must assume that this raft will be your home for weeks.

Boarding a Life Raft from the Water

With the life raft deployed and tied off alongside, the crew should try to remain on board for as long as possible. If the platform starts to sink rapidly, try to board the life raft directly, cutting the painter line once on board. If unable to board the raft directly, you will need to enter the water. Once in the water, get on board the life raft.

This is done by climbing on the boarding ladder or ramp, depending on the type of raft, and pulling yourself up using the straps attached to the raft for this purpose. Use the following procedure to board the life raft from the water:

Step	Action
1	Facing the entrance, use the buoyancy of worn PPE to spring up.
2	Legs together, kick.
3	Grab top tube, then straps inside raft to pull in.
4	Get help from people inside the raft.

Tasks Onboard a Raft

Upon boarding a raft, complete the following procedures as soon as possible:

Step	Action
1	Account for everyone and search for survivors.
2	If more than one raft is deployed, tie them together.
3	Check the physical condition of all people on board. Give first aid as necessary. Weather permitting, wash any oil or gasoline from clothing and body. These substances will not only burn skin but also pose a fire hazard. Additionally, they may be transferred from skin to the raft, deteriorating the rubber surfaces.
4	Salvage any floating equipment that may be useful. Inventory, stow, and secure all survival items. Ensure all items are securely attached to the raft to prevent them from being swept overboard.
5	If no longer attached to the vessel, deploy the sea anchor to reduce rate of drift and improve stability in heavy seas.
6	Check the raft for proper inflation and points of possible chafing where equipment may wear a hole in the buoyancy tubes.
7	Bail out any water that may have entered the raft.
8	Inflate the floor immediately.
9	In cold water, put on hypothermia protective clothing, if available. Rig the entrance cover; close when necessary. Add available material for additional layers of insulation. Avoid sitting in cramped positions for long periods. Move arms, feet, and legs as much as possible. Wiggle the toes inside boots. Keep feet as warm and dry as possible.
10	If other people are with you, huddle together for warmth.
11	Fresh water is supplied in 4.5-ounce packets and can be produced using the reverse osmosis pump supplied with the raft. You may begin drinking water as soon as you feel thirsty. However, rationing of food will be necessary. Most multiperson life rafts come equipped with food and water. The amount of food provided is enough to last for a twenty-four-hour period.

Life Raft Inventory

The equipment listed below is standard on board most life raft models. Every piece can be used to support survival in a variety of ways, and the applications of the pieces are limited only by your imagination.

Sea Anchor. A sea anchor is circular, cone shaped, fabric device with an open bottom. The sea anchor is used to keep drag on the raft and prevent significant drifting. This makes it easier for someone to swim to the raft should they fall overboard. The sea anchor automatically deploys when the raft is inflated.

Lifeline. The lifeline is a one-quarter-inch wide nylon webbing that encircles the floor or inside tube of the raft. The lifeline is used as an anchor point for attaching survival items or as a handhold during rough weather.

Nylon Cord. Fifty feet of Type-I nylon parachute cord is supplied with multiperson life rafts. This cord is used to tie survival items to the raft, or used as fishing line, or used for making snares.

Bailing Sponge. Bailing sponges are provided to soak up water collected on the floor of the raft, or they can be used to collect rainwater for drinking.

Drinking Water. Drinking water is provided in 4.5-ounce packets for use when no other clean water is available.

Reverse Osmosis Pump. The reverse osmosis pump is a simple, hand-operated water desalinator. To operate, place the strainer in the water and begin cycling the pump handle up and down until you get flow. Pump an additional two minutes (sixty-eighty strokes) while allowing the water to go overboard. This will give the preserving agents inside the membrane time to filter through. Then insert the freshwater hose into storage bag or directly into your mouth.

Forward Osmosis Filtration System. Converts saltwater or contaminated freshwater into a potable energy drink. Uses a nutrient syrup charge to draw water molecules out of a contaminated source across a filtration membrane into a catch bag requiring no pumping. This process removes 97 percent of salt from seawater as well as bacteria, virus, parasites, heavy metals, petroleum, and pesticides. One four-ounce syrup charge will produce seventeen ounces of drink from sixty ounces of 60-degree saltwater in five hours.

Pocket Compass. The pocket compass is encased in a cylindrical transparent plastic cover with a reference line scribed into the top. The compass is waterproof and should be protected from damage to maintain those properties. Fittings on the bottom of the compass case adapt it for wrist wear.

Hand Pump. Hand pumps are provided with all multiperson life rafts and are used to inflate the life raft center tube. The hand pump is also used to top off the tube pressures.

Dye Marker. The dye marker is used to attract the attention of rescue aircraft. Rapid dispersion of the dye is desirable, so you must agitate the packet vigorously in the water. For best results, the dye marker is used in calm to moderate seas. They dye is florescent green in color, visible for approximately two nautical miles from an altitude of three thousand feet, and it lasts for twenty to thirty minutes.

Space Blanket. The space blanket is a lightweight waterproof blanket made of aluminized plastic. The blanket is silver on one side and international orange on the other. The blanket can be used as a windbreak, sunshade, or poncho to provide warmth and protection. The blanket furnishes some radar reflectivity over water and may be used for signaling.

Water Storage Bag. The water storage bag is made of strong, transparent plastic. When using the bag to store items other than water, be sure to wrap those items prior to placing them into the bag. This will prevent contamination of the bag's interior and thus prevent spoilage of any water that may later be stored in the bag. Uses:

- Holds five quarts of freshwater.
- When two water bags are tied together, they can be used as water wings to assist swimming.
- Protecting miscellaneous items from water exposure.

Pocketknife. The pocketknife consists of a large knife blade, leather awl, can opener, and combination screwdriver–bottle opener. Great care should be taken when using the knife in a moving raft to prevent puncturing the raft fabric.

Drinking Water

Water is your most important need. With water alone, you can live for ten days or longer, depending on your will to live. When drinking water, moisten your lips, tongue, and throat before swallowing. Protect freshwater supplies from seawater contamination. Fix your daily water ration after considering the amount of water you have, the output of solar stills and desalting kit, and the number and physical condition of your party.

If you don't have water, don't eat. If your water ration is two liters or more per day, eat any part of your ration or any additional food that you may catch, such as birds, fish, and shrimp. The life raft's motion and anxiety may cause nausea. If you eat when nauseated, you may lose your food immediately. If nauseated, rest and relax as much as you can, and take only water.

To reduce your loss of water through perspiration, soak your clothes in the sea and wring them out before putting them on again. Don't overdo this during hot days when no canopy or sun shield is available. This is a trade-off between cooling and saltwater boils and rashes that will result. Be careful not to get the bottom of the raft wet.

Watch the clouds and be ready for any chance of showers. Keep the tarpaulin handy for catching water. If it is encrusted with dried salt, wash it in seawater. Normally, a small amount of seawater mixed with rain will hardly be noticeable and will not cause any physical reaction. In rough seas you cannot get uncontaminated fresh water.

At night, secure the tarpaulin like a sunshade, and turn up its edges to collect dew. It is also possible to collect dew along the sides of the raft using a sponge or cloth. When it rains, drink as much as you can hold.

When solar stills are available, read the instructions and set them up immediately. Use as many stills as possible, depending on the number of people in the raft and the amount of sunlight available. Secure solar stills to the raft with care. When desalting kits are available in addition to solar stills, use them only for immediate water needs or during long overcast periods when you cannot use solar stills. In any event, keep desalting kits and emergency water stores for periods when you cannot use solar stills or catch rainwater.

Drink the aqueous fluid found along the spine and in the eyes of large fish. Carefully cut the fish in half to get the fluid along the spine and suck the eye. If you are so short of water that you need to do this, then do not drink any of the other body fluids. These other fluids are rich in protein and fat and will use up more of your reserve water in digestion than they supply.

In arctic waters, use old sea ice for water. This ice is bluish, has rounded corners, and splinters easily. It is nearly free of salt. New ice is gray, milky, hard, and salty. Water from icebergs is fresh, but icebergs are dangerous to approach. Use them as a source of water only in emergencies.

Remember, do not drink seawater, urine, or alcohol. Do not smoke. And do not eat unless water is available.

Fishing

If you can catch fish, you will not die of hunger or thirst. The flesh of fish caught in the open sea is good to eat, cooked or raw, and liquid can be squeezed from the flesh or chewed out to quench your thirst. Life rafts and most personal survival kits designed for use at sea come packed with fishing equipment. If you don't have fishing gear, improvise. For fishing line, use shoelaces, paracord, or pieces of tarp or canvas. Unravel the threads and tie them together in short lengths in groups of three or more threads. Improvise hooks and grapples from pieces of plastic, wood, or metal. You can fashion lures by attaching a hook to any shiny piece of metal. Use grapples to hook seaweed. You may shake crabs, shrimp, or small fish out of the seaweed. Scoop small fish up with a net if they school near the raft. If you don't have a net, make one from available cloth. Hold the net under the water and scoop upward. If you catch small fish, you can cut them to make bait for larger ones. Use all the guts from birds and fish for bait.

When fishing, never tie a line to your finger, hand, or foot, or to your raft. If the line comes under strain with a fish on it, it may cut you or damage the raft. If you're with other survivors, let another individual hold the end of the bitter end of

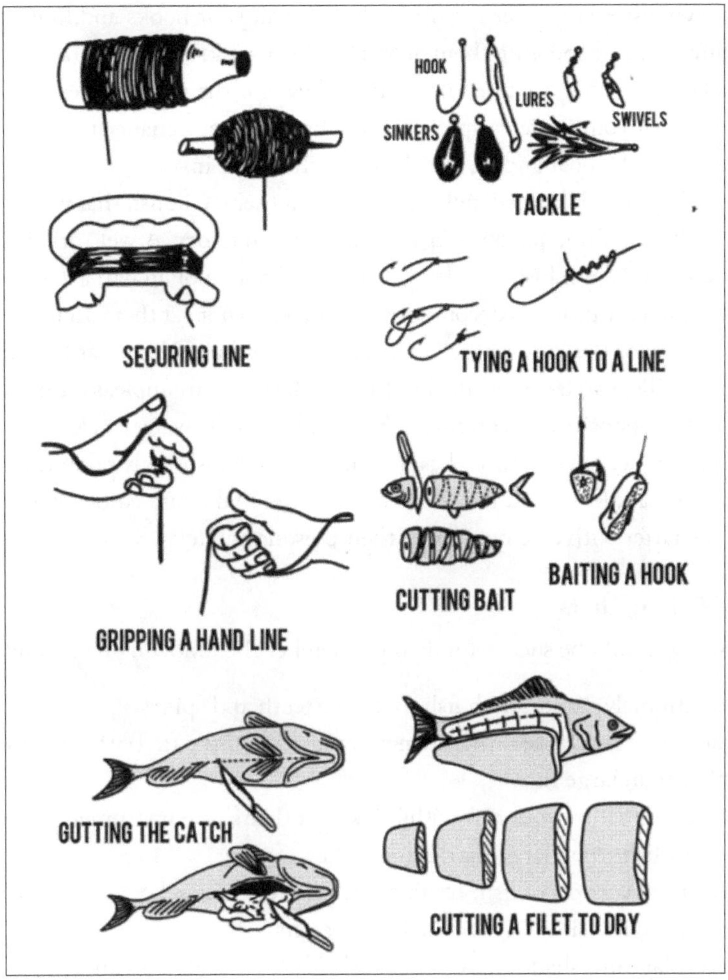

Survival fishing tactics

the fishing line. This gives you two chances to save your tackle and catch your fish if you lose your grip. Do not lean over or rush to the side of the boat when a fish is hooked, and don't let others do it. Try to catch small fish. Larger fish, like sharks, may break lines, carry off bait and tackle, injure you, or damage the raft.

Keep your bait moving in the water to make it look alive. Moving bait will be more productive than still bail. Wear gloves if available or use a cloth to handle any fish and avoid injury from sharp fins. Always be very careful how you lay your tackle down, especially after catching a fish. Clean your hooks and lines after use and secure them by spooling them onto a bottle or stick. Keep your tackle in a safe place or lash it in so it can't be lost overboard even if the raft capsizes. In a rubber raft, watch that your knife, grapple, and hooks never get a chance to prick, stab, or make a hole in the boat and let the air out or the water in.

Gut and bleed fish immediately after catching them. Cut fish that you do not eat immediately into thin, narrow strips and hang them to dry. A well-dried fish stays edible for several days. Fish not cleaned and dried may spoil in half a day. Fish with dark meat are very prone to decomposition. If you do not eat them all immediately, do not eat any of the leftovers. Use the leftovers for bait. Never eat fish that have pale, shiny gills, sunken eyes, flabby skin and flesh, or an unpleasant odor. Good fish show the opposite characteristics. Sea fish have a saltwater or clean fishy odor. Do not confuse eels with sea snakes that have an obviously scaly body and strongly compressed, paddle-shaped tail. Both eels and sea snakes are edible, but you must handle the latter with care because of their poisonous bites.

Helpful Fishing Hints

Your fishing should be successful if you remember the following important hints:

- Be extremely careful with fish that have teeth and spines.
- Cut a large fish loose rather than risk capsizing the raft. Try to catch small rather than large fish.
- Do not puncture your raft with hooks or other sharp instruments.
- Do not fish when large sharks are in the area.
- Watch for schools of fish; try to move close to these schools.
- Fish at night using a light. The light attracts fish.
- In the daytime, shade attracts some fish. You may find them under your raft.

Birds

Eat any birds you can catch. Sometimes birds may land on your raft, but usually they are cautious. You may be able to attract some birds by towing a bright piece of metal behind the raft. This will bring the bird within shooting range, provided you have a firearm. If a bird lands within your reach, you may be able to catch it. If the birds do not land close enough or land on the other end of the raft, you may be able to catch them with a bird noose or snare. Bait the center of the noose and wait for the bird to land. When the bird's feet are in the center of the noose, pull it tight.

Use all parts of the bird. Use the feathers for insulation, the entrails and feet for bait, and so on. Use your imagination.

Passing the Time

Sleep and rest are the best ways of enduring periods of reduced water and food intake. However, make sure that you have enough shade when napping during the day. If the sea is rough, tie yourself to the raft, close any cover, and ride out the storm as best you can. Relax is the key word—at least try to relax. It is not necessary to keep a "taut" ship with rigid regulations, but on the other hand, it is important that each person be given something to occupy their mind. Arguments, lowered morale, and possible violence may be avoided if each person has a specific duty. Normally the senior person will take control of the raft. In some cases, the condition or wishes of the senior person may make an elected skipper more suitable. The physical and emotional condition of the survivors living in close quarters will intensify the problems of living with other people. Select different duties for each individual, like keeping watch for rescue assets, fishing, practicing with the signal mirror, pumping water, dewatering the raft. You can even switch duties around to prevent boredom and pass the time.

Your morale is one of the most important factors effecting your well-being and chances of survival in a life raft. It is essential that you plan activities carefully so that you get adequate rest but at the same time keep busy during your waking hours. Keeping yourself occupied according to a schedule will not only help take care of your needs better, it will also keep you in good spirits.

Good leadership is the difference between groups that fight, steal from each other, commit suicide, become temporarily insane, or lose their essential supplies and a group that has survived for weeks on end without serious mental or morale difficulties. This kind of leadership starts on the beach—at your own unit.

Sharks

Whether you are in the water or in a boat or raft, you may see many types of sea life around you. Some may be more dangerous than others. Generally, sharks are the greatest danger to you. Other animals such as whales, porpoises, and stingrays may look dangerous but really pose little threat in the open sea.

Of the many hundreds of shark species, only about twenty species are known to attack humans. The most dangerous are the great white shark, the hammerhead, the mako, and the tiger shark. Other sharks known to attack humans include the gray, blue, lemon, sand, nurse, bull, and oceanic white tip sharks. Consider any shark longer than one meter dangerous.

There are sharks in all oceans and seas of the world. While many live and feed in the depths of the sea, others hunt near the surface. The sharks living near the surface are the ones you will most likely see. Their dorsal fins frequently project above the water. Sharks in the tropical and subtropical seas are far more aggressive than those in temperate waters. All sharks are basically eating machines. Their normal diet is live animals of any type, and they will strike at injured or helpless animals. Sight, smell, or sound may guide them to their prey. Sharks have an acute sense of smell, and the smell of blood in the water excites them. They are also very sensitive to any abnormal vibrations in the water. The struggles of a wounded animal or swimmer, underwater explosions, or even a fish struggling on a fishline will attract a shark.

Sharks can bite from almost any position; they do not have to turn on their side to bite. The jaws of some of the larger sharks are so far forward that they can bite floating objects easily without twisting to the side. Sharks may hunt alone, but most reports of attacks cite more than one shark present. The smaller sharks tend to travel in schools and attack en masse. Whenever one of the sharks finds a victim, the other sharks will quickly join it. Sharks will also eat a wounded shark as quickly as their prey.

Sharks feed at all hours of the day and night. Most reported shark contacts and attacks are during daylight, and many of these have been in the late afternoon. Some of the measures that you can take to protect yourself against sharks when you are in the water include staying with other swimmers. A group can maintain a 360-degree watch. A group can either frighten or fight off sharks better than one person. Always watch for sharks. Keep all your clothing on, including your shoes. Historically, sharks have attacked unclothed men first, mainly in the feet. Clothing also protects against abrasions should the shark brush against you. Avoid urinating. If you must, only do so in small amounts. Let it dissipate between discharges. If you must defecate, do so in small amounts and throw it as far away from you as possible. Do the same if you must vomit.

If a shark attack is imminent while you are in the water, splash and yell just enough to keep the shark at bay. Sometimes yelling underwater or slapping the water repeatedly will scare the shark away. Conserve your strength for fighting in case the shark attacks. If attacked, kick and strike the shark. Hit the shark on the gills or eyes if possible. If you hit the shark on the nose, you may injure your hand if it glances off and hits its teeth.

When you are in a raft and see sharks,

- Do not fish. If you have hooked a fish, let it go;
- Do not clean fish in the water;
- Do not throw garbage overboard;

- Do not let your arms, legs, or equipment hang in the water;
- Keep quiet and do not move around; and
- Bury all dead as soon as possible. If there are many sharks in the area, conduct the burial at night.

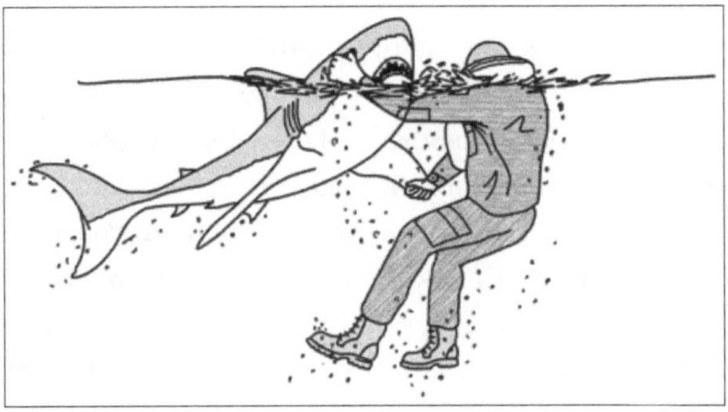

Fighting a shark

Beaching Techniques

If you find land, you must get ashore safely. Going ashore in a strong surf is dangerous. Take your time. Select your landing point carefully. Try to land on the lee side of an island or on a point of land jutting out into the water. Keep your eyes open for gaps in the surf line, and head for them. Avoid coral reefs and rocky cliffs. Either signal ashore for help or sail around and look for a sloping beach where the surf is gentle. As the raft nears the beach, ride in on the crest of a large wave. Paddle or row hard and ride into the beach as far as you can. Do not jump out of the raft until it has grounded, but once grounded, quickly get out and beach it. If you have a choice, do not land at night. If you have reason to believe that people live on the shore, lay away from the beach, signal, and wait for the inhabitants to come out and bring you in.

Swimming Ashore

If rafting ashore is not possible and you have to swim, wear shoes and at least one thickness of clothing. Use the sidestroke or breaststroke to conserve strength. If the surf is moderate, ride in on the back of a small wave by swimming forward with it. If you must land on a rocky shore, look for a place where the waves rush up onto the rocks. Avoid places where the waves explode with a high, white spray.

LAND SURVIVAL

On land the obstacles to overcome aren't so much physical as mental. Unlike ditching at sea, some people have made a home out of the geographical areas that you may find yourself stranded in. With varying degrees of effort, they adjusted to the demands of terrain, climate, and environment. Your problem is, because you did not expect to land there, you are not prepared to live there. First, set up a temporary shelter for protection from the elements. If fire is needed, start it at once. Have signaling equipment handy. Then relax and rest until you are over the shock of the mishap. Leave extensive preparations and planning until later.

After resting, organize the camp and appoint individuals to specific duties. Pool all food and equipment and place in the charge of one person. Prepare a more permanent shelter for protection from rain, hot sun, wind, cold, and insects. Collect all possible fuel. Look for a water supply and for animal and plant food. Prepare signals in order to be recognized from the air. You are the key person in the rescue. Help the search parties find you, and follow their instructions when you are sighted. Do not take chances that might result in injury. You will be easier to rescue if you are in one piece.

Shelter Construction

In any area, a shelter can be improvised from parts of wreckage and emergency equipment or from natural materials. The kind of shelter that is made depends on whether protection is needed from the outside environment and whether the camp is only for one or many days.

Improvised shelters

Shelter Tips

Following are some shelter tips to keep in mind.

- Do not make camp at the base of steep slopes or in areas where you run the risk of avalanches, floods, rock falls, or battering by winds.
- The life raft is very valuable in preparing an initial shelter. Overturned, inflated rafts can be propped up and used as a lean-to.
- You may also look for nature's "ready-made" shelters like rock shelves, caves, hollow trees, and bushes. These all provide adequate shelter under certain conditions. You can save yourself a great deal of time and effort if you make good use of them.
- In timberless country, choose a spot protected from wind and drifting snow. Remember that snow is a good insulator when making shelters. A simple snow cave or burrow can be made by digging into the side of a snowdrift and lining the hole with grass or brush. Snow caves must be ventilated. Vent holes are best placed at a 45-degree angle in the downwind side of a shelter, or an igloo.

Winter shelter

Signaling on Land

There are several ways to signal for help that you should consider if you are in a land survival situation. Following are just a few ways to attract the attention of rescuers:

- Keep snow and frost off surfaces to make sharp contrast with the surroundings.
- Build signal fires, using smoke by day, and flame by night. Add engine oil, rags soaked in oil, or pieces of rubber to make black smoke. Add green leaves, moss, or a little water to send up billows of white smoke.

- Spread out cargo parachutes, pine boughs, cowling panels, and so on, on the ground. Arrange them in a big geometric pattern, which will attract more attention than if they are scattered randomly.
- Use the sea dye marker by crushing it into powder and spreading it in the snow, or hold it in a river and let it spread down river.

Fire Making

Following are some safety guidelines for fire making.

- Do not build a fire inside your shelter.
- Do not pour fuel on a standing fire, even if it is only smoldering.
- Do not build the fire too big. Small fires require less fuel and are easier to control, and their heat can be concentrated.
- Clear away leaves, twigs, moss, and dry grass so that you do not start a grass or forest fire.
- Build a fire on a platform of logs or flat stones when the only place available for a fire is on snow, ice, or wet ground.
- Build fires against a rock or wall of logs that serve as a reflector to direct the heat into your shelter to maximize warmth and deflect winds.
- Carry some dry tinder with you in a waterproof container. Adding a little charcoal will improve it. Collect good tinder wherever and whenever it can be found.
- Dry damp tinder by placing it under your clothes against your skin.
- To split logs, whittle hardwood wedges, and drive them into cracks in the log with a rock or club; split wood burns more readily.
- Save matches, lighter fuel, or pyrotechnics for later use. A lit candle will burn longer.
- Use other common materials to start a fire when you lack matches, such as a cigarette lighter or pyrotechnics. Other materials are described below.

Magnesium Fire Starter Kit. This the easiest and most reliable way of making a fire without matches. The magnesium fire starter will produce a flash fire of great heat. Each starter kit comes with instructions on proper use.

Glass. Any convex lens two inches or more in diameter can be used in bright sunlight to concentrate the sun's rays on the tinder and start it burning.

Electric Spark. If you can salvage a live storage battery, direct a spark onto the tinder by scratching the end wires together to produce an arc. When using this method, take all necessary precautions to prevent receiving an electrical shock.

Types of Fire Material

The types of fire building material are described as follows:

Tinder. A dry material that ignites with little heat, just a spark, including

- Birch bark
- Shredded inner bark from cedar, chestnut, red elm trees
- Fine wood shavings
- Dead grass, ferns, moss, fungi
- Straw
- Sawdust
- Dead evergreen needles
- Punk (the completely rotten portion of dead logs or trees)
- Evergreen tree knots
- Bird down (fine feathers)
- Down seed heads (milkweed, dry cattails, bulrush, Canada thistle, goldenrod, dandelion)
- Dead palm leaves
- Lint from pocket and seams
- Charred cloth
- Waxed paper
- Cotton

Kindling. A readily combustible material added to the tinder to increase the temperature of the fire so that it will ignite less combustible material, including

- Small twigs
- Small strips of wood
- Split wood
- Heavy cardboard
- Wood that has been soaked or doused with flammable materials such as gasoline, oil, or wax

Fuel. A less combustible material that will burn slowly and steadily once it is well ignited, such as

- Dry standing wood and dry dead branches
- Dry inside (heart) of fallen tree trunk and large branches
- Green wood that is finely split
- Dry grasses twisted into bunches

- Dried animal dung
- Animal fats
- Coal, oil shale, or oily sand lying on the surface

Fire Lays

Fire lays are built to meet specific needs or uses.

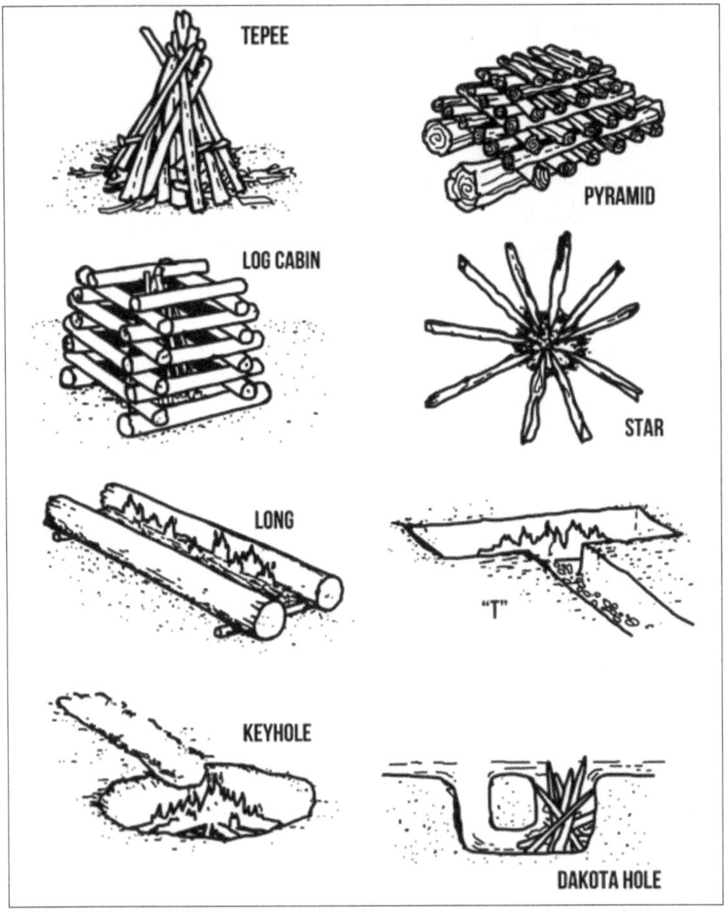

- Teepee fire or pyramid lay produces a concentrated heat source used primarily for cooking.
- Log cabin lay produces large amounts of light and heat to dry out wet wood, and provide coals for cooking, and so on.
- A Dakota fire hole is effective in high winds.

Lighting the Fire

Step	Action
1	Make a tinder bundle and place it near your fire lay.
2	Light your match (cigarette lighter or pyrotechnic), quickly grab the tinder, holding it slightly downward, and apply the flame.
3	As the tinder catches fire, turn the bundle in your hand until it is burning well.
4	Light the kindling in your fire lay.
5	Use the kindling to light your logs.

Clothing

Try to keep your clothing clean and in good repair. Clothing can keep you cool as well as warm and also protect you against sunburn, insects, pests, and scratches. Clean clothes are better insulators than dirty clothes, and they last longer. Try to keep your clothing and shoes dry by using a drying rack in front of a fire. Don't put your wet shoes too close to the fire or they will stiffen and crack. Think twice before you discard any clothing.

Cold Weather

Insulation combined with body heat is the secret of warmth. Insulation is largely determined by the combined thickness of all the garments worn. Below are just a few ideas that you should remember during survival situations in cold weather.

- When exerting yourself, cut down sweating by opening your clothes at the neck and wrists and loosening at the waist.
- Wear clothing loosely. Tight fits cut off circulation and increase danger of freezing.
- Keep ears covered. If necessary, improvise a scarf and headgear from parachute fabric to protect your neck and head.
- If you lose your boots or if your socks are wet and you have no spares, improvise footgear by wrapping parachute fabric lined with dry grass around your feet.
- Keep your clothing as dry as possible. Brush snow from your clothing before you enter your shelter or go near a fire.
- If you have fallen into water during the winter, excess water may be removed from your clothing by rolling in dry snow.

Hot Weather

Wearing clothing in hot weather is necessary to protect yourself from sunburn, heat, sand, and insects. Below are just a few ideas that you should remember during survival situations in hot weather.

- Keep your body and head covered during the daytime; you'll last longer on less water.
- Wear long pants and shirts with sleeves rolled down. Keep them loose and flapping.
- Wear a cloth neckpiece to cover the back of your neck from the sun. If you have no hat, make a headpiece.
- Wear a mosquito head net or tie a T-shirt around your head, especially at dawn and dusk in mosquito-infested areas.
- Dry your clothing before nightfall to avoid discomfort from cold.

Clothes that have been removed should be hung up. If laid on the ground, they may collect ants, scorpions, or snakes. Always check footgear and clothing for such "guests" before putting it on.

25

WELLNESS

The Coast Guard will train you to perform your military and professional duties, but your ability to perform them quickly and efficiently will depend upon the state of your physical and mental health. As a member of the Coast Guard, you will have the facilities and the opportunity to maintain a good standard of living and health. Even the smallest unit has facilities to provide nourishing meals, comfortable berthing, and all the amenities necessary to live in a comfortable and productive fashion. It is important that you take advantage of the healthful environment provided by the service and adopt good personal habits and standards. Good health habits and a positive attitude will make your job easier and improve relations with people.

Wellness is more than simply the absence of illness; it refers to your lifestyle habits and how you live your life. More than half of the diseases affecting us are caused by poor habits regarding tobacco, diet, exercise, and stress. To reach and maintain an optimal state of good health, you must actively pursue a healthy lifestyle. If you do, you are guaranteed to look better, feel better, play better, think better, work better, and live longer!

What does it mean to lead a healthy lifestyle? Most important is good nutrition or eating right, achieving and maintaining a high level of physical fitness, avoiding tobacco products and drinking alcohol in moderation or not at all, and lowering your risks for disease and long-term health problems. This means establishing personal development and fitness plans, periodically visiting your medical and dental providers, and taking advantage of work–life programs.

MASLOW'S HIERARCHY OF NEEDS

Psychologist Abraham Maslow's hierarchy of needs is a psychological theory concerning human motivation, depicted as five levels within a pyramid. Needs lower down in the hierarchy must be satisfied before individuals can attend to needs higher up, although behaviors can be motivated by several different needs. Life experiences, including divorce and loss of a job, may cause an individual to fluctuate between levels of the hierarchy. Meeting each of these needs contributes to your overall wellness. Because these needs are interrelated, understanding their factors is critical to meeting them and to moving up in the hierarchy and achieving not only peak performance but happiness and fulfillment.

1. Physiological needs: These are biological requirements for human survival—for example, air, food, drink, shelter, clothing, warmth, sex, sleep. If these needs are not satisfied, the human body cannot function optimally. Maslow considered physiological needs the most important, as all the other needs become secondary until these needs are met.
2. Safety needs: Once an individual's physiological needs are satisfied, the needs for security and safety become salient. People want to experience order, predictability, and control in their lives. These needs can be fulfilled by the family and society (e.g., police, schools, businesses, and medical care)—for example, emotional security, financial security (e.g., employment, social welfare), law and order, freedom from fear, social stability, property, health, and well-being (e.g., safety against accidents and injury).
3. Love and belongingness needs: After physiological and safety needs have been fulfilled, the third level of human needs is social and involves feelings of belonging. Belonging refers to a human emotional need for interpersonal relationships, affiliating, connectedness, and being part of a group. Examples of belonging include friendship, intimacy, trust, and acceptance, receiving and giving affection, and love.
4. Esteem needs are the fourth level in Maslow's hierarchy and include self-worth, accomplishment, and respect. Maslow classified esteem needs into two categories: (1) esteem for oneself (e.g., dignity, achievement, mastery, independence) and (2) the desire for reputation or respect from others (e.g., status, prestige). Maslow indicated that the need for respect or reputation is

most important for children and adolescents and precedes real self-esteem or dignity.

5. Self-actualization needs are the highest level in Maslow's hierarchy; they refer to the realization of a person's potential, self-fulfillment, seeking personal growth and peak experiences. Maslow describes this level as the desire to accomplish everything that one can, to become the most that one can be.

PERSONAL HYGIENE

Proper personal hygiene is essential in promoting good health. It is especially important in close living environments to keep bacteria, viruses, and illnesses at bay. Personal hygiene is simply keeping healthy by good cleanliness habits. You must take care of your grooming every single day. Poor personal hygiene is not only unhealthy for you, it can also expose your shipmates to many types of health risks.

Hand Washing

Good cleanliness habits begin with washing your hands. Frequently washing your hands will help to prevent the spread of germs and disease. If you are not able to wash your hands in the head, use waterless hand sanitizer. Wash your hands

- Before and after each trip to the head;
- Before and after each meal; and
- Any time you cough or sneeze in your hand.

Body Odor

There are many things you can do to avoid body odor, including

- Taking showers daily and after working out at the gym. While on board a cutter, you will be taking sea showers to conserve water. These are shortened showers to save time and water. This does not excuse you from maintaining proper hygiene;
- Using deodorant daily; and
- Changing clothing and underclothing daily.

Oral Health

The three most common dental diseases are tooth decay (caries), inflammation of the gum (gingivitis), and disease of the gum and bone surrounding the teeth (pyorrhea). They cause needless loss of teeth, for they all can be prevented or controlled. All Coast Guard personnel shall have an annual dental exam. This is essential for timely detection of tooth decay as well as any other type of mouth disease.

Cavities. There is no surefire way to prevent tooth decay, but it can be cut down greatly by regular, correct brushing and by reducing sweets in your diet. At the first sign of tooth decay, see a dentist. Untreated cavities can have serious long-term health consequences.

Gingivitis. Normal healthy gums are pale pink and firm in texture. If they are swollen or puffy, hang loosely about the teeth and bleed easily, you have gingivitis.

Pyorrhea. If gingivitis goes untreated, you may notice pockets or crevices between the tooth and gum. This is an indication of pyorrhea (sometimes called periodontitis). More teeth are lost from gingivitis and pyorrhea than from tooth decay. Most dental diseases result from poor mouth hygiene combined with the misuse of the toothbrush and lack of flossing.

Toothbrushes. Proper use of the toothbrush is most effective in control or prevention of dental diseases. Keep at least two brushes, so you can always use a dry one. When the bristles become soft or flattened, throw the brush away.

Brush your teeth regularly, after each meal if you can. Use dental floss at least once a day. The proper tooth-brushing technique will clean the teeth and gums, stimulate circulation in the gum tissues, and toughen the gums by the friction of the bristles but will not injure the teeth or gums. Brush with the mouth barely open and the muscles relaxed, which permits easier access to hard-to-get-at areas.

NUTRITION

Nutrition is the foundation of wellness. Since you literally are what you eat, a healthy lifestyle starts with your diet. To eat right and maintain a healthy weight,

- Eat a variety of foods;
- Limit your intake of total fat, animal fats, and cholesterol;
- Eat plenty of vegetables, fruits, and grain products;
- Use sugars in moderation;
- Use salt and sodium in moderation; and
- If you drink alcoholic beverages, do so in moderation.

Diet

The Coast Guard takes great care to serve good-tasting and well-balanced meals in its galleys and dining facilities, but the meals will do you little good unless you observe certain rules in your diet both on and off duty. Don't concentrate too much on one food group, such as meat or potatoes—or especially desserts! Do yourself, your strength and stamina, your general health, and your teeth and gums a favor by eating a balanced diet that provides the nutrients your body needs for peak performance. Three major nutrients the body needs are carbohydrates, protein, and fat.

Carbohydrates. Your digestive system changes carbohydrates into glucose or blood sugar. Your body then uses this sugar for energy. Carbohydrates are either simple or complex. Simple carbohydrates include sugars that are found in fruits, milk, and in processed sugars such as candy, syrups, and soft drinks. These sugars require little digestion and are broken down quickly by the body. Complex carbohydrates are found in grains, vegetables, breads, and legumes. They are starches and require more digestion, which satisfy you longer and give you energy over a longer period of time.

Protein. A protein provides a long-term source of energy. A complete protein source is one that provides all the essential amino acids. These are high-quality proteins such as meat, fish, cheese, and eggs. Incomplete proteins are low in one or more of the essential amino acids required for complete nutrition and must be combined with other sources to provide adequate amounts of the essential amino acids. Grains such as rice, pasta, and wheat bread contain proteins but must be combined with other incomplete proteins. For example, peanut butter on whole wheat bread would give you a complete protein.

Fats. Fats are composed of two components: glycerin and fatty acids. Glycerin is the energy source of fats, which your body metabolizes in much the same way as carbohydrates. The glycerin is broken down into sugars and used by your body for fuel. There are four types of fats: saturated, unsaturated, polyunsaturated, and hydrogenated fats.

- Saturated: Saturated fats are derived from animal products such as meat, butter, oil, and eggs. Saturated fats raise bad cholesterol levels (LDL).
- Monounsaturated Fats: These fats lower LDL levels without affecting good cholesterol (HDL) and are found in vegetables, nuts, canola, and olive oil.
- Polyunsaturated Fats: This type of fat tends to lower both LDL and HDL levels and are found in walnuts, sunflower seeds, and peanuts.
- Hydrogenated Fats: Hydrogenated fats are mostly found in processed foods such as margarine and vegetable shortening. The more solid and hydrogenated the fat, the more our LDL rises, but at the same time it lowers HDL, thereby greatly increasing the risk of coronary heart disease.

Five Major Food Groups

The U.S. Department of Agriculture defines five basic food groups. Each food group provides certain nutritional benefits, so foods from each group should be consumed each day.

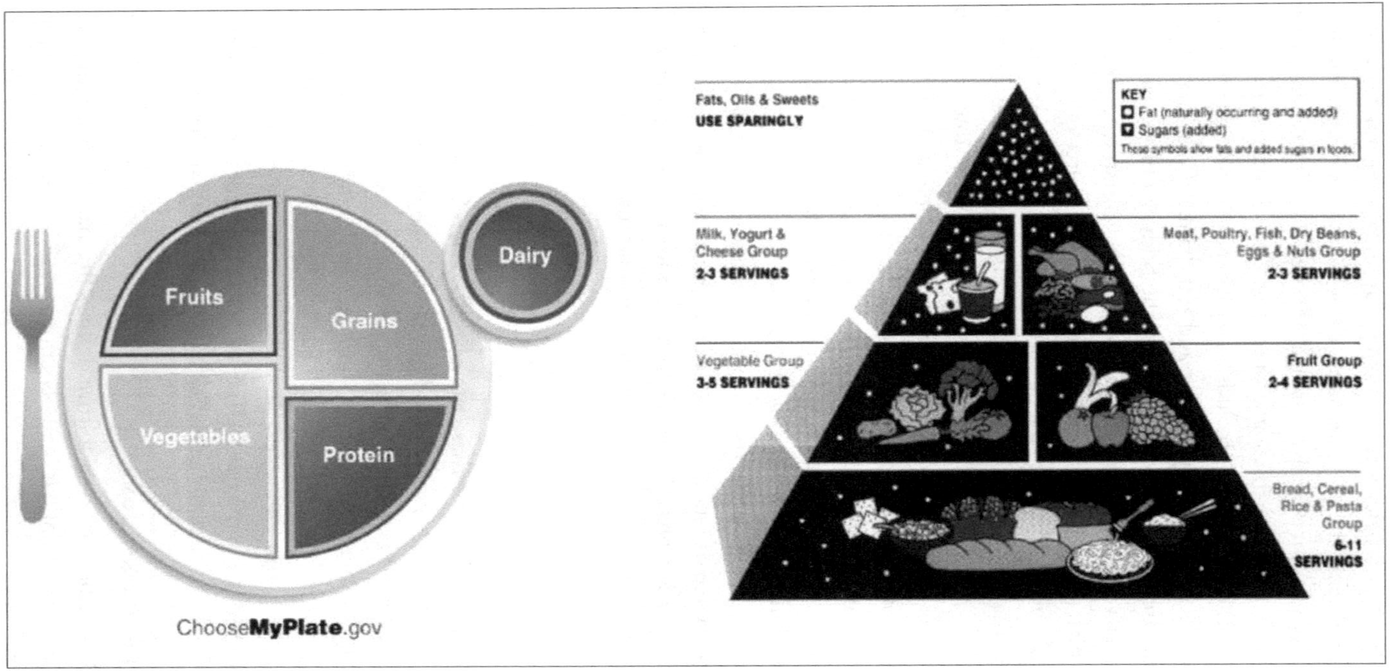

Fruits
Grains
Dairy
Vegetables
Protein
ChooseMyPlate.gov
Fats, Oils & Sweets
USE SPARINGLY
KEY
Fat (naturally occurring and added)
Sugars (added)
These symbols show fats and added sugars in foods.
Milk, Yogurt & Cheese Group
2-3 SERVINGS
Meat, Poultry, Fish, Dry Beans, Eggs & Nuts Group
2-3 SERVINGS
Vegetable Group
3-5 SERVINGS
Fruit Group
2-4 SERVINGS
Bread, Cereal, Rice & Pasta Group
6-11 SERVINGS

Nutrition

Proteins. Meat and poultry choices should be lean or low-fat, like 93 percent lean ground beef, pork loin, and skinless chicken breasts. Choose seafood options that are higher in beneficial fatty acids (omega-3s) and lower in methylmercury, such as salmon, anchovies, and trout. Vegetarian options in the protein group include beans, peas, lentils, nuts, seeds, and soy products.

Grains. Any food made from wheat, rice, oats, cornmeal, or barley is a grain product. Grains are divided into two subgroups, whole grains and refined grains.

- Whole Grains: Whole grains contain the entire grain kernel. Examples include whole wheat flour, oatmeal, and brown rice. Choosing whole grains provides health benefits. People who eat whole grains as part of their diet have a reduced risk of heart disease.
- Refined Grains: These grains have been milled, which removes the bran and germ. This gives the grain a finer texture and extends the shelf life but removes the dietary fiber. Examples include white flour, white bread, and white rice. To eat more whole grains, substitute whole wheat or multigrain bread instead of white bread or eat a whole grain cereal.

Vegetables. Any vegetable or 100 percent vegetable juice counts as a member of this group. Eating vegetables reduces your risk for heart disease and stroke and may protect against certain types of cancers. Vegetables are rich in fiber and lower in calories and may help to lower your overall calorie intake.

Vegetables are important sources of many nutrients including potassium, vitamin A, and vitamin C. To increase your intake of vegetables, try a salad for lunch, but go light on the salad dressing.

Fruits. Any fruit or 100 percent fruit juice counts as part of the fruit group. Individuals who eat more fruits and vegetables as part of their diet are likely to have a reduced risk of coronary heart disease, certain types of cancers, obesity, and type 2 diabetes. To increase your intake of fruits, at breakfast top your cereal with bananas or peaches, drink 100 percent orange juice, or mix fruits with low-fat yogurt.

Dairy. All fluid milk products and foods made from milk are part of this food group. Most dairy choices should be fat-free or low-fat. Intake of dairy products is linked to improved bone health and may reduce the risk of osteoporosis. The intake of dairy products is especially important to bone health during childhood and adolescence, when bone mass is being built, and is also associated with a reduced risk of cardiovascular disease and type 2 diabetes, and with lower blood pressure in adults. If you usually drink whole milk, switch gradually to fat-free milk to lower saturated fat and calories. If you drink cappuccinos or lattes, ask for them with fat-free skim milk.

Hydration

Hydration is basically taking in water to restore or maintain a balance of fluids. More than half your body weight is composed of water. Your body and its systems, which comprise cells, tissues, and organs, rely on water to survive, function correctly, and regulate body temperature. So it is important to fuel your body with water. Drinking water regularly helps keep you hydrated. We lose water when we urinate, sweat, and breathe. We even lose water through vomiting and diarrhea. If you do not replace this water, you will become dehydrated. Although dehydration can happen in warmer weather, it can also happen in cooler weather if fluid intake is insufficient. Becoming dehydrated can be a serious condition if not identified properly. You should know some basic symptoms of dehydration:

- Little or no urine output
- Urine that is darker than usual
- Extreme thirst
- No tears when crying
- Dizziness/light-headed feeling
- Confusion
- Headache

Never wait to act until you notice symptoms of dehydration. Dehydration is completely preventable by drinking plenty of water. You should drink before, during, and after exercise, even when the weather is cold. As a rule of thumb, drink at least two cups (sixteen ounces) of water before exercise to match fluids lost in sweat. Drink at least one cup of water every twenty to thirty minutes during exercise.

PHYSICAL FITNESS

There are five essential parts to being physically fit: aerobic capacity, muscle strength, muscle endurance, flexibility, and maintaining a healthy level of body fat. Each of these is important, and you cannot neglect any of them if you want to be fit. Physical fitness helps make you a more productive and capable member of the Coast Guard and a healthier and happier person in your private life.

Aerobic capacity is a measure of your body's endurance or stamina. Good endurance enables you to maintain work and play activities for extended periods of time. Aerobic capacity depends on the efficiency of your heart, lungs, and muscles to get and use oxygen. To increase your aerobic capacity, you must exercise with your large muscles in your legs and arms for at least thirty minutes at a time, at least three times per week. Aerobic exercises include running, jogging, cycling, stair-climbing, rowing, swimming, jumping rope, cross-country skiing, and fast walking. Aerobic exercise is also an effective way to burn calories, which is important if you want to reduce your body fat.

Muscle strength is a measure of the force you can produce, allowing you to lift, push, pull, and carry. These are all important in your Coast Guard duties, but they are also important in your recreation and home activities. The only way to build muscle strength is to exercise against resistance, either by lifting weights or by lifting your body weight against gravity by doing push-ups, pull-ups, calisthenics, and so on. Strength-training exercises not only help you build muscle but also help you lose fat. The more muscle mass you have, the higher your rate of metabolism and the faster you burn calories, even when you are asleep!

Muscle endurance is your ability to perform hard work over a prolonged period of time. Muscle endurance is important in your Coast Guard duties for both emergency situations and for operational requirements. The best muscle-endurance exercises are pull-ups, curl-ups, push-ups, parallel bar dips, and calisthenics. You can also build muscle endurance by lifting weights. But instead of using heavy weights and a small number of repetitions, as you would do to build muscle strength, you should use light weights and a high number of repetitions (fifteen to twenty) to build endurance.

Flexibility describes your ability to move your joints through a full range of motion. You need good flexibility to develop muscle strength; to prevent injuries to your muscles, tendons, ligaments, and joints; and to protect against back injuries. Proper stretching, both before and after exercise, is the most effective way to develop good flexibility.

Body Composition

Body composition refers to the relative amounts of fat and lean tissue in your body. The more fat you have, the harder it is on your joints and the harder your heart has to work. Trying to carry around an extra ten to twenty pounds of body fat is like walking with a ten- to twenty-pound pack on your back. Your goals should be to maximize your muscle mass and minimize your body fat. An ideal body composition for men is a body fat of 15 to 17 percent, while for women it is 18 to 22 percent. Try not to think of losing "weight" if you are too heavy; what you really want to lose is fat. The best way to lose fat and preserve muscle is to maintain a low-fat, healthy diet, perform aerobic exercise for at least thirty minutes three times per week, and perform strength-training and muscle-endurance exercises. If you're out of shape, start slowly and gradually increase frequency, intensity, and duration of exercise. Train regularly, vary the program, and understand plateaus. Train with a partner and use exercise cues, contracts, support from friends and family, and goal-setting strategies to help you stick with your program.

Body Mass Index Table

	Normal						Overweight					Obese										Extreme Obesity														
BMI	**19**	**20**	**21**	**22**	**23**	**24**	**25**	**26**	**27**	**28**	**29**	**30**	**31**	**32**	**33**	**34**	**35**	**36**	**37**	**38**	**39**	**40**	**41**	**42**	**43**	**44**	**45**	**46**	**47**	**48**	**49**	**50**	**51**	**52**	**53**	**54**
Height (inches)																		Body Weight (pounds)																		
58	91	96	100	105	110	115	119	124	129	134	138	143	148	153	158	162	167	172	177	181	186	191	196	201	205	210	215	220	224	229	234	239	244	248	253	258
59	94	99	104	109	114	119	124	128	133	138	143	148	153	158	163	168	173	178	183	188	193	198	203	208	212	217	222	227	232	237	242	247	252	257	262	267
60	97	102	107	112	118	123	128	133	138	143	148	153	158	163	168	174	179	184	189	194	199	204	209	215	220	225	230	235	240	245	250	255	261	266	271	276
61	100	106	111	116	122	127	132	137	143	148	153	158	164	169	174	180	185	190	195	201	206	211	217	222	227	232	238	243	248	254	259	264	269	275	280	285
62	104	109	115	120	126	131	136	142	147	153	158	164	169	175	180	186	191	196	202	207	213	218	224	229	235	240	246	251	256	262	267	273	278	284	289	295
63	107	113	118	124	130	135	141	146	152	158	163	169	175	180	186	191	197	203	208	214	220	225	231	237	242	248	254	259	265	270	278	282	287	293	299	304
64	110	116	122	128	134	140	145	151	157	163	169	174	180	186	192	197	204	209	215	221	227	232	238	244	250	256	262	267	273	279	285	291	296	302	308	314
65	114	120	126	132	138	144	150	156	162	168	174	180	186	192	198	204	210	216	222	228	234	240	246	252	258	264	270	276	282	288	294	300	306	312	318	324
66	118	124	130	136	142	148	155	161	167	173	179	186	192	198	204	210	216	223	229	235	241	247	253	260	266	272	278	284	291	297	303	309	315	322	328	334
67	121	127	134	140	146	153	159	166	172	178	185	191	198	204	211	217	223	230	236	242	249	255	261	268	274	280	287	293	299	306	312	319	325	331	338	344
68	125	131	138	144	151	158	164	171	177	184	190	197	203	210	216	223	230	236	243	249	256	262	269	276	282	289	295	302	308	315	322	328	335	341	348	354
69	128	135	142	149	155	162	169	176	182	189	196	203	209	216	223	230	236	243	250	257	263	270	277	284	291	297	304	311	318	324	331	338	345	351	358	365
70	132	139	146	153	160	167	174	181	188	195	202	209	216	222	229	236	243	250	257	264	271	278	285	292	299	306	313	320	327	334	341	348	355	362	369	376
71	136	143	150	157	165	172	179	186	193	200	208	215	222	229	236	243	250	257	265	272	279	286	293	301	308	315	322	329	338	343	351	358	365	372	379	386
72	140	147	154	162	169	177	184	191	199	206	213	221	228	235	242	250	258	265	272	279	287	294	302	309	316	324	331	338	346	353	361	368	375	383	390	397
73	144	151	159	166	174	182	189	197	204	212	219	227	235	242	250	257	265	272	280	288	295	302	310	318	325	333	340	348	355	363	371	378	386	393	401	408
74	148	155	163	171	179	186	194	202	210	218	225	233	241	249	256	264	272	280	287	295	303	311	319	326	334	342	350	358	365	373	381	389	396	404	412	420
75	152	160	168	176	184	192	200	208	216	224	232	240	248	256	264	272	279	287	295	303	311	319	327	335	343	351	359	367	375	383	391	399	407	415	423	431
76	156	164	172	180	189	197	205	213	221	230	238	246	254	263	271	279	287	295	304	312	320	328	336	344	353	361	369	377	385	394	402	410	418	426	435	443

Source: Adapted from *Clinical Guidelines on the Identification, Evaluation, and Treatment of Overweight and Obesity in Adults: The Evidence Report.*

FITT Principles

Your fitness plan should use the FITT principle to determine the ideal activity required to reach your fitness goals. FITT stands for the frequency, intensity, time, and type of activity. Train regularly and vary your program. If you struggle, use exercise cues, contracts, a partner, support from friends and family, and goal setting to help you stick with your program. Start new exercises and routines slowly and gradually increase frequency, intensity, and duration. The boxes below contain guidelines for using the FITT principles with your fitness plan

1. Name: Cardio-Respiratory Endurance (CRE)
   Frequency: 3 to 5 days per week
   Intensity: 55–64 percent max heart rate if unfit, 70–80 percent average, max 220
   Time: 20–60 min.
   Type: Aerobic endurance

2. Name: Muscular Strength / Muscle Endurance (MS/ME)
   Frequency: 2 to 3 days per week
   Intensity: resistance heavy enough to cause muscle fatigue, one set consisting of 8–10 reps each exercise
   Time: 9–10 strength exercises
   Type: focus on major muscle groups

3. Name: Flexibility (F)
   Frequency: 2 to 3 days per week
   Intensity: Stretch to the point of mild discomfort, not pain
   Time: Hold for 10–20 seconds
   Type: Focus on major joints

Moderate-intensity physical activity between 150 and 260 minutes per week will prevent weight gain and provide modest weight loss. Greater amounts of physical activity (>250 minutes per week) is associated with clinically significant weight loss and prevention of weight regain.

Stretches

Each day you perform aerobic or strength-training activities, take at least ten minutes to warm-up your ligaments, muscles, and tendons. Stretch the major muscles groups before and after your workouts. Hold stretches for ten to thirty seconds and repeat the stretch three to four times.

Boat Crew Physical Fitness Standards

All Coast Guard boat crew members are required to meet standards of physical fitness needed to ensure they have sufficient strength and endurance to safely perform their assigned duties during normal and adverse conditions. Meeting these standards is also an accepted way of complying with the weight standards required of all Coast Guardsmen. Knowing these standards will ensure that personnel are able to accurately gauge their level of fitness and make improvements where necessary. All sections of the physical fitness test must be completed one after the other within a reasonable amount of time, not to exceed thirty minutes between each section. If a section of the test is not completed successfully, the entire test must be completed again as a retest. Personnel who cannot meet the standard due to sickness, injury, recent pregnancy, and so on, should not perform boat crew duties for the safety of the member, crew, and public.

Males	Push-ups	Sit-ups	1.5-Mile Run	12-Minute Swim (yards)
Under 30	29	38	12:51	500
30 to 39	24	35	13:36	450
40 to 49	18	29	14:29	400
50 to 59	13	25	15:26	350
60+	10	22	16:43	300
Females	**Push-ups**	**Sit-ups**	**1.5-Mile Run**	**12-Minute Swim (yards)**
Under 30	15	32	15:26	400
30 to 39	11	25	15:57	350
40 to 49	9	20	16:58	300
50 to 59	9	16	17:55	250
60+	9	15	18:44	200

Note: Push-ups and sit-ups must be performed within one minute. Either the 1.5-mile run or the 12-minute swim may be performed to meet the standard.

STRESS

Stress is the everyday wear and tear on your body as you respond to the people, places, and things in your life. If you take good care of your body and make allowances for added stress, you can keep the wear and tear in check. While working in the fleet, you will be required to perform a variety of complex and at times stressful roles, some of which will result in lack of sleep, lack of food, extended periods on your feet, and extended periods in heavy weather. From such stressful situations, you will gain some important characteristics that the Coast Guard values in its members.

- Self-Knowledge: You will know more about yourself than you have known in the past. You will know your limits and how far you can be pushed.
- Self-Confidence: You will be able to resolve problems and situations with skill and ease because you will learn not to allow circumstances or emotions to control you. You will learn to evaluate and make solid decisions.
- Teamwork: This is one of the most important things that you will learn in the service. It is the backbone of the Coast Guard. Because we are a small service, everyone is important to the team.

Physiological Effects of Stress

Physiological effects of stress include increased awareness of aches and pains, increased perspiration, fatigue, dry mouth, shortness of breath, butterflies in the stomach, and loss of sleep.

Psychological Effects of Stress

Psychological effects of stress include mood changes, changes in appetite, withdrawal, loud outbursts, learning difficulty, poor concentration, irritability, guilt, and anger.

Coping with Stress

We operate best at an optimum stress level. Not having enough stress leaves us feeling bored and lethargic. Having too much stress damages our performance in the short term and can lead to unhappiness, exhaustion, or burnout in the long term. It may not be initially obvious what causes your stress, whether you are effective in controlling it, or what your optimum stress level is. Some ways to cope with stress to help you find your optimum levels include

- Keeping things in perspective;
- Understanding why stress occurs;
- Exercising and eating right;
- Making an action plan for stress; and
- Thinking positive!

You joined the Coast Guard to become a part of something bigger. There may be times when you are under a lot of stress. Realize that you are not alone. The services provided by the Employee Assistance Program are available for professional assistance with stress management.

ADDICTIVE BEHAVIORS

The Coast Guard defines substance abuse as the use of any substance (alcohol, caffeine, food, prescription or illicit drugs, etc.) by a member that causes problems or places the member's safety at risk. Abuse is the use of alcohol or drugs to a point that it interferes with duty, health, personal behavior, family, community, or safety. When people abuse drugs or alcohol, they are putting many things on the line. They can develop health problems, career problems, financial problems, family problems, and legal problems, among many others.

Alcohol

Alcohol is a well-recognized central nervous system depressant. It is one of the most frequently used and abused drugs in our society. Even small amounts of alcohol in the blood can seriously impair judgment, reflexes, and muscular control and can reduce the restorative effects of sleep. The level of alcohol in the body varies with the frequency and amount of alcohol intake, the length of time following cessation of drinking, and an individual's body weight. A zero-alcohol level is essential for duty personnel to meet the rigorous demands of operations. Detectable blood alcohol or symptomatic hangovers are causes for restricting personnel from operations. Although some people metabolize alcohol quicker than others, at least twelve hours must elapse following consumption of alcohol before reporting for duty, to allow an adequate margin of safety before resuming operations.

It is okay not to drink. Alcohol has been a part of our society for a very long time, and people have had problems with it for just as long. If you drink, do so responsibly. "It's okay not to drink" means not only is abstinence okay but also knowing your limits is okay. It means it is okay to stop after one or two drinks and not to feel pressure from your friends to drink more.

It also means that you, in turn, should not pressure your friends to drink more if they want to stop. The Coast Guard can provide you with professional help if you develop a problem with alcohol, but the best medicine is preventive medicine. Drink responsibly or not at all, and remember, it really is okay not to drink.

Alcoholic beverages are an accepted part of American life, and "social drinking" is considered normal behavior. The Coast Guard is a representative of our society. When a person's drinking creates problem situations, that person has an alcohol problem. The Coast Guard has a problem when alcohol problems cost the service time lost through absence, sickness, administration, and performance. A person drinking sensibly is usually not a problem. Think of the word "sense," and use good sense by keeping all your senses alert and limiting yourself accordingly. Don't drink on an empty stomach, when you are fatigued, when you are taking medication, or when you have to drive or operate machinery.

If a person develops an alcohol problem, that person should be encouraged to get professional help. Alcohol abuse training is not intended for "down and outers" only but also for less affected people. A person need not "hit bottom" to get help with an alcohol problem. That's why the Coast Guard offers help to its members. Some training programs even offer college credits to their participants.

The Coast Guard considers alcohol abuse a serious threat to job performance and safety. If you have an alcohol-related incident, your command will normally provide counseling for you. If there is a second alcohol incident, you will normally be processed for discharge. An alcohol incident is defined as alcohol abuse that has an outward detrimental effect on the abuser or others, such as involvement in any injury, loss of duty status, inability at any required time to properly perform duties, damage or loss of property, or violation of the UCMJ or federal, state, or local laws due to alcohol consumption.

Tobacco

The nicotine contained in tobacco is a quick-acting poison. Excessive smoking causes depression of the nervous system and impairment of vision. The carbon monoxide resulting from the combustion of tobacco is absorbed by the bloodstream in preference to oxygen, resulting in a lowering of altitude tolerance and irritation of the respiratory system.

Tobacco is one of the most harmful and addictive substances available in our society. It is responsible for more deaths and more disease than any other single agent. It is as addictive as heroin, so once you start, it is hard to get free of its hold. Cigarette smoking not only leads to heart disease and cancer, but it also causes you to have wrinkled skin, so you look prematurely old. Smokeless tobacco (dip, chew, etc.) can give you mouth and throat cancer while you are still young.

Caffeine

Contained in coffee, tea, and many soft drinks, caffeine can produce an adverse effect on the body. The amount of caffeine contained in just two cups of coffee appreciably affects the rates of blood flow and respiration. In small amounts, coffee can be considered a nervous system stimulant. Excessive amounts may produce nervousness, inability to concentrate, headaches, and dizziness. Individuals accustomed to daily intake of caffeine may develop headaches and experience a loss of sharpness if daily intake is stopped or significantly curtailed.

Drug Use

Illicit drug use and the non-medical abuse of drugs that are legally available undermines morale, mission performance, safety, and health and will not be tolerated in the Coast Guard. Illicit drugs include opium, heroin, cocaine, amphetamine, lysergic acid diethylamide (LSD), methamphetamine, phencyclidine, barbituric acid, and marijuana as listed in Article 112 of the UCMJ; substances listed on the schedule of controlled substances prescribed by the president for military personnel; and those substances listed in schedules I through V of the Controlled Substances Act (21

U.S.C. 812). Regardless of state measures to legalize certain controlled substances such as marijuana, the use, possession, manufacture, distribution, import, export, and introduction of illicit controlled substances onto military installations, vessels, vehicles, and aircraft remains illegal under the UCMJ and federal law. A service member who uses marijuana purportedly in compliance with state or foreign law is still in violation of the UCMJ. Service members who have had a drug incident are processed for discharge.

To eliminate drug use, commands randomly select members to submit urine samples, which are tested for evidence of drug use. Positive results from a confirmed urinalysis test constitute a drug incident, and the member will be processed for discharge. There is no second chance.

SUICIDE AWARENESS AND PREVENTION

Suicide is the act of intentionally killing oneself and is the second-leading cause of death among individuals between the ages of fifteen and thirty-four. More Americans died by suicide than in motor vehicle traffic-related injuries, and there are twice as many suicides each year as homicides. Death is just the "tip of the iceberg." More than 9.8 million adults report having serious thoughts of suicide, 2.7 million report making a suicide plan, and 1.4 million report a suicide attempt. Approximately 500,000 people per year require emergency medical treatment as a result of an attempted suicide. For every person who dies by suicide, more than 30 others attempt suicide. The suicide rate among males has remained approximately four times higher than among females. The Centers for Disease Control and Prevention states that suicide is the tenth leading cause of all deaths in the United States.

Danger Signs

It is difficult to identify individuals who are at the greatest risk for suicidal behaviors. Because of this difficulty, suicide prevention efforts are directed at improving and providing medical care for those already struggling with suicidal behaviors, intervening, educating, and maximizing efforts and resources for suicide prevention. Understanding the danger or warning signs is a start to understanding why someone might want to take their own life. It can help dispel the myths that suicide is a random act or results from stress alone. The importance of certain risk factors and their combination vary by age, gender, and ethnicity. An easy-to-remember mnemonic for recognizing the danger signs of suicidal behavior is the phrase "IS PATH WARM?"

IDEATION: Thoughts of suicide expressed, threatened, or written.
SUBSTANCE ABUSE: Increased or excessive alcohol or drug use.
PURPOSELESSNESS: Seeing no reason for living or having no sense of meaning or purpose in life.
ANXIETY: Feeling anxious, agitated, frequent nightmares, unable to sleep or sleeping all the time.
TRAPPED: Feeling trapped, like there is no way out.
HOPELESSNESS: Feeling hopeless about self, others, or the future.
WITHDRAWAL: Withdrawing from family, friends, usual activities, and society.
ANGER: Feeling rage or uncontrolled anger, seeking revenge for perceived wrongs.
RECKLESSNESS: Acting without regard for consequences, excessively risky behavior.
MOOD CHANGES: Experiencing dramatic changes in mood or having unstable moods.

Having a mood or a substance use disorder can greatly increase the risk for suicidal behaviors. Suicide rates are particularly high among individuals with mood disorders such as depression and bipolar disorder. Suicidal thoughts or behaviors are common among people with bipolar disorder, and suicide rates are estimated to be more than twenty-five times higher than the general population. Alcohol and drug abuse are second to mood disorders as the most frequent risk factors for suicidal behaviors. Alcohol was a factor in approximately one-third of suicides reported in sixteen states. More than 60 percent of suicidal deaths occur among individuals with mood disorders, and more than one in twenty adults have depression at any given time. Depression is also a contributing factor.

Responding to Suicide

Rarely is suicide spontaneous. It is usually thought out far in advance, and numerous warning signs are almost always given. Becoming aware of these clues and the severity of the person's problems can help prevent a tragedy. Many people convey their intentions directly with statements such as "I feel like killing myself," "I don't know how much longer I can take this," or "Life does not seem worth living anymore." It is important to listen to these cries for help. Tragically, suicide investigations frequently find people disbelieving a victim's statements regarding suicide or don't report it because they don't want to get the victim in trouble. If you make a promise to the victim not to tell, break it! Get the victim help! An easy-to-remember mnemonic for responding to the danger signs of suicidal behavior is ACE, Ask–Care–Escort.

ASK. This is often not an easy task, but it is essential that you do ask. It helps to role-play the question, or at least review in your mind how you will ask it. ASK the question directly: "Are you thinking about killing yourself?" or "You made a comment about ending your life that concerns me. Are you thinking about killing yourself?" No matter what the situation, follow up. Ask questions and get additional information if you can. Revisit risk factors and danger signs.

CARE. Be willing to listen and allow the person to express feelings in private. Talk openly about suicide. Show them you CARE. Use active listening skills, encourage person to seek help, be genuine and express optimism, do not promise confidentiality, and, as the situation warrants, remove or take possession of any means that could be used for self-injury.

ESCORT. Never leave the person alone unless you are convinced that they are not suicidal, and the person has convincingly denied that they are suicidal. If you have any doubts, ESCORT the person to the nearest emergency room, health care provider, or mental health care professional. Have a positive attitude and convey that you are genuinely going to help them. Use any means necessary to save a life. Remember, you do not have to deal with this yourself. For help, contact any of the following:

Chaplains
Medical hospitals/clinics/medical officers
Military Crisis Line (1-800-273-8255, press 1)
CG Regional Health, Safety, and Work–Life (HSWL)
CG Support Program (1-855-CG SUPRT)
National Suicide Prevention Lifeline (1-800-273-TALK)

SEXUALLY TRANSMITTED DISEASES

Sexually transmitted diseases (STD) were for many years "swept under the rug" and not mentioned in polite society. STDs are infectious diseases that can be transmitted only by intimate contact between two people. Anyone who is sexually active can catch an STD. The frequency of contact and number of sexual partners determine a person's risk of contracting STDs. Prostitutes or anyone who has frequent sexual contact with different partners are at high risk of contracting STDs and passing them on to others. All STDs can be transmitted from an infected person to an uninfected person through sexual intercourse. Syphilis can also be transmitted by a kiss if an infected person has an open sore on the lips. A woman can transmit syphilis to her unborn child or gonorrhea to her child at birth.

The incubation period, the time from contact until the first symptoms appear, varies for each STD: ten days to ten weeks for syphilis; two to ten days for gonorrhea and chancroid; longer times for others. Some types of gonorrhea, particularly in women, have no symptoms. A person who has become infected can transmit the disease to another before signs of infection appear. There is no way to tell that a person is not infected. Latent syphilis, the state in which clinical signs and symptoms of infection are absent, may appear as early as four years or as late as twenty years after infection. Among the infinite variety of results to be expected are destructive ulcers, disease of the heart or blood vessels, blindness, and insanity. Other kinds of STDs have other kinds of results, none of them pleasant.

STD control is a worldwide effort, but it all depends upon the individual. Medical department personnel will emphasize that individual abstinence from sexual intercourse is the one sure way to avoid STDs. Medical department personnel can also instruct in the effective use of condoms as a preventive measure.

There is no military punishment for contracting STDs, but it is misconduct to conceal an infection. It is important that you conduct your social life so as to avoid contracting STDs and to report for medical attention promptly at the first sign of infection. This is necessary for two reasons: to commence treatment to cure the infection, and to make certain that the person from whom the infection was received is identified and treated. Remember—STDs don't just happen. They can usually be cured, but prevention is a lot less trouble and pain.

Acquired Immune Deficiency Syndrome (AIDS)

AIDS is a deadly disease that breaks down the body's resistance. The cause of the disease is the human immune deficiency virus (HIV), which is transmitted through exposure to the blood, blood products, or semen of a previously exposed person. Means of exposure include sex, blood transfusions, and hypodermic needles. Not everyone who is exposed to the virus develops immunological deficiency or clinical illness, but the virus remains in the body for life. An individual who is exposed but who does not develop any illness can still transmit the virus to others. Once an individual develops immunological deficiencies, that person is prone to opportunistic illness, such as pneumonia and cancer. Victims die from these secondary illnesses rather than from the virus itself. Exposure to the virus can be prevented. The Coast Guard recognizes the severity of this illness and has initiated a testing program to detect exposure to the HIV virus. Applicants for regular or reserve service, officer or enlisted, or for the Coast Guard Academy who test HIV positive will be rejected from Coast

Guard service. Members of the Coast Guard and Coast Guard Reserve who test positive will be considered not fit for duty and evaluated by a medical review board. Members who test positive but do not demonstrate immunological deficiencies or clinical illnesses may be retained on active duty but will not be assigned to certain duty, including sea duty, isolated duty, aircrew duty, or small-boat crew duty. The Coast Guard constantly reviews its policy on AIDS as more is learned about the disease.

WORK–LIFE

Serving well requires single and married members to balance their military duties, personal and financial obligations, and the demands of an immediate or extended family. The purpose of the ongoing Work–Life Program is to strike a reasonable balance between the needs of the Coast Guard and the needs of our members and their dependents. The goal is to create a total quality environment by which the Coast Guard will achieve organizational excellence and continue to provide valued services to our members. Work–life balance enhances diversity, recruitment, retention, and the job performance of the Coast Guard's workforce. It builds support systems for military and civilian members as well as their families.

The Work–Life staff provides information that is vital to members and their families as they face the challenges and opportunities of Coast Guard life. The Work–Life Program offers the following support services.

Child Development Services Specialist

The child development services specialist is the point of contact and subject-matter expert for all active-duty Coast Guard members requesting a Family Child Care Certification and those seeking childcare services. They champion opportunities for maximizing the educational/academic success for children, and they identify state and local resources for children and families. They develop, promote, implement, and monitor the Coast Guard in-home childcare program.

Employee Assistance Program (CG SUPRT)

This program helps you resolve personal problems before they negatively affect your health, relationships with others, or job performance. CG SUPRT professionals can help you with marital and family problems, depression and anxiety, work-related concerns, career transition issues, personal growth, and development. In addition, CG SUPRT offers value-added life-balance services such as parenting and childcare, adoption, K–12 education, services for older adults, moving and relocation, and pet care.

This program can be accessed 24 hours a day, 365 days a year, by calling one toll-free number: 1-855-CG SUPRT (247-8778). After calling the toll-free number, you will speak with an experienced and specially trained counselor who will assist you in getting the help you need. Your participation in the program will be treated confidentially in accordance with state and federal laws. You may also visit the CG SUPRT website at www.CGSUPRT.com.

Employee Assistance Program Coordinator

The employee assistance program coordinator (EAPC) serves as the point of contact for all employee assistance issues and liaisons with the CG SUPRT program. The EAPC provides referrals to CG SUPRT for personal and life-skills counseling for active duty, civilians, and their families. Some types of assistance are suicide awareness and prevention training, workplace violence and threatening behavior prevention training and intervention, and critical incident stress management support.

Family Advocacy Specialist

The family advocacy specialist is a mental health professional who is trained to deal with family violence and crisis intervention. Family violence; child, elder, or spouse abuse; and child neglect are serious issues. The Coast Guard does not consider them to be private family matters. The purpose of this program is prevention, identification, and treatment. The goal of the program is to stop the violence, successfully rehabilitate the family, and retain the member.

Family Resource Specialist

The family resource specialist (FRS) helps Coast Guard members find childcare services, identify eldercare options, and respond to the special needs of their dependents. The FRS provides information on adoption resources as well as educational scholarships for Coast Guard dependents. The FRS is responsible for the enrollment of dependents who have special needs such as medical, physical, psychological, emotional, or educational disabilities. This is required to assist in assigning duty locations where services can meet the family's special needs.

Ombudsman

The ombudsman is an important resource who can refer families to services available within the community. The ombudsman is trained by the ombudsman coordinator. The ombudsman coordinator also ensures that the ombudsman is kept up to date on community services and events. What can an ombudsman do?

- Direct grievances or suggestions to the right department
- Pass important information to families
- Address military problems between member and command

Personal Financial Manager

The personal financial manager (PFM) is an accredited financial counselor and the designated subject-matter expert in all areas of personal finances. The PFM provides financial counseling, financial training, and financial information and referrals. Areas that are normally addressed with the PFM include goal setting and spending plans, transitional and security clearance challenges, debt reduction, credit report and identity theft assistance, Thrift Savings Plan questions, and retirement and estate planning strategies.

Transition Relocation Manager

The transition relocation manager helps active-duty members and their dependents during the relocation process. The transition relocation manager provides information and referral services on housing, the transfer process, overseas transfers, and spouse employment. This program is a vital link to ensure your move goes smoothly. In addition, the transition relocation manager also provides career and job skills information to service members transitioning from military service to the civilian workforce, such as interviewing techniques, resume writing, job search, and veteran's benefits.

Legal Assistance

Coast Guard legal assistance attorneys assigned to the Legal Service Command provide advice and counsel regarding personal legal issues to thousands of service members, dependents, and retirees each year at no cost. These issues may involve family law, estate planning, consumer law, landlord–tenant relations, immigration, or many other topics. They will interview, advise, and assist members of the Coast Guard and their dependents who have legal problems and, in certain cases, refer them to competent attorneys. All matters are treated confidentially and will not be disclosed without your permission.

Non-Appropriated Fund Activities

Most bases have an exchange, laundry, tailor shop, barber shop, snack bar, enlisted club, and so forth. These functions are called non-appropriated fund (NAF) activities because they are supported largely by their own profits rather than by government funds. Some of the profits are used by the command for recreational equipment for the crew, for crew parties, or for recognition items. Some of the profits are distributed to smaller units not having NAF activities in order that they may also benefit. You are entitled to use the exchanges, stores, and other NAF facilities at any Coast Guard, Army, Navy, Marine, Air Force, or Space Force installation.

Morale, Well-Being, and Recreation (MWR) Program

The MWR program offers recreation programs and leisure activities to military personnel, their families, and guests. MWR programs include sports activities and facilities such as bowling, golf, tennis, swimming, and stables; hobby facilities and activities including crafts, auto shop, and woodworking; general recreation and family support activities such as theaters, child development centers, campgrounds, and recreation areas; and military clubs, community centers, and food service facilities. Learn more about MWR programs near your unit and at locations you may be visiting at http://www.coastguardmwr.org.

Coast Guard Exchange System (CGX)

The Coast Guard Exchange system provides quality merchandise and services of necessity and convenience to Coast Guard personnel at competitive prices through exchange stores, gasoline stations, specialty stores, an online store and service activities including barber shops, laundries, and tailors. CGX is a military activity that operates as a business enterprise to maintain CGX locations in the United States and abroad and to provide supplemental funding to MWR programs. Learn more at http://www.shopcgx.com.

American Forces Travel

American Forces Travel is the only official U.S. Joint Services' MWR platform offering travel bookings for the military and government community. A portion of all bookings go back to support military quality of life programs, so members can feel confident knowing that their purchases are helping to fund events and activities made just for them. Learn more at http://www.americanforcestravel.com.

NONPROFITS

Many nonprofit organizations exist to provide support to active service members, their families, and veterans. Several are focused primarily on supporting members of the Coast Guard, while others support all of the armed forces. Such organizations are a great resource to help meet individual financial needs during crisis or for education, community building, and recognition of outstanding performance. Coast Guard–focused organizations include the following.

Coast Guard Foundation

Founded in 1969 to support Coast Guard families, bolster unit morale, honor the Coast Guard's mission at the national level, and more, the Coast Guard Foundation offers disaster relief, tragedy assistance, a Fallen Heroes Scholarship, procurement of fitness and outdoor recreation gear for units, the construction of community spaces, and tribute events such as Run to Remember. Learn more at http://www.coastguardfoundation.org.

Coast Guard Mutual Assistance

Coast Guard Mutual Assistance provides short-term financial assistance to members of the entire Coast Guard family who find themselves in unexpected financial need due to circumstances beyond their control. All such assistance is based on financial need. A wide range of financial assistance is available to members from Coast Guard Mutual Assistance through programs for emergency loans, debt management, and medical, housing, and educational assistance. Learn more at http://www.cgmahq.org.

Coast Guard Chief Petty Officers' Association / Coast Guard Enlisted Association

Established in 1969 to advance the interests of the United States Coast Guard Chief Petty Officer Corps, the association encourages integrity and fidelity to the service and fellowship within our ranks. In 1991 the Coast Guard Enlisted Association was chartered by and for junior enlisted personnel under the auspices of the Coast Guard Chief Petty Officers' Association. Over fifty chapters are now established in the continental United States, Alaska, and Hawaii. Through a strong and united network of over 10,000 members, these associations work to advance the interests of our membership through community service, scholarships, and fundraising and by partnering with military and veterans' organizations, both locally and nationally. They also represent members on Capitol Hill as a partner of The Military Coalition, promoting legislation to enhance the lives of past, present, and future members of the U.S. Armed Forces and their families. Learn more and join at http://www.uscgcpoa.org.

Coast Guard Tactical Law Enforcement Association

This association supports the families of seriously injured or killed deployable specialized forces with an emergency assistance fund and the Nate Bruckenthal Memorial Scholarship. Learn more at http://www.cgtlea.org.

Coast Guard Combat Veterans Association

Established in 1985 as a fraternal organization of active-duty, retired, reserve, and honorably discharged veterans of the U.S. Coast Guard who served in or provided direct support to combat situations recognized by an appropriate military award, this association is dedicated to making the public aware of the Coast Guard's service and participation in significant wartime events in United States history. It also provides illumination of the gravesite of U.S. Coast Guard Medal of Honor recipient Douglas Munro. Learn more at http://www.coastguardcombatvets.org.

American Red Cross

The American Red Cross Hero Care Center is available 24/7 to relay emergency communications regarding an immediate family member of a service member and to initiate a request for Red Cross emergency assistance for members of the military currently serving on active duty. The Red Cross role is to independently verify the emergency with a hospital, doctor's office, or funeral home and to enable the service member's commander to make an educated decision regarding emergency leave and then to provide transportation assistance or financial assistance, if needed. The American Red Cross facilitates emergency financial assistance on behalf of military aid societies. These aid societies determine the financial assistance package that will be offered—a grant or a loan. The Red Cross is the mechanism to expedite access to these financial resources. To speak to a Red Cross emergency communications specialist call (877) 272-7337 or visit https://www.redcross.org/get-help/military-families/emergency-communication.html.

26

ADMINISTRATION AND FINANCES

There are many benefits to serving in the Coast Guard, including some financial ones. Service members receive pay and benefits for the work and time sacrificed while serving their country. This chapter outlines the benefits you are entitled to by serving and the administrative processes needed to take full advantage of them.

SERVICEMEMBER, DEPENDENTS, AND BENEFICIARIES

It is essential that servicemembers update their Coast Guard service records when they gain or lose a family member or select a new beneficiary in order to ensure appropriate compensation, entitlements, and benefits. Certain family members, such as a spouse or child under twenty-one, are automatically entitled to dependency status. Other family members, such as parents, siblings, and adult children who are full-time students or incapacitated, require an initial dependency determination by the Defense Finance and Accounting Service.

MilConnect

MilConnect is a convenient self-service portal that allows Department of Defense (DoD) affiliates and beneficiaries to manage their benefits and records including DEERS enrollments, health care coverage, education benefits, and Servicemembers' Group Life Insurance (SGLI). Access the website at https://milconnect.dmdc.osd.mil/milconnect/.

Defense Enrollment Eligibility Reporting System

Update Defense Enrollment Eligibility Reporting System (DEERS) enrollment forms and other important papers, such as the SGLI election form, the designation of basic allowance for housing (BAH) form, and the emergency data form, immediately upon marriage, divorce, birth of a child, or death of a family member. Enrollment in DEERS requires you to present documents to verify identity and relationship to you. These documents may include original or certified copies of your Social Security card, government-issued photo ID, birth certificate, adoption papers, marriage certificate, divorce certificate, legal court order, or death certificate. To update DEERS records, contact Defense Manpower Data Center Support Office at 1-800-538-9552.

PAY AND COMPENSATION

Coast Guard members and their families are entitled to pay, access to the base commissary and exchange, family assistance, health care, a generous leave policy, travel opportunities, and—if the Coast Guard becomes a career—retirement.

Payslips

Each service member has access to an electronic copy of his or her payslip posted monthly to Direct Access. The payslip contains a monthly accounting of your leave and earnings. Members should read their payslip carefully each month.

Basic Pay

Active-duty members are paid twice a month. Paydays are the first (end-month) and the fifteenth (mid-month) of each month, or the business day preceding the first or the fifteenth, if either should fall on a federal holiday, Saturday, or Sunday. You should not set up automated payments or assume funds will be available prior to published paydays and should verify with your financial institutions as to the actual time funds are available in accounts.

There are many categories of pay; some are taxable, and others are not—usually "pay" is taxable and all "allowances" are not. Speak with your servicing yeoman or Servicing Personnel Office if you have any questions about your pay. Service members are only responsible for state taxes to their state of legal residence, not the state in which they are stationed. However, spouses can elect to use the legal residence of the service member for purposes of filing their state and local taxes. This law went into effect December 2018; consult a Coast Guard Legal Assistance office for assistance.

Basic Allowance for Subsistence

All active-duty personnel receive a tax-free food allowance each month, called basic allowance for subsistence (BAS). It is intended to offset costs for a member's meals and is not intended to offset the costs of meals for family members. The management of officer and enlisted BAS differs. Officers are authorized the officer BAS rate and are always expected to pay for government-provided meals. Enlisted members are authorized the enlisted BAS rate but, depending on their duty station assignment and government dining facility availability, can have the BAS amount adjusted to pay for government meals provided at the duty station at no cost to the member. At other duty locations, such as on board a Coast Guard cutter, enlisted personnel are required to use the onboard dining facility or galley so their BAS is deposited directly into the cutter's accounts vice paid to the member.

Housing Allowance

Coast Guard members on active duty entitled to basic military pay may be authorized a monthly BAH based on a member's pay grade, dependency status (with or without dependents), and duty station location. When authorized, BAH is paid in the continental United States (CONUS), District of Columbia, Alaska, and Hawaii. Outside of these locations, the housing allowance paid is called the overseas housing allowance (OHA). BAH is paid to offset the cost of private-sector housing when a member with or without dependents is not assigned to government-owned or -leased housing. BAH is not designed to cover all expenses because personal choices, such as education needs, commute time, or spouse employment, may determine housing needs. In BAH-payable locations, the cost of rental properties in each designated geographic location is surveyed annually to determine the correct housing allowance to enable members to afford suitable rental housing within a reasonable distance of the member's duty location. By design, BAH does not consider mortgage costs should a member decide to buy a home. Check with your new unit's Servicing Personnel Office or Housing Office prior to making any housing commitments.

Cost-of-Living Allowance

A cost-of-living allowance is paid to members residing in high-cost areas. It is based on your pay grade, years in service, whether you have dependents, and the location of your unit. Some members do not receive any cost-of-living allowance because the cost of living in the area they are assigned is not high.

Clothing Maintenance Allowance

Enlisted military members are paid a monthly allowance for the maintenance and replacement of required uniform items.

Sea Pay

All members assigned to cutters receive sea pay. Sea pay is an incentive pay that varies depending on your pay grade and years of sea service as well as the type of cutter you are assigned.

Special-Duty-Assignment Pay

Special-duty-assignment pay provides an additional monthly payment as both an incentive to retain enlisted personnel required to perform extremely demanding duties or duties demanding a large degree of responsibility and as an inducement to persuade qualified personnel to volunteer for such duties. For example, officers-in-charge receive special-duty-assignment pay to compensate for extremely long hours and the high degree of responsibility.

Hardship Duty

Hardship duty is compensation paid to members assigned outside the CONUS in locations where living conditions are substantially below the standard most members in the CONUS would generally experience. Rates are payable in increments of $50, $100, or $150 a month based on the level of hardship in a given area.

Hazardous-Duty Incentive Pay

Hazardous-duty incentive pay is authorized for members who, under orders, are required to participate in hazardous duties, such as aircraft tie-down crews on board major cutters. The pay is based on duty type and pay grade.

Family Separation Allowance

The family separation allowance provides compensation to those with dependents for added expenses incurred because of a forced family separation due to a dependent-restricted assignment or an underway patrol or temporary-duty period exceeding thirty days. After the initial thirty-day period, personnel are paid an allowance of $250 per month until the assignment ends.

Social Security

All service personnel participate in the Social Security program. A portion of your pay is deducted each month for this purpose. The amount of deduction is matched by the Coast Guard. These funds will earn Social Security benefits for you and your dependents.

BLENDED RETIREMENT SYSTEM

The fiscal year 2016 National Defense Authorization Act provides our military force with a modernized retirement plan built for retirement savings. The four parts to the Blended Retirement System (BRS) include

- Defined Benefit: Active duty are eligible after twenty years of service. Reservists are eligible after twenty years of service and beginning at age sixty. The full retirement pay annuity is calculated by taking 2.0 percent times years of service multiplied by basic pay.
- Defined Contribution: The DoD automatically contributes 1 percent of your basic pay to your Thrift Savings Plan (TSP) after sixty days of service. Matching contributions start at your third year of service through the completion of twenty-six years of service up to 5 percent government automatic and matching contributions. When eligible, you must invest 5 percent of your pay to get the full 5 percent from the government. See the chart below. You're fully vested (the money is yours to keep) at the beginning of three years of service and goes with you if you decide to leave the service.

You Contribute (%)	DoD Auto Contribution (%)	DoD Matches (%)	Total (%)
0	1	0	1
1	1	1	3
2	1	2	5
3	1	3	7
4	1	3.5	8.5
5	1	4	10

- Continuation Pay: Under BRS, you may be eligible to receive a one-time, midcareer bonus payment in exchange for an agreement to perform additional obligated service. It is payable between completion of eight years of service but before completion of your twelfth year of service. Each service publishes guidance related to continuation pay rates. Rates for continuation pay depend on a number of factors, including the specific needs of the service (e.g., retention rates, hard to recruit career fields). This one-time bonus is in addition to any career-field-specific incentives or retention bonuses you may receive during your career.

- Lump-Sum Option: Service members who retire from the military with at least twenty years (or twenty qualifying years in the Reserve) will have the opportunity to get a portion of their defined benefit as a lump sum upon retirement. That is, you can receive money up front as an advance on your own retired pay. Service members can elect to receive either a 25 percent or 50 percent discounted portion of their monthly retirement pay as a lump-sum payment, in exchange for reduced monthly retirement benefits.

LEAVE

Vacation time is referred to as annual leave in the military. Every service member earns two-and-a-half days of annual leave every month, totaling thirty days each year. Parental leave may be authorized in accordance with current instruction or policy. Not all time off in the Coast Guard is charged to annual leave. Travel time between duty stations, house hunting during a PCS (permanent change of station) move, and convalescent leave are examples of time off that is not charged to annual leave.

Emergency Leave

Emergency leave can be granted for grave illnesses, serious injury, or a death in the immediate family. Emergency leave comes out of a member's annual leave but does not require the same advance notice that annual leave does. District commanders and commanding officers may grant emergency leave to officers not to exceed thirty days at any one time, and to enlisted members not to exceed sixty days at any one time. If you are deployed and a family emergency arises, contact your command ombudsman, your chaplain, or the American Red Cross. They will notify your command of the emergency. If the emergency requires your presence at home and if you do not have enough saved annual leave, your commanding officer may approve advance leave. This leave will be earned back as you continue to serve.

LIBERTY

Liberty refers to the time when a member is not on duty, such as weekends and holidays. Regular liberty is a period not to exceed three days (seventy-two hours) commencing at the end of the normal working hours on a given day and expiring with the start of normal working hours on the next workday. Public holiday weekends and periods, which by the direction of the president of the United States, are extended to exceed seventy-two hours, are considered regular liberty periods.

Special liberty refers to time off during normal work hours that is not charged to leave. Commanding officers and officers-in-charge have the authority to combine special liberty with holidays, not to exceed four days. Special liberty may be awarded for especially hard work or outstanding job performance. Commanding officers may also grant special liberty to members who need to take care of pressing personal business. Four days of special liberty may be granted to any member experiencing the death of an immediate family member or legal guardian. This type of special liberty may be combined with normal liberty holidays or regular leave, at the request of the member.

All liberty is controlled by your command and is granted as workloads and operating schedules permit. Remember, liberty is a privilege, not a right.

Federal Holidays

The following government-approved holidays, established by law, are observed except when military operations prevent observation:

- New Year's Day, 1 January
- Martin Luther King Jr.'s Birthday, the third Monday in January
- Washington's Birthday, the third Monday in February
- Memorial Day, the last Monday in May
- Juneteenth, 19 June
- Independence Day, 4 July
- Labor Day, the first Monday in September
- Columbus Day, the second Monday in October
- Veterans Day, 11 November
- Thanksgiving Day, the fourth Thursday in November
- Christmas Day, 25 December
- Other, when directed by proper authority under law

When such holidays fall on a Saturday, the preceding Friday is also considered a holiday. When such holidays fall on a Sunday, the following Monday also will be considered a holiday. Holidays are charged as leave if they fall on or within the effective dates of leave. For example, if Veterans Day is Tuesday 11 November and you have requested leave Monday 10 November through Thursday 13 November, you will be charged four days leave for 10, 11, 12, and 13 November.

Combining Leave and Liberty

Two or more requests for leave that immediately precede and immediately follow a regular liberty period are not authorized. A member may not be on leave, immediately return to liberty status, then immediately resume leave status. Instead, the whole period of absence is charged as leave.

> LEAVE LIBERTY LEAVE: IS NOT AUTHORIZED.
> LIBERTY LEAVE LIBERTY: IS AUTHORIZED

HOUSING

The Coast Guard is committed to ensuring that Coast Guard members have access to adequate housing. Adequate housing is available through a number of avenues, including Coast Guard–owned housing, DoD housing, housing in the community, public–private venture housing, and government-leased housing. Assignment to Coast Guard–owned family and unaccompanied personnel housing (UPH) is mandatory when it is available and adequate. Assignment to leased housing is not mandatory. For locations outside the CONUS, where Coast Guard–owned family housing or UPH is not available, housing officers must refer members to DoD, Department of State, or other sponsoring agencies' housing offices. To ensure efficient use of Coast Guard–owned family housing and UPH, all members in receipt of PCS orders are required to submit an application for assignment to military housing, Form CG-5267, to their new unit's housing officer prior to receipt of final orders. This form is a requirement prior to making any housing arrangements at your new duty station. The housing office can advise you of your options and help you make the appropriate choices without conflicting with Coast Guard policy.

The housing officer is responsible for the overall management of all housing on a Coast Guard installation, including permanent party family housing and unaccompanied personnel housing. The housing officer and their staff are ready to support you in the transition to your new home and community. Do NOT make any housing arrangements prior to contacting the housing officer.

Community-Based Housing

Due to the remoteness of many Coast Guard assignments from established bases, most members are provided a housing allowance and are allowed to find a suitable home in the community. Members assigned to duty stations within the United States are provided a BAH, and members stationed outside of the United States are provided an OHA. Allowances vary based on a member's duty station, rank, and dependency status and are designed to cover an average member's rent, utilities, and renter's insurance. The housing office can help you find safe and adequate housing in the community. You must receive a release from mandatory assignment to Coast Guard housing before signing a lease. We recommend you review your lease with Coast Guard Legal Assistance prior to signing and have them explain anything you do not understand. Your lease should include a military clause that will allow you to terminate the lease without penalty if you receive PCS orders or leave active duty while the lease is in effect. The lease must clearly state who is responsible for payment of utilities and repairs. Complete a walk-through of the home before moving in to identify and document damages so you will not be held responsible when you move out.

Home Loans

The Department of Veterans Affairs offers home loans for qualified service members who wish to buy a home. These loans usually offer a lower interest rate and a lower down payment than commercial loans. Information on guaranteed home loans is available at your bank or credit union, at veteran service offices, and through real estate brokers. Please visit www.homeloans.va.gov for more information.

Personal Property Insurance

Your landlord's insurance policy and the Coast Guard will NOT cover damages or loss to your personal property. Therefore, whether you own your home, rent a home, or live in base housing, you need insurance to cover your personal belongings in case of fire, theft, wind, water, or other damages. Your housing allowance is calculated to cover a portion of your insurance costs.

MEDICAL

Free medical and dental care is provided all members of the Coast Guard on active duty. This care includes surgery and emergencies as well as routine immunizations, physical exams, and so forth. Dependents are eligible for medical care in the facilities of any of the armed services. Another medical care program is Tricare. Under this program dependents may receive treatment from civilian facilities. The government pays most of each Tricare bill, but the individual may also pay part of it. A dental plan is also available. You must change your primary care manager (PCM) when you arrive at your next duty station. You may change your dependent plan during the annual open enrollment period, typically mid-November to mid-December.

Emergencies

Call 911. In the event of an emergency: Do not get caught up in the details of which health plan you have. You do not need prior authorization. Contact your PCM within twenty-four hours or as soon as medically reasonable.

Tricare

The Department of Defense's worldwide health care program for active-duty and retired uniformed service members and their families, Tricare offers its beneficiaries different options for health care coverage that best suits the needs of the individual. While active-duty members are required to be enrolled in Tricare Prime, family members may choose a different option, depending on circumstances such as where they live, health status, and preference of health care providers. The most widely used plans for Coast Guard personnel and their family are described below.

Tricare Prime

Tricare Prime is a managed care option offering the most affordable and comprehensive coverage. It is similar to a civilian health maintenance organization. You must enroll to participate. Tricare Prime is available in the United States in designated Prime service areas. Check your zip code on the Tricare website to see if you live in a Prime service area. Tricare Prime provides comprehensive coverage including emergency care, outpatient visits, preventive care, hospitalization, maternity care, mental and behavioral health, and prescriptions.

With Tricare Prime you have an assigned PCM, either at a military hospital or clinic or from the Tricare network, who provides and directs most of your care. Your PCM will refer you to a specialist for care that he or she cannot provide and will coordinate with your regional contractor for authorization, find a specialist in the network, and file claims on your behalf. Active-duty service members and their families have no out-of-pocket costs for any type of care as long as care is received from the PCM or with a referral. All other beneficiaries pay annual enrollment fees and network copayments.

If you are an active-duty member or an activated reservist, you must enroll in Tricare Prime. All other eligible beneficiaries have the option to enroll or use Tricare Select. Tricare Prime offers fewer out-of-pocket costs than Tricare Select but less freedom of choice for providers as you must select a provider from the network. For more information, visit https://tricare.mil/Plans/HealthPlans/Prime.

Tricare Prime Remote

Tricare Prime Remote is a managed care option that provides the benefits of Tricare Prime in designated remote locations. You must enroll to participate. Once enrolled in Tricare Prime Remote, you may select a Tricare-authorized provider as your PCM from the Tricare network. If a network provider is not available, you may select any authorized provider as your PCM. All other Tricare Prime rules and conditions apply. For more information, visit https://tricare.mil/Plans/HealthPlans/TPR.

Tricare Select

Tricare Select is a fee-for-service plan available to all non-active-duty beneficiaries throughout the United States. Enrollment is required. Tricare Select enrollees may schedule appointments with any Tricare-authorized provider, network or non-network. Referrals are not required, but you may need prior authorization from your regional contractor for some services. Costs vary based on the sponsor's military status. After you have met an annual deductible, you are responsible to pay a cost share (or percentage) for covered services. For more information, visit https://tricare.mil/Plans/HealthPlans/TS.

Tricare Pharmacy Benefit

Tricare's pharmacy benefit is available to all eligible uniformed service members, retirees, and family members, including beneficiaries aged sixty-five and older. The Tricare Pharmacy Program provides outpatient prescription drugs to more than 9.6 million individuals. Express Scripts, which administers the Tricare pharmacy benefit, provides home delivery, retail, and specialty pharmacy services. The Tricare Pharmacy Program is designed to provide the medications you need, when you need them, in a safe, convenient, and cost-effective manner. Tricare Pharmacy Home Delivery is the least expensive option when not using a military treatment facility pharmacy. You can receive up to a ninety-day supply of maintenance medications delivered directly to your home. All prescriptions dispensed through military treatment facilities, Tricare Pharmacy Home Delivery, and retail network pharmacies are rigorously checked for accuracy and against your Tricare prescription history for potential drug interactions. The Tricare retail pharmacy network includes most of the major national chains, over 100 regional chains, and more than 20,000 independent community pharmacies. If you are unsure if your current pharmacy is a Tricare retail network pharmacy, ask your pharmacist. For more information, visit www.tricare.mil/pharmacy or www.express-scripts.com/.

Federal Employees Dental and Vision Insurance Program

Family members of active-duty service members who are enrolled in a Tricare Health Plan can enroll in vision and dental plans under this program.

Dental

Dental coverage is separate from Tricare's medical coverage. The Tricare Dental Program is a voluntary dental program available to eligible active-duty and family members as well as National Guard and Reservists and their family members. United Concordia is the Tricare Dental Program contractor. The amount covered by the plan ranges from 100 percent to 50 percent, depending on the type of dental service. Additional information is available at www.tricare.mil/dental/.

Supplemental Health Insurance

Tricare supplements are insurance policies or other health benefit plans offered by a private entity that is specifically designed to supplement Tricare benefits. They usually pay whatever is left after Tricare has paid its share of the cost of covered health care and supplies. Supplementary coverage of this sort doesn't affect Tricare payments. Be cautious! Before you purchase any supplemental insurance, carefully consider which plan is best suited to your individual needs. The cost of these plans may vary depending on the services offered.

COMMISSARY AND EXCHANGE PRIVILEGES

After receiving a military ID card, the holder can begin to benefit from its advantages. Two of the most frequently used benefits are access to the commissary and the Coast Guard Exchange System (CGX). The commissary is the military version of a grocery store. Many of the same goods found in a local store are available at a reduced cost. Commissaries apply a small surcharge that pays for operating the store; however, there is no sales tax on items purchased at the commissary. The CGX is similar to a department store but is operated by the military. Prices are competitive with civilian stores, and you can normally pay by credit card, check, or cash. Sales tax is not collected on items sold through the exchange. Base stores also honor coupons. The profits from CGX go into the Morale, Well-Being, and Recreation program. Military families can save between 10 and 30 percent using these stores, due to the service's volume buying and federal tax-free status. These stores are strictly for service personnel and family members, and a military ID must be shown to make purchases. There are strict rules regarding shopping privileges, but the one stressed most is do not buy for anyone who is not a military family member unless it is a real gift. Purchases can also be made online at https://shopcgx.com. This is especially helpful to those families who do not live close to a military base.

SERVICEMEMBERS' GROUP LIFE INSURANCE

Servicemembers' Group Life Insurance (SGLI) was established to provide life insurance protection to members of the armed forces at a reasonable cost. There are currently separate plans to cover the Coast Guard members and family members. For coverage of active-duty members, your options are

- Decline coverage;
- Elect a reduced level of coverage in multiples of $50,000; or
- Elect a higher level of coverage up to a maximum of $400,000.

By law, when you first enter active duty, you are automatically insured for $400,000 coverage at a cost of $25.00/month.

Coast Guard Reserve members are entitled to SGLI. The Coast Guard automatically pays premiums for all its drilling members, regardless of pay category, unless the member specifically declines the coverage using Form SGLV-8286. If you are drilling in a non-pay status and wish to retain your SGLI coverage, you must begin sending your monthly premiums directly to the Pay and Personnel Center in Topeka, Kansas, or you risk losing this low-cost benefit.

Family Servicemember Group Life Insurance (FSGLI) is a program extended to the spouses and dependent children of members insured under the SGLI program that provides up to a maximum of $100,000 of coverage for spouses and $10,000 of coverage for dependent children. Premium rates will vary and are automatically deducted from the service member's pay each month. Once separated from the Coast Guard, your coverage under SGLI and your family's coverage under FSGLI will continue for 120 days after your separation without any further payment of premiums.

TRANSFERS

The Coast Guard requires frequent moves. The Relocation Assistance Program in the Office of Work–Life (www.dcms.uscg.mil/worklife) is available to help you with your move and provide you with information on your new home. Your sponsor at the new unit can help with location-specific information.

e-Resume and Assignment Process

During the assignment process, a shopping list of available positions is published. Members use this list to construct their e-Resume. The e-Resume is a communication tool that allows members to express their assignment desires. When creating an e-Resume, members should consider professional growth opportunities, assignment priority, evaluations, past assignments, sea/shore ratio for rating, and personal objectives. Members are encouraged to list as many positions as possible on their e-Resume, beginning with the positions they most desire. Ultimately, assignments are determined by considering multiple factors: the needs of the service, the needs of the unit, and the needs/desires of the member.

Once selected for a position, members will receive orders. Service members are encouraged to seek career counseling from their assignment officer.

Sponsor Program

The Sponsor Program was developed to help you to smoothly transition to your new unit. Your sponsor is your primary means of communicating to your new unit. They can provide information regarding your new assignment, duties, and expectations. Sponsors serve as a personal liaison to the community, housing, commissaries, markets, public transportation, carpools, and other local functions.

Overseas Duty

An overseas tour can be an exciting and rewarding experience when service and family members are properly prepared for it. The more you know about your overseas duty station, the fewer problems you will have during the transfer. The Coast Guard has many resources for information on your overseas destination that will ensure your tour is an enjoyable one. If stationed in a foreign country, passports are a requirement. Ensure that you work with the command to receive up-to-date information.

Overseas Screening

The overseas screening process ensures that necessary support services are available at your new duty station and that all medical and educational needs can be met. Make sure to address during the screening process any medical or special support needs that your family requires. Problems that are not addressed can create unnecessary hardships and result in an early return. Service members are screened for financial stability and responsibility, compliance with weight standards, alcohol or substance abuse or dependency, performance and conduct, and medical readiness.

Military One Source

When you receive PCS orders, you are entitled to shipment or storage of your household goods and personal effects, or both, if necessary. Service members and their families use the Defense Personal Property System, known as DPS, which can be accessed through www.militaryone source.mil/. This website is a one-stop shop where members can learn about and manage their move process. Training videos and presentations on the website explain how to navigate through the move process. To help transportation officials gauge the overall success of DPS and to identify the best providers, customers are encouraged to fill out the DPS Customer Satisfaction Survey after delivery of their household goods. The survey is also available online at Move.mil.

Government Bill-of-Lading Move

For a government bill-of-lading (GBL) move, the government contracts with a carrier to pack, pick up, deliver, unpack, and place your household goods at your residence. They also inventory your goods. You or your designated agent must be present for packing and pickup or delivery and are responsible for ensuring that the mover packs and inventories your property and that each item has a condition code. It takes an average of ten to fourteen days to ship household goods within the CONUS. Outside CONUS, expect delivery in about forty to sixty days. This type of move is a MUST for anyone going overseas to Alaska, Hawaii, Guam, or Puerto Rico. Authorized items include

- All personal property and affects;
- Snowmobiles, motorcycles, and boats under fourteen feet without trailers;
- Firearms within CONUS; however, you must comply with local laws;
- Overseas shipments require compliance with that country's laws; and
- A second vehicle with limited reimbursement when a Coast Guard member drives themselves and dependents in one car.

The transportation of perishable, combustible, flammable, or explosive items is prohibited.

Personally Procured Move

In a personally procured move, you move your household goods, and the government pays you an incentive of 95 percent of the cost of a GBL move. You may be eligible to receive an advance payment for the cost of the rental, gas, and packing material. You will need to provide a copy of the rental estimate. Prior written permission must be obtained from your Transportation Office before the move. You must have two weight tickets of the vehicle—one empty and one full/loaded from origin. Some locations where you may weigh vehicles include public scales, scrap yards, truck stops, and lawn and garden centers. You are responsible for submitting all receipts associated with the move with the claim (gas, oil, etc.). All of the restrictions of the GBL method also apply to the personally procured move.

Non-Temporary Storage

This method of shipment is used most often when a member is assigned to a unit where quarters are provided furnished and the member will not be living on the economy. In this instance, the government would contract with a moving carrier to pack, pick up, and place in storage at the government-approved warehouse nearest to where the household goods are located, until such time as the member receives new PCS orders or the requirement for non-temporary storage expires. Members use this method of shipment with household goods while attending a class A school.

Privately Owned Conveyance

In order to ship a privately owned conveyance (POC; also known as a POV, privately owned vehicle), you must be in receipt of PCS orders to a unit outside the CONUS or be medically unable to drive the vehicle to the next duty assignment within the CONUS. You may only ship one vehicle outside the CONUS. The DoD pamphlet "Shipping Your POC" has a complete listing of all ports that service overseas assignments. To facilitate shipment, you will drive your vehicle to the nearest port that services the location of your current or future duty station. You must have an entry approval authorization from the command outside the CONUS in order for the vehicle to be shipped. This approval is done by the Personnel Office and is usually a message. You are the on-scene inspector. If things are proceeding incorrectly, you must notify the transportation officer.

Dislocation Allowance

The purpose of dislocation allowance is to partially reimburse members who have dependents for expenses incurred in relocating your household on a PCS move and who do not occupy government quarters. If you have dependents relocating with you, you may request an advance dislocation allowance. This money is to help with expenses associated with setting up your residence for you and your dependents.

Mutual Exchange of Station and Unilateral Transfers

It is not always possible to match duty stations with your personal preference. Personnel are sent where they are needed most. Mutual exchanges make it possible for people to be sent to stations they prefer, but because there is no advantage to the Coast Guard in such a transfer, it must be made at no expense to the government. The person wanting the transfer must find someone at the other station with the same rating and experience who will "mutual" with them—swap duty stations with both paying their own expenses. If a mutual exchange cannot be made, a unilateral change of station may be requested. Two conditions for a unilateral transfer are that the person's unit exceeds its allowance for the member's rating or that the command recommends that no immediate replacement would be required if the request were granted.

Humanitarian Assignments

These are special assignments made to help a member handle a hardship of such a nature that it cannot be taken care of by emergency leave. Such transfers are usually for four months, but in special cases a PCS would be considered. A PCS humanitarian assignment is at government expense.

TRAVEL

Travel orders are issued via Direct Access for PCS, class "A" and "C" schools, and temporary-duty (TDY) assignments. When in receipt of orders, it is your responsibility as well as the unit approving/authorizing official's (AO) to read all order notes in their entirety, ensure proper travel and entitlement counseling occurs before the travel begins, and certify that the accurate reimbursement of travel entitlements is paid upon completion. You are obligated to exercise the same care and regard when incurring government reimbursable expenses as would a prudent person traveling at personal expense. You should schedule entitlement counseling with your AO or administration office as soon as possible.

Amendments

Orders can be amended in many circumstances to accommodate personal preferences—for example, when you prefer to drive your personal vehicle rather than fly commercial air as directed by the orders, although you will only be authorized reimbursement up to the cost of a government transportation request. To obtain an amendment, speak with your AO or unit administrator who, if in concurrence, will normally provide a memorandum to be attached to your original orders authorizing the change.

Government Travel Charge Card Program (GTCC)

This program provides travelers with a safe, effective, convenient, and commercially available method to pay for expenses associated with official travel. The GTCC includes individually billed accounts and centrally billed accounts (CBA). Coast Guard policy is that the GTCC should be used to pay for ALL official travel expenses while on temporary duty or temporary additional duty. Personal use of the travel card or using the travel card to pay for someone else's travel expenses is prohibited. The use of the travel card for non-official expenses may result in disciplinary actions. Payment of the GTCC bill is the responsibility of the traveler, who will then be reimbursed by the government. A CBA is a variation of the travel card, whose payment is the responsibility of the government. Use of a CBA is limited and must have prior approval.

Booking Travel

Schedule transportation and lodging as early as possible in an effort to maintain fiscal responsibility. Unnecessary or unjustified delays in making reservations precludes you from using the least expensive transportation and accommodations available. CWTSato Travel is the sole travel agent for the Coast Guard, and use of its Travel Management Center is mandatory. All reservations through CWTSato are made using your issued GTCC or, if authorized on your orders, a CBA. Learn more at https://www.dcms.uscg.mil/Our-Organizatiton/Assistant-Commandant-for-Human-Resources-CG-1/Personnel-Service-Center-PSC/BOPS/PSC-BOPS-R/GOVTrvl.ADTRAV/.

POV Mileage

A mileage allowance for using a POV for local, TDY, and PCS travel is reimbursed as a rate per mile in lieu of reimbursement of actual POV operating expenses.

Government Lodging and Meals

While TDY to a U.S. installation where adequate government quarters and meals are available, the standard government meal rate or proportional meal rate is authorized. The standard government meal rate is the authorized meal rate for installations where meals are served at "à la carte" prices. Current government meal rates can be found at https://www.defensetravel.dod.mil/site/faqmeal.cfm.

Commercial Lodging and Meals

The full per diem rate is authorized for duty in locations where government lodging and meals are not available. Per diem is the daily allowance for traveler lodging, meals, and incidental expenses. The General Services Administration establishes per diem rates for destinations within the CONUS; the State Department for foreign destinations; and the DoD for Alaska, Hawaii, and all U.S. territories. On the first and last travel day, travelers are only eligible for 75 percent of the total per diem rate for their TDY travel location. Per diem rates are broken down to account for payment of three meals a day, incidental expenses, and lodging costs. An incidental expense includes the fees and tips given to porters, baggage carriers, hotel staff, and staff on ships. Current per diem rates can be found at https://www.defensetravel.dod.mil/site/perdiemCalc.cfm.

Travel Claims

Work with your AO and unit administrator to file a travel claim upon return to your parent unit, or every thirty days when on long-term orders. Note that for any official temporary travel destination, you must provide a receipt to substantiate your claimed travel expenses for lodging and receipts for any authorized expenses over $75, or provide a reason acceptable to your agency explaining why you are unable to provide the necessary receipt.

PERSONAL FINANCE

It is important for all Coast Guard members to develop a basic understanding of personal finances to build a secure future. How we handle our finances and the decisions we make about money will significantly impact our quality of life. We all want to be financially secure and enjoy the good things that life has to offer; yet most of us have little formal training in financial management to help us attain our goals successfully. Contact your personal financial manager (PFM) at your Health, Safety, and Work–Life (HSWL) Regional Practice or your unit's command financial specialist to learn what type of financial counseling, financial education (including free classes), and other financial resources are available in your area. Additionally, the CG SUPRT Financial Wellness program can provide you one-on-one money coaching, financial webinars, and much more. To access CG SUPRT online visit www.cgsuprt.com. Click on My CG SUPRT and log in. Additionally, there is a CG SUPRT app available for download for your smartphone or device. You can also visit www.dcms.uscg.mil/financial/ for additional information and resources.

Financial Planning

We all want enough money for everything we desire, but the reality is most of us have to be mindful of our spending habits. By establishing a sound financial plan and committing yourself to following it, you can achieve your financial goals. The biggest step is to start. Take the time to develop a financial plan that will provide security for your family in case you face an emergency. You can secure your financial future by budgeting your money, not overspending, and investing wisely in opportunities that will make your money grow. Coast Guard policy states that service members are responsible for paying their debts in a timely manner. Incurring debts beyond your ability to pay is evidence of irresponsibility and may jeopardize your security clearance status, duty assignment, and qualification for reenlistment or extension of enlistment. In cases that are more serious, indebtedness may become grounds for disciplinary action or other-than-honorable discharge.

You can start on the road to improved personal finances by following these six steps:

1. Set financial goals: Life goals can include buying a home, saving for your child's education, or planning for retirement. The process involves gathering relevant financial information, setting short-term, medium-term, and long-term goals, examining your current financial status, and coming up with a strategy or plan to meet these goals given your current situation and future plans.
2. Create and follow a spending plan: A spending plan allows you to see where your money is going by documenting all sources of spending and making adjustments as needed to meet your financial goals.
3. Save for emergencies: Most financial experts agree you should have between three and six months of expenses saved in an emergency fund. This safety net will cover unexpected financial burdens in the event of an emergency, thus reducing the need to obtain necessary funds by drawing on high-interest credit cards or loans.
4. Adequately insure your family: Life is full of risk and uncertainty. Insurance plays a key role in mitigation of this risk. For example, life insurance provides financial assistance to the family of the insured in the event of the insured's death. Other types of insurance, such as homeowners, auto, and health, provide you with protection from unexpected financial loss.
5. Save and invest 10–15 percent pretax income: Saving 10–15 percent of your pretax income in a retirement savings plan can provide you with a source of income in addition to your military pension and Social Security, helping to build financial stability and independence at retirement.
6. Have an estate plan: An estate plan, which provides for your heirs in the event of your death, is covered in depth later in this chapter.

Thrift Savings Plan

The Thrift Savings Plan (TSP) is an excellent way to save for retirement. A range of investment options is offered. Contributions to your TSP account yield tax benefits and savings; it is safe and offers flexible options for contributions and fund growth. A TSP account can be started with a monthly contribution as little as a 1 percent of your base pay or as much as the TSP's monetary cap, which varies yearly. Your contribution and fund election can be adjusted as your base pay increases. Visit www.tsp.gov for more information. Service members who joined the military after January 2018 are automatically enrolled into the Blended Retirement System (BRS), which consists of automatic and matching contributions into TSP, a pension after twenty years of service, a midcareer bonus, and a lump-sum option at retirement. All service members who have completed two years of service are considered fully vested. For more information on BRS, please visit the local PFM at your HSWL Regional Practice or http://www.dcms.uscg.mil/brs.

Credit

In today's economy, borrowing on credit can be a necessity. However, careless use of credit can make each purchase cost more over the long run, once interest charges are added. If possible, pay in cash to avoid high finance charges. Using credit unwisely is a dangerous trap. Be cautious. Consumers can request, view, and print their free credit report at www.annualcreditreport.com. This is the only authorized website to request your free report. The report is also available by calling toll-free at (877) 322-8228 or by writing to Annual Credit Report Request Service, P.O. Box 105281, Atlanta, GA 30348-5281.

Banks and Credit Unions

Choosing a bank or credit union is a matter of personal choice. Often your first financial decision is to choose a banking institution. So it is important to select one that meets your needs and has a good reputation. Banks are for-profit institutions. Generally interest rates on loans tend to be higher and interest paid on deposit accounts tend to be lower than credit unions. Credit unions are nonprofit organizations owned by their members. Interest rates on loans tend to be lower, and interest paid on deposit accounts tend to be higher. When choosing a bank or credit union, you should consider the following:

- Convenience: Are ATMs and branch locations easily accessible?
- Fees: What charges are associated with the type of banking you plan to do?
- Balances: What are the minimum balance requirements?

- Direct deposit: Does the bank / credit union offer early access to direct deposit of active-duty pay, which allows you to access your pay sooner?
- Special programs: Are there any special military saving programs that will help you boost your savings?
- Mobile banking: What services are available online?

There are a number of services provided by banks and credit unions, such as

- Interest on deposits: Deposit interest rates are paid by financial institutions to account holders. These guaranteed fixed interest rates tend to be smaller than the variable returns of other investment vehicles.
- Automatic teller machines (ATM): These enable customers to withdraw cash, make deposits, transfer funds, and make account inquiries at any time.
- Online and mobile banking: This enables customers to execute bank transactions via computer or mobile devices electronically. Conducting transactions online may offer a great deal of convenience; however, there is risk of exposing your personal data and account information. Avoid unsecure Wi-Fi.
- Cashier checks: Cashier checks are checks that the bank issues and guarantees and are usually required for large purchases.
- Overdraft protection: An overdraft occurs when money is withdrawn from an account and the available balance is not sufficient to cover the amount. Fees may be charged on overdrafts.
- Automatic and web bill pay: Automatic bill pay allows you to set up recurring payments to pay your bills each month. Web bill pay allows you to pay your bills electronically online or via paper check issued by the bank or credit union.
- Credit and debit cards: With credit cards, you are borrowing money in exchange for repayment at a later date. The companies expect you to pay in full or make minimum payments, which often come with fees and interest charges. Debit cards draw against money in your account and therefore do not charge fees or interest.

Fundamentals of Debt and Credit

Success or failure with money depends much more on appropriate behavior than it does on the amount of money you have. Using credit responsibly is the result of good planning. Conversely, inappropriate behavior, lack of planning, or emotional spending can lead to serious financial trouble, which may result in being discharged from the Coast Guard.

Credit can be a valuable part of your long-term financial plan. It may be used for large purchases, such as a home, new car, or college education. Some make the mistake of viewing credit as a license to spend. Poor spending decisions can leave you deeply in debt and damage your credit rating for years. Boost your overall financial well-being by using credit responsibly. Practice the following healthy habits to improve your credit rating and limit unnecessary debt:

- Pay bills on time. Do not skip payments.
- Strive to pay off credit cards in full each month.
- Do not apply for credit you do not need.
- Keep credit card and loan information in a safe, secure location to reduce the risk of identity theft.
- Keep your receipts and compare charges when billing statements arrive. Notify your bank or credit agency if there is a discrepancy.

Credit Report

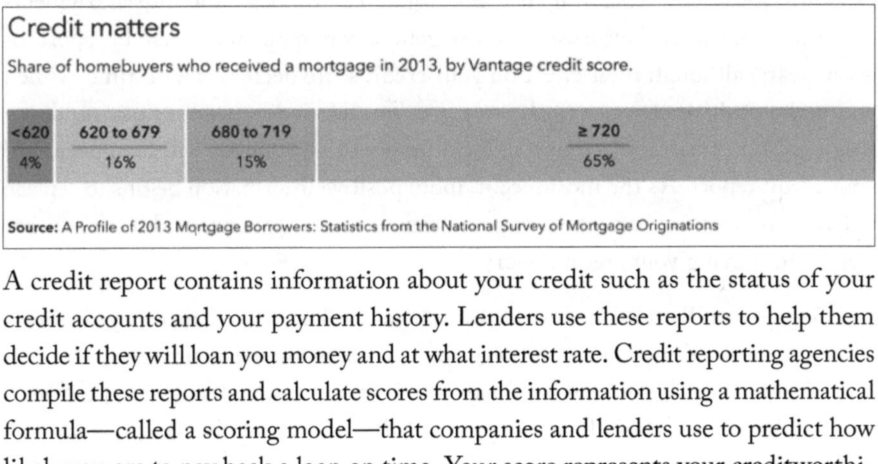

A credit report contains information about your credit such as the status of your credit accounts and your payment history. Lenders use these reports to help them decide if they will loan you money and at what interest rate. Credit reporting agencies compile these reports and calculate scores from the information using a mathematical formula—called a scoring model—that companies and lenders use to predict how likely you are to pay back a loan on time. Your score represents your creditworthiness, and lenders use this to determine who qualifies for loans, what interest rate you will pay, and what your credit limit should be. Your credit report shows how much credit you are using, how well you pay your debts, who is inquiring about your credit, and information on bankruptcies or tax liens. Credit scores range from 300 to 850 depending on the credit scoring model. Most lenders consider people with scores above 660 to be fair credit risks. Scores below 600 may indicate credit problems. The median FICO score nationwide is 721. The best rates go to borrowers with credit scores in the mid-700s or above.

It's important to understand that you don't have just one credit score. There are three major credit reporting companies—Equifax, Experian, and TransUnion—and each maintains a separate report. You have the right to a free copy of your credit report once per year from each of the three companies at www.annualcreditreport.com. You can also call (877) 322-8228.

If you haven't checked your credit recently, now is the time to do so. You'll have concrete information to make the best decisions and find out if there are any errors on your credit report that may be lowering your credit scores. You'll also see which areas you may be able to improve. Checking your own credit won't hurt your credit scores.

Improving Your Credit

Building or rebuilding your credit won't happen overnight. Beware of anyone who claims that they can "fix" your credit quickly for a fee. This could be a credit repair scam. There are no shortcuts to building a good credit history as it takes time and making regular payments to your accounts. Bankruptcy, foreclosures, and defaults can lower your credit score by up to a few hundred points. Late and missed payments also lower your score. Negative items can generally remain on your credit report for seven years, although their effect on your credit score decreases with time. When you make positive changes to the way you handle credit—such as paying down balances and consistently paying bills on time—these changes will be reflected in your credit report. As the most recent, more positive information begins to replace older, more negative information, your credit scores will improve. Here are some tips for improving your credit history:

- Pay your bills on time, every time. It may take six months to a year, or more, of making consistent, on-time payments to raise your credit scores significantly. If you've missed payments, get current and stay current. For credit cards, it's best if you can pay the full amount, but if you can't, at least make the minimum payment on time.
- Know your limits. Don't get close to your credit limit, the maximum amount you can borrow on a credit card. Credit scoring models look at how close you are to being "maxed out," so try to keep your balances low in proportion to your overall credit limit. A good target is no more than 30 percent of your total credit limit. For example, if your credit limit is $1,000, you'd want to limit your use to $300 or less at any time.
- Think before you cancel old cards. You might think that a good way to improve your credit history is to cut up your credit cards and throw them away. If you are tempted to overspend, cutting up your cards or storing them

in a safe, inaccessible place for a while might well be a good idea. But that doesn't necessarily mean you want to actually cancel the account. Pay down your balance each month, and when it gets to zero, you can decide whether to leave the account open or close it. If you leave the account open, the extra available credit may help to improve your credit score. However, it's important to check the statements every month to make sure there aren't any charges you don't recognize. Identity thieves sometimes target unused accounts. If the card has an annual fee, the best choice is usually to close the account once you've paid off the balance. Don't pay to keep open a card that you're not using.

- Apply for new credit slowly. You need to use credit—borrowing and repaying money using credit cards or other loans—in order to build your credit history. However, applying for several new credit cards or loans in a short period of time can hurt your credit scores. Taking on too much new debt at the same time is risky for you and for lenders. Take things one step at a time. When you apply for a new loan or credit card, demonstrate at least six months of on-time payments for that account before you apply for another account. Never apply for more accounts than you can reasonably manage—it's more important to manage the credit you already have responsibly than it is to apply for new credit.

Financing a Major Purchase

Making a major purchase requires skill and careful, thorough research and planning to get the best deal and keep your budget balanced. Stop major purchases from becoming major problems by understanding the following:

- Additional costs matter. What else will you have to pay? As an example, cars require maintenance, repairs, gas, and insurance. Other big purchases are similar.
- Things change. Will today's purchase still make sense in a few years? For example, if you buy a new car and are transferred to a cutter, you may not be driving for up to six months a year.
- Shop around. Shopping around forces you to slow down your decision and find the best price before purchasing. Cash or credit. Paying cash is smarter than financing. Just don't use all of your cash. Keep your emergency fund.

Loans

Like credit cards, loans vary depending on the institution offering them. When shopping for a loan, compare the following:

- Annual percentage rate (APR): This represents the rate of interest you are charged plus any fees.
- Loan term (length): This is the amount of time you have to pay the loan in full. Longer-term loans have lower monthly payments, but you will pay more interest.
- Prepayment penalties: These are fees for paying the loan in full before the due date.

It is easier to commit to a loan with a longer term when the lower monthly payment looks attractive. Compare the total amount of interest over the entire life of the loan before deciding.

Filing Taxes

When the service member is preparing to file federal and state taxes, they can also retrieve their military W-2 from Direct Access. If a spouse or family member is filing a tax return for the service member, then a power of attorney form needs to be completed. For assistance with tax planning or basic tax-related questions, contact www.CGSUPRT.com or the designated PFM at your HSWL Regional Practice. CG SUPRT offers free tax filing through the CG SUPRT website.

Predatory Financial Services

Located near many U.S. military bases are questionable businesses that consumer advocates describe as being predatory. These businesses offer services such as payday loans; rent-to-own furniture, electronics, and appliances; check cashing; pawn options; and advance tax refunds. While the services offered may seem very attractive, they come at a high cost and often trap the consumer in an endless cycle of debt. Often service members are not equipped with the financial education to protect themselves against these predators. The issue has received so much attention that there is a division of the Consumer Financial Protection Agency (www.consumerfinance.gov) aimed at curbing such practices directed toward military service members. It is important to the financial well-being of service members and their families to educate themselves on these practices and learn to avoid them.

Payday Lending

Payday lending (also called a "cash advance") is the practice of using a post-dated check or electronic checking account information as collateral for a short-term loan. To qualify, borrowers need only to provide personal identification, a checking account, and income from a job or government benefits, such as Social Security or disability payments. Research shows that the payday lending business model is designed to keep borrowers in debt—not to provide one-time assistance during a time of financial need. For more information and to keep from falling prey to the "debt traps" associated with payday lending, contact the designated PFM at your HSWL Regional Practice.

Identity Theft

Identity theft is a crime in which an impostor obtains key pieces of your personal information such as your Social Security number or driver's license number and uses it for their personal gain. Be sure to take advantage of the option to review your credit report for free, and review it for skeptical activity at https://www.annualcreditreport.com/. For tools and resources to protect you from identity theft or if your identity has been stolen, visit https://www.identitytheft.gov/.

If you think someone else is using your Social Security number, report the problem to the Social Security Administration at (800) 772-1213. You can also check your earnings record by calling the above number or by making a request online at https://www.ssa.gov/. Your PFM at your HSWL Regional Practice or your nearest legal assistance office may also be able to assist with ID theft and fraud cases if you are a victim of this crime.

LEGAL PROTECTIONS

A variety of legal protections are available to service members and their families to limit interest rates, prevent evictions, and terminate leases upon being ordered to a new duty station.

Service Member Civil Relief Act

The Service Member Civil Relief Act offers financial and legal protections for active-duty service members and their families. Some protections include

- Reduced interest rates (applies to before active-duty debt or liabilities; can be challenged by creditor);
- Evictions prevented (depending on monthly rent amount);
- Postponed foreclosures;
- Terminated residential or automobile leases (depending on PCS or deployment orders); and
- Protection in civil court proceedings.

There are several other protections that are not listed above. For more details, contact your nearest legal assistance office.

Military Lending Act (MLA)

This federal regulation limits creditors from charging more than 36 percent APR on certain loans such as payday loans, vehicle title loans, and tax refund anticipation loans. These three types of loans have high interest rates coupled with short payback terms. The MLA also prohibits mandatory arbitration clauses, mandatory allotments to repay the loan, and prepayment penalties. The MLA applies to active-duty, Reservists (activated thirty days or more), and their family members enrolled in DEERS. If your rights have been violated under the MLA, you may file a complaint with the Consumer Financial Protection Bureau at www.consumerfinance.gov/complaint.

Military Spouse Residency Relief Act

The Military Spouse Residency Relief Act pertains to residency, taxes, and voting. It allows a military spouse to retain the same place of residency that their service member has established so long as the military spouse has also established residency in the same state. You've probably lived in a number of interesting places during your service member's career and might like to claim residency in one of those places, but this law does not permit you to choose any state you desire for residency. Not all residency requirements are the same, so spouses will need to check the specifications for their particular state. Ways to establish residency may include voting, paying taxes, owning a home, and registering a vehicle in that state. This act is complicated but can be helpful once you understand its basic benefits. Seek assistance from a military legal assistance office to get information that pertains to your particular situation.

OFF-DUTY / OUTSIDE EMPLOYMENT

Personnel on active duty are in a twenty-four-hour duty status, and their military duties take precedence at all times. However, with approval of your CO/OIC, you can hold a part-time job or work for yourself in off-duty hours. If you do such work, you must remember that even though you are on leave or liberty, you are subject to recall and duty at any time. You are also subject to all conditions and prohibitions defined by the Standards of Ethical Conduct set forth in 5 CFR 2635 and COMDTINST M5370.8B, and in the Military Civil and Dependent Affairs Manual, COMDTINST M1700.1.

ESTATE PLANNING

Estate planning is the process of arranging the management and disposition of your estate to ensure your heirs are cared for according to your wishes following your death or incapacitation. You may think that an estate is just for rich people. It is not. Everyone has an estate. Your estate comprises everything you own: your car, home, savings and checking accounts, and personal possessions. A good estate plan answers important questions ahead of need:

- To whom should your personal possessions and property pass?
- How should your assets such as savings, investments, real estate, and insurance proceeds be used in the event you are incapacitated or deceased?
- Who will care for your children?
- Who will oversee your finances and health care options if you are unable?

Estate plans should include

- Last Will and Testament: This legal document dictates your wishes about the disposition of your assets after your death. Without a will, state law governs how property will be distributed and who will take care of your children.
- Power of Attorney: This legal document gives a person the authority to act on your behalf on legal and financial matters if you are unable to handle your own affairs.
- Living Will: If you are seriously injured or suffering a terminal illness, this document allows you to describe which medical treatments you do and do not want. You may also designate someone to make your medical decisions for you if you are unable to do so.
- Servicemembers' Group Life Insurance: Your policy will outline who receives your life insurance upon your death.
- Funeral and Burial Arrangements: This document will ensure your final wishes are carried out.

Contacting an estate planning attorney or your installation's legal office is a good first step in getting together a will and other pieces of an estate plan.

27

CAREER DEVELOPMENT

Service in the Coast Guard offers you many career opportunities that provide value to the nation. The career path and community of work you select will depend, in large part, on your own aptitudes and interests. Once in an enlisted rating, you can advance as far as master chief petty officer or can pursue a commission and seek promotion to the flag ranks. Advancement is highly competitive, and only the best-qualified progress. For those who enlist for only a single tour of duty, much of the training and many of the skills you will acquire are highly marketable in the private sector.

The Coast Guard is also ready to help you attain your educational goals, whether a college degree or a professional certification. Accreditation of Coast Guard training, funding of coursework through tuition assistance, and programs such as the Post-9/11 GI Bill can be combined to fully fund your goal and allow you to achieve it in a reduced timeline.

RECRUIT TRAINING

Your service in the Coast Guard will begin in Cape May, New Jersey. After being accepted into the Coast Guard and sworn in, you will be ordered to report to Training Center, Cape May, New Jersey. Here, in eight weeks, you must learn to make the difficult change from school and home life to a strict military routine. You will probably be wishing you were anywhere else but at Coast Guard boot camp at first, but so will almost everyone else. Your company commanders and instructors are among the most professional members of the U.S. Coast Guard and are there to help you get organized and learn to do things the Coast Guard way.

One of the biggest problems the average recruit has on entering military service is learning how to live with a large group of people and become a team member. Recruit training is designed to assist you in solving those problems, but you must help yourself by being tolerant of other people's peculiarities since you probably have some of your own.

While you are in recruit training, you will learn many things, some of which will be taught in formal classes, and others you will gradually absorb in your daily work. Recruit training is an entry-level instructional program designed to provide you with fundamental skills and knowledge so you can be productive in a military environment. You will learn the various skills required in the Coast Guard: military courtesies, drills and ceremonies, military justice and code of conduct, leadership and supervision, seamanship, uniforms, ranks and rates, career development, first aid and survival, fitness, wellness and quality of life, Coast Guard history, traditions and values, safety, damage control, small arms, Coast Guard assets and missions, administration and personal finance, watch-standing, security, and communications. You will also learn the main essential of good citizenship—getting along with others.

FIRST DUTY ASSIGNMENT

Upon completion of recruit training, each graduate is permitted to take up to five days regular leave. This "recruit leave" may provide you time to visit your family and friends back home for a few days. They may base their entire opinion of the Coast Guard on what you do and how you act and display the training you have received, so make sure you leave them with a good impression.

First duty assignments may be either to sea duty or shore duty. Sea duty could mean a cutter, icebreaker, buoy tender, patrol boat, or any of many other vessels. Shore duty could be at a boat station, aids-to-navigation team, or sector. Wherever you go, you will find new friends and have exciting experiences.

ADVANCEMENT

There are five means of advancement in the enlisted rates, two applicable to nonrated personnel and three to those who have a rating.

- Nonrated advancements: Recruits are normally advanced to seaman apprentice or fireman apprentice (E-2) upon graduation from basic training. Six months later, they may be advanced to seaman or fireman (E-3) by their commanding officers if they have maintained good conduct and satisfactory performance and completed the enlisted professional military education requirement (EPME) and enlisted performance qualifications (EPQ) for E-3.

- "A" school advancements: Graduates of "A" schools will be advanced to petty officer third class once they have completed the EPQs for E-4 and have the required six months' service in paygrade E-3.
- Competitive advancements: Advancement to E-5 and above is achieved through completion of Enlisted Rating Advancement Training System (ERATS) requirements and competition of the Servicewide Examination (SWE).
- Meritorious advancements: Coast Guard senior leadership identify and advance the most talented, experienced, and proficient members based on demonstrated performance and ability.
- Master Chief Advancement Panel: This panel recommends and ranks eligible active-duty senior chief petty officers for advancement to master chief petty officer for all ratings, except the musician rating.

Individual Development Plan

The individual development plan (IDP) is a personally tailored action plan used to communicate your short- and long-term personal and career goals to your chain of command. Your IDP supports you in reaching your aspirations in terms of education, promotion and advancement, professional accreditation, and career opportunities. When completing the IDP, understand that it is one of the best tools available for your supervisors to understand your motivations and goals so they can aid you in building the expertise, confidence, and self-esteem to have a successful, fulfilling career and life. Take time to complete this plan and use it to guide conversations with your supervisors about career development and progression.

Coast Guard Business Intelligence

The Coast Guard employs a web-based readiness tool known as CGBI, or Coast Guard Business Intelligence, that allows individual members and Coast Guard senior leadership to view and analyze a vast amount of personal readiness data. This data is important as it will help determine our ability to respond to our many missions and operations. You are responsible to ensure that your readiness data in CGBI is up to date at all times. There are four views in CGBI: Enterprise, Unit, Personal, and Cubes and Reports.

The Personal view is designed to show your readiness to be deployed. There are two different tabs you need to be familiar with, Compliance and Skills. Each of these tabs contains colored icons that indicate your compliance status. The Compliance tab displays your immunization, influenza, evaluations or marks, weigh-ins, dental, periodic health assessment (PHA), and medical tests. If you have any colored icon other than green in a specific category, then you must take action to overcome this deficiency. Discuss the information with your supervisor and develop an action plan to address any item that appears in red or yellow. The Skills tab displays your personal information, rank and position history, education, training, mandated training, tasks, competencies, language, and pipeline skills.

Much of the information displayed in CGBI is imported into Direct Access to aid algorithms and board and panel decision-making on your future advancements and assignments. It is essential to keep this information up to date and engage with your command and administrative staff to correct any discrepancies.

ASVAB / AFCT

The Armed Services Vocational Aptitude Battery (ASVAB) is a Department of Defense test used as a preliminary classification and screening device administered only to applicants through Military Entrance Processing Sites. The Coast Guard uses the ASVAB for screening applicants for enlistment into the Coast Guard, for class "A" schools, and for Officer Candidate School (OCS). If the scores you received on the ASVAB do not qualify you for the class "A" school you are interested in, you can retest by taking the Armed Forces Classification Test (AFCT) through your unit's education services officer. If you desire to raise your ASVAB test scores, you must take the AFCT in its entirety. The AFCT results then replace all ten ASVAB subtest scores in your personnel record, even if the AFCT scores are lower. Your most current score is the only one that counts. There must be thirty days between the first time you take the test; after that there must be at least six months between tests. Your command may waive up to five points on your score so you can attend "A" school, but any waiver is only granted when two or more scores are added together.

"A" SCHOOL

Class "A" schools are formal training courses by which nonrated personnel become rated petty officers. These schools provide the basic technical knowledge and skills required for entry-level petty officers. Graduates are apprentices who will initially require on-the-job training under close supervision or further formal class "C" schools to carry out their duties. Nonrates are required to wait a minimum of four months after reporting on board their first unit to apply for "A" school, although the waiting period is waived for critical ratings. E-2s with more than one year in a pay grade are not eligible. In order to attend an "A" school, you must have attained minimum scores in your classification tests, have a record of good conduct, and be recommended by your command. Your commanding officer's recommendation will be based on observation of your work and attitude by your commanding officer, executive officer, and division and petty officers. Following are additional requirements:

- These ratings require normal hearing: BM, IS, ME, AMT, AET, and AST.
- These ratings require normal color perception: BM, DV, IS, GM, OS, EM, ET, MK, HS, IT, PA, ME, AET, AMT, and AST.
- These ratings require eligibility for a secret clearance: GM, OS, ET, IT, IS, PA, ME, AET, AMT, and AST.
- This rating requires eligibility for a top secret clearance: IS.
- These ratings require U.S. citizenship: AET, AMT, AST, ET, EM, GM, IS, IT, OS, and PA.

Assignment to "A" school may be made upon completion of recruit training, but most attend after serving on board a cutter or at a station for a specific amount of time. The Coast Guard conducts the following "A" schools:

Aviation Technical Training Center, Elizabeth City, North Carolina	
Avionics electrical technician (AET)	26-week course
Aviation maintenance technician (AMT)	24-week course
Aviation survival technician (AST)	24-week course
Training Center, Petaluma, California	
Electronics technician (ET)	29-week course
Food service specialist (FS)	12-week course
Health services technician (HS)	23-week course
Information systems technician (IT)	29-week course
Operations specialist (OS)	17-week course
Storekeeper (SK)	8-week course
Yeoman (YN)	8-week course
Training Center, Yorktown, Virginia	
Boatswain's mate (BM)	14-week course
Damage controlman (DC)	15-week course

Training Center, Yorktown, Virginia (*continued*)	
Electrician's mate (EM)	19-week course
Gunner's mate (GM)	13-week course
Intelligence specialist (IS)	14-week course
Machinery technician (MK)	13-week course
Marine science technician (MST)	10-week course
Coast Guard Maritime Law Enforcement Academy at the Federal Law Enforcement Training Center (FLETC), Charleston, South Carolina	
Maritime enforcement specialist (ME)	10-week course
Defense Information School, Ft. Meade, Maryland	
Public affairs specialist (PA)	12-week course
Naval Diving and Salvage Training Center, Naval Support Activity Panama City, Florida	
Diver (DV)	17.5-week course

Rating Apprenticeship Program

The Rating Apprenticeship Program (RAP) courses provide a performance-based training pipeline as an alternative to full duration "A" school with the opportunity for personnel to return to the unit they were assigned to at the time orders to school were received. Boatswain's mate (BM), maritime enforcement specialist (ME), aviation maintenance technician (AMT), and avionics electrical technician (AET) ratings all have RAP courses.

- BM RAP is a six-week alternative to the existing fourteen-week "A" school for members who already hold a boat crew member competency and meet all BM "A" school requirements.
- ME RAP is a 7.5-week alternative to the existing 10-week "A" school for members who already hold a boarding team member competency and meet all ME "A" school requirements.
- AMT and AET RAP candidates are assigned to an air station where they complete twelve weeks of structured on-the-job-training and distance education through e-learning lessons and reading assignments, weekly discussions, virtual classroom, and teleconference meetings. Once complete with the first phase of "A" school at their unit, students travel to the Aviation Technical Training Center in Elizabeth City, North Carolina, where they will complete an additional twelve weeks of technical training.

Striker Program

Striking involves completion of E-4 rating performance qualifications (RPQ) at a field unit vice an "A" school under the supervision of a professional development coach. FORCECOM determines what ratings are eligible for striking and suspends these programs when "A" schools' throughput falls below capacity. Currently all striker programs are suspended.

"C" SCHOOL

Class "C" schools provide advanced and specialized skills and knowledge to perform a task, or group of tasks, required by a specific billet or unit type. Eligibility requirements vary by course and program and are available through the Training Quota Management Center CGPortal site. Applications for class "C" school training are submitted via electronic training request in Direct Access by the unit training officer.

PERSONAL QUALIFICATION STANDARD

The Personal Qualification Standard (PQS) is a qualification system wherein certification of a minimum level of competency is required prior to qualifying to perform a specific job such as boat crew member or boarding team member. School attendance, such as class "C" school, may satisfy specific portions of PQS books, but all are completed at the unit level. When all PQS items for a competency are signed off, the trainee is considered qualified. To perform the job requires certification of the member by their command, normally after successful completion of a board and check ride or practical skills assessment administered by the unit's Qualification Examining Board.

ENLISTED RATING ADVANCEMENT TRAINING SYSTEM

Graduates of "A" schools will be advanced to petty officer third class once they have completed the EPQs for E-4 and have the required six months' service in pay grade E-3. Those not advanced on graduation will be given a school-assigned designator and classified as a "rated nonrate" to allow them to work in their new rating. They will be promoted when the above requirements have been satisfied and they are recommended by their commanding officer.

Requirements for advancement to E-5 and higher pay grades are established by the Enlisted Rating Advancement Training System (ERATS), which defines performance requirements for each rating and pay grade, helps members meet those requirements through training and performance support, and assesses performance to determine whether members possess the knowledge for advancement to the next higher grade. ERATS has two types of performance requirements that serve as the basis for all enlisted advancement training programs and assessments: RPQ and EPME and EPQ. RPQs are the core tasks of a rating. EPME are general leadership and management tasks associated with enlisted grades. RPQs and EPQs are housed on the ERATS CGPortal page.

The RPQ standard is the document used for completing rating advancement requirements and contains instructions, advancement requirements, RPQs, references, and supplemental performance guides. Competencies required for advancement in certain ratings and grades are called core competencies. Core competencies are promulgated via ALCOAST messaging and outlined in the RPQ standard. The RPQ standard also contains instructions for the professional development coach on how to mentor, track progress, and evaluate RPQ performance. The coach trains, evaluates, and documents RPQ completion in the RPQ standard. Rating force master chiefs and rating knowledge managers manage a rating-specific knowledge repository providing additional performance support resources not otherwise captured, and a rating community of practice. Upon completion of all RPQs for a pay grade, a rating competency code is entered into the training management tool and Direct Access to indicate eligibility to receive a Servicewide Examination (SWE).

Servicewide Examination

The SWE is an assessment held annually in May that tests both RPQ and EPME knowledge to rank enlisted members in the same rating who have completed all eligibility requirements and are competing for advancement. Candidates become eligible to advance and sit for the SWE after they

- Complete the RPQs for the next pay grade;
- Complete the EPQs for the next pay grade;
- Fulfill sea time requirements of the next pay grade;
- Serve a specified number of months in their current grade in the Coast Guard (TIG, time in grade);
- Serve a specific number of years in service in the Coast Guard (TIS, time in service);
- Meet service weight standards; and
- Receive a recommendation for advancement from their commanding officer or officer-in-charge.

All eligibility requirements except TIS and TIG must be met before an individual may be recommended by the command. TIS and TIG must be completed prior to terminal eligibility dates established for each advancement cycle. A Personal Data

Extract will be made available in Direct Access confirming SWE eligibility and points for awards, sea time, TIS, and TIG. If your Personal Data Extract contains errors, notify your unit administrator.

The difficulty of the SWE is set so that a minimally competent candidate will score 70 percent or higher on the SWE. Usually, 80 percent of the candidates should be minimally competent; however, the subject matter expert will make that determination.

Beginning in May 2022 the SWE is administered annually in May for active-duty personnel in pay grades E-5 through E-8 and in October for Reserve forces. Previously it was held in May and November. The results are announced eight to ten weeks after the SWE is administered. All candidates who participate in the SWE will receive a profile letter. The profile letter indicates the overall performance of the candidate. Section scores are provided to identify those areas that the candidate may use to study for the next SWE, if applicable.

An advancement eligibility list is published shortly after the profile letters are mailed. The rank order is based on the final multiple, a number computed from the SWE score, average performance evaluations for a given period of time, TIS, TIG, points for certain medals and awards, and sea time. This list includes the names of all candidates who are eligible for advancement. Names that fall above the cutoff are guaranteed advancement. Names below the cutoff may be advanced but should plan to compete in the next SWE. The number of candidates to be advanced is based upon a forecast of servicewide vacancies made by the Coast Guard Personnel Service Center's Enlisted Personnel Management Division. The forecast for each rate varies with each examination cycle. Candidates must remain qualified and be recommended for advancement by the command each SWE cycle.

Chief Petty Officer

The transition from E-6 to E-7 is the most critical in the Coast Guard enlisted career as E-7s transition into a role with unparalleled influence both up and down the chain of command. E-6 members above the published cutoff for advancement to E-7 must attend the Chief Petty Officers Academy, the Air Force Senior Non-Commissioned Officer Academy, or the Navy Senior Enlisted Academy prior to advancement. Once advanced, E-7s must complete the Chiefs Call to Initiation to earn the title "Chief."

E-8 members above the published cutoff for advancement to E-9 must attend the Senior Enlisted Leadership Course prior to advancement. The Master Chief Advancement Panel convenes annually to select eligible active-duty senior chief petty officers for advancement to E-9. This panel was established to serve as a talent management tool that incorporates qualitative information found in a member's record, with a focus on professionalism, leadership, and performance. This panel replaces the E-9 SWE.

Chief Warrant Officer

Every enlisted rating has a normal path of advancement to a commission as a chief warrant officer (CWO). The following is the normal path for all enlisted ratings to the twenty CWO specialties:

Enlisted Rating	CWO Specialty
AET/AMT	Aviation engineering (AVI)
AST	Aviation engineering (AVI) Marine safety specialist deck (MSSD)
BM	Boatswain (BOSN) Marine safety specialist deck (MSSD) Marine safety specialist response (MSSR)
CS	Finance and supply (F&S)
DC	Material maintenance (MAT) Marine safety specialist engineering (MSSE) Marine safety specialist response (MSSR)
DV	Diver (DIV)
EM	Marine safety specialist engineering (MSSE) Naval engineering (ENG)
ET	Electronics (ELC)
GM	Weapons (WEPS)
HS	Medical administration (MED)
IS	Intelligence systems specialist (ISS)
IT	Information systems management (ISM)
IV	Criminal investigator (INV)
ME	Maritime law enforcement specialist (MLES)
MK	Naval engineering (ENG) Marine safety specialist engineering (MSSE) Marine safety specialist response (MSSR)
MST	Marine safety specialist deck (MSSD) Marine safety specialist response (MSSR)

(continued)

Enlisted Rating	CWO Specialty
OS	Operations systems specialist (OSS)
PA	Public information (INF)
SK	Finance and supply (F&S)
YN	Personnel administration (PERS)

Candidates cannot have more than twenty-six years of active-duty service when applying and must meet all published eligibility criteria. The commanding officer's recommendation is the most important step in the CWO appointment process, as the recommendation assists the service in affirming the candidate's mental, moral, physical, and professional qualifications, taking into consideration the totality of the applicant's career, not just the time observed while assigned to the unit.

An appointment board composed of commissioned officers rank order the applicants considered best qualified to fill the vacancies in the various CWO categories. Selected candidates are promoted to W-2 with promotion potential to W-4.

Those at the rank of W-3 and above are eligible to apply to the CWO to lieutenant selection panel in order to fill occupational specialty vacancies at the O-3 level. Applicants must have completed at least thirteen years but not more than twenty-six years of active duty for retirement at the time of application. Selectees are offered appointments as temporary officers and incur three years of obligated service.

OFFICER CANDIDATE SCHOOL

Civilians and current enlisted members of the Coast Guard may apply for selection as a Coast Guard officer through the Officer Candidate School (OCS) program. OCS is a highly specialized, seventeen-week course of instruction in leadership, management, navigation, law enforcement, and military subjects at the U.S. Coast Guard Leadership Development Center in New London, Connecticut. OCS classes are convened twice a year, in January and in the summer. Two programs, Reserve Commission (OCS-R) and Temporary Commission (OCS-T) are available. Currently serving Coast Guard members may be eligible to compete for either and should review the eligibility requirements for OCS-R and OCS-T to determine which program they are eligible for. In the final stages of OCS, officer candidates compete for positions that will be available to them upon graduation and commissioning. Initial assignments will be within one of the four main officer operational specialties, including afloat, aviation, prevention, and response. Many junior officers typically receive subsequent assignments in the career field of their initial assignment. Opportunities exist to receive training and qualifications in other specialties and to pursue a subsequent assignment in a different career field or a subspecialty such as intelligence, human resources, engineering and C4IT (command, control, communication, computers, and information technology), international affairs, legal, recruiting, financial management, and training.

On commissioning, which occurs upon graduation from OCS, members will be discharged from their enlisted status and commissioned as a Coast Guard ensign with an initial three-year active-duty obligation. Many such temporary officers eventually receive permanent commissions after meeting service requirements and after being selected for promotion to lieutenant. For more information and to apply talk to a recruiter or visit http://www.gocoastguard.com.

COAST GUARD ACADEMY

The U.S. Coast Guard Academy in New London, Connecticut, is a top military college granting bachelor of science degrees in one of nine engineering or professional majors. Graduates earn a commission as an ensign in the Coast Guard.

Enlisted members who have not reached their twenty-third birthday may apply to the Coast Guard Academy. To be eligible, they must be unmarried without dependents, have demonstrated a responsible use of GTCC and personal credit, have Armed Forces Qualification Test scores of 75 or higher with a minimum App Review score of 52, and have completed high school or college basic calculus and either physics or chemistry. To learn more and apply, visit http://www.uscga.edu.

COLLEGE STUDENT PRE-COMMISSIONING INITIATIVE

Students currently enrolled, accepted for enrollment, or pending enrollment in a full-time bachelor's degree program at an accredited college or university designated as a minority-serving institute with the desire to complete college and receive a guaranteed commission as an officer in the U.S. Coast Guard can apply for College Student Pre-Commissioning Initiative (CSPI). This means that a student currently pursuing an associate's degree at a junior college or community college who plan to continue with their bachelor's degree may be eligible for CSPI if they are accepted for enrollment at a qualifying college.

Students who are accepted into this scholarship program will be enlisted into the U.S. Coast Guard, complete basic training during the summer, and receive full funding for up to two years of college. Funding includes not only payment of tuition, books, and fees but a full-time Coast Guard salary, housing allowance, and medical benefits.

During the school year, students participate in a minimum of sixteen hours a month of Coast Guard activities and are supervised by the recruiter-in-charge at their recruiting office. During school breaks, CSPI students receive training and exposure to Coast Guard missions and units. After completing their junior year of college, CSPI students attend a three-week leadership training course in New London, Connecticut. Following college graduation, CSPI students attend the seventeen-week OCS in New London. Upon successful completion of OCS, graduates receive a commission as an ensign (O-1) and an assignment in one of the many Coast Guard mission fields. To learn more and apply, visit http://www.gocoastguard.com.

LEADERSHIP DEVELOPMENT FRAMEWORK

The development of Coast Guard leaders at all levels is a service priority. A complex organization such as the Coast Guard doesn't just "run." It requires people who decide what must be done, how and when it will be done, and who will do it. These people may be petty officers, department heads, chiefs, commanding officers, or admirals, but they all have one common responsibility—to lead their teams, often in challenging and sometimes extreme circumstances. Every Coast Guardsman must understand and employ the basic principles of leadership. These principles are essential to success when operating in the uncompromising environment of the sea. The Coast Guard considers leadership the most important criteria for advancement in all pay grades. Leadership makes the difference between a good, well-functioning cutter, unit, or organization and a poor one. A well-led unit is characterized by commitment to a unified vision; a strong sense of teamwork and pride, high morale, and motivated personnel. Successful leaders do two things well—they get the job done and at the same time develop their subordinates by allowing them to contribute, grow, and mature as military professionals.

The Coast Guard has developed a formal Leadership Development Framework, which outlines twenty-nine leadership competencies, or the knowledge, skills, and expertise the Coast Guard expects of its leaders. While there is some overlap in these competencies, they generally fall within four broad categories: Leading Self, Leading Others, Leading Performance and Change, and Leading the Coast Guard. Together, these four leadership categories and their elements are instrumental to career success, and you should work to develop your capacity and skills in each of them.

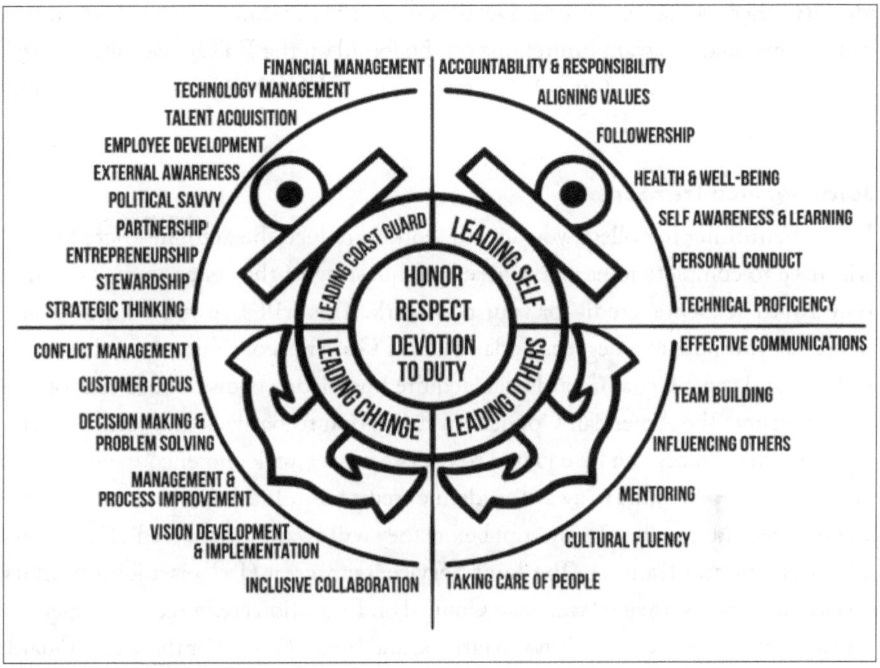

Leadership competencies

You will not become a good leader overnight. While for a few people leadership comes naturally, for most it takes time. The Coast Guard begins this developmental process by having you be a follower first. As you advance in your career, you will be put in positions of increasing responsibility and be given more leadership opportunities. Remember that today's master chiefs once were seaman recruits. Leadership and learning go hand in hand, and developing the capacity to lead increases your capacity to serve.

VOLUNTARY EDUCATION

The Voluntary Education Department within the Education and Training Quota Management Center (ETQC) and unit-level education services officers (ESO) provide services to assist you with professional growth through higher education throughout your Coast Guard career. You are strongly encouraged to seek ESO guidance when deciding which college or university maximizes military training and credits-by-exam and provides the best value for tuition dollars. ESOs can also help identify other sources of funding. Details on specific program requirements, tuition caps, forms, and program limitations can be found on the ETQC website, https://www.forcecom.uscg.mil/Our-Organization/FORCECOM-UNITS/ETQC/, and by speaking with your ESO.

Joint Service Transcript

Prior to enrolling in college you should work to reduce the amount of credits you will have to complete to earn a degree by requesting of the college you select that they award academic credit for your past work. This will save you both time and money as you pursue a degree. Today's Coast Guard is continuously and expertly trained, and most Coast Guardsmen acquire knowledge equivalent to that of college students. The American Council on Education recognizes that some military learning experiences can be equated to college coursework and encourages college and university personnel to award academic credit for military learning experiences. Colleges require an official transcript before they will award college credit for military advancement and training. The Joint Service Transcript (JST) is a joint military service transcript showing American Council on Education credit recommendations for the Army, Coast Guard, Navy, Marines, and Space Force. For the Coast Guard, the JST is available for active-duty, Reserve, civilian, spouses of active duty, and Auxiliary members. Your JST will list military training history, rate history, military correspondence course history, college credit examinations, and certifications.

DANTES

The Defense Activity for Non-Traditional Education Support (DANTES) provides funds to active-duty and Reserve service members and their spouses in order to complete standardized American College Testing (ACT), Scholastic Assessment Test (SAT), College Level Examination Program (CLEP), and DANTES Subject Standardized Tests (DSST) exams. ACT and SAT scores may be needed for college and various program enrollments, while successful CLEP and DSST exams will result in the award of academic credits toward a degree. Prior to testing, determine if the college or university you're enrolled in awards credit for CLEP and DSST exams. DANTES only funds fees for the first attempt on all CLEP and DSST titles. Examinees may retest on a self-paid basis after a required waiting period. For more information, contact your ESO or visit https://www.dantes.doded.mil/examinations/funding-and-reimbursement-eligibility/funding-eligibility.html#sthash.8HB1WDoX.bY6f7acu.dpbs.

Degree Plan

Establish a plan for successfully completing any certificate or degree requirements. Engage your ESO to assist with this plan and submit your JST to your academic institution to maximize your military service and training to reduce the number of credits required to complete the certificate or degree. The plan will list all courses required to complete the academic program and are typically provided by the college and university on your student portal page. Degree plans are mandatory for those using tuition assistance.

Tuition Assistance

Tuition assistance (TA) is government funding for higher education that is authorized for active and Selected Reserve (SELRES) service members who meet the following eligibility criteria:

- Satisfactory progress toward completion of unit quals;
- Satisfactory progress toward watch station quals;
- Satisfactory proficiency of craft;
- Satisfactory conduct during the six months prior; and
- Satisfactory progress toward participation standards if member is SELRES.

TA is authorized for courses leading to the first certificate, associate's, bachelor's, and master's degrees. TA is not authorized for members pursuing additional degrees at the same or lower level regardless of how they received their first degree. TA is not authorized for any degree above a master's degree, including juris doctor, doctoral degrees, and similar terminal degrees. Members who have a bachelor's, associate's, or no degree, may use TA to pursue their first certificate, at any academic level; however, members who currently hold a master's degree or higher are not authorized to use TA to pursue a certificate.

Schools authorized to receive TA funding must participate and have a signed memorandum of understanding with the DoD. The authorized school list can be found at https://www.dodmou.com.

Pending annual allocation of funding, TA will fund tuition costs not to exceed a set annual cap and set amounts per semester, quarter, or clock hour. Reimbursement of TA will be required from the service member if the student does not successfully complete the courses. For reimbursement, a successful course completion is defined as a grade of a C or higher for undergraduate courses, a B or higher for graduate courses, and a pass for pass/fail grades.

Other Coast Guard sources of funding and reimbursement programs for education expenses can be found on the ETQC Web page. Federal student aid loans obtained by Coast Guard members for the purpose of funding their education will not be considered as a duplication of benefit and not subtracted when calculating the use of TA.

To apply, engage with your ESO to submit a command-approved TA application with a degree plan uploaded to ETQC no fewer than fourteen days prior to the class start date.

Post-9/11 GI Bill

The Post-9/11 GI Bill provides up to thirty-six months of benefits to pay for higher education or job training. Benefits under the bill include paid tuition and fees, money for housing, funding for books and supplies, and moving costs needed to attend school. Veterans who have served at least ninety days of active-duty service after 10 September 2001 and have received an honorable discharge qualify for the Post-9/11 GI Bill. To qualify for the full benefit, a veteran must have served at least three years of active duty after 10 September 2001 or been awarded a Purple Heart on or after 11 September 2001. If your service ended before 1 January 2013, your benefits will expire fifteen years after your last separation date from active service, and you must use all of your benefits by that time or you will lose whatever is left. If your service ended on or after 1 January 2013, your benefits won't expire thanks to the Forever GI Bill—Harry W. Colmery Veterans Educational Assistance Act. If you're a qualified service member, you can also transfer all thirty-six months or a portion of your Post-9/11 GI Bill benefits to a spouse or child.

To use your benefits you must apply through the Department of Veterans Affairs at http://www.va.gov.

Coast Guard Foreign Language Program

The purpose of the Coast Guard Foreign Language Program is to identify a strong pool of foreign language–qualified military members, both active duty and Reserve, to support Coast Guard missions. The foreign languages used by the Coast Guard in support of Coast Guard missions are listed in an annual ALCOAST authorizing interpreters and linguists at specific units. To qualify as a unit interpreter, a military member must take the Defense Language Proficiency Test, which serves as the standard measure of language proficiency for members of the armed forces. It measures listening and reading skills in a foreign language. You must score at least a 2 in listening and a 2 in reading to qualify. The Foreign Language Program is funded to provide foreign language proficiency pay, which members may receive simultaneously with special-duty assignment pay. Neither civilians nor Auxiliarists are eligible.

COAST GUARD CREDENTIALING OPPORTUNITIES ON-LINE (CG COOL)

The Coast Guard Voluntary Credentialing Program established the CG COOL Portal to allow all Coast Guard service members to find information on credentials, certifications, and licenses related to their military occupations and duties. Many civilian occupations require credentials that identify their owner as meeting specific professional and technical standards. By enabling credentialing of military personnel, any transition into civilian employment is more seamless. CG COOL identifies military occupation-related credentials, highlights potential knowledge gaps between military training and civilian credentialing requirements, and provides guidance to assist in filling those gaps. The Coast Guard will reimburse eligible personnel for voluntary credentialing expenses, including certification fees, licensing exam fees, maintenance fees, renewals, and membership dues for those credentials listed on CG COOL that are directly related to their ratings or service experience. Eligible members include enlisted active duty and SELRES service members. Eligible service members may only apply for one credential per fiscal year. To learn more, engage with your ESO and visit http://www.cool.osd.mil.

SEPARATION

To assist those members who do not choose to make the Coast Guard a career as well as those who are retiring, the Coast Guard provides transition and relocation managers (TRM) at its Health, Safety, and Work–Life offices to help make the process of transitioning to a new career easier. Services include access to computerized job banks, resume-writing assistance, help with the interviewing process, and more. All separating or retiring members must receive pre-separation counseling from a TRM, their Servicing Personnel Office, or their command. The TRM can provide you with additional information about transition assistance and pre-separation counseling. See https://www.dcms.uscg.mil/TAP.

Discharge Types

There are several types of discharges for enlisted personnel separating from the service, to include several of a punitive nature. The kind of discharge you receive may affect you for the rest of your life. Eligibility for veterans' preference in federal employment, for payments for a service-connected disability, for a pension, and for many other benefits and privileges will depend upon the type of discharge you receive. Civilian employers may refuse to hire a person with the "wrong" kind of discharge. Do all you can to earn a discharge under honorable conditions.

Form DD-214, Report of Discharge, contains separation/discharge information that can be used by potential employers for screening veterans, determining eligibility of VA benefits, and reenlistment. Here are the seven blocks (box 23–29) that make up the character of discharge on the "long" or "undeleted" DD-214 form.

Character of Service

The "character of your service" determines your eligibility for VA benefits. There are six types of "discharge of character" listed on military discharges. The DD-214 must have a service characterization of anything other than dishonorable to be considered a veteran.

Honorable Discharge. This is the highest discharge a military member can receive. It indicates the service member performed duties well, faithfully executed the mission, and was an asset to the branch of the military where the member served.

General Discharge under Honorable Conditions. This type of military administrative discharge is motivated by different things depending on the branch of service. The overall conduct of the military member may have been exemplary in some areas, but other areas of misconduct or failure to adapt to the military environment may have resulted in such a discharge. The separation paperwork for these military discharges may be quite specific about the reasons for the discharge, so while it's not stigmatized the same as a dishonorable discharge, the general discharge under honorable conditions may still hurt the veteran in some ways where a DD Form 214 Report of Discharge is required for employment or other reasons.

Other Than Honorable (OTH) Discharge. This is the most severe of the administrative discharges that do not require a court-martial. Reasons for the OTH discharge may depend on the severity of the offenses, how a particular branch of the military has traditionally handled such issues, and other variables. Security violations, trouble with civilian authorities, assault, drug possession or various degrees of drug violations, or other problems could all potentially motivate an OTH discharge. The OTH discharge should be considered to be a barrier to future military service.

Bad Conduct Discharges. A bad conduct discharge comes as the result of a court-martial and may be followed by prison time depending on the nature and severity of the conduct. This type of military discharge is not considered an administrative one and is a barrier to future military service.

Dishonorable Discharge. This is the most punitive of all military discharges and is the result of a court-martial. Desertion, murder, fraud, and other crimes performed in uniform can result in court-martial proceedings that lead to a dishonorable discharge. No military benefits or future military service is possible with a military discharge characterized as dishonorable.

Other Military Discharges. A new recruit who cannot complete basic training, adapt to the military environment while in recruit training or "A" school, or otherwise is unable or unwilling to complete the initial phases of training before moving out of training and into "permanent party" status would be given an entry-level discharge or entry-level separation, depending on the branch of military service. These separations generally happen before the new recruit has served more than 180 days. These are not considered "good" or "bad" discharges, but the recruit is not considered a veteran, and those receiving entry-level separations are not eligible for benefits.

Sometimes a situation may require the separation of a new recruit or permanent party military member "for the convenience of the government." This type of discharge is done at the discretion of the service and is not considered a common or routine practice.

A medical discharge may be given to service members who become sick or injured to the point where military duty is no longer possible based on a medical evaluation of the medical condition. This process can be lengthy and may or may not be appealed depending on a variety of factors.

RETIREMENT

To qualify for retirement, a member must serve for at least twenty years (exceptions may be made in disability cases). Retirement pay depends on the length of active-duty service, pay grade, and the retirement plan that is in place at the time a member enters the service. There are several classes of retirement, including compulsory retirement at the age of sixty-two, voluntary retirement after twenty or more years' service, involuntary retirement after twenty or more years' service, and retirement for disabilities incident to service. In time of war or national emergency, retired personnel may be called to active duty.

A member who remains on active duty until retirement receives retired pay, commissary and exchange privileges, "space available" use of recreation and club facilities, a survivor benefit (if elected), reduced-cost medical insurance, burial rights, and more. Members leaving active-duty service are eligible to receive transition assistance.

Coast Guard Retiree Services Program

All retirees, including their family members and survivors, are entitled to certain rights, benefits, and privileges and are subject to responsibilities arising from these entitlements. Retirees, as part of the total Coast Guard workforce, remain capable of making life-long contributions to Coast Guard missions. To ensure their contributions continue, retirees must be kept apprised of changing programs, services, and policies. The Coast Guard Retiree Services Program serves the Coast Guard military retirees and eligible surviving family members by keeping the lines of communication open to provide retiree information, retiree benefits advice, and services. Additional information is available at https://www.dcms.uscg.mil/retiree.

Veterans Affairs

The U.S. Department of Veterans Affairs offers many benefits for uniformed service members, veterans, spouses, children, and parents of a deceased or disabled service member or veteran. Benefits include compensation, pension, education and training, employment assistance, life insurance, home loans, health care, and burial. Visit http://ebenefits.va.gov to learn about and apply for your benefits. You may also visit the nearest VA regional office or call 1 (800)-MyVA411 (800-698-2411).

Abbreviations

AFCT	Armed Forces Classification Test
AIDS	acquired immune deficiency syndrome
AIS	automated identification system
ALMIS	Asset Logistics and Maintenance Information System
AMVER	automated mutual assistance vessel rescue
ANB	aids-to-navigation boat
AO	approving/authorizing official
APR	air-purifying respirator
ASVAB	Armed Services Vocational Aptitude Battery
ATON	aids to navigation
AWS	Alert Warning System
BAH	basic allowance for housing
BAS	basic allowance for subsistence
BDC	boat deck captain
BMOW	boatswain's mate of the watch
BRS	Blended Retirement System
BU	bow-loading buoy boat
BUSL	buoy utility stern loading
CBA	centrally billed account
CB-OTH	cutter boat—over the horizon
CBR	chemical, biological, or radiation
CBRN	chemical, biological, radiation, or nuclear
CBRNE	chemical, biological, radiological, nuclear, and high-yield explosive
CGAS	Coast Guard Air Stations
CGBI	Coast Guard Business Intelligence
CG COOL	Coast Guard Credentialing Opportunities On-Line

CGOne	Coast Guard Network
CGX	Coast Guard Exchange
CIWS	close-in weapon system
CLEP	College Level Examination Program
CONUS	continental United States
CPB	coastal patrol boat
CPR	cardiopulmonary resuscitation
CSPI	College Student Pre-Commissioning Initiative
DANTES	Defense Activity for Non-Traditional Education Support
DCC	damage-control center
DCMS	deputy commandant for mission support
DCO	deputy commandant for operations
DECON	decontamination
DEERS	Defense Enrollment Eligibility Reporting System
DMB	datum marker buoy
DoD	Department of Defense
DPS	Defense Personal Property System
DR	dead reckoning
DSC	Digital Selective Calling
DSF	deployable specialized forces
DSST	DANTES Subject Standardized Tests
EAPC	employee assistance program coordinator
EBL	electronic bearing line
EEBD	emergency escape breathing device
EEZ	exclusive economic zone
EPME	enlisted professional military education
EPQ	enlisted performance qualification
e-QIP	Electronic Questionnaire for Investigations Processing
ERATS	Enlisted Rating Advancement Training System
ESO	education services officer
ETQC	Education and Training Quota Management Center
FITT	frequency, intensity, time, and type of activity
FFE	firefighter ensemble
FOIA	Freedom of Information Act
FOUO	For Official Use Only
FPCON	Force Protection Conditions

GBL	government bill-of-lading
GCS	gun computer system
GE	general emergency
GMDSS	Global Maritime Distress and Safety System
gpm	gallons of water per minute
GQ	general quarters
GTCC	Government Travel Charge Card Program
HELP	heat escape lessening posture
HF	high frequencies
HIV	human immunodeficiency virus
HSWL	Health, Safety, and Work–Life
ICW	Intracoastal Waterway
IDLH	immediately dangerous to life or health
IDP	individual development plan
IRR	Individual Ready Reserve
JST	Joint Service Transcript
kn	knot
lat	latitude
LED	light-emitting diode
LNG/LPG	liquefied natural gas / liquefied petroleum gas
long	longitude
LOP	line of position
LORAN	long-range navigation
LRI	long-range interceptor
LRS	long-range surveillance
MCPO	master chief petty officer of the Coast Guard
medevac	medical evacuation
Mk	mark
MLA	Military Lending Act
MMSI	Maritime Mobile Service Identification
MODU	mobile offshore drilling unit
MPC	maintenance and procedure card
MSRT	maritime security response team
MWR	Morale, Well-Being, and Recreation (program)
NAC	National Agency Check
NACI	National Agency Check and Inquiries

NAF	non-appropriated fund
nm	nautical mile
NOAA	National Oceanic and Atmospheric Administration
NSC	national security cutter
NTAS	National Terrorism Advisory System
OCS	Officer Candidate School
ODU	operational dress uniform
OHA	overseas housing allowance
OOD	officer of the deck
OPC	offshore patrol cutter
OPSEC	operational security
OSC	on-scene coordinator
OTH	other than honorable
PCM	primary care manager
PCS	permanent change of station
PDW	personal defense weapon
PFD	personal flotation device
PKP	Purple-K powder
PLB	personal locator beacon
POV	privately owned vehicle
POW	prisoner of war
PPE	personal protective equipment
PQS	Personal Qualification Standard
PSC	polar security cutter
PSU	port security unit
PWCS	Ports, Waterways, and Coastal Security
RAP	Rating Apprenticeship Program
RM	risk management
RPQ	rating performance qualification
SAP	search action plan
SAR	search and rescue (n); search-and-rescue (adj)
SAR	supplied-air respirator
SBU	Sensitive but Unclassified information
SCBA	self-contained breathing apparatus
SCI	Sensitive Compartmented Information
SDB	service dress blue / service dress blue "Bravo"

SDS	safety data sheet
SELRES	Selected Reserve
SGLI	Servicemembers' Group Life Insurance
SIPRNet	Secret Internet Protocol Router Network
SMC	SAR mission coordinator
SRU	search-and-rescue unit
SS	square search
STCW	Standards of Training, Certification, and Watchkeeping
STD	sexually transmitted disease
SWE	Servicewide Examination
TA	tuition assistance
TACLET	tactical law enforcement team
TIG	time in grade
TIS	time in service
TRM	transition and relocation manager
TSP	Thrift Savings Plan
UAS	unmanned aerial system
UCMJ	Uniform Code of Military Justice
UHF	ultra-high frequencies
UPH	unaccompanied personnel housing
USCG	United States Coast Guard
VDS	video distress signal
VHF	very high frequencies
VRM	variable range marker
VS	sector search
WAGB	icebreaker
WCC	waterways commerce cutter
WHEC	high-endurance cutter
WLB	seagoing buoy tender
WLBB	Great Lakes icebreaker
WLI	inland buoy tender
WLIC	inland construction tender
WLL	working load limit
WLM	coastal buoy tender
WLR	river tender
WMD	weapons of mass destruction

WMEC	medium-endurance cutter
WMSL	maritime security cutter, large
WMSM	offshore patrol cutter
WPB	patrol boat
WPC	fast-response cutter
WTGB	icebreaking tugs
WYTL	harbor tugs
XTE	cross-track error

APPENDIX A
Code of Conduct

The 12 January 2016 capture of ten U.S. Navy sailors by Iran served as an unwanted reminder for everyone wearing the uniform to review their obligations under the Code of Conduct. The Code of Conduct is an ethical guide. Its six articles deal with your chief concerns as a prisoner of war (POW); these concerns become critical when you must evade capture, resist while a prisoner, or escape from the enemy.

ARTICLE I

I am an American, fighting in the forces which guard my country and our way of life. I am prepared to give my life in their defense.

All men and women in the armed forces have the duty, at all times and under all circumstances, to oppose the enemies of the United States and support its national interests. In training or in combat, alone or with others, while evading capture or enduring captivity, this duty belongs to each American defending our nation regardless of circumstances.

ARTICLE II

I will never surrender of my own free will. If in command, I will never surrender the members of my command while they still have the means to resist.

As an individual, a member of the armed forces may never voluntarily surrender. When isolated and no longer able to inflict casualties on the enemy, members of the armed forces have an obligation to evade capture and rejoin friendly forces.

Only when evasion by an individual is impossible and further fighting would lead only to death with no significant loss to the enemy should one consider surrender. With all reasonable means of resistance exhausted and with certain death the only alternative, capture does not imply dishonor.

The responsibility and authority of a commander never extends to the surrender of a command to the enemy while the command has the power to fight and evade.

When isolated, cut off, or surrounded, a unit must continue to fight until relieved or able to rejoin friendly forces through continued efforts to break out or evade the enemy.

ARTICLE III

If I am captured I will continue to resist by all means available. I will make every effort to escape and aid others to escape. I will accept neither parole nor special favors from the enemy.

The duty of a member of the armed forces to use all means available to resist the enemy is not lessened by the misfortune of captivity. A POW is still legally bound by the Uniform Code of Military Justice and ethically guided by the Code of Conduct. Under provisions of the Geneva Convention, a prisoner of war is also subject to certain rules imposed by the captor nation. When repatriated, a prisoner of war will not be condemned for having obeyed reasonable captor rules, such as sanitation regulations. The duty of a member of the armed forces to continue to resist does not mean a prisoner should engage in unreasonable harassment as a form of resistance. Retaliation by captors to the detriment of that prisoner and other prisoners is frequently the primary result of such harassment.

The Geneva Convention recognizes that a POW may have a duty to attempt escape. In fact, the Geneva Convention prohibits a captor nation from executing a POW simply for attempting escape. Under the authority of the senior official (often called the senior-ranking officer, or SRO), a POW must be prepared to escape whenever the opportunity presents itself. In a POW compound, the senior POW must consider the welfare of those remaining behind after an escape. However, as a matter of conscious determination, a POW must plan to escape, try to escape, and assist others to escape.

Contrary to the spirit of the Geneva Convention, enemies engaged by U.S. forces since 1950 have regarded the POW compound as an extension of the battlefield. In so doing, they have used a variety of tactics and pressures, including physical and mental mistreatment, torture, and medical neglect to exploit POWs for propaganda purposes, to obtain military information, or to undermine POW organization, communication, and resistance.

Such enemies have attempted to lure American POWs into accepting special favors or privileges in exchange for statements, acts, or information. Unless it is essential to the life or welfare of that person or another prisoner of war or to the success of efforts to resist or escape, a POW must neither seek nor accept special favors or privileges.

One such privilege is called parole. Parole is a promise by a prisoner of war to a captor to fulfill certain conditions—such as agreeing not to escape nor to fight again once released—in return for such favors as relief from physical bondage, improved food and living conditions, or repatriation ahead of the sick, injured, or longer-held prisoners. An American POW will never sign nor otherwise accept parole.

ARTICLE IV

If I become a prisoner of war, I will keep faith with my fellow prisoners. I will give no information nor take part in any action which might be harmful to my comrades. If I am senior, I will take command. If not, I will obey the lawful orders of those appointed over me and will back them up in every way.

Informing or any other action to the detriment of a fellow prisoner is despicable and is expressly forbidden. Prisoners of war must avoid helping the enemy identify fellow prisoners who may have knowledge of particular value to the enemy and who may, therefore, be made to suffer coercive interrogation.

Strong leadership and communication are essential to discipline. Discipline is the key to camp organization, resistance, and even survival. Personal hygiene, camp sanitation, and care of the sick and wounded are imperative. Officers and noncommissioned officers of the United States must continue to carry out their responsibilities and exercise their authority in captivity. The senior, regardless of service, must accept command. This responsibility and accountability may not be evaded.

If the senior is incapacitated or is otherwise unable to act, the next senior person will assume command. Camp leaders should make every effort to inform all POWs of the chain of command and try to represent them in dealing with enemy authorities. The responsibility of subordinates to obey the lawful orders of ranking American military personnel remains unchanged in captivity.

The Geneva Convention Relative to Treatment of Prisoners of War provides for election of a "prisoners' representative" in POW camps containing enlisted personnel but no commissioned officers. American POWs should understand that such a representative is only a spokesman for the actual senior-ranking person. Should the enemy appoint a POW chain of command for its own purposes, American POWs should make all possible efforts to adhere to the principles of Article IV.

As with other provisions of this code, common sense and the conditions of captivity will affect the way in which the senior person and the other POWs organize to carry out their responsibilities. What is important is that everyone support and work within the POW organization.

ARTICLE V

When questioned, should I become a prisoner of war, I am required to give name, rank, service number, and date of birth. I will evade answering further questions to the utmost of my ability. I will make no oral or written statements disloyal to my country and its allies or harmful to their cause.

When questioned, a prisoner of war is required by the Geneva Convention and this code to give name, rank, service number (Social Security number), and date of birth. The prisoner should make every effort to avoid giving the captor any additional information. The prisoner may communicate with captors on matters of health and welfare and additionally may write letters home and fill out a Geneva Convention "capture card."

It is a violation of the Geneva Convention to place a prisoner under physical or mental duress, torture, or any other form of coercion in an effort to secure information. If under such intense coercion a POW discloses unauthorized information, makes an unauthorized statement, or performs an unauthorized act, that prisoner's peace of mind and survival require a quick recovery of courage, dedication, and motivation to resist anew each subsequent coercion.

Actions every POW should resist include making oral or written confessions and apologies, answering questionnaires, providing personal histories, creating propaganda recordings, broadcasting appeals to other prisoners of war, providing any other material readily usable for propaganda purposes, appealing for surrender or parole, furnishing self-criticisms, and communicating on behalf of the enemy to the detriment of the United States, its allies, its armed forces, or other POWs.

Every POW should also recognize that any signed confession or any statement may be used by the enemy as false evidence that the person is a "war criminal" rather than a POW. Several countries have made reservations to the Geneva Convention in which they assert that a "war criminal" conviction deprives the convicted individual of prisoner-of-war status, removes that person from protection under the Geneva Convention, and revokes all rights to repatriation until a prison sentence is served.

Recent experiences of American prisoners of war have proved that, although enemy interrogation sessions may be harsh and cruel, one can resist brutal mistreatment when the will to resist remains intact.

The best way for a prisoner to keep faith with country, fellow prisoners, and self is to provide the enemy with as little information as possible.

ARTICLE VI

I will never forget that I am American, fighting for freedom, responsible for my actions, and dedicated to the principles which made my country free. I will trust in my God and in the United States of America.

A member of the armed forces remains responsible for personal actions at all times. A member of the armed forces who is captured has a continuing obligation to resist and to remain loyal to country, service, unit, and fellow prisoners.

Upon repatriation, POWs can expect their actions to be reviewed, both regarding the circumstances of capture and their conduct during detention. The purpose of such reviews is to recognize meritorious performance as well as to investigate possible misconduct. Each review will be conducted with due regard for the rights of the individual and consideration for the conditions of captivity; captivity of itself is not a condition of culpability.

Members of the armed forces should remember that they and their dependents will be taken care of by the appropriate service and that pay and allowances, eligibility, and procedures for promotions and benefits for dependents will continue while the service member is detained. Service members should make sure that their personal affairs and family matters (such as pay, powers of attorney, current will, and provisions for family maintenance and education) are properly and currently arranged. Failure to do so can create a serious sense of guilt for a POW and place unnecessary hardship on family members.

The life of a prisoner of war is hard. Each person in this stressful situation must always sustain hope and resist enemy indoctrination. Prisoners of war standing firm and united against the enemy will support and inspire one another in surviving their ordeal and in prevailing over misfortune with honor.

APPENDIX B

Rank Insignia of U.S. Armed Forces

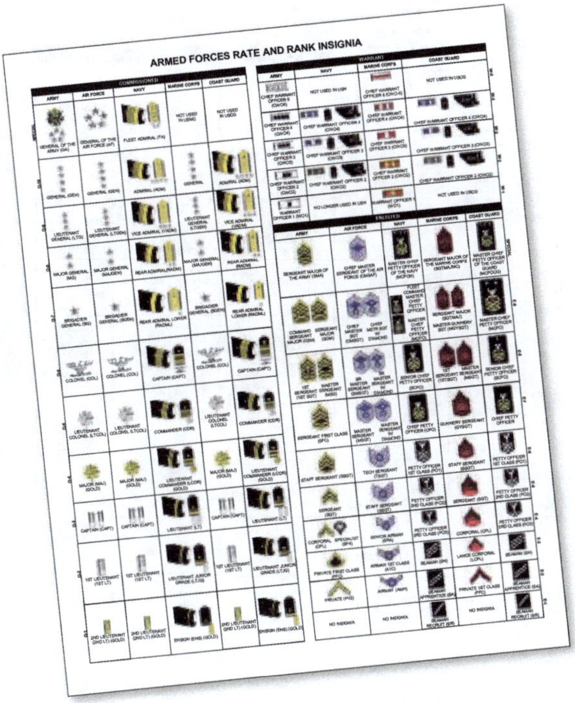

ARMED FORCES RATE AND RANK INSIGNIA

COMMISSIONED

	ARMY	AIR FORCE	NAVY	MARINE CORPS	COAST GUARD
SPECIAL	GENERAL OF THE ARMY (GA)	GENERAL OF THE AIR FORCE (AF)	FLEET ADMIRAL (FA)	NOT USED IN USMC	NOT USED IN USCG
O-10	GENERAL (GEN)	GENERAL (GEN)	ADMIRAL (ADM)	GENERAL	ADMIRAL (ADM)
O-9	LIEUTENANT GENERAL (LTG)	LIEUTENANT GENERAL (LTGEN)	VICE ADMIRAL (VADM)	LIEUTENANT GENERAL (LTGEN)	VICE ADMIRAL (VADM)
O-8	MAJOR GENERAL (MG)	MAJOR GENERAL (MAJGEN)	REAR ADMIRAL(RADM)	MAJOR GENERAL (MAJGEN)	REAR ADMIRAL (RADM)
O-7	BRIGADIER GENERAL (BG)	BRIGADIER GENERAL (BGEN)	REAR ADMIRAL LOWER (RADML)	BRIGADIER GENERAL (BGEN)	REAR ADMIRAL LOWER (RADML)
O-6	COLONEL (COL)	COLONEL (COL)	CAPTAIN (CAPT)	COLONEL (COL)	CAPTAIN (CAPT)
O-5	LIEUTENANT COLONEL (LTCOL)	LIEUTENANT COLONEL (LTCOL)	COMMANDER (CDR)	LIEUTENANT COLONEL (LTCOL)	COMMANDER (CDR)
O-4	MAJOR (MAJ) (GOLD)	MAJOR (MAJ) (GOLD)	LIEUTENANT COMMANDER (LCDR) (GOLD)	MAJOR (MAJ) (GOLD)	LIEUTENANT COMMANDER (LCDR) (GOLD)
O-3	CAPTAIN (CAPT)	CAPTAIN (CAPT)	LIEUTENANT (LT)	CAPTAIN (CAPT)	LIEUTENANT (LT)
O-2	1ST LIEUTENANT (1ST LT)	1ST LIEUTENANT (1ST LT)	LIEUTENANT JUNIOR GRADE (LTJG)	1ST LIEUTENANT (1ST LT)	LIEUTENANT JUNIOR GRADE (LTJG)
O-1	2ND LIEUTENANT (2ND LT) (GOLD)	2ND LIEUTENANT (2ND LT) (GOLD)	ENSIGN (ENS) (GOLD)	2ND LIEUTENANT (2ND LT) (GOLD)	ENSIGN (ENS) (GOLD)

WARRANT

ARMY	NAVY	MARINE CORPS	COAST GUARD	
CHIEF WARRANT OFFICER 5 (CWO5)	NOT USED IN USN	CHIEF WARRANT OFFICER 5 (CWO-5)	NOT USED IN USCG	W-5
CHIEF WARRANT OFFICER 4 (CWO4)	CHIEF WARRANT OFFICER 4 (CWO4)	CHIEF WARRANT OFFICER 4 (CWO4)	CHIEF WARRANT OFFICER 4 (CWO4)	W-4
CHIEF WARRANT OFFICER 3 (CWO3)	CHIEF WARRANT OFFICER 3 (CWO3)	CHIEF WARRANT OFFICER 3 (CWO3)	CHIEF WARRANT OFFICER 3 (CWO3)	W-3
CHIEF WARRANT OFFICER 2 (CWO2)	CHIEF WARRANT OFFICER 2 (CWO2)	CHIEF WARRANT OFFICER 2 (CWO2)	CHIEF WARRANT OFFICER 2 (CWO2)	W-2
WARRANT OFFICER 1 (WO1)	NO LONGER USED IN USN	WARRANT OFFICER 1 (WO1)	NOT USED IN USCG	W-1

ENLISTED

ARMY	AIR FORCE	NAVY	MARINE CORPS	COAST GUARD	
SERGEANT MAJOR OF THE ARMY (SMA)	CHIEF MASTER SERGEANT OF THE AIR FORCE (CMSAF)	MASTER CHIEF PETTY OFFICER OF THE NAVY (MCPON)	SERGEANT MAJOR OF THE MARINE CORPS (SGTMAJMC)	MASTER CHIEF PETTY OFFICER OF THE COAST GUARD (MCPOCG)	SPECIAL
COMMAND SERGEANT MAJOR (CSM); SERGEANT MAJOR (SGM)	CHIEF MASTER SGT (CMSGT); CHIEF MSTR SGT W/ DIAMOND	FLEET COMMAND MASTER CHIEF PETTY OFFICER; MASTER CHIEF PETTY OFFICER (MCPO)	SERGEANT MAJOR (SGTMAJ); MASTER GUNNERY SGT (MGYSGT)	MASTER CHIEF PETTY OFFICER (MCPO)	E-9
1ST SERGEANT (1ST SGT); MASTER SERGEANT (MSG)	SR MASTER SERGEANT (SMSGT); SR MASTER SERGEANT W/ DIAMOND	SENIOR CHIEF PETTY OFFICER (SCPO)	1ST SERGEANT (1STSGT); MASTER SERGEANT (MSGT)	SENIOR CHIEF PETTY OFFICER (SCPO)	E-8
SERGEANT FIRST CLASS (SFC)	MASTER SERGEANT (MSGT); MASTER SERGEANT W/ DIAMOND	CHIEF PETTY OFFICER (CPO)	GUNNERY SERGEANT (GYSGT)	CHIEF PETTY OFFICER	E-7
STAFF SERGEANT (SSGT)	TECH SERGEANT (TSGT)	PETTY OFFICER 1ST CLASS (PO1)	STAFF SERGEANT (SSGT)	PETTY OFFICER 1ST CLASS (PO1)	E-6
SERGEANT (SGT)	STAFF SERGEANT (SSGT)	PETTY OFFICER 2ND CLASS (PO2)	SERGEANT (SGT)	PETTY OFFICER 2ND CLASS (PO2)	E-5
CORPORAL (CPL); SPECIALIST (SP4)	SENIOR AIRMAN (SRA)	PETTY OFFICER 3RD CLASS (PO3)	CORPORAL (CPL)	PETTY OFFICER 3RD CLASS (PO3)	E-4
PRIVATE FIRST CLASS (PFC)	AIRMAN 1ST CLASS (A1C)	SEAMAN (SN)	LANCE CORPORAL (LCPL)	SEAMAN (SN)	E-3
PRIVATE (PV2)	AIRMAN (AMN)	SEAMAN APPRENTICE (SA)	PRIVATE 1ST CLASS (PFC)	SEAMAN APPRENTICE (SA)	E-2
NO INSIGNIA	NO INSIGNIA	SEAMAN RECRUIT (SR)	NO INSIGNIA	SEAMAN RECRUIT (SR)	E-1

APPENDIX C

International Alphabet Flags and Pennants

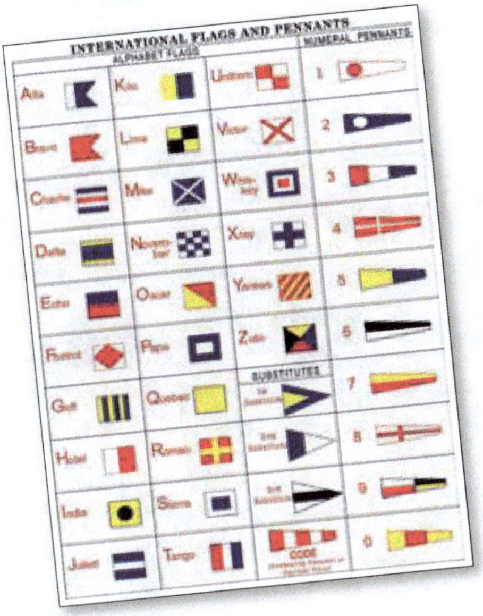

INTERNATIONAL FLAGS AND PENNANTS

APPENDIX D

IALA Maritime Buoyage System Region A

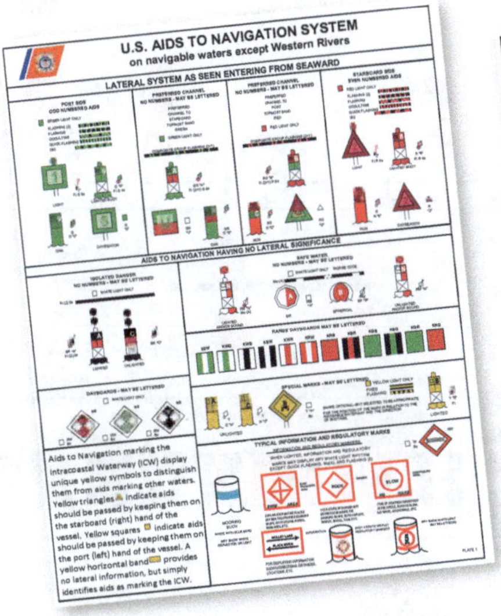

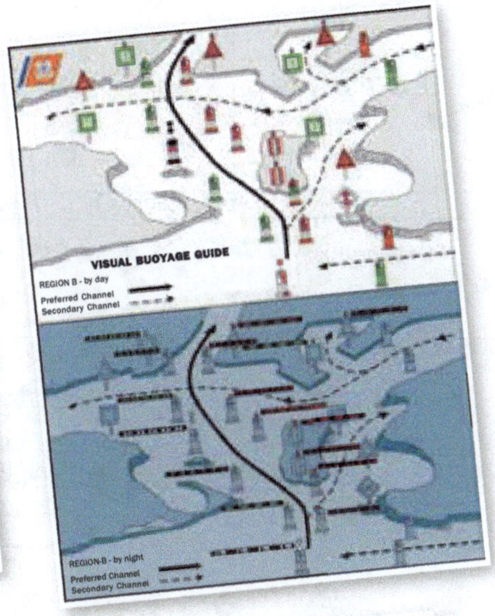

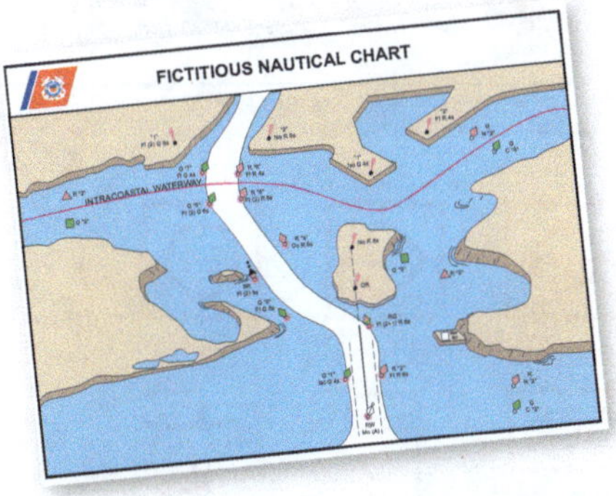

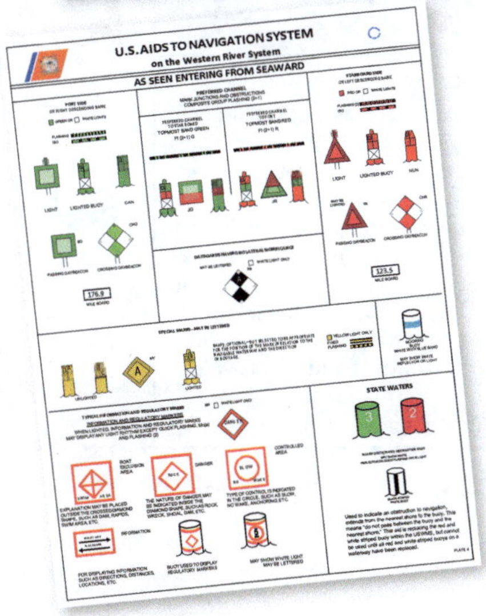

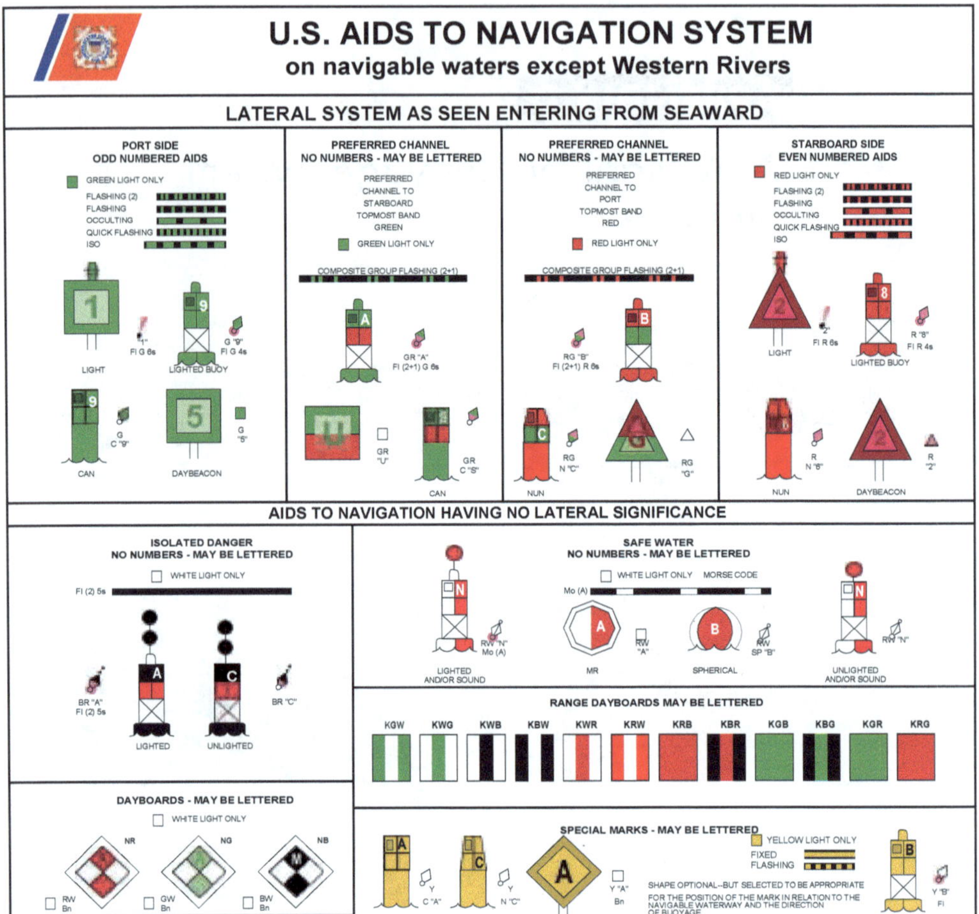

Aids to Navigation marking the Intracoastal Waterway (ICW) display unique yellow symbols to distinguish them from aids marking other waters. Yellow triangles ▲ indicate aids should be passed by keeping them on the starboard (right) hand of the vessel. Yellow squares ■ indicate aids should be passed by keeping them on the port (left) hand of the vessel. A yellow horizontal band ▬ provides no lateral information, but simply identifies aids as marking the ICW.

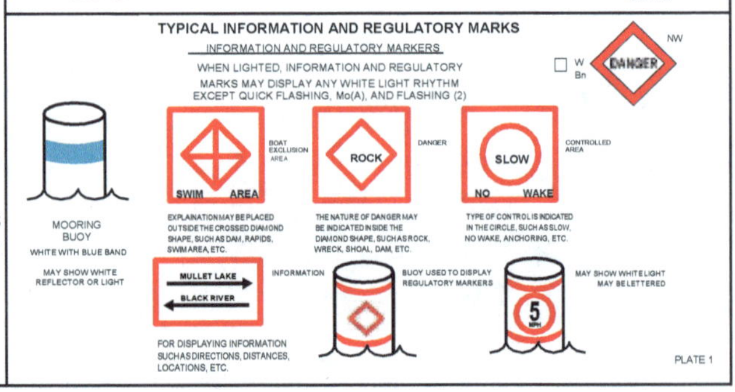

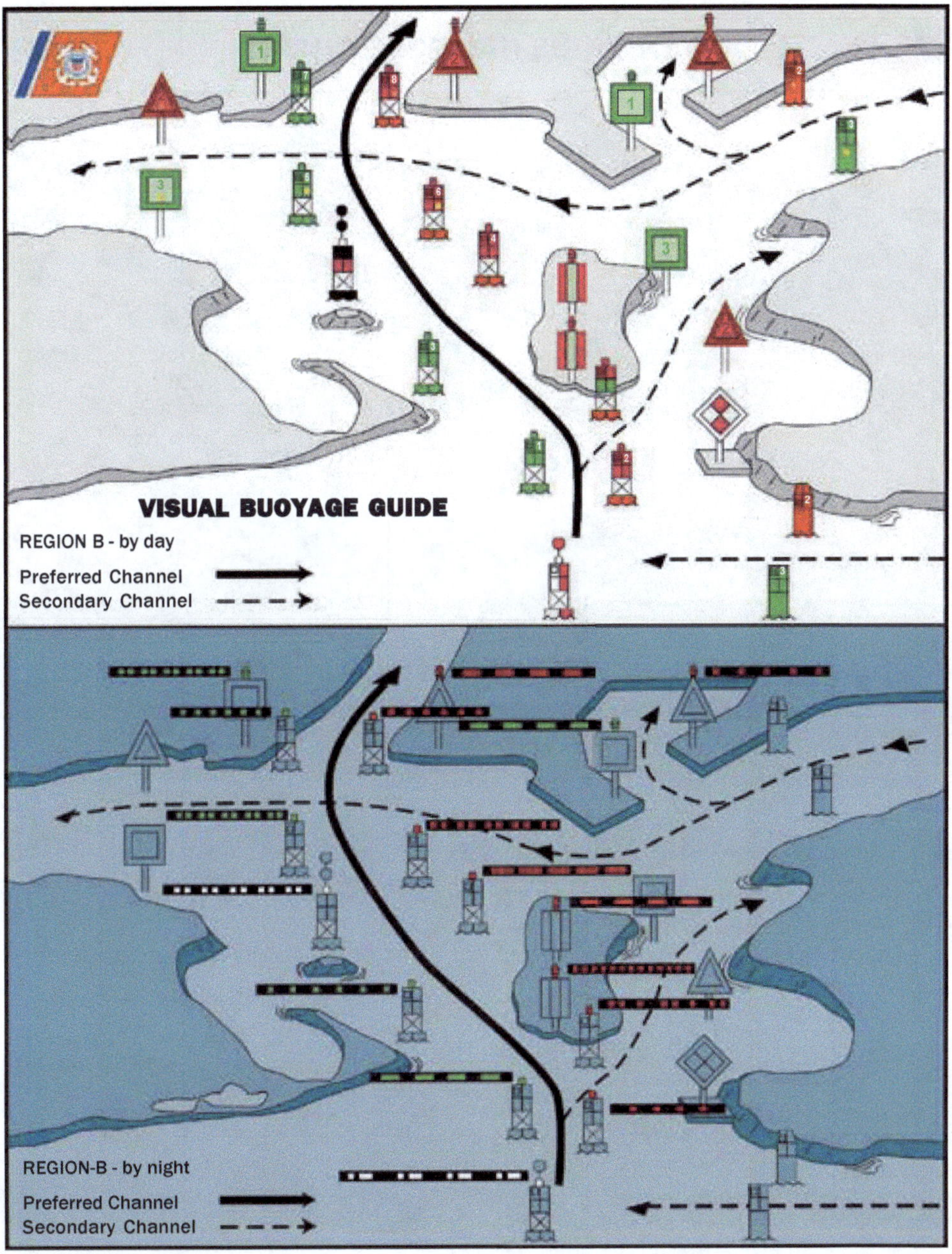
VISUAL BUOYAGE GUIDE
REGION B - by day
Preferred Channel
Secondary Channel
REGION-B - by night
Preferred Channel
Secondary Channel

FICTITIOUS NAUTICAL CHART

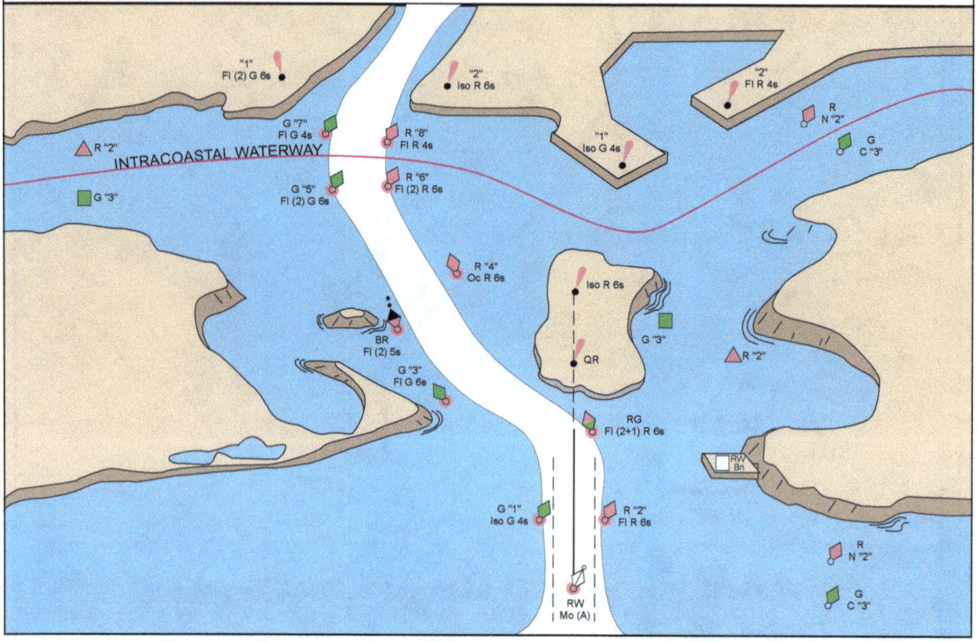

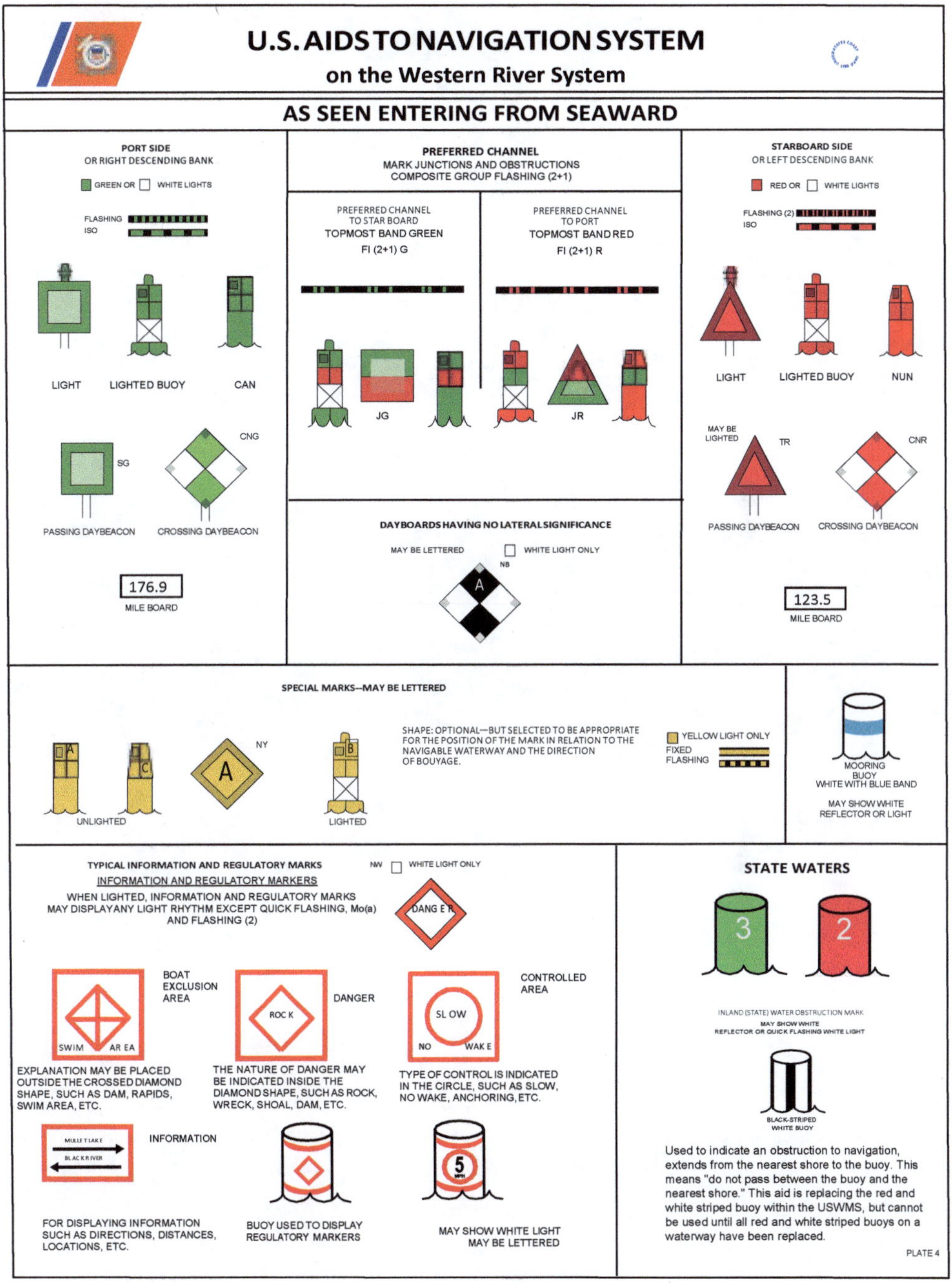

U.S. AIDS TO NAVIGATION SYSTEM
on the Western River System
AS SEEN ENTERING FROM SEAWARD
PORT SIDE
OR RIGHT DESCENDING BANK
GREEN OR WHITE LIGHTS
FLASHING
ISO
LIGHT
LIGHTED BUOY
CAN
SG
CNG
PASSING DAYBEACON
CROSSING DAYBEACON
176.9
MILE BOARD
PREFERRED CHANNEL
MARK JUNCTIONS AND OBSTRUCTIONS
COMPOSITE GROUP FLASHING (2+1)
PREFERRED CHANNEL
TO STAR BOARD
TOPMOST BAND GREEN
Fl (2+1) G
PREFERRED CHANNEL
TO PORT
TOPMOST BAND RED
Fl (2+1) R
JG
JR
DAYBOARDS HAVING NO LATERAL SIGNIFICANCE
MAY BE LETTERED
WHITE LIGHT ONLY
NB
A
STARBOARD SIDE
OR LEFT DESCENDING BANK
RED OR WHITE LIGHTS
FLASHING (2)
ISO
LIGHT
LIGHTED BUOY
NUN
MAY BE LIGHTED
TR
CNR
PASSING DAYBEACON
CROSSING DAYBEACON
123.5
MILE BOARD
SPECIAL MARKS--MAY BE LETTERED
A
C
NY
A
B
UNLIGHTED
LIGHTED
SHAPE: OPTIONAL—BUT SELECTED TO BE APPROPRIATE FOR THE POSITION OF THE MARK IN RELATION TO THE NAVIGABLE WATERWAY AND THE DIRECTION OF BOUYAGE.
YELLOW LIGHT ONLY
FIXED
FLASHING
MOORING BUOY
WHITE WITH BLUE BAND
MAY SHOW WHITE REFLECTOR OR LIGHT
TYPICAL INFORMATION AND REGULATORY MARKS
NW
WHITE LIGHT ONLY
INFORMATION AND REGULATORY MARKERS
WHEN LIGHTED, INFORMATION AND REGULATORY MARKS MAY DISPLAYANY LIGHT RHYTHM EXCEPT QUICK FLASHING, Mo(a) AND FLASHING (2)
DANG E R
BOAT EXCLUSION AREA
SWIM
AR EA
EXPLANATION MAY BE PLACED OUTSIDE THE CROSSED DIAMOND SHAPE, SUCH AS DAM, RAPIDS, SWIM AREA, ETC.
DANGER
ROC K
THE NATURE OF DANGER MAY BE INDICATED INSIDE THE DIAMOND SHAPE, SUCH AS ROCK, WRECK, SHOAL, DAM, ETC.
CONTROLLED AREA
SL OW
NO
WAK E
TYPE OF CONTROL IS INDICATED IN THE CIRCLE, SUCH AS SLOW, NO WAKE, ANCHORING, ETC.
INFORMATION
FOR DISPLAYING INFORMATION SUCH AS DIRECTIONS, DISTANCES, LOCATIONS, ETC.
BUOY USED TO DISPLAY REGULATORY MARKERS
5
MAY SHOW WHITE LIGHT
MAY BE LETTERED
STATE WATERS
3
2
INLAND (STATE) WATER OBSTRUCTION MARK
MAY SHOW WHITE REFLECTOR OR QUICK FLASHING WHITE LIGHT
BLACK-STRIPED WHITE BUOY
Used to indicate an obstruction to navigation, extends from the nearest shore to the buoy. This means "do not pass between the buoy and the nearest shore." This aid is replacing the red and white striped buoy within the USWMS, but cannot be used until all red and white striped buoys on a waterway have been replaced.
PLATE 4

APPENDIX E

Conversion and Timing Charts

Timing Chart

Distance (nm)	Speed (kts)					
	3	4	8	10	15	20
0.5	10:00	6:00	3:45	3:00	2:00	1:30
1.0	20:00	12:00	7:30	6:00	4:00	3:00
1.5	30:00	18:00	11:15	9:00	6:00	4:30
2	40:00	24:00	15:00	12:00	8:00	6:00
2.5	50:00	30:00	18:45	15:00	10:00	7:30
3.0	60:00	36:00	22:30	18:00	12:00	9:00
3.5		42:00	26:15	21:00	14:00	10:30
4.0		48:00	30:00	24:00	16:00	12:00
4.5		54:00	33:45	27:00	18:00	13:30
5.0		60:00	37:30	30:00	20:00	15:00
6.0			45:00	36:00	24:00	18:00
7.0			52:30	42:00	28:00	21:00
8.0			60:00	48:00	32:00	24:00

Note: Time to complete one leg (t) in minutes and seconds

Initial Track Spacing (NM)

Search Object	Good Conditions Wind < 15 kts Seas < 3 ft	Poor Conditions Wind > 15 kts Seas > 3 ft
PIW	0.1*	0.1*
< 15 ft	0.5	0.2
> 15 ft	1.0	0.5

* or > 0.1 depending on SRU's minimum navigational accuracy and maneuvering capability

GPS Conversion Chart

Decimal Minutes	Seconds	Decimal Minutes	Seconds	Decimal Minutes	Second
.17	01	.350	21	.683	41
0.33	02	.367	22	.700	42
0.50	03	.383	23	.717	43
0.67	04	.400	24	.733	44
0.83	05	.417	25	.750	45
.100	06	.433	26	.767	46
.117	07	.450	27	.783	47
.133	08	.467	28	.800	48
.150	09	.483	29	.817	49
.167	10	.500	30	.833	50
.183	11	.517	31	.850	51
.200	12	.533	32	.867	52
.217	13	.550	33	.883	53
.233	14	.567	34	.900	54
.250	15	.583	35	.917	55
.267	16	.600	36	.933	56
.283	17	.617	37	.950	57
.300	18	.633	38	.967	58
.317	19	.650	39	.983	59
.333	20	.667	40	1.000	60

APPENDIX F

Maritime Jurisdiction

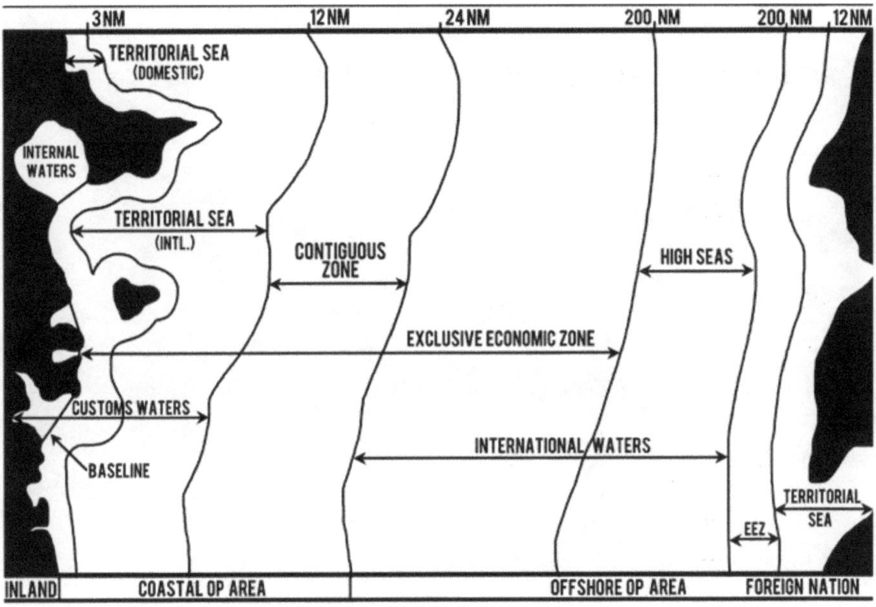

APPENDIX G

Risk Assessment

USCG Afloat Risk Assessment			
Mission:		**Date:**	
Step 1: Identify, Assess, & Mitigate Risk Elements			
Instructions: To determine the level of risk for each element below, estimate the risk level based on the Low/Medium/High scale. If your perceived rating is Medium or High, explore mitigations. Draw a line through the risk zone that corresponds to the mitigated risk level and document the perceived risk(s) and mitigation(s) in the space provided.	**Rate Risk Zone**		
Planning - Enough time and information to conduct thorough pre-mission planning. Consider: B-0 response, completeness of mission information and of on-scene details. NOTES/MITIGATIONS:	Complete *L*	Partial *M*	None *H*
Event - Refers to mission complexity. Consider: non-standard mission profile, coordinating multi-agency/nationality, language barriers, not performed often, etc. NOTES/MITIGATIONS:	Low *L*	Moderate *M*	Extreme *H*
Asset – Crew – Proper number and skill set for the mission. Consider: time at unit, familiarity w/OP area, fatigue, u/w time, crew selection, adequate supervision, etc. NOTES/MITIGATIONS:	Excellent *L*	Marginal *M*	Poor *H*
Asset – Cutter/Boat Resources – Proper number and operational characteristics for mission. Consider: operational thresholds/limitations, status of equipment, etc. NOTES/MITIGATIONS:	Ideal *L*	Restrictions *M*	Limitations *H*
Communications/Supervision - Ability to maintain comms throughout mission. Consider: availability/quality of internal w/command and external w/customer. NOTES/MITIGATIONS:	Excellent *L*	Partial *M*	None *H*
Environment - External conditions surrounding mission. Consider: weather, night/day, sea state, currents, water temp, air temp, visibility, etc. NOTES/MITIGATIONS:	Ideal *L*	Marginal *M*	Extreme *H*
**Other (Unit Specific Element):*	*L*	*M*	*H*
**Other (Unit Specific Element):*	*L*	*M*	*H*
Step 2: Determine Overall Risk Level			
Consider: 1) the rating for each element above, 2) the importance of the element for mission execution, and 3) how elements may interact. Rate the perceived **Overall Risk Level** when considering this information. Circle the risk zone (***Low, Medium, or High***) that corresponds to your perceived overall risk level:	**Low**	**Medium**	**High**
PEACE elements are required per COMDTINST 3500.3A. Additional unit specific elements are permitted.			

USCG Afloat Risk Assessment

Step 3: Determine Risk vs. Gain: Do gains warrant the risk?

Step 3a. Enter the ***Overall Risk Level*** (*Step 2 on prior page*) in the **RISK** box below (*Low, Medium, or High*).

Step 3b. Review the definitions for Gain below and enter the level in the **GAIN** box below. (*Low, Medium, or High*).

Level of Gain

- **Low** – Situation with unclear benefits or a low probability for providing concrete results.
  Examples: passenger transport, non-critical logistics missions, and public affairs demonstrations.
- **Medium** – Situation that provides immediate and real benefits.
  Examples: saving property, protecting the environment, deterring illegal operations.
- **High** – Situation that provides immediate and real benefits that if ignored could result in loss of life.
  Examples: Urgent SAR and MEDEVACs.

[ ] Vs. [ ]

RISK (Low, Med, High) **GAIN** (Low, Med, High)

Step 3c. Use the **Risk vs. Gain** values from above and follow the column and row until they cross. The intersecting point is the recommended action.
Example, if Risk is 'low' and Gain is 'medium', the recommendation is: "Accept the Mission. Continue to monitor Risk Factors, if conditions or mission changes".

Risk vs. Gain	High Gain	Medium Gain	Low Gain
Low Risk	**Accept the Mission.** Monitor Risk Factors and re-evaluate if conditions or mission/activities change.	**Accept the Mission.** Monitor Risk Factors and re-evaluate if conditions or mission/activities change.	**Accept the Mission.** Monitor Risk Factors and re-evaluate if conditions or mission/activities change.
Medium Risk	**Accept the Mission.** Monitor Risk Factors and employ Controls when available. Re-evaluate if conditions or mission change.	**Accept the Mission.** Monitor Risk Factors and employ Controls when available. Re-evaluate if conditions or mission change.	**Accept the Mission Only with Command Endorsement** Communicate Risk vs. Gain to Chain of Command. Implement Controls and continuously evaluate conditions and mission for change.
High Risk	**Accept the Mission Only with Command Endorsement.** Communicate Risk vs. Gain to Chain of Command. Implement Controls and monitor Risk Factors. Continuously evaluate conditions and mission change.	**Accept the Mission Only with Command Endorsement.** Communicate Risk vs. Gain to Chain of Command. Implement Controls and monitor Risk Factors. Continuously evaluate conditions and mission change.	**DO NOT Accept the Mission.** Communicate to Chain of Command. Wait until Risk Factors change or Controls are available to warrant Risk exposure.

NOTES:

USCG Ashore Risk Assessment				
Mission:			**Date:**	

Step 1: Identify, Assess, & Mitigate Risk Elements			
Instructions: To determine the level of risk for each element below, estimate the risk level based on the Low/Medium/High scale. If your perceived rating is Medium or High, explore mitigations. Draw a line through the risk zone that corresponds to the mitigated risk level and document the perceived risk(s) and mitigation(s) in the space provided.	**Rate Risk Zone**		
Planning - Enough time and information to conduct thorough pre-mission planning. Consider: B-0 response, completeness of mission information and of on-scene details. **NOTES/MITIGATIONS:**	**Complete** *L*	**Partial** *M*	**None** *H*
Event - Refers to mission complexity. Consider: non-standard mission profile, coordinating multi-agency/nationality, language barriers, not performed often, etc. **NOTES/MITIGATIONS:**	**Low** *L*	**Moderate** *M*	**Extreme** *H*
Asset – Crew – Proper number and skill set for the mission. Consider: time at unit, familiarity w/OP area, fatigue, u/w time, crew selection, adequate supervision, etc. **NOTES/MITIGATIONS:**	**Excellent** *L*	**Marginal** *M*	**Poor** *H*
Asset – Cutter/Boat Resources – Proper number and operational characteristics for mission. Consider: operational thresholds/limitations, status of equipment, etc. **NOTES/MITIGATIONS:**	**Ideal** *L*	**Restrictions** *M*	**Limitations** *H*
Communications/Supervision - Ability to maintain comms throughout mission. Consider: availability/quality of internal w/command and external w/customer. **NOTES/MITIGATIONS:**	**Excellent** *L*	**Partial** *M*	**None** *H*
Environment - External conditions surrounding mission. Consider: weather, night/day, sea state, currents, water temp, air temp, visibility, etc. **NOTES/MITIGATIONS:**	**Ideal** *L*	**Marginal** *M*	**Extreme** *H*
****Other (Unit Specific Element):***	*L*	*M*	*H*
****Other (Unit Specific Element):***	*L*	*M*	*H*

Step 2: Determine Overall Risk Level

Consider: 1) the rating for each element above, 2) the importance of the element for mission execution, and 3) how elements may interact. Rate the perceived **Overall Risk Level** when considering this information. Circle the risk zone (***Low, Medium, or High***) that corresponds to your perceived overall risk level:

Low	Medium	High

****PEACE elements are required per COMDTINST 3500.3A. Additional unit specific elements are permitted.***

USCG Ashore Risk Assessment

Step 3: Determine Risk vs. Gain: Do gains warrant the risk?

Step 3a. Enter the ***Overall Risk Level*** (*Step 2 on prior page*) in the **RISK** box below (*Low, Medium, or High*).

Step 3b. Review the definitions for Gain below and enter the level in the **GAIN** box below. (*Low, Medium, or High*).

Level of Gain

- **Low** – Situation with unclear benefits or a low probability for providing concrete results.
  Examples: passenger transport, non-critical logistics missions, and public affairs demonstrations.
- **Medium** – Situation that provides immediate and real benefits.
  Examples: saving property, protecting the environment, deterring illegal operations.
- **High** – Situation that provides immediate and real benefits that if ignored could result in loss of life.
  Examples: Urgent SAR and MEDEVACs.

RISK (Low, Med, High)	Vs.	GAIN (Low, Med, High)

Step 3c. Use the **Risk vs. Gain** values from above and follow the column and row until they cross. The intersecting point is the recommended action.
Example, if Risk is 'low' and Gain is 'medium', the recommendation is: "Accept the Mission. Continue to monitor Risk Factors, if conditions or mission changes".

Risk vs. Gain	High Gain	Medium Gain	Low Gain
Low Risk	**Accept the Mission.** Monitor Risk Factors and re-evaluate if conditions or mission/activities change.	**Accept the Mission.** Monitor Risk Factors and re-evaluate if conditions or mission/activities change.	**Accept the Mission.** Monitor Risk Factors and re-evaluate if conditions or mission/activities change.
Medium Risk	**Accept the Mission.** Monitor Risk Factors and employ Controls when available. Re-evaluate if conditions or mission change.	**Accept the Mission.** Monitor Risk Factors and employ Controls when available. Re-evaluate if conditions or mission change.	**Accept the Mission Only with Command Endorsement** Communicate Risk vs. Gain to Chain of Command. Implement Controls and continuously evaluate conditions and mission for change.
High Risk	**Accept the Mission Only with Command Endorsement.** Communicate Risk vs. Gain to Chain of Command. Implement Controls and monitor Risk Factors. Continuously evaluate conditions and mission change.	**Accept the Mission Only with Command Endorsement.** Communicate Risk vs. Gain to Chain of Command. Implement Controls and monitor Risk Factors. Continuously evaluate conditions and mission change.	**DO NOT Accept the Mission.** Communicate to Chain of Command. Wait until Risk Factors change or Controls are available to warrant Risk exposure.

NOTES:

USCG Aviation Risk Assessment

Mission:		Date:	

Step 1: Identify, Assess, & Mitigate Risk Elements

Instructions: To determine the level of risk for each element below, estimate the risk level based on the Low/Medium/High scale. If your perceived rating is Medium or High, explore mitigations. Draw a line through the risk zone that corresponds to the mitigated risk level and document the perceived risk(s) and mitigation(s) in the space provided.	**Rate Risk Zone**		
Planning - Enough time and information to conduct thorough pre-mission planning. Consider: B-0 response, completeness of mission information and of on-scene details. **NOTES/MITIGATIONS:**	**Complete** *L*	**Partial** *M*	**None** *H*
Event - Refers to mission complexity. Consider: non-standard mission profile, coordinating multi-agency/nationality, language barriers, not performed often, etc. **NOTES/MITIGATIONS:**	**Low** *L*	**Moderate** *M*	**Extreme** *H*
Asset – Pilots – Proper number and skill set for the mission. Consider: time at unit, familiarity w/OP area, fatigue, u/w time, crew selection, adequate supervision, etc. **NOTES/MITIGATIONS:**	**Excellent** *L*	**Marginal** *M*	**Poor** *H*
Asset – Aircrew – Proper number and skill set for the mission. Consider: time at unit, familiarity w/OP area, fatigue, u/w time, crew selection, adequate supervision, etc. **NOTES/MITIGATIONS:**	**Ideal** *L*	**Restrictions** *M*	**Limitations** *H*
Asset – Airframe/Resources – Proper number and operational characteristics for mission. Consider: operational thresholds/limitations, status of equipment, etc. **NOTES/MITIGATIONS:**	**Excellent** *L*	**Partial** *M*	**None** *H*
Communications/Supervision - Ability to maintain comms throughout mission. Consider: availability/quality of internal w/command and external w/customer. **NOTES/MITIGATIONS:**	**Ideal** *L*	**Marginal** *M*	**Extreme** *H*
Environment - External conditions surrounding mission. Consider: weather, night, illum, mountain terrain, alternate airfields, on-scene cover, etc. **NOTES/MITIGATIONS:**	*L*	*M*	*H*
****Other (Unit Specific Element):***	*L*	*M*	*H*

Step 2: Determine Overall Risk Level

Consider: 1) the rating for each element above, 2) the importance of the element for mission execution, and 3) how elements may interact. Rate the perceived **Overall Risk Level** when considering this information. Circle the risk zone (***Low, Medium, or High***) that corresponds to your perceived overall risk level:

Low	Medium	High

**PEACE elements are required per COMDTINST 3500.3A. Additional unit specific elements are permitted.*

USCG Aviation Risk Assessment

Step 3: Determine Risk vs. Gain: Do gains warrant the risk?

Step 3a. Enter the ***Overall Risk Level*** (*Step 2 on prior page*) in the **RISK** box below (*Low, Medium, or High*).

Step 3b. Review the definitions for Gain below and enter the level in the **GAIN** box below. (*Low, Medium, or High*).

Level of Gain

- **Low** – Situation with unclear benefits or a low probability for providing concrete results.
  Examples: passenger transport, non-critical logistics missions, and public affairs demonstrations.
- **Medium** – Situation that provides immediate and real benefits.
  Examples: saving property, protecting the environment, deterring illegal operations.
- **High** – Situation that provides immediate and real benefits that if ignored could result in loss of life.
  Examples: Urgent SAR and MEDEVACs.

[ ]	Vs.	[ ]
RISK **(Low, Med, High)**		**GAIN** **(Low, Med, High)**

Step 3c. Use the **Risk vs. Gain** values from above and follow the column and row until they cross. The intersecting point is the recommended action.
Example, if Risk is 'low' and Gain is 'medium', the recommendation is: "Accept the Mission. Continue to monitor Risk Factors, if conditions or mission changes".

Risk vs. Gain	High Gain	Medium Gain	Low Gain
Low Risk	**Accept the Mission.** Monitor Risk Factors and re-evaluate if conditions or mission/activities change.	**Accept the Mission.** Monitor Risk Factors and re-evaluate if conditions or mission/activities change.	**Accept the Mission.** Monitor Risk Factors and re-evaluate if conditions or mission/activities change.
Medium Risk	**Accept the Mission.** Monitor Risk Factors and employ Controls when available. Re-evaluate if conditions or mission change.	**Accept the Mission.** Monitor Risk Factors and employ Controls when available. Re-evaluate if conditions or mission change.	**Accept the Mission Only with Command Endorsement** Communicate Risk vs. Gain to Chain of Command. Implement Controls and continuously evaluate conditions and mission for change.
High Risk	**Accept the Mission Only with Command Endorsement.** Communicate Risk vs. Gain to Chain of Command. Implement Controls and monitor Risk Factors. Continuously evaluate conditions and mission change.	**Accept the Mission Only with Command Endorsement.** Communicate Risk vs. Gain to Chain of Command. Implement Controls and monitor Risk Factors. Continuously evaluate conditions and mission change.	**DO NOT Accept the Mission.** Communicate to Chain of Command. Wait until Risk Factors change or Controls are available to warrant Risk exposure.

NOTES:

Index

Page numbers in italics indicate figures.

About the Author

A native of Charleston, West Virginia, **Phillip Null** enlisted in the U.S. Coast Guard on the morning of 11 September 2001 and served on active duty for more than twenty-one years before retiring in 2022 as a chief boatswain's mate. An accomplished operator across the service's full range of missions, Chief Null held qualifications as an officer in charge, boarding officer, master training specialist, and coxswain on fourteen boat types ranging from lifeboats to airboats with advanced heavy weather, tactical, pursuit, ice rescue, and aids to navigation competencies. During his career he sortied on missions along the Pacific, Atlantic, Gulf, and Great Lakes coasts and deployed overseas multiple times to locales including the Caribbean, Europe, Africa, and Asia.

Chief Null is a graduate of American Military University and Red Rocks Community College with degrees in Emergency and Disaster Management. He is also a member of the Sons of the American Revolution National Society, the Coast Guard Chief Petty Officers Association, the Coast Guard Tactical Law Enforcement Association, the National Tactical Officers Association, the Naval Order of the United States, and the U.S. Naval Institute.

An award-winning writer, photographer, and videographer, he has been published in the U.S. Naval Institute *Proceedings*, *The Counter Terrorist*, *Tactical Weapons*, *Survivor's Edge*, and *PoliceOne*, to name a few. He is also the creator and host of *They Had to Go Out* podcast, which tells the dangerous, daring, and epic sea stories of U.S. Coast Guard veterans in support of the National Coast Guard Museum. Today he continues to work at sea.

Suggestions and feedback regarding this book are welcomed. E-mail the author at phillip.null@gmail.com.